Disorders of Movement

Disorders of Movement

Clinical, Pharmacological and Physiological Aspects

Edited by

N. P. QUINN
Department of Clinical Neurology, Institute of Neurology, London, UK

P. G. JENNER
Department of Pharmacology, King's College London, London, UK

ACADEMIC PRESS
Harcourt Brace Jovanovich, Publishers
London   San Diego   New York   Berkeley
Boston   Sydney   Tokyo   Toronto

Academic Press Limited
24-28 Oval Road
London NW1

US edition published by
Academic Press Inc.
San Diego, CA 92101

ISBN 0-12-569685-X

This book is printed on acid-free paper.⊗

Typeset by Lasertext, Stretford, Manchester
Printed in Great Britain at T.J. Press (Padstow) Ltd, Cornwall

Contributors

Y. Agid   Laboratoire de Médecine Expérimentale et Clinique de Neurologie et Neuropsychologie, Hôpital de la Salpêtrière, 47 Bd de l'Hôpital-75634, Paris Cedex 13, France

A. Agnoli   First Clinic of Neurology, Department of Neurological Sciences University "La Sapienza", Viale dell'Universita 30, 00185 Roma, Italy

J. Artieda   Department of Neurology, Clinica Universitaria, Apartado 192 31080 Pamplona, Spain

P. J. Bedard   Laboratoire de Neurobiologie, Hôpital de l'Enfant-Jésus, 1401 18e Rue, Quebec (QC), Canada G1J 1ZA

J. Benavidès   Laboratoires d'Etudes et de Recherches Synthelabo (L.E.R.S.), Biomedical Pharmacology, 31 Avenue Paul Vaillant Couturier, 92220 Bagneux, France

R. Benecke   Neurologische Universitatsklinik, Moorenstr. 5, 4000 Dusseldorf, FRG

A. Berardelli   V Clinica Neurologica, Dipartimento de Scienze Neurologiche, Universita di Roma "La Sapienza", Viale dell'Universita n.30, Roma, Italy

R. Boucher   Laboratoire de Neurobiologie, Hôpital de l'Enfant-Jésus, 1401 18e Rue, Quebec (QC), Canada G1J 1ZA

C. Brazell   Merck, Sharp and Dohme Neuroscience Research Centre, Terlings Park, Eastwick Road, Harlow, Essex CM20 2QR, UK

R. G. Brown   Medical Research Council Human Movement and Balance Unit, Queen Square, London WC1N 3BG, UK

A. Carta   First Clinic of Neurology, Department of Neurological Sciences University "La Sapienza", Viale dell'Universita 30, 00185 Roma, Italy

C. J. Carter   Laboratoires d'Etudes et de Recherches Synthélabo (L.E.R.S.), Biomedical Pharmacology, 31 Avenue Paul Vaillant Couturier, 92220 Bagneux, France

L. Cleeves   MRC Neuro-Otology Unit, National Hospital, Queen Square, London WC1N 3BG, UK

A. Clow   Bernhard Baron Memorial Research Laboratories, Queen Charlotte's

Hospital, London W6 0XG, UK

R. J. Coleman Department of Neurology, Institute of Psychiatry, de Crespigny Park, London SE5 8AF, UK

B. Costall Postgraduate School of Studies in Pharmacology, University of Bradford, Bradford, West Yorkshire BD7 1DP, UK

P. A. Cullis Department of Neurology, Wayne State University School of Medicine, Detroit, Michigan, USA

B. L. Day MRC Human Movement and Balance Unit, Institute of Neurology, Queen Square, London WC1N 3BG, UK

T. Di Paulo Laboratoire de Neurobiologie, Hôpital de l'Enfant-Jésus, 1401 18e Rue, Quebec (QC), Canada G1J 1ZA

A. M. Domeney Postgraduate School of Studies in Pharmacology, University of Bradford, Bradford, West Yorkshire, BD7 1DP, UK

A. Dubois Laboratoires de'Etudes et de Recherches Synthelabo (L.E.R.S.), Biomedical Pharmacology, 31 Avenue Paul Vaillant Couturier, 92220 Bagneux, France

S. B. Dunnett Department of Experimental Psychology, University of Cambridge, Downing Street, Cambridge CB2 3EB, UK

R. C. Duvoisin Department of Neurology, University of Medicine and Dentistry of New Jersey, Robert Wood Johnson Medical School, New Brunswick, New Jersey, USA

J. S. Elston Department of Neuro-Ophthalmology, The National Hospital for Nervous Diseases, Queen Square, London WC1N 3BG, UK

S. Fahn Neurological Institute, 710 West 168th Street, New York, NY 10032, USA

L.J. Findley MRC Neuro-Otology Unit, National Hospital, Queen Square, London WC1N 3BG, UK

C. R. Gardner Merck, Sharp and Dohme Research Laboratories, West Point, PA, USA

W. R. G. Gibb Department of Neurology, Guy's Hospital, St. Thomas Street, London SE1 9RT, UK

F. Grandas Servicio de Neurologia, Hospital Provincial de Madrid, Doctor Esquerdo 46, 28007 Madrid, Spain

M. Hallett National Institute of Neurological and Communicative Disorders and Stroke, National Institutes of Health, Bldg 10, Rm 5N226, Bethesda, MD 20892, USA

A. E. Harding Institute of Neurology, Queen Square, London WC1N 3BG, UK

E. Hirsch Laboratoire de Médecine Expérimentale et Clinique de Neurologie et Neuropsychologie, Hôpital de la Salpêtrière, 47 Bd de l'Hôpital 75634, Paris Cedex 13, France

S.-C. G. Hui Department of Pharmacology, Faculty of Medicine, University of Hong Kong, Hong Kong

F. Javoy-Agid   Laboratoire de Médecine Expérimentale et Clinique de Neurologie et Neuropsychologie, Hôpital de la Salpêtrière, 47 Bd de l'Hôpital-75634, Paris Cedex 13, France

P. Jenner   Department of Pharmacology, King's College London, London SW3 6LX, UK

J. B. Kulisevsky   Neurology Service, Hospital Clinic i Provincial, University of Barcelona, Villarroel 170, 08036 Barcelona, Spain

A. E. Lang   Movement Disorders Clinic, Toronto Western Hospital, Toronto, Canada

K. L. Leenders   Clinical Head, PET Group PSI, Paul Scherrer Institut, Vormals SIN, CH-5234 Villigen, Switzerland

A. J. Lees   The National Hospital for Nervous Diseases, Queen Square, London WC1N 3BG, UK

P. N. Leigh   Department of Neurology, Institute of Psychiatry, De Crespigny Park, London SE5 8AF, UK

G. P. Luscombe   Boots Pharmaceuticals Research Department, Nottingham NG2 3AA, UK

J. J. Maccabe   The Neurosurgical Unit, The Maudsley Hospital, de Crespigny Park, London SE5 8AZ, UK

J. M. Martínez-Lage   Department of Neurology, Clinica Universitaria, Apartado 192, 31080 Pamplona, Spain

E. Melamed   Department of Neurology, Beilinson Medical Centre, Petah Tiqva 49 100, Israel

R. J. Naylor   Postgraduate School of Studies in Pharmacology, University of Bradford, Bradford, West Yorkshire, BD7 1DP, UK

M. Nomoto   Department of Neurology, Institute of Psychiatry, de Crespigny Park, London SE5 8AF, UK

T. G. Nygaard   The Department of Neurology, Columbia University College of Physicians and Surgeons, Neurological Institute of New York, 710 West 168th Street, New York, NY 10032, USA

J. A. Obeso   Department of Neurology, Clinica Universitaria, Apartado 192 31080 Pamplona, Spain

W. Oertel   Neurologische   Klinik,   Ludwig-Maximilians-Universitat Munchen, D-8000 Munchen 70, GDR

C. W. Ogle   Department of Pharmacology, Faculty of Medicine, University of Hong Kong, Hong Kong

D. Parkes   University Department of Neurology, King's College Hospital School of Medicine and Dentistry and Institute of Psychiatry, London SE5 8AF, UK

J. S. Perlmutter   Division of Radiation Sciences, Barnes Hospital Plaza, 510 S. Kingshighway, St Louis, MO 63110, USA

N. P. Quinn   Department of Clinical Neurology, Institute of Neurology, Queen Square, London WC1N 3BG, UK

J. C. Rothwell   MRC Human Movement and Balance Unit, Institute of Neurology, Queen Square, London WC1N 3BG, UK

M. Ruberg   Laboratoire de Médecine Expérimentale et Clinique de Neurologie et Neuropsychologie, Hôpital de la Salpêtrière, 47 Bd de l'Hôpital 75634, Paris Cedex 13, France

S. Ruggieri   First Clinic of Neurology, Department of Neurological Sciences, University "La Sapienza", Viale dell'Universita 30, 00185 Roma, Italy

S. M. Stahl   Merck, Sharp and Dohme Research Laboratories, Harlow, Essex, UK

R. Stell   The University Department of Clinical Neurology, The National Hospital for Nervous Diseases, Queen Square, London WC1N 3BG, UK

F. Stocchi   First Clinic of Neurology, Department of Neurological Sciences, University "La Sapienza", Viale dell'Universita 30, 00185 Roma, Italy

D. Tarsy   Section of Neurology, Department of Medicine, New England Deaconess Hospital, 110 Francis Street, Boston, MA 02215, USA

J. A. Temlett   Department of Neurology, Institute of Psychiatry, de Crespigny Park, London SE5 8AF, UK

P. D. Thompson   University Department of Clinical Neurology, The National Hospital for Nervous Diseases, Queen Square, London WC1N 3BG, UK

E. S. Tolosa   Neurology Service, Hospital Clinic i Provincial, University of Barcelona, Villarroel 170, 08036 Barcelona, Spain

P. C. Walker   The National Spasmodic Torticollis Association, Royal Oak, Michigan, USA

G. M. Zentner   Department of Pharmaceutical Chemistry, University of Kansas, Lawrence, KS 66046, USA

Contents

PART IV   OTHER MOVEMENT DISORDERS

35 Tics   495
A. J. Lees

36 Classification of Tremor   505
L. J. Findley and L. Cleeves

37 Orthostatic Tremor   521
J. C. Rothwell

38 Long-latency Reflexes   529
M. Hallett

Foreword

I was delighted to have been invited to write a foreword to this volume on movement disorders in honour of Professor David Marsden. Without question, David Marsden is one of the most distinguished living British neurologists, and indeed his contributions to our understanding of disorders of movement have assured for him a well-deserved and enviable international reputation. Since he became Professor of Neurology at King's College Hospital and the Institute of Psychiatry in 1972, he has pursued with total dedication a programme of research into Parkinson's disease and other disorders of movement, using clinical, physiological, biochemical, pharmacological and other relevant methods of study in a truly multidisciplinary approach which has borne considerable fruit in improving our understanding of this group of disorders and in being able to recommend for our patients much improved treatment. In his research he has deployed his own personal and considerable expertise in basic neuroanatomy, neurophysiology and neurochemistry, but also his encyclopaedic knowledge of neuropharmacology, and it therefore came as no surprise to his colleagues when in 1983 he was elected to a Fellowship of the Royal Society in acknowledgement of his outstanding scientific achievements. But above all David Marsden, despite his distinction and his scientific skills, has continued to be an outstanding and able doctor, able to deploy in patient care and management those virtues of the able communicator, demonstrating compassion and human understanding which have made him a true polymath and one of the acknowledged leaders of our profession.

It is a testimony to David's outstanding distinction that so many national and international experts should have come together to pay tribute to his memory by contributing to this volume on movement disorders. A glance at the list of contributors shows the regard in which he is held in countries throughout the world. The editors, Dr. Quinn and Professor Jenner, are to be congratulated in bringing together such an outstanding team and in editing so skilfully a major manuscript which will, in my opinion, stand as a permanent testimony to David Marsden's clinical and scientific expertise, but will also serve as a definitive reference text for all of those working in clinical neurology, neurophysiology, neurochemistry and neuropharmacology who wish to have an update on basic scientific and clinical aspects of disorders of movement. Writing as an editor of a series of monographs on "Major Problems in Neurology", I have to confess that I was considering the possibility of commissioning a volume on movement disorders in the series. Having surveyed this manuscript, I am now entirely satisfied that no such monograph is needed. I regard it as a privilege to be able to join in the congratulations offered by the contributors to David Marsden for his personal contributions to the field, combined with our good wishes to him in his new career as head of the Academic Department of Neurology at the National Hospital, Queen Square, where we know that he will continue, with the aid of his colleagues, to make outstanding contributions to the whole field of neurology, but above all to his chosen research interest in disorders of movement.

JOHN WALTON
Oxford

Preface

David Marsden was appointed Senior Lecturer and Consultant Neurologist at the Institute of Psychiatry and King's College Hospital in 1970, and appointed to the first Chair of Neurology at these institutes in 1972, at the age of 34. Over the next 15 years until he took up the Chair at the Institute of Neurology in October 1987, he built up a highly successful, productive and friendly academic department of neurology. Many British neurologists owe their interest in neurology, and in particular movement disorders, to the enthusiasm, interest and scientific approach that he exemplified. Many other medical and non-medical research workers were also drawn from all over the world to be trained by, and work with, him. These were the "Denmark Hill fellows", so named after the road in Camberwell that bisects the two institutions. This explains our choice of cover for this volume, on which the globe is juxtaposed with a picture of a Camberwell Beauty butterfly.

After organizing a symposium in honour of David Marsden in 1987, we felt that a more permanent tribute to him from his colleagues was called for, and set about inviting contributions that would do justice to his mastery not only of clinical aspects of movement disorders, but also their pharmacology and pathophysiology. Such is the breadth of talent on which we have been able to call that our greatest difficulty has been in narrowing down the list of authors. Almost all of the contributors have either spent time in the Department, or collaborated in various research projects. Indeed Gerald Stern, in chairing a session at the meeting, was speaking only part in jest when he estimated that he was one of the few people present who had never written a paper with David Marsden!

We are grateful to all the contributors who responded with such enthusiasm to this special tribute, and congratulate them on producing such authoritative and up-to-date chapters. We hope that this volume will serve as a valuable reference source for all those interested in movement disorders, and record something of the remarkable achievements arising from David Marsden's conception and stewardship of the Denmark Hill School of Movement Disorders.

NIALL QUINN
PETER JENNER

PART I

Parkinson's Disease

1

Is there a Parkinson's Disease?

Roger C. Duvoisin

INTRODUCTION

Implicit in the current usage of the term "Parkinson's disease" is the concept of a specific morbid entity with a particular albeit as yet unknown etiology. As will be shown here, that concept is widely held today by physicians especially concerned with parkinsonism. The concept may be traced back nearly 200 years to James Parkinson's initial description of the "Shaking Palsy" (Parkinson, 1817). It was confirmed by the nineteenth century founders of modern neurology. Its persistence inevitably influences the search for an etiology as well as diagnostic classification of the various parkinsonian syndromes recognized today. It is therefore necessary that we critically examine the question whether there is indeed a particular morbid entity which merits the designation *Parkinson's disease*. This brief essay is devoted to examining that question.

HISTORICAL BACKGROUND

James Parkinson had sought to limit the use of the name "Shaking Palsy" specifically to the clinical entity which he so eloquently described. He

DISORDERS OF MOVEMENT: CLINICAL, PHARMACOLOGICAL
AND PHYSIOLOGICAL ASPECTS   ISBN 0-12-569685-X

speculated on its etiology and on the location within the brain of the lesion; he appealed to morbid anatomists to search for its cause and nature—not its *causes* but its *cause*, in the singular. He seems to have thought of the disorder as a single nosologic entity and to have expected a particular cause would be found.

Charcot (1880), Gowers (1893) and other nineteenth century neurologists added additional clinical features, delineating in considerable detail its signs and symptoms, and painting in the process a picture which seemed sufficiently consistent and circumscribed to represent a specific morbid entity. As a result, by the end of the century, paralysis agitans or Parkinson's disease was widely recognized as a distinctive nosologic entity even though the site of the lesion and the nature of the pathology remained undefined and its etiology a mystery. Speculating about possible etiologies, Charcot spoke of it as a "neurosis", meaning thereby a disorder without a material pathology, and Gowers (1902) developed the notion of "abiosis", but neither suggested that the disorder might be a syndrome of varied etiology.

With minor additions and refinements recognized subsequently, such as the occasional occurrence of dysautonomia, mild defects of ocular motility and depression, the concept of Parkinson's disease bequeathed to us by the nineteenth century neurologists persists essentially unchanged today. The manifestations comprise the classic triad of tremor, rigidity and bradykinesia plus characteristic attitudes and postures and certain autonomic symptoms including seborrhea, episodic diaphoresis and chronic constipation, the whole pursuing a slowly progressive course extending 20–30 years or more. Evidence of involvement of cerebellar or pyramidal systems, peripheral neuropathy, amyotrophy or ophthalmoplegia, is inconsistent with the concept of Parkinson's disease and suggests some other disorder such as progressive supranuclear palsy or multiple system atrophy, disorders which have been defined by their own distinctive pathology.

Subsequent generations of neurologists searching for its anatomical substrate also assumed that Parkinson's disease was a unitary nosologic entity. Tretiakoff (1919), for example, en lieu of providing any clinical details, simply indicated that some of the cases he studied post-mortem had "typical" parkinsonism. The implication is clear that his contemporaries knew what that meant. Hassler (1938), who 20 years later in the most detailed study of the morbid anatomy of parkinsonism done up to that time confirmed Tretiakoff's observations on the constancy of the nigral lesion and the presence of Lewy bodies, also provided sparse clinical details.

The abrupt advent of encephalitis lethargica in the early years of the present century made it clear that there could be different etiologic types of parkinsonism. The extraordinary range of clinical manifestations seen in the chronic phase of that remarkable illness, the phase of *post-encephalitic parkinsonism*, so broadened concepts of parkinsonism that a variety of disorders came to be regarded as "forms" of parkinsonism. Thus, the lacunar state was renamed *arteriosclerotic parkinsonism* (Critchley, 1929). Similarly, the sequelae of carbon monoxide, manganese and other intoxications came to be considered forms of parkinsonism. The parkinsonism which prevailed prior to the epidemics of encephalitis lethargica and had been so well defined by nineteenth century clinicians became generally known as *idiopathic parkinsonism*.

The recognition of multiple forms of parkinsonism inevitably called into question the existence of a specific morbid entity which could be termed paralysis agitans or *Parkinson's disease*. As Lhermitte and Cornil (1921) put it over 60 years ago: "we must either admit that there exists, in addition to multiple parkinsonian syndromes, an authentic Parkinson's disease with particular lesions and a characteristic evolution and symptomatology or we must say that there is no Parkinson's disease just as there is no hemiplegic disease or pseudo-bulbar disease". The question has not yet been answered.

Although a "characteristic evolution and symptomatology" was generally recognized at the time, there was no agreement regarding "particular lesions". In the absence of such agreement the dilemma remained unresolved and leading authorities disagreed. A half-century later, Merritt (1955), for example, reflected a widely held view in stating in his textbook that "the symptom-complex identified by Parkinson stems from a variety of lesions". Others believed that the symptoms, signs and pattern of progression clearly distinguished paralysis agitans from other forms of parkinsonism. Walshe (1955) characteristically commented that to lump all the forms under one designation encouraged "slovenly clinical observation". In practice, a compromise prevailed: the term *idiopathic parkinsonism*, was widely used with the implicit understanding that it was more or less synonymous with paralysis agitans and Parkinson's disease as it was known prior to the epidemics of encephalitis lethargica, yet left open for future investigation the possibility of diverse etiologies and nosologic heterogeneity.

The deliberately implied uncertainty of the term *idiopathic parkinsonism* encourages the persistence of an inconsistent nomenclature and as a result we are left today with a confused terminology which complicates thought on the nature and possible etiology of the "parkinsonisms". Illogical expressions such as "post-encephalitic Parkinson's *disease*" and "arteriosclerotic Parkinson's *disease*" are commonly encountered in the current literature. We may read of the heterogeneity of Parkinson's disease as if the disorder could be etiologically heterogenous and still be considered a *disease*. Some authors straining for clarity have resorted to "*classical* idiopathic parkinsonism", but this awkward expression compounds the problem by suggesting that there might be some other kinds of idiopathic parkinsonism. More recently the expression "Lewy body parkinsonism" has become current, implying that the clinician can reliably identify that disorder and predict the pathology.

THE "PARTICULAR" LESION

Agreement on the "particular lesion" was very gradually reached by the 1960s despite uncertainty regarding the nosologic unity of Parkinson's disease. The early observations of Tretiakoff and Hassler were subsequently confirmed by Greenfield and Bosanquet (1953), Hartog-Jager and Bethlem (1960) and many others. Thus at the Second International Parkinson Symposium in 1963, Earle (1966) was able to report a general consensus among his fellow neuropathologists that nigral cell loss was the "indispensable change".

Significantly, agreement on the nigral lesion was reached before the nigro-striatal dopaminergic projection was demonstrated by the Falk–Hillarp histologic technique (Anded et al., 1964) and the fact appreciated that nigral lesions resulted in striatal dopamine depletion (Sourkes and Poirier, 1966). That demonstration gave meaning to the nigral lesion of parkinsonism and complemented earlier discovery of the striatal depletion of dopamine in both idiopathic and post-encephalitic parkinsonism (Ehringer and Hornykiewicz, 1960). It explained why different morbid entities could produce similar manifestations and how "supranigral" parkinsonism was possible. The clinical syndrome could be understood as a pathophysiologic state reflecting dysfunction of the nigro-striatal dopaminergic system which might result from various lesions of that structure. Parkinsonism was, as Horneykiewicz (1973) noted, the "syndrome of striatal dopamine depletion".

Earle (1966) also reported agreement that "in general" Lewy bodies were associated with paralysis agitans and neurofibrillary tangles with post-encephalitic parkinsonism. Within a few years, striato-nigral degeneration (Adams et al., 1964) and progressive supranuclear palsy (Steele et al., 1964) were distinguished by differences in the cellular morphology of the neuronal lesion and by the topographical distribution of lesions, not by their clinical features. More recent students of the pathology of the disease such as Jellinger (1986) and Forno (1982) have affirmed the constancy of the nigral lesion and emphasized the association of the Lewy body with Parkinson's disease. Thus the Lewy body has come to be seen as the "hallmark" of Parkinson's disease and the expression "Lewy body parkinsonism" has become current as a designation of the morbid entity Parkinson's disease.

THE VIEWS OF CONTEMPORARY AUTHORITIES

To assess how widely these views are held among clinicians who work in the field, I sought the views of 50 recognized leading authorities on

Parkinson's disease in 10 countries with the aid of a simple questionnaire. Anonymity was promised to assure candid replies. Respondents were asked to check "agree", "uncertain" or "disagree" for each of the following five statements, shown here with the results:

	Agree	Uncertain	Disagree	Total
1. Parkinson's disease is a specific morbid entity, albeit as yet of unknown cause.	31	2	0	33
2. Although Lewy bodies may occur in other conditions, if there are no Lewy bodies, it isn't Parkinson's disease.	24	3	7	34
3. Parkinson's disease and idiopathic parkinsonism are synonymous terms.	24	2	5	34
4. The term "arteriosclerotic Parkinson's disease" is acceptable.	1	3	29	33
5. The term "arteriosclerotic parkinson*ism*" is preferable.	16	11	6	33

A surprising 31 (94%) of the 33 respondents agreed with the first proposition. We may confidently conclude that there is wide acceptance among authorities in the field of the view that Parkinson's disease is a specific morbid entity. However, there was less agreement regarding the underlying pathology. Only 24 respondents agreed outright with the second and third propositions. Another two disagreed but qualified their responses by observing that Lewy bodies may be missed at post-mortem examination. Accepting these two responses as affirmative, we can increase the number of respondents agreeing that the presence of Lewy bodies defines Parkinson's disease to 26 (76%) and reduce the number who were uncertain or disagreed to 8 (24%).

The dissenting views are important and show that there is serious doubt among these experts that the Lewy body does indeed correspond to the clinical concept of Parkinson's disease. As several respondents noted the Lewy body is not unique to Parkinson's disease. One lone respondent believed that clinically typical cases of Parkinson's disease without Lewy bodies should not be rejected in view of our meagre understanding of their significance.

The great majority agreed that the terms "idiopathic parkinsonism" and "Parkinson's disease" were synonymous. This may be deemed consistent with the view that there is a specific disorder which we may call Parkinson's disease, especially since the dissenters regarded the former term as less specific. However, the number of dissenters indicates poor agreement on the relative meaning of these terms.

It is reassuring that only one authority considered the term "arteriosclerotic Parkinson's disease" acceptable but strange that only half agreed with the reciprocal proposition that "arteriosclerotic parkinsonism" was the preferable term. This inconsistency was explained by a number of respondents who noted that they did not believe such an entity exists. The question remains open, therefore, among leading thinkers in this field whether the parkinson-like syndromes found in association with multiple cerebral infarctions, the lacunar state, Binswanger's disease and other types of cerebral vascular disease merit the appellation "parkinsonism".

THERE IS NO PARKINSON'S DISEASE

Although there is wide acceptance of the view that Parkinson's disease is a specific morbid entity, there is significant disagreement on the meaning of the Lewy body, its presumed morphologic correlate. There is also appreciable confusion in the use of diagnostic terms. These results are consistent with the historical perspective reviewed above indicating that our concept of Parkinson's disease is based primarily on clinical observation. It was established in the nineteenth century, long before the anatomical substrate of parkinsonism was understood, in the absence of any clue to etiology and long before the various morbid entities which comprise the spectrum of parkinsonism were recognized. It is a purely clinical concept.

Unfortunately few pathological studies have attempted a strict detailed clinico-pathologic correlation. Where the attempt has been made to correlate clinical features with morbid anatomy, the results have been imperfect. For example, Tretiakoff (1919) found Lewy bodies in only six of the 10 cases he considered "typical" and in one of the two cases who were atypical. The few clinical details he provided make it clear that these two cases were very atypical indeed! Hassler (1938) found Lewy bodies in 10 of his 23 cases: only five of these had clinically typical Parkinson's disease; whilst five clearly had post-encephalitic parkinsonism! Subsequent studies have usually described findings in larger series of cases collected over periods of many years and do not allow such an analysis. My own recent experience has also been disappointing: two of five patients who had Lewy body disease at post-mortem had not been thought to have typical Parkinson's disease in life (Sage et al., unpublished data). One had been suspected of having progressive supranuclear palsy but never developed the characteristic ocular palsy. The other had a severe dementia with minimal akinetic parkinsonism and was thought to have Alzheimer's disease but on post-mortem study had brain stem plus diffuse cerebral cortical Lewy bodies. I tentatively estimate that in roughly one-quarter of the patients thought by experienced clinicians to have typical Parkinson's disease Lewy bodies will not be found at post-mortem study and conversely, Lewy bodies will be found in about one-quarter of the patients thought not to have Lewy body disease. The total of positive and negative "misdiagnosis" is thus roughly 50%! Our clinical concept does not appear to correspond closely to a particular pathology.

Several morphologically distinct disorders can produce the clinical picture we have seen for the past century as "Parkinson's disease", some frequently, some occasionally, and others rarely. This complex circumstance may usefully be illustrated in terms of set theory as shown in Fig. 1.1. The range of clinical manifestations corresponding to the clinician's vision of Parkinson's disease is an area in which several distinct morbid entities, each represented by a circle, overlap. The major portion of the circle representing Lewy body disease is shown lying within the area corresponding to the concept of Parkinson's disease but also extending beyond it. Other entities overlap to a lesser extent, some only minimally. Only a few patients with one of these entities may fit within the circle and they may fit only for a few years in their clinical course.

The most important entity is a chronic progressive disorder which we may term "Lewy body disease" with Lewy bodies confined mainly to the brain stem. It corresponds best to our clinical concept of Parkinson's disease and accounts for the great majority, perhaps 60–70% of patients clinically diagnosed as having Parkinson's disease. In a small percentage of patients with Lewy body disease the inclusions are widely dispersed throughout the cerebral cortex and often associated with Alzheimer-type neurofibrillary

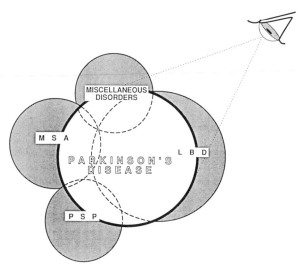

Fig. 1.1 Diagrammatic illustration using the principles of set theory to show how the historic clinical concept or vision of Parkinson's disease overlaps a number of discrete morbid entities.

tangles (Gibb *et al.*, 1987). These patients present primarily a chronic progressive dementia with some parkinsonian features which may easily be overlooked! Some would consider this a different entity but the existence of intermediate cases with numerous Lewy bodies in the brain stem and a few in the temporal lobe cortex (Forno *et al.*, 1978) suggests a continuum representing one common pathogenesis (Yoshimura, 1983). Thus the circle diagrammatically representing Lewy body disease is shown extending somewhat beyond the area corresponding to the clinical concept of Parkinson's disease.

The correspondence of Lewy bodies with our historic concept of Parkinson's disease is further weakened by recent observations of typical Lewy body pathology in patients who cannot reasonably be considered to have the same pathogenesis. For example, Lewy bodies have been seen in the substantia nigra of a patient with the juvenile onset dystonia-Parkinson syndrome (Yokochi *et al.*, 1984), recently renamed "dopa-responsive dystonia" (Nygaard *et al.*, 1988). This is clearly on clinical and genetic grounds a different disorder. In addition, typical brain stem Lewy body pathology was found in a recently described kindred of autosomal dominant parkinsonism, clinically atypical in its young adult onset and lack of tremor (Golbe *et al.*, 1988). This is clearly on clinical and genetic grounds still another morbid entity. Thus three circles might be required to represent these three types of Lewy body disease in the figure. The possibility that there are still more cannot be excluded.

Probably the second most important cause of Parkinson's disease is progressive supranuclear palsy. My colleagues and I have elsewhere indicated that it may account for up to 12% or more of the patients clinically diagnosed as having Parkinson's disease (Duvoisin *et al.*, 1987). Next in numerical importance we can place the various forms of olivopontocerebellar atrophy (Duvoisin, 1987) and striato-nigral degeneration which some would lump together under the broader designation of "multiple system atrophy" (Oppenheimer, 1984). These are again a large heterogenous group of disorders, some dominantly inherited and others apparently sporadic. We have become increasingly aware since the advent of levodopa therapy that these disorders can closely mimic Parkinson's disease for several to many years in their evolution and are commonly diagnosed as such (Forno, 1982). They may account for as many as 5 or 6% of patients so classified. Some cases of spinocerebellar degeneration may also present a clinical picture with parkinsonian features predominating at least for some time in their course. Rarely, Hallervorden–Spatz disease of late onset may mimic Parkinson's disease. There are many more disorders which on rare occasions may present the clinical manifestations of Parkinson's disease, and doubtless some which have not yet been classified.

From the foregoing, we may reasonably doubt that there is a Parkinson's disease and we may question the usefulness of the concept and the term. Clearly we need to reconsider the nosology and diagnostic classification of the "parkinsonisms" and to redefine our terms. "Idiopathic parkinsonism" has the advantage of long usage but appears too non-specific. A possible solution might be to use the general term "parkinsonism" as a clinical descriptor, then adding a qualifying phrase such as "probable" or "possible" Lewy body disease to indicate with appropriate tentativeness the clinician's estimate of the likely underlying morbid entity. A more specific nomenclature must necessarily await future progress in clarifying clinico-pathologic correlations and elucidating etiologies.

REFERENCES

Adams, R., Van Bogaert, I. and Van Der Eecken, H. (1964). *J. Neuropath. Exp. Neurol.* **23**, 219–259.

Anden, N. E., Carlsson, A., Dahlstrom, A., Fuxe, K., Hillarp, N.-A. and Larson, K. (1964). *Life Sci.* **3**, 523–530.

Charcot, J. M. (1880). *Leçons sur les maladies du système nerveux faites á la Salpêtrière*, 4th edn, collected and published by Bourneville, p. 186. Delahaye & Lecrosnier, Paris.

Critchley, M. (1929). *Brain* **52**, 23–83.

Duvoisin, R. C. (1987). In *Movement Disorders 2* (edited by C. D. Marsden and S. Fahn) pp. 249–269. Butterworths, London.

Duvoisin, R. C., Golbe, L. I. and Lepore, F. E. (1987). *Can. J. Neurol. Sci.* **14**, 547–554.

Earle, K. M. (1966). *J. Neurosurg.* **24**, 247–249.

Ehringer, H. and Hornykiewicz, O. (1960). *Klin. Wschr.* **38**, 1236–1239.

Forno, L. S. (1982). In *Movement Disorders* (edited by C. D. Marsden and S. Fahn), pp. 25–30. Butterworths, London.

Forno, L. S., Barbour, P. J. and Norville, R. L. (1978). *Arch. Neurol. (Chicago)* **35**, 818–822.

Gibb, W. R., Esiri, M. M. and Lees, A. J. (1987). *Brain* **110**, 1131–1153.

Golbe, L. I., Miller, D. C. and Duvoisin, R. C. (1988). *Ann. Neurol.* **24**, 151–152A (abstract).

Gowers, W. R. (1893). *A Manual of Diseases of the Nervous System*, 2nd edn, pp. 636–657. Blakiston, Philadelphia.

Gowers, W. R. (1902). *Lancet* i, 1003–1007.

Greenfield, J. G. and Bosanquet, F. (1953). *J. Neurol. Neurosurg. Psychiat.* **16**, 213–226.

Hartog-Jager, W. A. and Bethlem, J. (1960). *J. Neurol. Neurosurg. Psychiat.* **23**, 283–290.

Hassler, R. (1938). *J. Psychol. Neurol.* **18**, 387–476.

Hornykiewicz, O. (1973). *Proc. Fed. Am. Soc. Exp. Biol.* **32**, 183–190.

Jankovic, J., Kirkpatrick, J. B., Blomquist, K. A., Langlais, P. J. and Bird, E. D. (1985). *Neurology* **35**, 227–234.

Jellinger, K. (1986) *Adv. Neurol.* **45**, 1–18.
Jellinger, K. and Neumayer, E. (1972). *Z. Neurol.* **203**, 105–118.
Lhermitte, J. and Cornil, L. (1921). *Rev. Neurol.* **28**, 587–592.
Merritt, H. H. (1955). In *A Textbook of Neurology*, pp. 432–441. Lea & Febiger, Philadelphia.
Nygaard, G. N., Marsden, C. D. and Duvoisin, R. C. (1988). *Adv. Neurol.* **50**, 377–384.
Oppenheimer, D. R. (1984). In *Greenfield's Neuropathology*, 4th edn (edited by J. H. Adams, J. A. N. Corsellis and L. W. Duchen), pp. 699–747. W. J. Arnold, London.
Parkinson, J. (1817). *An Essay on the Shaking Palsy.* Sherwood, Neely and Jones, London.
Sourkes, T. L. and Poirier, L. J. (1966). *J. Neurosurg.* **24**, 194–195.
Steele, J. C., Richardson, J. C. and Olszewski, J. (1964). *Arch. Neurol. (Chicago)* **10**, 333–359.
Tretiakoff, C. (1919). *Contribution a l'Etude de l'Anatomie du Locus Niger de Soemmering.* MD Thesis, Université de Paris.
Walshe, F. M. R. (1955). In *James Parkinson 1755–1824* (edited by M. Critchley), pp. 245–268. MacMillan, London.
Yokochi, M., Narabayashi, H., Iizuka, R. and Nagatsu, T. (1984). *Adv. Neurol.* **40**, 407–413.
Yoshimura, T. (1983) *J. Neurol.* **229**, 17–32.

2

Functional Organization of the Basal Ganglia

P. N. Leigh

INTRODUCTION

In the first three decades of this century pathologists and clinicians, pre-eminently the Vogts and Kinnier Wilson, laid the foundations for a theoretical understanding of basal ganglia function in relation to movement disorders. "It is essential in discussing the physiology of the basal ganglia to think anatomically" wrote Wilson in 1912. At that time, "the currently familiar concept of the corpus striatum as a mechanism consisting of striopallidal links... and a complex pallidal discharge, not projecting further caudally than the substantia nigra" was established (Obersteiner, 1896, quoted by Mettler, 1968) but the pathological basis of Parkinson's disease had yet to be defined (Tretiakoff, 1919; Hassler, 1937; Greenfield and Bosanquet, 1953). The known anatomy of the striatum provided few indications that it might be implicated in "complex" functions such as motor planning and cognition. In the Croonian lectures in which he distilled his thoughts on basal ganglia function and symptomatology, Wilson (1925) commented: "When we remember the histological simplicity and comparative structural homogeneity of the corpus striatum, in contrast with the greater dimensions, much more intricate cytoarchitectonic complexity, and far wider connexions of the

DISORDERS OF MOVEMENT: CLINICAL, PHARMACOLOGICAL
AND PHYSIOLOGICAL ASPECTS   ISBN 0-12-569685-X

rolandic motor cortex, the idea of attributing all these disturbances to striatal disease, and of crowding corresponding 'centres' into that ganglion, becomes nothing short of ludicrous". Cautious and critical in his interpretation of clinical phenomena (he referred scathingly to the trend at that time to see all motor symptoms "through an encephalitic fog") Wilson elected to stress "positive" symptoms such as rigidity and tremor as characteristic of the striatal syndrome, and concluded that "striatal symptomatology and striatal function must be comparatively simple". To a large extent, re-evaluation of this reductionist position, which was nevertheless crucial for the development of ideas on basal ganglia function, followed advances in understanding of striatal anatomy and biochemistry, redirecting basic scientists and clinicians to examine more complex aspects of basal ganglia function, such as their role in cognition, emotion and affective disorders (Nauta, 1979; Divac, 1983; Iversen, 1984; Gotham et al., 1986; Brown and Marsden, 1988; Starkstein et al., 1988). Wilson's notion of the striatum as a morphologically homogeneous structure is no longer tenable, and our notions of "simple" and "complex" are clearly inadequate as analytical tools for constructing a theoretical model of basal ganglia function. Nevertheless, Wilson's dictum that we should "think anatomically" has never been more relevant or productive of useful hypotheses (e.g. DeLong and Georgopoulos, 1981; Marsden, 1982). I will review some of the major advances in understanding of basal ganglia anatomy, and attempt to relate them to the pathophysiology of movement disorders. If the cartography of the basal ganglia is more complex than it was 15 years ago, major misapprehensions have been corrected (for example, we now know that the basal ganglia do not project back upon the primary motor and sensory areas as had been thought, but upon premotor areas, including the supplementary motor area) and for the clinician, neurologist and psychiatrist alike, study of basal ganglia anatomy is increasingly rewarding as techniques for investigating the function of discrete parts of the basal ganglia in man and experimental animals become more powerful.

NEW CONCEPTS OF BASAL GANGLIA ORGANIZATION

Recent interest has focused upon three main aspects of basal ganglia organization. First, relationships between the cerebral cortex and the neostriatum have been reassessed. Second, recognition of the compartmental organization of the striatum has generated a new anatomy of the striatum, which has allowed correlations to be made between the development, connectivity, neurochemistry, and (at a more speculative level) functional attributes of the striatum and of its input and output systems (Mensah, 1977; Graybiel and Ragsdale, 1978; Graybiel, 1984; Penney and Young, 1986;

references in Table 1). Finally, striatal outflow pathways have now been defined with great precision, using more sensitive tracing techniques. The functional anatomy of these pathways has been studied in a variety of rodent models, mapped by single cell recordings in awake, behaving monkeys and by ^{14}C-2-deoxyglucose (2-DG) autoradiography in monkeys with experimental movement disorders very similar to those occurring in man. The subthalamic nucleus, long known to be important in the pathogenesis of movement disorders, has received renewed attention; and interest has also focused upon what might be regarded as its satellite nucleus, the pedunculopontine nucleus (PPN) in the midbrain reticular formation, particularly since PPN neurones are lost in PD (Jellinger, 1988), suggesting that dysfunction of this nucleus may influence the clinical features of PD.

RELATIONSHIPS BETWEEN CEREBRAL CORTEX AND NEOSTRIATUM: CORTICO-STRIATAL PROJECTIONS

The striatum receives inputs from all parts of the cerebral cortex so far examined. Kemp and Powell (1970) showed that these inputs were topographically arranged (Fig. 2.1). They observed that the motor cortex

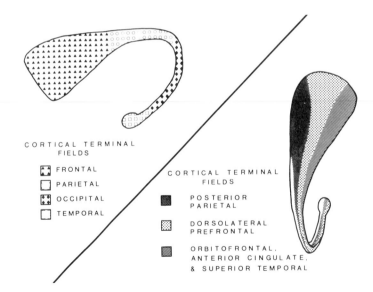

Fig. 2.1 Comparison of observations on the topographical organization of cortico-striatal projections from (left) Kemp and Powell (1970) and (right) Selemon and Goldman-Rakic (1985). (Reproduced, with permission, from Selemon and Goldman-Rakic, 1985.)

projects mainly upon the putamen, and that the supplementary motor area gives rise to a bilateral projection to the striatum. A close relationship between the putamen and primary sensorimotor and supplementary motor cortex, and between the caudate nucleus and "association cortex" has been confirmed in subsequent studies (Kunzle, 1975, 1977, 1978; Jones et al., 1977; Goldman and Nauta, 1977; Selemon and Goldman-Rakic, 1985), and the predominance of motor impairments in Parkinson's disease may reflect greater dopamine depletion in the putamen compared with the caudate nucleus (Bernheimer et al., 1973). Anatomical studies also revealed the patch-like nature of cortico-striatal terminals, and a similar pattern is seen in the pattern of thalamo-striatal and (during development) nigro-striatal dopamine inputs (Olson et al., 1972; Kalil, 1978; Royce, 1978; Graybiel, 1986). Kemp and Powell (1970) noted considerable overlap within the striatum of projections from different cortical areas. Recent studies have modified the details of this scheme, particularly with regard to allocortical areas, which project to the ventral striatum and nucleus accumbens (Heimer et al., 1982; Heimer and Wilson, 1975) and so provide a link between motor and cognitive-affective functions of the basal ganglia. Kemp and Powell (1970) also noted that projections from discrete cortical areas terminated in striatal zones restricted in their antero-posterior and medio-lateral extent (Fig. 2.1). In fact, cortical inputs to the striatum are probably more completely (although not totally) segregated than was apparent from the work of Kemp and Powell. It is now clear that individual cortical regions project upon longitudinally and medio-laterally arranged "strips" of striatum, encompassing varying domains of the caudate nucleus and putamen (Selemon and Goldman-Rakic, 1985; Fig. 2.1).

Using two sensitive anterograde tracing techniques to detect striatal projections from interconnected areas of cortex Selemon and Goldman-Rakic (1985) have now shown that even when terminal fields from such areas converge on one part of the striatum, they tend to interdigitate rather than intermix (Fig. 2.2).

In certain areas such as the head of the caudate nucleus, however, there may be a true intermingling of inputs, e.g. those from prefrontal and parietal areas (Selemon and Goldman-Rakic, 1985). The anatomical basis for convergence and integration of information flow (long assumed to be a key function of the striatum) may therefore lie elsewhere than in the arrangement of cortical inputs to the striatum.

These findings have major functional implications, but many questions remain to be answered before their significance is fully understood. Are adjacent but "segregated" cortico-striatal inputs in fact linked by inter-neurones, or by the dendrites of medium spiny neurones, the major projection cells of the striatum? What is the full extent and nature of convergence of

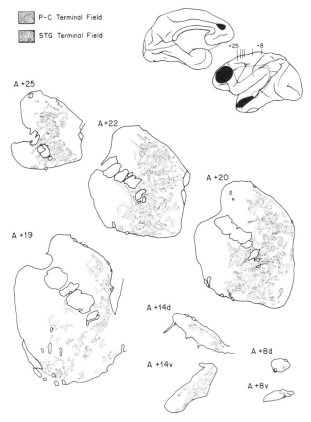

Fig. 2.2 Diagram illustrating the interdigitation of projections originating in the prefrontal cortex (red) and superior temporal cortex (black). HRP was injected into frontal cortex and tritiated amino acids were injected into temporal cortex, and adjacent sections reacted histochemically for localization of HRP and processed as autoradiograms, respectively. (Reproduced, with permission, from Selemon and Goldman-Rakic, 1985.) (This figure appears in colour in the plate section facing page 360.)

functionally related areas of cortex upon a given striatal domain? What kind of integrative processing occurs in such domains? The striosome (patch)– matrix organization of the striatum may provide some of these answers.

NEUROCHEMICAL ANATOMY OF THE STRIATUM

The recognition that the striatum is composed of neurochemically distinct compartments (Butcher and Hodge, 1976; Pert *et al.*, 1976; Graybiel and Ragsdale, 1978) has stimulated a far-reaching re-evaluation of striatal organization. Acetylcholinesterase (ACHE) histochemistry of striatum in many species, including man, reveals a complex system of ACHE-rich and ACHE-poor zones, the former referred to as matrix, and the latter termed striosomes or patches (see Graybiel, 1983, 1984, 1986 for reviews). Subsequent work has revealed that a mosaic or patch-like organization applies to most neurochemical systems in the striatum (Table 2.1), although it is not necessarily identical in different species. It is now clear that some perikarya

Table 2.1 Neurochemical characteristics of neostriatal compartments

	Neostriatal compartment	
	Striosomes (patches)	Matrix
1. Neurochemical marker		
	Low ACHE	High ACHE (1)
	High u-opiate receptor binding	Low u-opiate receptor binding (2, 3)
	Low ChAT	High ChAT (4)
	Low muscarinic-receptor binding	High muscarinic-receptor binding (5)
	Low dopamine D2-receptor binding	High dopamine D2-receptor binding (6)
	Dopamine "islands" (7, 8)	—
	DARRP-32 positive cell clusters (9)	—
	Low SOM IR	High SOM IR (10)
	Low NADPH-D activity	High NADPH-D activity (11)
	High ENK IR	Low ENK IR (12)
	High NT IR	Low NT IR (13)
	High SP IR	Low SP IR (14)
	High DYN IR	Low DYN IR (12)
	High GAD	Low GAD (15)
2. Connectivity		
Inputs	Cortico-striatal inputs (e.g. PF, limbic)	Cortico-striatal inputs (e.g. sensorimotor)
	Dopamine "island" system (7)	"Diffuse" dopamine system (8, 17).
	Ventral A9 and DA cells in ventral SN	VTA (18, 19). Thalamic inputs (3, 20, 21)
Outputs	SNC (10)	GP and SNR (10, 22)

References: 1, Graybiel and Ragsdale (1978); 2, Pert *et al.* (1976); 3, Herkenham and Pert (1981); 4, Graybiel *et al.* (1986); 5, Nastuk and Graybiel (1985); 6, Joyce *et al.* (1986); 7, Graybiel (1984); 8, Graybiel (1986); 9, Fuxe *et al.* (1986); 10, Gerfen (1984); 11, Sandell *et al.* (1986); 12, Graybiel and Chesselet (1984); Goedert *et al.* (1983); 14, Bolam *et al.* (1988); 15, Graybiel *et al.* (1983); 16, Donoghue and Herkenham (1986); 17, Olson *et al.* (1972); 18, Gerfen *et al.* (1987); 19, Herkenham *et al.* (1984); 20, Kalil (1978); 21, Royce (1978); 22, Graybiel *et al.* (1979).

ACHE, acetylcholinesterase; ChAT, choline acetyltransferase; DARRP-32, dopamine- and adenosine 3′, 5′-monophosphate-regulated phosphoprotein; DYN, dynorphin; ENK, enkephalin; GAD, glutamic acid decarboxylase; GP, globus pallidus; IR, immunoreactive; NADPH-D, NADPH-diaphorase; NT, neurotensin; SNC, substantia nigra pars compacta; SNR, substantia nigra, pars reticulata; SP, substance P; PF, prefrontal; SOM, somatostatin; VTA, ventral tegmental area. (Modified, with permission, from Leigh, 1987.)

(e.g. those of cholinergic and somatostatin-immunoreactive interneurones) may be found in both compartments, whereas their processes can be identified mainly in the matrix (Graybiel *et al.*, 1986; Bolam *et al.*, 1988). Graybiel *et al.* (1986) have suggested that cholinergic modulation of striatal function occurs mainly within the extra-striosomal matrix, from which derives the major part of nigro- and pallido-thalamic outputs influencing the cortex. Output neurones are also arranged so as to allow both segregation of information within striosomal compartments and communication between compartments. Bolam *et al.* (1988) have shown that medium spiny neurones in striosomes fall into two classes, one class comprising neurones whose dendritic arbors are apparently uninfluenced by the boundary of striosomes as defined by SP immunoreactivity, and one class composed of neurones whose dendritic arbors do not cross striosomal boundaries. Dendrites of medium spiny neurones located in the matrix likewise may either cross, or in other cases, respect striosomal boundaries. These findings provide a morphological basis for information flow from striosome to matrix, and vice versa. This raises the possibility that the level of incoming activity and on-going "background activity" in cholinergic and somatostatin interneurones may influence the extent of striosome–matrix co-activation or inhibition (Bolam *et al.*, 1988), so determining the degree of integration or segregation in a given functional state. On clinical grounds one would suspect that dopamine and acetylcholine are of crucial importance in this respect.

The relationship between dopamine and the patch–matrix arrangement of the striatum has been reviewed by Graybiel (1986). In the adult mammalian brain, dopamine terminals are relatively homogeneously distributed through the striatum. In the foetal or newborn animal, however, dopamine markers are clustered in well-defined islands, which are *rich* in ACHE and these can be related to the ACHE-*poor* zones (striosomes) of the adult brain (Graybiel, 1984, 1986). In the cat the change from ACHE-rich zones in the developing striatum to ACHE-poor zones in the adult occurs in the first 2–4 weeks of life (Graybiel, 1984) but the functional significance of this is unknown.

The scheme shown in Table 2.1 simplifies a highly complex system, since the pattern of striosomes and matrix differs between the dorsal and ventral striatum, and between species (Graybiel, 1986; Semba *et al.*, 1987). Nevertheless, there is indirect evidence that the striosome–matrix organization has functional significance. First, in a microdialysis model, acetylcholine releases dopamine from striosomes but not from the matrix compartment (Gauchy *et al.*, 1987). Second, systemic apomorphine reduces glucose utilization in striosomes, and increases it in matrix areas, whereas haloperidol has the opposite effect on striosomes but fails to alter glucose utilization in the matrix (McCulloch *et al.*, 1983). Third, single-cell recordings in the putamen reveal groups of functionally related neurones, clustered within a

100–500 μm span, which would be appropriate for the dimensions of striatal compartments and it has been suggested that these putaminal cell groups might correspond to functional columns within the cerebral cortex (Crutcher and DeLong, 1984). Finally, it is not surprising that selective loss of striatal neurones in disease is reflected in selective changes in striatal compartments. Ferrante *et al.* (1986) observed that substance P and enkephalin-like IR disappeared from the dorsal caudate nucleus and putamen in advanced Huntington's disease whereas a patch–matrix pattern of ACHE activity was discernible throughout the striatum, even in advanced disease. High ACHE regions were reduced in area, but low ACHE areas were of normal dimensions. These findings correlate with the loss of output neurones, comprising populations of medium spiny neurones in which GABA coexists with SP or enkephalin associated with relative preservation of cholinergic and somatostatin–NPY interneurones, although some of the ACHE activity is derived from nigral and thalamic afferents to striatum (Ferrante *et al.*, 1986; Kowall *et al.*, 1987). As yet, little information exists on alterations of striatal compartments in other disorders associated with striatal pathology, but this should be a rewarding area of clinico-pathological research in the near future.

SYNAPTIC ORGANIZATION WITHIN THE STRIATUM

Much progress has been made in mapping details of neuronal connectivity within the striatum, and current knowledge has been reviewed by Bolam (1984) and Beal and Martin (1986). The known synaptic inputs to type 1 medium spiny neurones which comprise 70–80% of striatal neurones, and which represent the major output neurones of the striatum, are summarized schematically in Fig. 2.3. These neurones are thought to be GABAergic, but in different populations of medium spiny neurones GABA appears to coexist with substance P, enkephalin or dynorphin (Vincent *et al.*, 1982; Chesselet and Graybiel, 1983a; Aronin *et al.*, 1984; Morelli and Di Chiara, 1984; Penney *et al.*, 1986; Semba *et al.*, 1987). As we have seen, some populations of output neurones have different targets. A striking feature of the striato-fugal projection is the localization of enkephalin in GPe, and SP in GPi, suggesting that striatal enkephalin containing cells project to the external pallidum whereas SP containing neurones project to the internal pallidum, although both SP and enkephalin cells project in addition to the substantia nigra (Haber and Elde, 1981; Chesselet and Graybiel, 1983a,b).

It will be seen that glutamatergic neocortical inputs terminate on the tips of dendritic spines, whereas dopamine inputs impinge upon the stem of the spines, and other neurotransmitters contact more proximal parts of the dendritic tree of perikaryon. Impulse flow from the cortex is presumably subject to modification by dopamine and other neurotransmitters, including

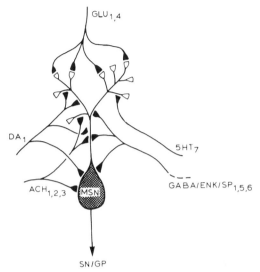

Fig. 2.3 Diagrammatic representation of synaptic inputs upon a typical striatal medium spiny neurone. ACH, acetylcholine; DA, dopamine; ENK, enkephalin; GABA, glutamic acid decarboxylase; GLU, glutamate; GP, globus pallidus; SN, substantia nigra; SP, substance P. References: 1, Bolam (1984); 2, Wainer et al. (1983); 3, Izzo and Bolam (1988); 4, Somogyi et al. (1981); 5, Somogyi et al. (1982); 6, Bolam and Izzo (1988); Pasik et al. (1981).

GABA and several neuropeptides, via recurrent collaterals of medium spiny neurones and axons of GABAergic interneurones (Bolam, 1984). Thalamo-striatal and cortico-striatal inputs appear to terminate upon different subsets of medium spiny neurones (Dube et al., 1988). For clinicians, the relationship between dopamine and acetylcholine in the striatum is of crucial importance. There is no direct evidence to date that dopamine axons synapse directly on cholinergic neurones. However, both dopamine and cholinergic terminals impinge upon medium spiny neurones (Izzo and Bolam, 1988) and it may be that the major synaptic interaction between dopamine and acetylcholine occurs there rather than at the membrane of the cholinergic neurone. The diversity and complexity of neurochemical inputs to medium spiny neurones at present precludes the creation of an adequate model for striatal processing of cortical information, but activity in striatal output pathways is more accessible to experiment, and allows us to draw tentative conclusions on the outcome, if not the nature, of striatal information processing.

STRIATAL OUTPUT PATHWAYS

Striatal cartography has been revised in the light of refinements in anatomical tracing techniques, and new concepts are relevant to the study of clinical problems. First, whereas all areas of cortex project to the striatum, striatal outflows via the pallidum, substantia nigra and thalamus directly influence restricted regions of cortex. In the primate these include frontal areas comprising the supplementary motor area, dorsolateral cortex, frontal eye fields, lateral orbitofrontal area, and anterior cingulate cortex (Percheron *et al.*, 1984; Strick, 1985; Alexander *et al.*, 1986) but not primary motor cortex, as had been proposed in earlier schemes (Kemp and Powell, 1970, 1971). Second, there is increasing evidence that topographical segregation of cortico-striatal terminals within the striatum is at least partly maintained in the output pathways. Schell and Strick (1984) and Evarts and Wise (1984) have drawn attention to the close relationship between the basal ganglia and the supplementary motor area (SMA), and between the cerebellum and the primary motor area, such that interactions between these motor systems occur via transcortical rather than subcortical links (Fig. 2.4) and parallel thalamic channels convey information from cerebellum and basal ganglia to cortical areas. DeLong and Georgopoulos (1981) and Alexander *et al.* (1986) have taken this further to suggest that the basal ganglia may be organized

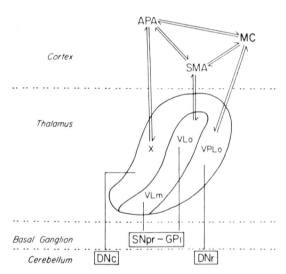

Fig. 2.4 Summary of anatomical relationships between cerebellar and basal ganglia efferents and motor and premotor areas. This diagram illustrates: (1) the pathway from caudal portions of the deep cerebellar nuclei (DNc), to area X and the arcuate premotor area (APA); (2) the pathways from the pars reticulata of the substantia nigra (SNpr) and the internal segment of the globus pallidus (GPi), to VLm and VLo and the supplementary motor area (SMA); (3) the pathway from rostral portions of the deep cerebellar nuclei (DNr), to VPLo and the motor cortex (MC); and (4) the reciprocal connections between the MC, APC, and SMA. (Reproduced, with permission, from Schell and Strick, 1984.)

as parallel, segregated functional circuits with the implication that these circuits might have the capacity to preserve specificity of information at each level of the system (Fig. 2.5).

If it is now possible to see how the striosome–matrix organization of the striatum may provide the necessary level of anatomical heterogeneity to permit segregation of different functional inputs within the striatum and at the same time to allow interaction between such inputs, it is less easy to see how such specificity can be maintained in the pallidal and nigral output stations. Percheron *et al.* (1984) have shown that dendritic arbors of neurones from macaque pallidum form broad flat discs orientated at right angles to incoming striato-pallidal fibres, such that as few as 5–6 overlapping discs cover the lateral border of the pallidum, through which pass all striatal axons. Bearing in mind the funnel shape of the pallidal system, they estimate that up to 500 000 striatal axons might pass through one of the most medial pallidal dendritic discs. The organization of nigral neurones with respect to striatal inputs is similar. Although convergence of striatal information is a striking feature of these output stations, Percheron *et al.* (1984) note that some pallidal neurones receive only sensorimotor or associative information, so the notion of the basal ganglia being composed of a restricted number of "parallel pathways" is not necessarily in conflict with these anatomical observations. Furthermore, physiological evidence supports preservation of somatotopic and functional specificity in the pallidum (DeLong and Georgopoulos, 1981; Georgopoulos *et al.*, 1981, 1983) and in the substantia nigra (Hikosaka and Wurtz, 1983a). The latter showed that some nigro-collicular neurones selectively altered their rate of discharge prior to memory-contingent saccades (i.e. saccades towards the position of a remembered but no longer visible stimulus) indicating that nigral GABAergic output neurones have the capability to respond selectively to specific visuo-spatial and "cognitive" information. This dependence of neuronal activity upon cognitive or functional state may be a general feature of basal ganglia systems, for similar phenomena have been observed in SMA, frontal eye field, striatal, and pallidal neurones (Evarts and Wise, 1984).

SUBSTANTIA NIGRA, PALLIDUM AND SUBTHALAMIC NUCLEUS

The functions of striato-nigral and striato-pallidal outflows have been the subject of intensive study, although much of the work has centred on animal models such as circling behaviour induced by manipulating basal ganglia function with lesions or drugs, and these may be of limited relevance to the situation in primates (Pycock, 1980). It is generally agreed that striatal GABAergic projection neurones inhibit the activity of GABAergic pallidal

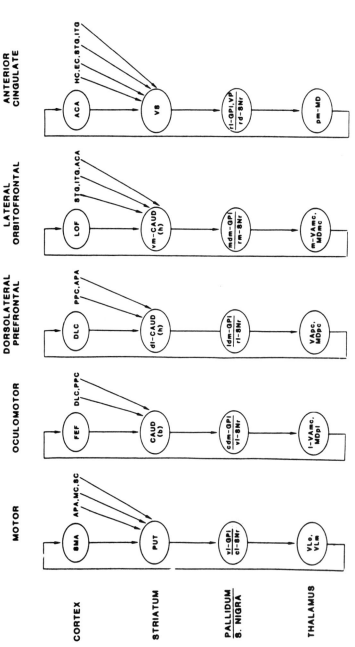

Fig. 2.5 Proposed organization of basal ganglia–thalamocortical circuits as a series of parallel pathways engaging specific regions of the cortex, striatum, substantia nigra and thalamus. Abbreviations: ACA, anterior cingulate area; APA, arcuate motor area; CAUD, caudate nucleus, (b) body, (h) head; DLC, dorsolateral prefrontal cortex; EC, entorhinal cortex; FEF, frontal eye fields; GPi, internal segment of the globus pallidus; HC, hippocampal cortex; ITG, inferior temporal gyrus; LOF, lateral orbitofrontal cortex; MC, motor cortex; MDmc, medialis dorsalis pars magnocellularis; MDpc, medialis dorsalis pars parvocellularis; MDpl, medialis dorsalis pars paralamellaris; PPC, posterior parietal cortex; PUT, putamen; SC, somatosensory cortex; SMA, supplementary motor area; SNr, substantia nigra pars reticulata; STG, superior temporal gyrus; VAmc, ventralis anterior pars magnocellularis; VApc, ventralis anterior pars parvocellularis; VLm, ventralis lateralis pars medialis; VLo, ventralis lateralis pars oralis; VP, ventral pallidum; VS, ventral striatum; cl-, caudolateral; cdm-, caudal dorsomedial; dl-, dorsolateral; l-, lateral; ldm-, lateral dorsomedial; m-, medial; mdm-, medial dorsomedial; pm-, posteromedial; rd-, rostrodorsal; rl-, rostrolateral; rm-, rostromedial; vm-, ventromedial; vl-, ventrolateral. (Reproduced, with permission, from Alexander *et al.*, 1986.)

and substantia nigra pars reticulata (SNR) neurones which show high rates of spontaneous activity, with the result that pallidal and nigral target areas (i.e. thalamus, tectum and reticular formation) become disinhibited (Dray, 1979; Chevalier et al., 1981; Chevalier and Deniau, 1984; Deniau and Chevalier, 1984). There is convincing evidence from the work of Hikosaka and Wurtz (1985a,b) that this mechanism, which is well documented in the rat, applies also to the primate basal ganglia. Furthermore, abnormalities of eye movements which follow manipulation of GABA function in the primate SNR and superior colliculus mirror abnormalities detected in parkinsonian patients (White et al., 1983). The targets of GPi and SNR GABA neurones include the ventral and dorsomedial thalamic nuclei (Fig. 2.5), the superior colliculus and adjacent reticular formation, and the pedunculopontine nucleus. SNR has been considered a caudal extension of the internal pallidal segment (DeLong and Georgopoulos, 1979) concerned particularly with orienting movements of the head and neck, and with control of eye movements via major outputs to the superior colliculus and reticular formation (Sprague, 1975; Kilpatrick et al., 1982; Leigh et al., 1983; Hikosaka and Wurtz, 1985a,b).

New information on the pathophysiology of striato-fugal pathways in parkinsonism comes from 2-DG uptake studies in monkeys rendered parkinsonian by MPTP (Mitchell et al., 1986). In these monkeys, increased uptake of 2-DG was observed in both pallidal segments and in the ventro-anterior and ventro-lateral thalamus, with reduction of 2-DG uptake in the subthalamic nucleus (Table 2.2). These changes were interpreted as reflecting increased activity in striato-pallidal and pallido-thalamic neurones, with decreased activity in pallido-subthalamic neurones, since the increased uptake of 2-DG reflects mainly activity in terminals rather than neuronal perikarya (Crossman, 1987). This implies that the consequence of striatal dopamine deficiency in this paradigm is hyperactivity of striato-pallidal neurones, which in turn would be expected to inhibit pallido-thalamic neurones, although this should lead to a *decrease* rather than increase in 2-DG uptake in thalamic relay nuclei. An alternative explanation, which would account for the observed changes in 2-DG uptake, is that inhibition of pallido-subthalamic neurones due to striato-pallidal hyperactivity results in increased activity of excitatory subthalamo-pallidal inputs upon a population of pallidal thalamic projecting neurones which may therefore continuously and inappropriately inhibit thalamic neurones so interfering with the normal operation of motor programmes (Marsden, 1982; Hikosaka and Wurtz, 1985a, b; Crossman, 1987). Surprisingly, increased uptake of 2-DG was not seen in SNR, perhaps reflecting the selective involvement of a subpopulation of striatal neurones which project to the pallidum (Feger and Crossman, 1984; Parent et al., 1984) and underlining the association between parkinsonism and putamino-pallidal as opposed to caudato-nigral dysfunction (Crossman, 1987). The

Table 2.2 Summary of alterations of 2-DG uptake determined by autoradiography in experimental movement disorders in primates (for references see text).

| Movement disorder | Model | Neostriatum | Changes in regional 2-DG uptake in basal ganglia areas | | | | | VA/VL thalamus | Effects on thalamo-cortical neurones |
			GP$_e$	GP$_i$	STN	SNR	PPN		
Parkinsonism	MPTP (unilateral)	↑	↑	—	→	—	↑	↑	Suppression (increased inhibition)
	MPTP (bilateral)	—	↑	↑	→	—	↑	↑	
Contralateral hemichorea	GABA antagonists into ventral putamen or GP$_E$	—	—	→	↑	—	→	→	Activation (decreased inhibition)
Contralateral hemiballism	GABA antagonist into STN	—	→	→	—	—	—	→	Activation (decreased inhibition)

Abbreviations: 2-DG, ^{14}C-2-deoxyglucose; GP$_e$, lateral pallidal segment; GP$_i$, medial pallidal segment; PPN, pedunculopontine nucleus; SNR, substantia nigra, pars reticulata; STN, subthalamic nucleus; VA/VL thalamus, ventroanterior and ventrolateral thalamic nuclei.

importance of STN is underlined by the well-known association of lesions of the subthalamic nucleus (STN) with hemiballism (Martin, 1927). Its major connections are summarized in Fig. 2.6. Some electrophysiological studies (e.g. Perkins and Stone, 1980, 1981) suggested that STN neurones were inhibitory upon the pallidum and by implication, upon SNR, since single STN neurones send branching fibres to the pallidum and SNR (Deniau *et al.*, 1978; Van der Kooy and Hattori, 1980). The balance of evidence now favours an excitatory influence, possibly mediated by glutamate, although there may be populations of inhibitory and excitatory neurones within STN (Hammond *et al.*, 1978; Kitai *et al.*, 1985; Kita and Kitai, 1987; Crossman *et al.*, 1988). In a reappraisal of the functional anatomy of STN, Kita and Kitai (1987) point out that STN neurones "contact those GP and SNR neurones which are influenced by topographically similar regions of the neostriatum". Furthermore, STN may be able to activate pallidal and nigral neurones as a result of its direct input from the neocortex, *before* impulse flow through striato-pallido-nigral pathways leads to their inhibition (Kita and Kitai, 1987). STN therefore occupies a key position in basal ganglia circuitry.

This has been confirmed recently in experiments which have direct bearing upon understanding of movement disorders. Injection of GABA antagonists into the monkey STN can induce contralateral dyskinesia virtually identical

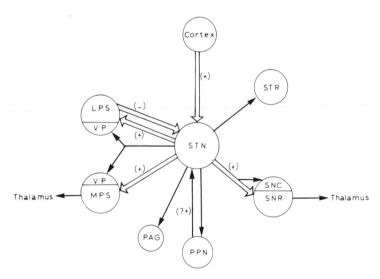

Fig. 2.6 Connections of the subthalamic nucleus (STN) showing putative excitatory (+) and inhibitory (−) action between STN and neocortex, lateral pallidal segment (LPS), medial pallidal segment (MPS), ventral pallidum (VP), substantia nigra (SNR and SNC) and pedunculopontine nucleus (PPN).

to hemiballismus seen in human subjects (Crossman *et al.*, 1980; Jackson and Crossman, 1984). Evidence from 2-DG studies suggests that hemiballistic movements are associated with decreased uptake in both segments of the pallidum, in STN, and in the ventral thalamic nuclei (Mitchell *et al.*, 1985a, b) indicating decreased impulse flow in the subthalamo-pallidal tract and in the pallido-thalamic pathways. This is in contrast to the situation in parkinsonism, where there is increased uptake in the pallidal segments and thalamic nuclei, as discussed above. The notion that decreased impulse flow in the (excitatory) subthalamo-pallidal pathway is of critical importance in the pathophysiology of dyskinesias is supported by the observation that injection of the endogenous glutamate receptor antagonist kynurenic acid into the lateral aspect of GPi elicits contralateral choreiform or hemiballistic movements (Robertson *et al.*, 1987; Crossman *et al.*, 1988).

To summarize, in parkinsonism, pallido-thalamic outputs are hyperactive (presumably driven by excitatory STN inputs) resulting in excessive inhibition of thalamo-cortical neurones, whereas in chorea and hemiballism pallido-thalamic pathways are underactive, resulting in abnormal activation of thalamo-cortical neurones. These findings indicate that excitatory neuro-transmission (perhaps glutamatergic) subthalamofugal pathways is likely to be important in human movement disorders. Of course, the analogy between these experimental movement disorders and human movement disorders is a little misleading, since bradykinesia, which is not seen in experimental hemichorea, is a feature of PD and Huntington's disease, as well as a variety of other disorders (Thompson *et al.*, 1988). Penney and Young (1986) have suggested that in PD there is loss of excitatory dopaminergic input to the striatal neurones projecting to GPi and SNR. They also propose that the distinguishing feature between PD and Huntington's disease would be selective or predominant involvement of the striatal projections to the pallidum. In PD there would be increased activity of STN neurones projecting to GPi (and SNR), and because the ultimate effect would be to inhibit thalamo-cortical circuits, there would be difficulty switching to new behaviours or "learnt motor plans" (Marsden, 1982). This would be in accord with the conclusion that there is excessive activation of some subthalamo-pallidal pathways in PD. Penney and Young (1986) point out that disruption (albeit by different mechanisms) of striatal throughput could account for rigidity and bradykinesia in PD and Huntington's disease. Choreiform movements in the latter could be explained by early selective loss of striatal matrix neurones projecting to GPe, leading to disinhibition of GPe neurones, and excessive inhibition of subthalamic neurones, the effect of which would be to reduce inhibitory drive upon thalamo-cortical neurones, so impairing suppression of unwanted movements. In keeping with this notion, the 2-DG data (Table 2.2) suggest that there is decreased impulse flow in the subthalamo-pallidal pathways in experimental chorea and hemiballismus (Crossman *et al.*, 1988). The consequence would be impaired suppression of unwanted movements, and, depending upon the degree of selective striatal cell loss, the clinical picture would include "parkinsonian" features.

Some caution is needed in drawing conclusions concerning the pathophysiology of movement disorders from changes in focal 2-DG uptake, but further experiments of this kind are likely to provide important new insights into the pathogenesis of movement disorders. We lack a satisfactory model of dystonia, but evidence from brain imaging in patients with hemidystonia suggests that the syndrome may be associated with lesions in the pallidum, ventral thalamus or upper brain stem (Marsden *et al.*, 1985; Leenders *et al.*, 1986). Dystonia is related in some instances to striatal dopamine deficiency (Leenders *et al.*, 1986) so it is possible that the clinical manifestations of dystonia are related to selective involvement of certain striatal neurones or output systems.

BASAL GANGLIA ANATOMY AND FUNCTION: FUTURE PERSPECTIVES

Study of basal ganglia anatomy cannot by itself generate usefully specific hypotheses about the functions of these structures or the mechanisms of movement disorders. Fortunately, techniques now exist to correlate physiological and pathological events with features of basal ganglia structure, both in experimental animals and in man. It has to be admitted that the relevance of newer concepts of basal ganglia organization to our understanding of movement disorders remains speculative, although evidence that the striosome–matrix arrangement has functional importance in experimental animals is now compelling, and work on the selective involvement of one or other component in disease may provide insight into the pathogenesis of movement disorders (Ferrante *et al.*, 1986; Penney and Young, 1986). The question as to precisely how dopamine and acetylcholine function in the striatum remains largely unresolved, although intracellular recordings show that dopamine may be either inhibitory or excitatory upon striatal neurones, so change of one or other action may generate different abnormalities of movement, or cognitive function. The notion that the basal ganglia are organized around multiple segregated parallel pathways (Alexander *et al.*, 1986) provides testable hypotheses and raises fundamental questions about the nature of convergence of striatal inputs at the level of the pallidum and substantia nigra (Percheron *et al.*, 1984) which will have to be taken into account if we are to achieve a satisfactory model of basal ganglia information processing.

Despite the wealth of anatomical, biochemical and physiological detail which has accrued from animal studies over the last 15 years, one can argue that careful study of patients with movement disorders exemplified by Wilson (1912, 1925), Schwab and England (1958), Martin (1967) and Marsden (1982), amongst many others, has furnished information on basal ganglia function of equal or greater importance. In arguing that the essential defect in Parkinson's disease is "the inability of patients to automatically execute learnt motor plans", Marsden (1982, 1984) summarizes a concept of basal ganglia function which is certainly compatible with the evidence reviewed above: "The multiple inputs from all areas of cerebral cortex into the striatum provide moment to moment information on the external environment and the wishes and feelings of the individual. At any moment, such inputs may alter the execution of the motor plan, in response to changes in the environment or aims of the subject. Since the workings of the basal ganglia are regarded as subconscious, such adjustments would be automatic. The internal workings of the striopallidal complex, with the subthalamus, could digest this mass of information and reduce it to appropriate signals relayed via pallidal output to thalamus and thence to premotor (and other frontal) cortex, so as to adjust the form and sequence of motor programmes... Viewed in this light, the Parkinsonian patient's motor deficit is reduced to inability to specify the detailed accuracy of the motor programmes and to automatically run their sequence, switching from one programme to another as required."

These difficulties in switching from one programme to another and in scaling the amount of muscle activity to the task in question, detected in parkinsonian patients, may well be related to the effects of increased inhibition of thalamo-cortical neurones as predicted by electrophysiological and 2-DG studies. It seems likely that the diversity of symptoms and signs in movement disorders may be related to involvement of topographically and functionally discrete populations of striatal, pallidal, subthalamic and nigral neurones, and that changes in the activity of such areas in human movement disorders may be revealed in the near future by increasingly refined nuerophysiological and *in vivo* imaging techniques.

REFERENCES

Alexander, G. E., DeLong, M. and Strick, P. L. (1986) *Ann. Rev. Neurosci.* 9, 357–381.
Aronin, N., Difiglia, M., Graveland, G. A., *et al.* (1984) *Brain Res.* 300, 376–380.
Beal, M. F. and Martin, J. B. (1986) *Ann. Neurol.* 20, 547–565.
Bernheimer, H., Birkmayer, W., Hornykiewicz, O. and Seitelberger, F. (1973) *J. Neurol. Sci.* 20, 415–455.

Bolam, J. P. (1984) In *Functions of the Basal Ganglia*, CIBA Foundation Symposium No. 107 (edited by D. Evered and M. O'Connor), pp. 30–42. Pitman, London.

Bolam, J. P. and Izzo, P. N. (1988) *J. Comp. Neurol.* **269**, 219–234.

Bolam, J. P., Izzo, P. N. and Graybiel, A. M. (1988) *Neuroscience* **24**, 853–875.

Brown, R. G. and Marsden, C. D. (1988) *Neuroscience* **25**, 363–387.

Butcher, L. L. and Hodge, G. K. C. (1976) *Brain Res.* **106**, 223–240.

Chesselet, M.-F. and Graybiel, A. M. (1983a) *Life Sci.* **33**, 37–40

Chesselet, M.-F. and Graybiel, A. M. (1983b) *Neurosci. Abstr.* **9**, 16.

Chevalier, G. and Deniau, J. M. (1984) *Neuroscience* **12**, 427–439.

Chevalier, G., Deniau, J. M., Thierry, A. M. and Feger, J. (1981) *Brain Res.* **213**, 253–263.

Crossman, A. R. (1987) *Neuroscience* **21**, 1–40.

Crossman, A. R. and Jackson, A. (1984). *J. Physiol. (Lond)* **350**, 36P.

Crossman, A. R., Mitchell, I. J., Sambrook, M. A. and Jackson, A. (1988) *Brain* **111**, 1211–1233.

Crutcher, M. D. and DeLong, M. (1984) *Exp. Brain Res.* **53**, 233–243.

DeLong, M. and Georgopoulos, A. (1979) *Adv. Neurol.* **24**, 131–140.

DeLong, M. and Georgopoulos, A. (1981) In *Handbook of Physiology, the Nervous System*, Vol. 2 (edited by J. M. Brookhart and V. B. Mountcastle), pp. 1017–1061. American Physiological Society, Bethesda, MD.

Deniau, J. M. and Chevalier, G. (1984). In *Functions of the Basal Ganglia*, CIBA Foundation Symposium No. 107 (edited by D. Evered and M. O'Connor), pp. 48–63. Pitman, London.

Deniau, J. M., Hammond, C., Chevalier, G. and Feger, J. (1978). *Neurosci. Lett.* **9**, 117–121.

Divac, I. (1983) *Neuroscience* **10**, 1151–1155.

Donoghue, J. P. and Herkenham, M. (1986) *Brain Res.* **365**, 397–403.

Dray, A. (1979). *Neuroscience* **4**, 1407–1439.

Dube, L., Smith, A. D. and Bolam, J. P. (1988) *J. Comp. Neurol.* **267**, 455–471.

Evarts, E. V. and Wise, S. P. (1984) In *Functions of the Basal Ganglia*, CIBA Foundation Symposium No. 107 (edited by D. Evered and M. O'Connor), pp. 83–102. Pitman, London.

Feger, J. and Crossman, A. R. (1984). *Neurosci. Lett.* **49**, 7–12.

Ferrante, R. J., Kowall, N. W., Richardson, E. P., Bird, E. D. and Martin, J. B. (1986) *Neurosci. Lett.* **71**, 283–288.

Fuxe, K., Agnati, L. F., Harfstrand, A., Ogren, S.-O. and Goldstein, M. (1986) In *Recent Developments in Parkinson's Disease* (edited by S. Fahn *et al.*), pp. 17–32. Raven Press, New York.

Gauchy, C., Kemel, M. L., Desban, M. and Glowinski, J. (1987) In *Proceedings of International Symposium on the Basal Ganglia*, University of Leeds, July 1987.

Georgopoulos, A. P., DeLong, M. R. and Crutcher, M. D. (1983) *J. Neurosci.* **3**, 1586–1598.

Gerfen, C. R. (1984) *Nature* **311**, 461–464.

Gerfen, C. R., Baimbridge, K. G. and Thibault, J. (1987) *J. Neurosci.* **7**, 3935–3944.

Goedert, M., Mantyh, P. W., Hunt, S. P. and Emson, P. C. (1983) *Brain Res.* **274**, 176–179.

Goldman, P. S. and Nauta, W. J. H. (1977) *J. Comp. Neurol.* **171**, 369–386.

Gotham, A.-M., Brown, R. G. and Marsden, C. D. (1986) *J. Neurol. Neurosurg. Psychiat.* **49**, 381–389.

Graybiel, A. M. (1983) In *Progress in Brain Research* (edited by J.-P. Changeux, J.

Glowinski, M. Imbert and F. E. Bloom), pp. 247–255. Elsevier Science Publishers, Amsterdam.

Graybiel, A. M. (1984) *Neuroscience* **13**, 1157–1187.

Graybiel, A. M. (1986) In *Recent Developments in Parkinson's Disease* (edited by S. Fahn *et al.*), pp. 1–16. Raven Press, New York.

Graybiel, A. M. and Chesselet, M. F. (1984) *Proc. Natl. Acad. Sci. USA* **81**, 7980–7984.

Graybiel, A. M. and Ragsdale, C. W. (1978) *Proc. Natl. Acad. Sci. USA* **75**, 5723–5726.

Graybiel, A. M., Ragsdale, C. W. and Moon Edley, S. (1979) *Exp. Brain Res.* **34**, 189–195.

Graybiel, A. M., Chesselet, M.-F., Wu, J.-Y., Eckenstein, F. and Joh, T. E. (1983) *Neurosci. Abstr.*, **9**, 14.

Graybiel, A. B., Baughman, R. W. and Eckenstein, F. (1986) *Nature* **323**, 625–627.

Greenfield, J. G. and Bosanquet, F. D. (1953). *J. Neurol. Neurosurg. Psychiat.* **16**, 213–226.

Haber, S. N. and Elde, R. P. (1981) *Neuroscience* **6**, 1291–1297.

Hammond, C., Deniau, J. M., Rouzaire-Dubois, B. and Feger, J. (1978). *Neurosci. Lett.* **9**, 171–176.

Hassler, R. (1937) *J. Psychol. Neurol. (Lpz)* **48**, 387–476.

Heimer, L. and Wilson, R. D. (1975). In *Golgi Centennial Symposium* (edited by M. Santini), pp. 177–193. Raven Press, New York.

Heimer, L., Switzer, R. D. and Van Hoesen, G. W. (1982) *Trends Neurosci.* **5**, 83–87.

Herkenham, M. and Pert, C. B. (1981) *Nature* **291**, 415–418.

Herkenham, M., Moon Edley, S. and Stuart, J. (1984) *Neuroscience* **11**, 561–593.

Hikosaka, O. and Wurtz, R. H. (1983a) *J. Neurophysiol.* **49**, 1268–1284.

Hikosaka, O. and Wurtz, R. H. (1983b) *J. Neurophysiol.* **49**, 1285–1301.

Hikosaka, O. and Wurtz, R. H. (1985a) *J. Neurophysiol.* **53**, 266–291.

Hikosaka, O. and Wurtz, R. H. (1985b) *J. Neurophysiol.* **53**, 292–308.

Iversen, S. D. (1984) In *Functions of the Basal Ganglia*, CIBA Foundation Symposium No. 107 (edited by R. Evered and M. O'Connor), pp. 183–195. Pitman, London.

Izzo, P. N. and Bolam, J. P. (1988) *J. Comp. Neurol.* **269**, 219–234.

Jackson, A. and Crossman, A. R. (1984). *Neurosci Lett.* **46**, 41–45.

Jellinger, K. (1988) *J. Neurol. Neurosurg. Psychiat.* **51**, 540–543.

Jones, E. G., Coulter, J. D., Burton, H. and Porter, R. (1977) *J. Comp. Neurol.* **173**, 53–80.

Joyce, J. N., Douglas, W. S. and Marshall, J. F. (1986) *Proc. Natl. Acad. Sci. USA* **83**, 8002–8006.

Kalil, K. (1978) *Brain Res.* **140**, 333–339.

Kemp, J. M. and Powell, T. P. S. (1970) *Brain* **93**, 525–546.

Kemp, J. M. and Powell, T. P. S. (1971) *Phil. Trans. Roy. Soc. London (Ser. B)* **262**, 441–457.

Kilpatrick, I. C., Starr, M. S., Fletcher, A., James, T. A. and MacLeod, N. K. (1982) *Exp. Brain Res.* **40**, 45–54.

Kita, H. and Kitai, S. T. (1987). *J. Comp. Neurol.* **260**, 435–452.

Kitai, S. T., Nakanishi, H. and Kita, H. (1985). *Soc. Neurosci. Abstr.* **110**, 4.

Kowall, N. W., Ferrante, R. J. and Martin, J. B. (1987) *Trends Neurosci.* **10**, 24–29.

Kunzle, H. (1975) *Brain Res.* **88**, 195–210.

Kunzle, H. (1977) *Exp. Brain Res.* **30**, 481–492.

Kunzle, H. (1978) *Brain Behav. Evol.* **15**, 185–234.

Leenders, K. L., Frackowiak, R. J. J., Quinn, N., Brooks, D., Sumner, D. and Marsden, C. D. (1986) *J. Movement Disorders* **1**, 51–58.

Leigh, P. N. (1987) In *Parkinson's Disease. Clinical and Experimental Advances* (edited by F. Clifford Rose), pp. 13–17. John Libbey, London, Paris.

Leigh, P. N., Reavill, C., Jenner, P. and Marsden, C. D. (1983) *J. Neural Trans.* **58**, 1–41.

Marsden, C. D. (1982) *Neurology* **32**, 514–539.

Marsden, C. D. (1984) In *Functions of the Basal Ganglia*, CIBA Foundation Symposium No. 107 (edited by D. Evered and M. O'Connor), pp. 225–236. Pitman, London.

Marsden, C. D., Obeso, J. A., Zarranz, J. J. and Lang, A. E. (1985) *Brain* **108**, 463–483.

Martin, J. P. (1927) *Brain* **50**, 637–651.

Martin, J. P. (1967) *The Basal Ganglia and Posture*. Pitman Medical Publishing Co., London.

McCulloch, J., Kirkham, K., MacPherson, H. and Sharkey, J. (1983) *J. Cereb. Blood Flow Metab.* **3** (suppl), S258–259.

Mensah, P. L. (1977) *Brain Res.* **137**, 53–66.

Mettler, F. (1968) In *Diseases of the Basal Ganglia. Handbook of Neurology*, Vol. 6 (edited by P. J. Vinken and G. W. Bruyn), pp. 1–55. North Holland Pub. Co., Amsterdam.

Mitchell, I. R., Jackson, A., Sambrook, M. A. and Crossman, A. R. (1985a) *Brain Res.* **339**, 346–350.

Mitchell, I. R., Sambrook, M. A. and Crossman, A. R. (1985b). *Brain* **108**, 421–438.

Mitchell, I. R., Cross, A. J., Sambrook, M. A. and Crossman, A. R. (1986) *J. Neural Trans.* Suppl. 20, 41–46.

Morelli, M. and Di Chiara, G. (1984) *Neuropharmacology* **23**, 847.

Nastuk, M. A. and Graybiel, A. M. (1985) *J. Comp. Neurol.* **237**, 176–194.

Nauta, W. J. H. (1979) *Neuroscience* **4**, 1875–1881.

Olson, L., Seiger, A. and Fuxe, K. (1972) *Brain Res.* **44**, 283–288.

Parent, A., Bouchard, C. and Smith, Y. (1984) *Brain Res.* **303**, 385–390.

Pasik, P., Pasik, T., Pecci Saavedra, J. and Holstein, G. R. (1981) *Anat. Rec.* **199**, 194A.

Penney, G. P., Afsharpour, S. and Kitai, S. T. (1986) *Neuroscience* **17**, 1011–1045.

Penney, J. B. and Young, A. B. (1986) *J. Movement Disorders* **1**, 3–16.

Percheron, G., Yelnik, J. and Francois, C. (1984) In *The Basal Ganglia* (edited by J. S. McKenzie, R. E. Kemm and L. N. Wilcock), pp. 87–105. Pitman, London.

Perkins, M. N. and Stone, T. W. (1980). *Exp. Neurol.* **68**, 500–511.

Perkins, M. N. and Stone, T. W. (1981). *Quart. J. Exp. Physiol.* **66**, 225–236.

Pert, C. D., Kuhar, M. J. and Snyder, S. H. (1976) *Proc. Natl. Acad. Sci. USA* **73**, 3729–3733.

Pycock, C. J. (1980) *Neuroscience* **5**, 461–514.

Robertson, R. G., Farmery, S. M., Sambrook, M. A. and Crossman, A. R. (1987). In *Proceedings of the International Symposium on the Basal Ganglia*. University of Leeds, July 1987. (Abstract).

Royce, G. J. (1978) *Brain Res.* **146**, 145–150.

Sandell, J. H., Graybiel, A. M. and Chesselet, M. F. (1986) *J. Comp. Neurol.* **243**, 326–334.

Schell, G. R. and Strick, P. L. (1984) *J. Neurosci.* **4**, 539–560.

Schwab, R. S. and England, A. C. (1958) *J. Chron. Dis.* **8**, 488–509.

Selemon, L. D. and Goldman-Rakic, P. S. (1985) *J. Neurosci.* **5**, 776–794.

Semba, K., Fibiger, H. C. and Vincent, S. R. (1987) *Can. J. Neurol. Sci.* **14(B)**, 386–394.

Somogyi, P., Bolam, J. P. and Smith, A. D. (1981) *J. Comp. Neurol.* **195**, 567–584.

Somogyi, P., Priestley, J. V., Cuello, A. C., Smith, A. D. and Takagi, H. (1982) *J. Neurocytol.* **11**, 779–807.

Sprague, J. M. (1975) In *Neuroscience Research Symposium Bulletin*, Vol. 13 (edited by D. Inge and J. M. Sprague), pp. 204–213.

Starkstein, S. E., Brandt, J., Folstein, S., *et al.* (1988) *J. Neurol. Neurosurg. Psychiat.* **51**, 1259–1263.

Strick, P. L. (1985). *Behav. Brain Res.* **18**, 107–123.

Thompson, P. D., Berardelli, A., Rothwell, J. C., Day, B. L., Dick, J. P. R., Benecke, R. and Marsden, C. D. (1988). *Brain* **111**, 223–244.

Tretiakoff, C. (1919) Thèse, Paris, No. 293.

Van de Kooy, D. and Hattori, T. (1980). *J. Comp. Neurol.* **192**, 751–768.

Vincent, S. R., Hokfelt, T., Christensson, I. and Terenius, L. (1982). *Eur. J. Pharmacol.* **85**, 251–252.

Wainer, B. H., Bolam, J. P., Clarke, D. J., *et al.* (1983) *Soc. Neurosci. Abstr.* **9**, 963.

White, O. D., Saint-Cyr, J. A. and Sharpe, J. A. (1983) *Brain* **106**, 571–588.

Whittier, J. R. and Mettler, F. A. (1949) *J. Comp. Neurol.* **90**, 319–372.

Wilson, S. A. K. (1912) *Brain* **36**, 427–492.

Wilson, S. A. K. (1925) *Lancet* **ii**, 268–276.

Zahm, D. S., Zaborsky, L., Alones, V. E., *et al.* (1985) *Brain Res.* **325**, 317–321.

3

The Pathology of Parkinsonian Disorders

W. R. G. Gibb

INTRODUCTION

This chapter will address Parkinson's disease and other degenerative parkinsonian disorders (the Parkinson-plus syndromes), thus excluding secondary or symptomatic disorders such as drug-induced, post-encephalitic and pugilistic parkinsonism. There are three principal aims in studying the pathology of these degenerative disorders. First, the identification of appropriate features for pathological diagnosis with the purpose of improving clinico-pathological correlation and clinical diagnosis. Second, the investigation of aetiology by providing pathological data for the purposes of descriptive epidemiology. Third, the study of specific pathological features with the aim of elucidating pathogenetic mechanisms. There is space here only to concentrate on the first of these objectives, thus considering a pathological and clinico-pathological description of these disorders.

In most degenerative parkinsonian disorders the pathological diagnosis can be made by macroscopic examination of the sliced brain, although microscopic examination allows definitive classification. Bradykinesia and muscular rigidity are believed to result from degeneration in the nigro-striatal system. The substantia nigra invariably shows neuronal depletion, but the striatum shows significant damage in only selected disorders. Nerve

DISORDERS OF MOVEMENT: CLINICAL, PHARMACOLOGICAL
AND PHYSIOLOGICAL ASPECTS   ISBN 0-12-569685-X

cell loss and gliosis in other brain regions are responsible for differences between the various clinical syndromes. The unique combinations of areas affected contribute to the pathological diagnosis. However, in practice a wide and detailed examination of the brain can be laborious if undertaken at the outset. The pathological changes in some areas may be equivocal in degree, and establishing the diagnosis in this way can sacrifice considerable material which might otherwise be used for research investigations. In many of these disorders the degenerate substantia nigra contains neuronal inclusions. These are of diverse types and are often sufficiently characteristic and specific to suggest the correct diagnosis. This emphasizes the advantage of thoroughly examining the substantia nigra as the first step in microscopic examination. If necessary only a relatively small quantity of tissue need be prepared, because the inclusions lie scattered in the nigra and can be rapidly identified. In some instances it is diagnostically useful to quantify them, and they can also be studied in an attempt to unravel pathogenetic mechanisms. Neuronal inclusions imply different kinds of injury and probably represent specific products, or non-specific by-products of degeneration, rather than indicating attempts at regeneration. Once the substantia nigra has been studied then the distribution of cerebral lesions can be established to confirm the diagnosis, if necessary without disrupting too much material.

Clinical series of patients believed to have Parkinson's disease are contaminated by undiagnosed cases of multiple system atrophy (MSA), Steele–Richardson–Olszewski syndrome (SRO; progressive supranuclear palsy) and so on. The converse can also occur. For example, the unusual and therefore unexpected occurrence of autonomic failure in Parkinson's disease may lead to a mistaken diagnosis of MSA (Shy–Drager syndrome). Even careful clinical scrutiny may fail to identify as many as 10–15% of patients with parkinsonian disorders other than Parkinson's disease. The true relative frequency of these disorders is not exactly known because of bias inherent in referrals for pathological study, but clinical and pathological surveys suggest figures of approximately 70–75% for Parkinson's disease, 10–15% for MSA, 5–10% for SRO and 1–5% for corticobasal degeneration (CBD) and other degenerative parkinsonian diseases.

PARKINSON'S DISEASE

History and diagnosis

Approximately a century after James Parkinson speculated that pathological changes in Parkinson's disease resided in the upper cervical spinal cord and medulla oblongata three important developments took place. Professor Brissaud suggested that damage to the substantia nigra, as recently seen in a patient at the Salpêtrière with a midbrain tuberculoma, might also be the cause of the bradykinesia and rigidity of Parkinson's disease (Brissaud, 1895). Secondly, Friederich Lewy working on cases of Parkinson's disease in Berlin in 1912 recognized a particular kind of inclusion body in nerve cells of the dorsal vagal nucleus and nucleus basalis (Lewy, 1912). Lastly, Constantin Tretiakoff working in Paris in 1919 implied that these "Lewy bodies" were specific to Parkinson's disease (Tretiakoff, 1919). He found concentric hyaline inclusions in the substantia nigra in six of nine typical cases of Parkinson's disease and in one of three "catatonic" cases, but he was unable to find them in a variety of other disorders. Sadly, owing to difficulties in the clinical diagnosis of various parkinsonian syndromes, the significance of these inclusions has remained a matter of dispute until recently.

Historically the next major event was the endemic spread of encephalitis lethargica in Europe, North America and other parts of the world between 1917 and 1925. This had a considerable impact on the understanding of the pathology of Parkinson's disease because approximately half of the survivors developed a parkinsonian disorder, and many patients died. One of the main pathological changes consisted of severe nerve cell loss in the substantia nigra. Unfortunately, the clinical diagnosis of encephalitis lethargica was complicated by a coincident pandemic of influenza accompanied by feverish and lethargic symptoms which started in Italy in 1918 spreading through the rest of Europe in 1919 and lasting until 1925 (Von Economo, 1931). In patients presenting with a parkinsonian syndrome the aetiological role of a flu-like illness was often unclear. Furthermore some physicians were aware of previous outbreaks resembling encephalitis lethargica, such as the nona epidemic, which was probably an encephalitic illness present in Italy in 1890–1891 (Von Economo, 1931). Understandably they were tempted to believe that all cases of Parkinson's disease were essentially post-encephalitic. A diagnosis of post-encephalitic parkinsonian syndrome was therefore positively encouraged even to the extent of assuming previous subclinical encephalitis in virtually all parkinsonian patients with an earlier age of onset than usual. Lewy bodies were frequently identified in the substantia nigra of patients labelled as post-encephalitic and were often not found in apparently idiopathic cases, so that most neuropathologists did not attach much significance to them. In contrast the post-encephalitic parkinsonian syndrome helped to confirm that the substantia nigra was an important site of pathology in parkinsonian disorders, although it was not until Hassler's work of 1938 that a long-standing dispute about nigral versus striatal pathology was mostly resolved (Hassler, 1938). Hassler consistently found lesions in the substantia nigra, but not in the pallidum, in 10 patients with Parkinson's disease in whom pallidal abnormalities had previously been reported by the Vogts (Vogt and Vogt, 1920).

There was little further progress until 1952 when Beheim-Schwarzbach (1952) reported her studies of material in the Vogt collection, including that reported by Hassler (1938). She turned to the locus coeruleus where "Masson-positive vacuoles" (cytoplasmic inclusions with a red staining core) were found in 16 parkinsonian cases with the typical nigral pathology of Parkinson's disease identified by Hassler. These Lewy bodies were not found in 13 post-encephalitic patients or in seven cases with "supranigral" disease. Most other studies at this time failed to find Lewy bodies in all cases of Parkinson's disease, but Lewy bodies were less often seen in apparently post-encephalitic cases (Alvord, 1958; Lipkin, 1959), which were now more reliably distinguished from Parkinson's disease on account of their long survival following a major encephalitic illness.

Forno and Alvord (1971) demonstrated that patients presenting with largely unclassified parkinsonian disorders presented a variety of pathologies. Among 67 patients, 22 (33%) had neither Lewy bodies nor Alzheimer tangles in the substantia nigra, 29 (43%) had Lewy bodies and 16 (24%) had tangles. In contrast to the findings of Beheim-Schwarzbach most studies showed that the exclusion of patients clinically atypical for Parkinson's disease left a residue of 10–15% of patients without nigral Lewy bodies (Lipkin, 1959; Stadlan et al., 1965). To determine whether Lewy bodies occur in every case of Parkinson's disease it is necessary to show, first, that every parkinsonian case without Lewy bodies shows a distribution of neuronal loss and gliosis that differs from that of Parkinson's disease. For example, one of the fundamental distinguishing points of striatonigral degeneration (SND) is involvement of the striatum. Second, Lewy bodies must occur with a certain statistically definable frequency so as to show a clear difference in the pathology between Parkinson's and non-Parkinson's disease patients.

We therefore collected a group of 99 patients, with a clinical diagnosis of Parkinson's disease, that were referred for pathological study (Gibb and Lees, 1989). Retrospective scrutiny of clinical documents led to the identification of features incompatible with typical Parkinson's disease in 21 patients, who were therefore excluded. Eight were believed to have had encephalitis lethargica, four were drug-induced, three had early dementia, four had cerebellar signs, one had motor neurone disease, and one had a "vascular" onset. Of the 78 remaining patients, 74 showed Lewy bodies in the substantia nigra. One of these also had MSA. The four cases without Lewy bodies had the distribution of pathology seen in MSA (two cases) and post-encephalitic parkinsonian syndrome (two cases), but none had demonstrated positive clinical evidence for these diagnoses. For example, the two patients with post-encephalitic parkinsonian syndrome did not have a documented history of encephalitis. They had parkinsonian symptoms for 22 and 43 years before death, disease durations which are compatible with Parkinson's disease.

In 110 cases of Parkinson's disease (comprising the 73 uncomplicated cases with an additional 37) cytoplasmic Lewy bodies occurred with a median frequency of one every 29 pigmented cells, but with a positive skew so that in some cases they were as sparse as one every 200–300 cells (Gibb and Lees, 1989). Logarithmic adjustment of these data showed that in Parkinson's disease Lewy bodies were highly unlikely to occur less often than once every 330 pigmented cells. A crucial point in Parkinson's disease is that neuronal loss in horizontal cross-sections of the substantia nigra is never extreme in degree, and rarely severe, even with long survival. Pigmented cells are reduced to between 20 and 30% of normal so that a 7-μm thick, unilateral section taken from the mid-substantia nigra (level of the oculomotor nerve) which normally contains approximately 900 pigmented cells, has 100–300 cells. Consequently there are always enough cells in two unilateral sections from the mid-substantia nigra to clarify whether Lewy bodies are present or not. If they are not found within two sections or 330 pigmented cells then the diagnosis is highly unlikely to be Parkinson's disease. Therefore, the greatest power of the Lewy body is the exclusion of Parkinson's disease if it cannot be found this easily. If Lewy bodies occur very infrequently then the subject probably had either incidental (presymptomatic) Lewy body disease, coincidental MSA or SRO, or CBD or Hallervorden–Spatz disease (HSD), which are associated with Lewy bodies in 15–40% of cases. The finding of very infrequent Lewy bodies, despite considerable nigral cell depletion, is therefore insufficient to diagnose Parkinson's disease. Furthermore a positive pathological diagnosis depends on identifying 60% or greater nigral cell loss, and the exclusion of other parkinsonian disorders by the absence of other cerebral lesions showing nerve cell loss and gliosis. In the locus coeruleus, Lewy bodies are seen in about twice as many neurons as in substantia nigra, so this can be helpful in diagnosis. However, examination of the locus coeruleus cannot be substituted for examination of the substantia nigra, because it is still necessary to identify significant nigral nerve cell loss for the diagnosis of Parkinson's disease.

Another useful diagnostic point in Parkinson's disease is selective neuronal vulnerability within the substantia nigra. Hassler (1937) identified an internal anatomy in the nigra and established that certain cell groups, for example Spez and Sped, were preferentially destroyed in Parkinson's disease (Hassler, 1938). His work was accurate and meticulous, but proved too detailed to apply in practice. The nigra is constructed from constant and well-defined arrangements of pigmented nerve cells, with the most dominant cell group in the mid-nigra consisting of a ventro-lateral cell strip corresponding to region α defined by Olszewski and Baxter (1954), and in part to groups Spez and Sped of Hassler. In Parkinson's disease cells of this group, starting from the lateral part, are the first to succumb, so by the time the patient dies the area is devoid of nerve cells (Gibb and Lees, 1988a). Remaining parts of the nigra show a constant and ordered pattern of cell degeneration. Nerve cell counts in whole sections of the nigra therefore obscure these specific regional differences. This selective neuronal degeneration within the nigra is not specific to Parkinson's disease, but may be specific to diseases associated with Lewy bodies, such as Hallervorden–Spatz (see below). We have recently observed that neuronal groups in the nigra also vary in their melanin content, such that cells of the ventrolateral part contain least melanin. The susceptibility of nuclear groups throughout the nigra negatively correlates with their melanin content, so it is the least melanized populations which preferentially degenerate. This may have important implications for pathogenesis.

If Lewy bodies are the major diagnostic hallmark of Parkinson's disease, and if there is a small but definite variation in their frequency between cases, to what extent do they correlate with clinical variables and define clinical differences? Numerical assessment of Lewy bodies in relation to nerve cell numbers, including patients with a propensity for multiple inclusions per cell, does not relate to a particular age at onset, disease duration or age at death (Gibb and Lees, 1989).

What is the age range of patients shown pathologically to have Parkinson's disease? Bernheimer et al. (1973) reported patients with Lewy body pathology ranging in age from 20 to 87 years at the time of symptom onset. A recent pathological study compared 12 patients with disease onset at 40 years or less and 22 with onset at 70 years or more (age limits of 23 and 87 years) for the purpose of establishing whether the basic disease process was the same at these extremes of age (Gibb and Lees, 1988b). The young-onset cases showed a median duration of survival that was 12 years longer, and accompanied by 24% greater nigral cell loss, than old-onset cases. The nigral morphology and extent of disease in extranigral sites were otherwise identical, thus failing to support claims that Parkinson's disease is a different process at the extremes of age.

Idiopathic Lewy body disease

The uniform and consistent pathology in Parkinson's disease at the level of the substantia nigra and locus coeruleus justify using the term (idiopathic) Lewy body–Parkinson's disease to emphasize this pathological substratum. The term (idiopathic) Lewy body disease has been introduced as a more inclusive description of the pathological process as well as the various clinical syndromes that result (Gibb, 1987). In order to define and justify this title it is necessary to consider the distribution of Lewy bodies in Parkinson's disease, the occurrence of cerebral Lewy bodies in healthy persons, and the existence of other clinical states (dementia or autonomic failure) with a similar pathological picture.

Distribution of Lewy bodies in Parkinson's disease

The Lewy bodies in Parkinson's disease are widely distributed, but specific to certain nuclear groups. They can be found in every part of the autonomic nervous system, namely the hypothalamus, the Edinger–Westphal, salivatory and dorsal vagal nuclei, the intermediolateral columns of the spinal cord, sympathetic ganglia, and myenteric plexi of the oesophagus, colon and rectum. In addition to the substantia nigra and locus coeruleus they are found in the nucleus basalis and thalamus in virtually all cases, in the parahippocampus and temporal lobe neocortex in about one-third and in raphé and pedunculopontine nuclei. There are also many areas which never contain Lewy bodies, such as the striatum. They are not described in anterior horn motor neurons in Parkinson's disease, despite the rare occurrence of anterior horn cell disease with Lewy body Parkinson's disease (Delisle et al., 1987; Gibb et al., 1989a). In one of these cases Lewy bodies were present in grey matter of the spinal cord, but were not found in the few remaining anterior horn cells (Delisle et al., 1987). This is unlike all other pathological lesions in Lewy body Parkinson's disease where neuronal degeneration and death is always associated with Lewy bodies. Another unique quirk of anterior horn cells is that they contain Lewy bodies in various rare sporadic and familial cases of motor neurone disease (Gibb, 1986; Kato et al., 1988), sometimes associated with a parkinsonian disorder, but without Lewy bodies in the substantia nigra (Schmitt et al., 1984).

It is not clear what factor is common to the specific nerve cells damaged in Parkinson's disease, but the distribution of Lewy bodies and nerve cells is moderately specific to medium- and large-sized monoaminergic and cholinergic neurons, for example the small pigmented neurons of the arcuate nucleus of the hypothalamus are spared. If Lewy bodies are found at any extranigral site other than anterior horn cells, it is virtually certain that they will be found without difficulty in the substantia nigra.

In view of the widespread occurrence of Lewy bodies in the autonomic nervous system and the high proportion of Parkinson's disease patients in which they are found in peripheral ganglia, they can be found in surgically excised specimens of oesophagus, colon, rectum or sympathetic ganglia (Stadlan et al., 1965; Qualman et al., 1984; Kupusky et al., 1987). It is worth remembering that if these tissues have to be removed in patients with parkinsonian disorders, then the finding of Lewy bodies would make

Parkinson's disease by far the most likely diagnosis. While it might be encouraging to know this for prognosis, rectal biopsies have not found a role in diagnosis because myenteric ganglion cells and Lewy bodies are not sufficiently numerous. There is no firm evidence that Lewy bodies and nerve cell loss in these locations contribute substantially to dysphagia, constipation or postural hypotension, which could result from centrally located autonomic or nigrostriatal dysfunction.

Justifiably Lewy bodies are considered to be markers of progressive nerve cell degeneration, because nerve cell loss has been documented for most locations in which they are found. In Parkinson's disease cell loss does not generally occur in the absence of Lewy bodies and Lewy bodies themselves form as one of the earliest pathological changes before neuronal fallout is detectable. Although they are associated with progressive neuronal degeneration we do not know whether they represent a regenerative or degenerative product.

Lewy bodies in non-parkinsonian persons

In 1952 Lewy bodies were found in the locus coeruleus in 10 non-parkinsonian persons aged 59–100 years, from a group of 30 of different ages (Beheim-Schwarzbach, 1952). Subsequent studies involving Japanese, European and American populations reported that the frequency of "incidental Lewy bodies" (Forno, 1969) in the substantia nigra varied between 6.8% (Lipkin, 1959) and 20.9% (Hamada and Ishii, 1963) for those aged over 60 years. Only two of 837 persons dying younger than 60 years showed nigral Lewy bodies. Some of these studies included a few patients with early Parkinson's disease, so we have re-estimated the prevalence of Lewy bodies. In a group of 308 non-parkinsonian persons the prevalence rates for Lewy bodies were 3.6% at 50–59 years, 4.4% at 60–69 years, 5.6% at 70–79 years, 11.5% at 80–89 years and 11.1% at 90–99 years (Gibb, 1988a). Subsequent work suggests that some of these figures may still overestimate the true frequency, but only by a small degree. Considering both the published data and our current work we would suggest figures of 1–2% at ages 50–59, 4–5% at ages 60–69, 6–7% at ages 70–79 and 10% at ages 80–89 years.

The crucial point about incidental Lewy body disease is that the distribution of Lewy bodies in the nervous system is the same as in Parkinson's disease (Woodward, 1962). The neuronal morphology (Lewy bodies and pale bodies) in the substantia nigra and elsewhere, and the progressive neuronal degeneration are also identical (Gibb and Lees, 1988a). The important difference is that neuronal loss in the substantia nigra is mild, and less than the approximate 60% depletion required for the development of parkinsonian features. The existence of a presymptomatic population is also supported by biochemical data suggesting an 80% loss of striatal dopamine by the time of first symptom onset (Bernheimer et al., 1973). Pathologically the disease processes are therefore identical, the fundamental difference being the milder subthreshold loss of nigral neurons in incidental cases.

Other clinical states

If Lewy body disease is defined by a specific distribution of Lewy body–neuronal degeneration then it is justified to include other clinical conditions, such as dementia and autonomic failure, which have the same underlying pathology. These patients mostly have parkinsonism as well, but may present an isolated picture of dementia or autonomic failure.

Dementia attributable to Lewy body disease is associated with multiple Lewy bodies in the cerebral cortex, and Lewy body–neuronal degeneration in the nucleus basalis, the relationship between the two being complex (Gibb, 1989). When Lewy bodies are very numerous in the cerebral cortex, as in fewer than 5% of patients with Parkinson's disease, then the term "cortical Lewy body dementia" is used. Another term, "diffuse Lewy body disease", was introduced to describe patients with cortical Lewy body dementia (Yoshimura et al., 1980) or autonomic failure (Kono et al., 1976), and is now applied specifically to patients with cortical Lewy body dementia. However the term is inaccurate as there is no evidence that the distribution of lesions in cortical Lewy body dementia is more severe or widespread at the subcortical level, and Lewy bodies never occur "diffusely" at any level of the nervous system. The reason for the variability in emphasis on cortical and autonomic involvement in Lewy body disease is unknown.

Patients with dementia associated with cortical Lewy bodies with or without Parkinson's disease usually show cortical features such as apraxia, agnosia, aphasia and spatial disorientation (Gibb et al., 1987). Confusional states, delusions, visual and auditory hallucinations also occur. The cortical dementia is similar to that in Alzheimer's disease and in some instances there are also cortical Alzheimer changes including neurofibrillary tangles, plaques and granulovacuolar degeneration, but these do not necessarily indicate an association between Lewy body and Alzheimer pathologies (Gibb et al., 1989b). The relationship between cortical Lewy bodies and Lewy body disease is, in general, a continuum. Lewy bodies are found in the medial temporal cortex of one-third of patients with Parkinson's disease (Gibb and Lees, 1987). In approximately 5% of such brains there are modest numbers in the parahippocampus, temporal, frontal, anterior cingulate and insular cortex, and in a smaller number of patients the spectrum includes a wider distribution, possibly involving parietal and occipital cortex as well. Cortical Lewy bodies are found only in specific locations and cells; they never occur in the hippocampus and are most frequent in deeper cortical layers of frontal, temporal and limbic cortex.

Autonomic failure in Lewy body disease has been associated with a greater than average loss of intermediolateral column cells (Oppenheimer, 1980), but other sites in the autonomic system may be involved. Examples of Lewy body autonomic failure are provided by Fichefet *et al.* (1965) who described a woman of 71 years who developed postural hypotension followed by Parkinson's disease. She died a year from the onset when Lewy bodies were found in cervical, thoracic and lumbar sympathetic ganglia, as well as the substantia nigra. A year later Johnson *et al.* (1966; Case 1) described a 62-year-old man who developed autonomic failure and died 4 years later without symptoms of Parkinson's disease. Lewy bodies were described in the substantia nigra, intermediolateral columns, and cervical and thoracic sympathetic ganglia.

The direct clinical consequences of Lewy body–neuronal degenerations in most other sites are not known. One area of recent interest includes disturbances of oesophageal motility, defaecation and micturition, which are dependent on central dopaminergic function (Christmas *et al.* 1988; Mathers *et al.* 1988; Kempster *et al.* 1988).

Other Lewy body diseases

Lewy bodies occur in other parkinsonian disorders either as an incidental finding with a prevalence similar to that of their age-specific prevalence in the population (equivalent to a figure including presymptomatic and symptomatic Parkinson's disease), or as a more specific pathological feature. Parkinsonian disorders associated with Lewy bodies at roughly the same frequency as in the general population include MSA, SRO, Guam parkinsonism–dementia complex and post-encephalitic parkinsonian syndrome. Some other parkinsonian disorders are more closely associated with Lewy bodies, the association being evident from the greater than incidental frequency of Lewy bodies, the occurrence of Lewy bodies in more than one affected family member, or an age at death of less than 50 years. These disorders include CBD, HSD, some familial olivopontocerebellar atrophies, juvenile hereditary parkinsonian syndrome, rare young-onset cases of Alzheimer's disease, ataxia–telangiectasia, and finally motor neurone disease, where Lewy bodies are confined to anterior horn cells. Lewy bodies are reported most frequently in CBD (approximately 40% of cases) and HSD (approximately 15% of cases). In some instances they are very sparse. This is incompatible with Parkinson's disease and serves as an important point of distinction. Unlike Parkinson's disease, Lewy bodies are not invariable in these disorders, although it is quite possible that they will be consistently found in juvenile hereditary parkinsonian syndrome (Segawa disease, dopa-responsive dystonia) (Nygaard and Duvoisin, 1986), and in all affected members of autosomal dominant parkinsonian disorder with (Muenter *et al.*, 1986) and without dementia (Golbe *et al.*, 1988).

One area of great clinical controversy concerns a possible association between Alzheimer's disease and Parkinson's disease. A recent pathological study examined the age-specific prevalence of Lewy bodies in the substantia nigra in 273 controls without Parkinson's disease or dementia and in 121 cases of "Alzheimer's disease" (Gibb *et al.*, 1989b). There was a non-significant excess of Lewy bodies in the Alzheimer's disease group (14.0% compared to 7.8% over age 60 years). However, cortical plaques and tangle estimations were lower in the Alzheimer cases with Lewy bodies compared to those without, whereas cortical choline acetyltransferase (ChAT) activities were similar. This suggests that Lewy body neuronal degeneration in the nucleus basalis contributes to the deficit of cortical ChAT, but not to the cortical Alzheimer pathology. Persons with otherwise borderline or presymptomatic Alzheimer's disease may thus cross the clinical threshold to develop dementia if Lewy body disease is also present. In general, there is no good evidence for a greater than chance association between the two processes in elderly subjects. However, coexistent Lewy body and Alzheimer pathology has exceptionally been reported in eight young patients with age at onset of a parkinsonian syndrome or dementia between 22 and 47 years, and age at death between 28 and 56 years (Kosaka *et al.*, 1973; Kayano *et al.*, 1980; Okeda *et al.*, 1982; Gibb *et al.*, 1987, Case 2; Popovitch *et al.*, 1987; Delisle *et al.*, 1987; Gibb *et al.*, 1989b, two cases). Lewy bodies also coexisted with neurofibrillary tangles in a patient aged 46 years with Hallervorden–Spatz disease (Eidelberg *et al.*, 1987).

Neuronal morphology in Parkinson's disease

Although the hallmark of Parkinson's disease is the Lewy body, other neuronal inclusions and morphologies are found, namely the granular pale body, and neuronal degeneration resembling the Lewy body or pale body (Fig. 3.1). It is imperative that none of these features are taken as absolutely diagnostic of Parkinson's disease, because of the occurrence of Lewy bodies in other disorders. Additionally there are other neuronal morphologies, such as eosinophilic granules, and inclusions in corticobasal degeneration, which also resemble pale bodies. Lewy bodies are found within neuronal cell bodies, their processes, or they lie free in the neuropil. Their precise morphology varies between each of their locations in the nervous system, but the basic constituents are similar.

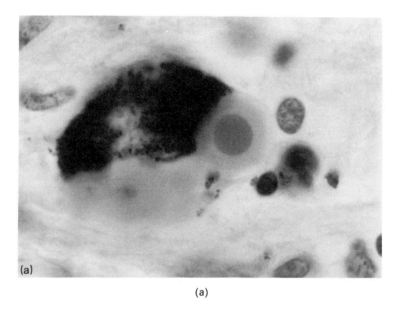

(a)

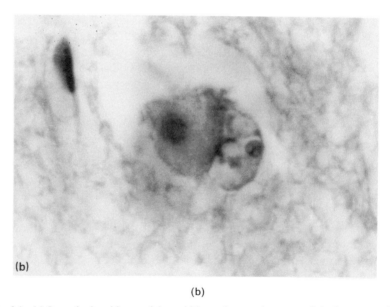

(b)

Fig. 3.1 (a) Lewy body with core lying within a pigmented neuron of the locus coeruleus. Haematoxylin and eosin stain (H and E), × 4480. (b) Lewy body in the temporal cortex. H and E, × 5600.

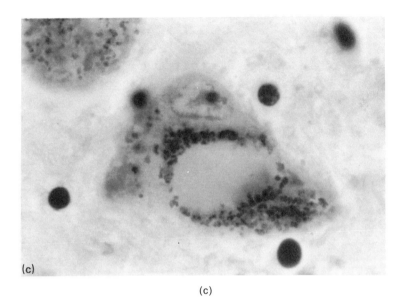

(c)

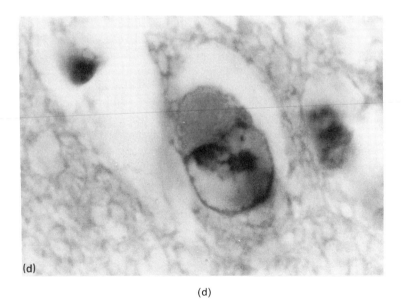

(d)

Fig. 3.1 – *continued* (c) Pale body surrounded by melanin granules in pigmented neuron of the substantia nigra. H and E, ×4480. (d) Pale inclusion in the temporal cortex. H and E, ×5600.

Ultrastructurally the "body" is composed of filamentous matter, often radially orientated, thinning out at the periphery or "halo" to melanin granules and a variety of organelles including dense core vesicles, mitochondria and lysosomes. Lewy bodies with central densities (cores) contain a central, amorphous, electron-dense debris of uncertain composition. It has been suggested that sphingomyelin is a component of the Lewy body, as well as a variety of polypeptide components such as those common to neurofilament protein and ubiquitin (Kuzuhara *et al.*, 1988). The pale body fails to stain with conventional histological techniques, including haematoxylin and eosin and silver impregnations, so that it appears as a non-staining inclusion. Ultrastructurally it consists of a featureless, vacuolated area devoid of mitochondria and filament (unpublished observations).

STRIATONIGRAL DEGENERATION

In the same way that Lewy bodies can be specifically associated with parkinsonian disorders other than Parkinson's disease, varying degrees of striatonigral degeneration (SND) are found in HSD, CBD and Huntington's chorea. However, the term SND usually refers to one specific pathological entity causing parkinsonism included under the title MSA (see below).

In general macroscopic examination of the brain in Parkinson's disease shows only depigmentation of the substantia nigra and locus coeruleus, whereas in SND there is also a distinctive atrophy and brown-green discoloration of the putamen, which is occasionally so mild at the time of death that it cannot be seen with the naked eye. Pontocerebellar atrophy may also be found. Microscopic examination of the substantia nigra can show varying degrees of neuronal loss, ranging from mild to extreme, depending on the relative degree of associated cerebellar involvement (see below). However, in contrast to the moderate or severe depletion found in Parkinson's disease the damage is often severe and extensive, so that few cells remain intact. This provides a histological clue to the diagnosis. In addition, unlike the generally indolent, but selective degeneration in Parkinson's disease, there is no pattern of selectivity within the substantia nigra, and the rate of cell destruction is faster, as judged by greater quantities of extraneuronal melanin and greater numbers of glial cells. Another histological clue to the diagnosis is the failure to find Lewy bodies at an appropriate frequency for Parkinson's disease, except in the occasional case, despite finding sufficient nigral cells to examine. SND differs from most other degenerative parkinsonian disorders in not being associated with a distinctive nigral neuronal morphology as yet detected by conventional light microscopy. It is notable for this very reason. However, Lewy bodies can be found at a frequency which is not definitely greater than in non-parkinsonian persons. In a total of 119 autopsied patients Lewy bodies were seen in 14.7% in the 50–59 year age group and 7.8% in the 60–69 year age group, or 11.8% overall (Gibb, 1988b). The most important additional positive histological feature is the severe degree of striatal, especially putamenal, nerve cell loss. The putamen may be totally gliosed without remaining neurons, although striatal and nigral atrophy does not always progress in parallel. This striatal atrophy was first described in 1961 in three patients with a parkinsonian syndrome, in one patient also associated with cerebellar signs (Adams et al., 1961, 1964). At the same time Shy and Drager (1960) described a patient, who, in addition to parkinsonian and cerebellar features, had autonomic failure. Pathological examination showed striatal and nigral cell loss and gliosis, with marked cell depletion in intermediolateral columns and milder changes in many other parts of the autonomic nervous system. It also emerged that some previous reports of patients dying with cerebellar degeneration described elements of the same pathological picture. This included nerve cell loss in pontine nuclei, inferior olives, and the cerebellar pyramidal cell layer (olivopontocerebellar atrophy), as well as SND, and degeneration of the autonomic nervous system. The three clinical presentations corresponding to these areas of maximal cell loss are a cerebellar syndrome, a parkinsonian disorder and autonomic failure, respectively. A greater or lesser degree of pathology in all three main systems is usually found at autopsy. Nerve cell loss is also found in the locus coeruleus, subthalamic nucleus, dorsal vagal nucleus, medial vestibular nucleus and sympathetic ganglia. Fibres of the middle cerebellar peduncle and olivocerebellar pathway are substantially depleted.

The term MSA was introduced to describe this disease in an attempt to provide a rational classification of diseases which are "merely the expressions of neuronal atrophy in a variety of overlapping combinations" (Graham and Oppenheimer, 1969). At the time its diagnostic limits were not defined because of limited experience of the disorder. The original definition covered "a primary degeneration of the nervous system, familial or sporadic, with onset in middle life, and affecting a selection of the following structures: striatum, pigmented nuclei (substantia nigra and locus coeruleus), pontine nuclei, inferior olives, cerebellar Purkinje cells, and dorsal vagal and vestibular nuclei" (Oppenheimer, 1988). Degeneration of intermediolateral columns was added to the list and it was later suggested that the definition should include patients with motor neurone disease. However, despite statements such as "often there is wasting and fasciculation of distal limb muscles with EMG evidence of anterior horn cell degeneration" (Côté, 1984), involvement of motor neurons in MSA has not yet been adequately documented. MSA is not associated with significant dementia, and pathology in the cortex or nucleus basalis is not well described. Clinically and pathologically reported cases of MSA are almost invariably sporadic and non-familial. Familial diseases with an element of striato-nigral degeneration are best considered part of a separate disorder. Despite its ambiguity the term MSA is now indelible and should be used to describe a sporadic disease usually manifest as a parkinsonian disorder, but occasionally with prominent cerebellar signs or autonomic failure.

Certain aspects of MSA are now better appreciated. It is more frequent than previously imagined, most commonly presenting with a parkinsonian syndrome (striato-nigral degeneration), with a relative frequency of approximately one for every 10 cases of Parkinson's disease. Patients are aged 40–75 years at onset, with a mean age of 55 years. The disease duration is usually less than 10 years with a mean of 5 years. At presentation the clinical picture is usually one of symmetrical bradykinesia and rigidity, with no or minimal rest tremor, although rest and intention tremors occur. The response to L-dopa is usually absent or poorly sustained. Occasionally a beneficial response can be complicated by mild dyskinesia, but probably never by severe fluctuations in response. At present it is not clear how closely the disease can, in its early stages, mimic Parkinson's disease in terms of its L-dopa response. In time it progresses more rapidly than Parkinson's disease, the response to L-dopa fails completely, and cerebellar features or autonomic failure may supervene. Our pathological observations suggest that within MSA the lesions in some patients are relatively confined to SND, with moderate or severe nigral degeneration and very mild or moderate neuronal loss in the striatum. It is in these patients that clinical differentiation from Parkinson's disease is most difficult.

STEELE–RICHARDSON–OLSZEWSKI SYNDROME

Clinical and pathological features of this disorder were formerly identified as being distinct from post-encephalitic parkinsonian syndrome and vascular disease in 1963 (Richardson et al., 1963; Steele et al., 1964), although its features were sufficiently unusual for isolated cases to have been described early in the twentieth century. The most distinctive feature on macroscopic examination of the brain is shrinkage of the midbran with loss of definition of the colliculi and red nucleus. On computed tomography, the first corresponding change to occur in SRO is a reduction of the antero-posterior diameter of the midbrain to less than 15 mm, in comparison with the normal range of 18 ± 1.3 mm (Duvoisin et al., 1987). The subthalamic nucleus shows a loss of definition and the substantia nigra and locus coeruleus are depigmented.

Microscopic examination of the substantia nigra often shows an extreme degree of neuronal depletion, as in SND, although a few patients will show only moderate (75%) neuronal loss. The nigral feature unique to SRO is the mildly basophilic, globose (spool-like), neurofibrillary tangle, which is about as numerous as the Lewy body in Parkinson's disease. These are also found in the locus coeruleus, raphé nuclei and many other locations showing neuronal loss. The two distinguishing morphological aspects of these inclusions are their globular shape, and more characteristically, their spool-like texture (Fig. 3.2a). They appear to consist of finely entwined filaments often arranged in parallel, so their structure resembles a ball of string. Isolated globular tangles can be found in the substantia nigra in Guam parkinsonism–dementia complex, post-encephalitic parkinsonian syndrome, and possibly Alzheimer's disease, although tangles in these disorders are generally known as Alzheimer tangles and are flame shaped. Microscopic study of flame-shaped tangles, and probably of the occasional globular tangle that coexists, reveals a coarse, rope-like texture of fibrils. These tangles are more argyrophilic than tangles in SRO and show coarse basophilic strands when stained with H and E. Thus, the faintly fibrillar, globose tangle of SRO is specific to the condition and usually distinct from the Alzheimer tangle. Ultrastructurally it consists of 15 nm straight filaments, but occasional paired helical filaments characteristic of the Alzheimer tangle, or twisted paired fibrils with unusual periodicities are seen. Conversely straight filaments can be seen in Alzheimer tangles admixed with paired helical filaments. Despite the differing basic structure of tangles in SRO and Alzheimer's disease their immunocytochemical reactions seem identical. They both contain antigenic determinants of neurofilaments, and of Alzheimer neurofibrillary tangles, and they have special affinity for antibodies recognizing phosphorylated epitopes.

The main distribution of nerve cell loss in SRO includes the locus coeruleus, periaqueductal grey area, superior colliculus, pontine tegmentum, subthalamic nucleus, red nucleus, Edinger–Westphal, trochlear and abducens nuclei, raphé nuclei, nucleus basalis, dentate nucleus and parts of the spinal cord, including a modest dropout of anterior horn cells. Significant neocortical pathology is not typical, but can occur coincidentally.

The frequency of Lewy bodies in the substantia nigra or locus coeruleus in SRO, as in SND, is similar to the population frequency. A total of 55 cases reported in the literature (excluding Japanese papers) and 26 personal cases provide five patients with Lewy bodies in the brain stem. These data give a prevalence of Lewy bodies of 6.2% at an approximate mean age of 64 years, or 8.3% for patients aged over 60 years at death. Some of the Japanese reports were written for the purpose of documenting the association (Mori *et al.*, 1986), but there is insufficient evidence to suggest that the pathologies when concurrent represent a greater than chance association.

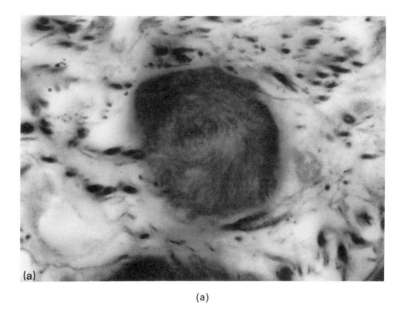

(a)

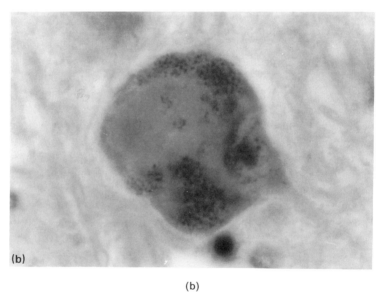

(b)

Fig. 3.2 (a) Globose neurofibrillary tangle in the substantia nigra in SRO. Bielschowsky silver stain, × 3500. (b) Corticobasal inclusion in a pigmented nigral neuron in CBD, containing a slightly basophilic centre and a few melanin granules. H and E, × 4480.

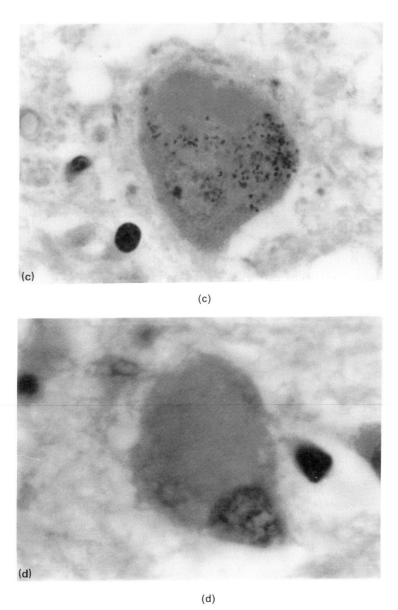

(c)

(d)

Fig. 3.2 – *continued* (c) Nigral inclusion body in Pick's disease. H and E, × 4480. (d) Pick cell in the parietal cortex in CBD. H and E, × 5600.

Pathological studies in patients aged 45–70 years at onset of disease, with mean age of 58 years and mean duration of 5.5 years, have identified the most common early symptoms of SRO as unsteady gait with frequent falls, visual difficulties with difficulty focusing and double vision, and personality change consisting of irritability, forgetfulness, emotional lability, mental slowing and depression (Gibb, 1988c). These features with the supranuclear gaze palsy, axial rigidity and pseudobulbar palsy are well known, but these characteristic signs may not develop in the first few months or years of the condition, and a number of patients die even without opthalmoplegia (Dubas et al., 1983). Modest L-dopa responses are well documented. Limb dystonia has recently been reported in eight of 30 clinically diagnosed cases, in four of which it preceded ophthalmoplegia (Rafal and Friedman, 1987). Two other similar patients presented with upper limb dystonia, before developing gaze paresis, and one of these showed marked widening of convexity sulci on computed tomography (Léger et al., 1987). In the absence of pathology it is not clear how many patients with bradkinesia–rigidity, supranuclear gaze palsy and limb dystonia have SRO, CBD or even MSA.

CORTICOBASAL DEGENERATION

Corticodentatonigral degeneration with neuronal achromasia was described in three adults presenting to Massachusetts General Hospital in 1967 (Rebeiz et al., 1967, 1968). In very recent years a few similar cases have been suspected on clinical grounds, but few have been confirmed pathologically (Gibb et al., 1989c). The name of CBD has now been adopted by those clinicians who suspect such patients. This brief name remains appropriately descriptive and excludes the now redundant term neuronal achromasia.

Macroscopic examination shows cortical atrophy mostly affecting the anterior parietal region and posterior frontal lobes, in combination with depigmentation of the substantia nigra and, occasionally, visible depigmentation of the locus coeruleus.

The neuronal morphology in the substantia nigra is more complex than in other parkinsonian disorders and mimics that of PD, SRO and Pick's disease (Gibb et al., 1989c). The most frequent inclusion, occurring in every case, is weakly basophilic and has a faintly fibrillar structure, often containing entwined melanin granules (Fig. 3.2b). It is as numerous as the globose tangle in SRO and the Lewy body in Parkinson's disease. It superficially resembles the globose tangle of SRO, and some dense basophilic forms resemble the nigral Pick body of Pick's disease (Fig. 3.2d). It is unique to corticobasal degeneration and has therefore been called the corticobasal inclusion. Ultrastructurally it contains fibrillar elements that require further clarification. Immunocytochemical work using RT97, RS18 and 147, which recognize epitopes on the 210k neurofilament protein, show positive reactions for RT97, but weak or absent staining for RS18 and 147 (Gibb et al., 1989c). These reactions mirror those of neurofibrillary tangles and Pick bodies (Probst et al., 1983). The inclusions also frequently occur in the locus coeruleus, raphé nuclei and in some other subcortical nuclei. Another inclusion in CBD is non-staining, pale and surrounded by melanin granules. Most of these probably represent morphological variations of the corticobasal inclusion, although some closely resemble the pale body of Parkinson's disease. Furthermore, a proportion of cases (approximately 40%, or 3/8 cases examined by the author) have sparse nigral Lewy bodies, and are therefore associated with other Lewy body pathology in the substantia nigra and locus coeruleus, including pale bodies. Pale inclusions, pale bodies and Lewy bodies in the substantia nigra in CBD can therefore lead to confusion with Parkinson's disease. This diagnostic trap is avoided by attending to the usually infrequent numbers of Lewy bodies in CBD (less than one every 300 pigmented nerve cells), and the presence of corticobasal inclusions.

Neuronal loss and gliosis are mostly found in the substantia nigra, locus coeruleus, midbrain tectum and tegmentum, parietal cortex, thalamus, and lentiform, subthalamic and red nuclei. Rather less neuronal loss occurs at other locations including frontal and temporal cortex. In atrophic parts of the cerebral cortex, especially in pre- and post-central gyri, there is usually severe neuronal loss accompanied by cortical and subcortical gliosis. In these areas swollen or Pick cells are characteristic (Fig. 3.2d). The immunocytochemical reactions of swollen cells seen in one probable case of CBD, a man dying aged 71 years with a parkinsonian disorder and dementia, were identical to those of Pick cells in Pick's disease (Dickson et al., 1986). A number of neurons in cortical and subcortical regions show chromatolytic-like changes.

Diagnostic pathological features of CBD are nerve cell loss in the parietal cortex and midbrain tegmentum, and corticobasal inclusions with neuronal degeneration in the substantia nigra.

The most common clinical features in seven pathologically verified cases (Rebeiz et al., 1968; Scully et al., 1985; Gibb et al., 1989c) were postural instability, dysarthria, focal limb dystonia, supranuclear gaze palsy, short-stepped gait, bradykinesia, muscular rigidity and intention tremor. However, the most characteristic and distinctive features are focal limb dystonia, supranuclear gaze palsy, bradykinesia–rigidity, parietal lobe signs (constructional difficulties) and the alien hand sign. The presence of three or more of these features is strongly suggestive of CBD, although can also be seen in rare cases of SRO and MSA. Damage in the parietal cortex explains some of the cortical features, that in the tectal regions the supranuclear gaze palsy, and damage in the substantia nigra the parkinsonian features.

HALLERVORDEN–SPATZ DISEASE

Hallervorden–Spatz disease (HSD) is increasingly recognized as a cause of a parkinsonian disorder in adults, although it more commonly affects persons in the first and second decades (Jankovic *et al.*, 1985). The diagnostic pathological features are degeneration of the globus pallidus and substantia nigra zona reticulata associated with axonal swellings and iron pigment deposition. However, reticulata pathology may often not be sufficiently clear-cut from that in the zona compacta to justify its use as an objective diagnostic criterion. Degeneration with axonal swellings also occurs in the nucleus gracilis and cuneatus, cerebral cortex and other locations.

Three recently described adult women showed a variant of HSD with marked neurofibrillary tangle pathology (Eidelberg *et al.*, 1987). They were mentally retarded, aged between 31 and 42 years at onset, and they had an average 6-year duration of an illness that comprised dementia with parkinsonian features. Their classification is based on the finding of pallidonigral degeneration with pigmentation and axonal pathology. Neurofibrillary tangles were present in the neocortex, hippocampus, nucleus basalis, substantia nigra, locus coeruleus, as well as in the thalamus, subthalamic nucleus and other sites.

Lewy bodies are reported in the substantia nigra in approximately 15% (11 of 80) of reported cases of HSD in the same profusion and distribution as in Parkinson's disease (Gibb, 1987). In such cases they are often present in the locus coeruleus, dorsal vagal nucleus, nucleus basalis, cerebral cortex and spinal cord. The 11 Lewy body cases were aged between 2 and 27 years at onset and 17 and 38 years at death. Other late-onset cases include one who developed a parkinsonian syndrome at age 25 years, and died aged 38 years without dementia (Alberca *et al.*, 1987). The brain showed pallidonigral degeneration with spheroids, and Lewy bodies were present in the substantia nigra and locus coeruleus. An additional patient with cerebral Lewy bodies was one of those described by Eidelberg *et al.* (1987). She had mild mental retardation and then at age 42 developed progressive dementia with parkinsonian and pyramidal features and died 6 years later. There were profuse cortical and subcortical neurofibrillary tangles with Lewy bodies and tangles in the substantia nigra.

CONCLUSION

Over the past 30 years, pathological study of progressive parkinsonian disorders has defined a number of different entities. Further pathological study is required on large numbers of clinically "typical" patients so that clinical and pathological diagnostic criteria can be improved. All "atypical" parkinsonian states should also be studied to improve clinical diagnosis, treatment and estimates of prognosis, and permit accurate neuropathological and, hopefully, aetiological classification of these disorders.

REFERENCES

Adams, R., Bogaert, van L. and Van Der Eecken, H. (1961) *Psychiat. Neurol.* **142**, 219–259.

Adams, R. D., Bogaert, van L. and Eecken, H. V. (1964) *J. Neuropathol. Exp. Neurol.* **24**, 584–608.

Alberca, R., Rafel, E., Chinchon, I., Vadillo, J. and Navarro, A. (1987) *J. Neurol. Neurosurg. Psychiat.* **50**, 1665–1668.

Alvord, E. C. (1958) In *Pathogenesis and Treatment of Parkinsonism* (edited by W. S. Fields), pp. 161–186. CC Thomas, Springfield, Illinois.

Beheim-Schwarzbach, D. (1952) *J. Nerv. Ment. Dis.* **116**, 619–632.

Bernheimer, H., Birkmayer, W., Hornykiewicz, O., Jellinger, K. and Seitelberger, F. (1973) *J. Neurol. Sci.* **20**, 415–455.

Brissaud, E. (1895). Leçons sur les Maladies Nerveuses (Salpêtrière 1893–1894). Paris.

Christmas, T.J., Chapple, C.R., Lees, A.J. *et al.* (1988) *Lancet* **ii**, 1451–1453.

Côté, L. J. (1984) In *Autonomic Nervous System. Merritt's Textbook of Neurology* (edited by L. P. Rowland), 617 pp. Lea and Febiger, Philadelphia.

Delisle, M. B., Gorce, P., Hirsch, E., Hauw, J. J., Rascol, A. and Bouissou, H. (1987) *Acta Neuropathol.* **75**, 104–108

Dickson, D. W., Yen, S.-H., Suzuki, K. I., Davies, P., Garcia, J. H. and Hirano, A. (1986) *Acta Neuropathol.* **71**, 216–223.

Dubas, F., Gray, F. and Escourolle, R. (1983) *Rev. Neurol.* **139**, 407–416.

Duvoisin, R. C., Golbe, L. I. and Lepore, F. E. (1987) *Can. J. Neurol. Sci.* **14**, 547–554.

Eidelberg, D., Sotrel, A., Joachim, C., Selkoe, D., Forman, A., Pendlebury, W. W. and Perl, D. P. (1987) *Brain* **110**, 993–1013.

Fichefet, J. P., Sternon, J. E., Franken, L., Dermanet, J. C. and Vanderhaeghen, J. J. (1965). *Acta Cardiol.* **20**, 332–348.

Forno, L. S. (1969) *J. Am. Geriat. Soc.* **17**, 557–575.

Forno, L. S. and Alvord, E. C. (1971) In *Recent Advances in Parkinson's Disease* (edited by F. H. McDowell and C. H. Markham), pp. 120–161. Blackwells, Oxford.

Gibb, W. R. G. (1986) *Neuropathol. Appl. Neurobiol.* **12**, 223–234.

Gibb, W. R. G. (1987) In *Recent Developments in Parkinson's Disease*, Vol. II (edited by S. Fahn, C. D. Marsden, D. B. Calne and M. Goldstein), pp. 1–13. Macmillan Healthcare, New Jersey.

Gibb, W. R. G. (1988a) In *Parkinson's Disease and Movement Disorders* (edited by J. Jankovic and E. Tolosa). pp. 205–223. Urban and Schwarzenberg, Baltimore.

Gibb, W. R. G. (1988b) In *Autonomic Failure* (edited by R. Bannister), pp. 484–497. Oxford University Press, Oxford.

Gibb, W. R. G. (1988c) *Postgrad. Med. J.* **64**, 345–351.

Gibb, W. R. G. (1989) *Br. J. Psychiat.* **154**, 596–614.

Gibb, W. R. G. and Lees, A. J. (1987) *Lancet* **i**, 861.

Gibb, W. R. G. and Lees, A. J. (1988a) *J. Neurol. Neurosurg. Psychiat.* **51**, 745–752.

Gibb, W. R. G. and Lees, A. J. (1988b) *Neurology*, **38**, 1402–1406.
Gibb, W. R. G. and Lees, A. J. (1989) *Neuropathol. Appl. Neurobiol*, **15**, 27–44.
Gibb, W. R. G., Esiri, M. M. and Lees, A. J. (1987) *Brain* **110**, 1131–1153.
Gibb, W. R. G., Luthert, P. J., Janota, I. and Lantos, P. L. (1989a) *J. Neurol. Neurosurg. Psychiat*, **52**, 185–192.
Gibb, W. R. G., Mountjoy, C. Q., Mann, D. M. A. and Lees, A. J. (1989a) *J. Neurol. Neurosurg. Psychiat*, in press.
Gibb, W. R. G., Luthert, P. J. and Marsden, C. D. (1989c) *Brain*, in press.
Golbe, L. I., Duvoisin, R. C. and Miller, D. C. (1990) *Ann. Neurol.* In press.
Graham, J. G. and Oppenheimer, D. R. (1969) *Quart. J. Med.* **32**, 28–34.
Hamada, S. and Ishii, T. (1963) *Adv. Neurol. Sci.* **7**, 184–186.
Hassler, R. (1937) *J. Psychol. Neurol.* **48**, 1–55.
Hassler, R. (1938) *J. Psychol. Neurol.* **48**, 387–476.
Jankovic, J., Kirkpatrick, J. B., Blomquist, K. A., Langlais, P. J. and Bird, E. D. (1985) *Neurology* **35**, 227–234.
Johnson, R. H., Lee, de J., Oppenheimer, D. R. and Spalding, J. M. K. (1966) *Quart. J. Med.* **35**, 276–292.
Kato, T., Katagiri, T., Hirano, A., Sasaki, H. and Arai, S. (1988) *Acta Neuropathol.* **76**, 208–211.
Kayano, T., Funada, N., Okeda, R., Kojima, T., Miki, M. and Iwama, H. (1980) *Neuropathology* **1**, 27–28.
Kempster, P. A., Lees, A. J., Crichton, P., Frankel, J. and Shorvon, P. (1989) *Movement Disorders*, **4**, 47–52.
Kono, C., Matsubara, M. and Inagaki, T. (1976) *Neurol. Med.* **4**, 568–570.
Kosaka, K., Shibayama, H., Kobayashi, H., Hoshino, T. and Iwase, S. (1973). *Psychiat. Neurol. Jap.* **75**, 18–34.
Kupusky, W. J., Grimes, M. M., Sweeting, J., Bertsch, R. and Côté, L. J. (1987) *Neurology* **37**, 1253–1255.
Kuzuhara, S., Mori, H., Izumiyama, N., Yoshimura, M. and Ihara, Y. (1988) *Acta Neuropathol.* **75**, 345–353.
Léger, J. M., Girault, J. A. and Bolgert, F. (1987) *Rev. Neurol.* **143**, 140–142.
Lewy, F. H. (1912) In *Handbuch der Neurologie III*, pp. 920–933. Springer, Berlin.
Lipkin, L. E. (1959) *Am. J. Pathol.* **35**, 1117–1133.
Mathers, S.E., Kempster, P.A., Swash, M. and Lees, A.J. (1988). *J. Neurol. Neurosurg. Psychiat.* **51**, 1503–1507.
Mori, H., Yoshimura, M., Tomonaga, M. and Yamanouchi, H. (1986) *Acta Neuropathol.* **71**, 344–46.
Muenter, M. D., Howard, F. M., Okazaki, H., Forno, L. S., Kish, S. J. and Hornykiewicz, O. (1986) *Neurology* **36**, suppl. 1, 115
Nygaard, T. G. and Duvoisin, R. C. (1986) *Neurology* **36**, 1424–1428.
Okeda, R., Kayano, T., Funata, N., Kojima, T., Miki, M. and Iwama, H. (1982) *Brain Nerve* **38**, 761–767.
Olszewski, J. and Baxter, D. (1954) *Cytoarchitecture of the Human Brain Stem.* S. Karger, New York.
Oppenheimer, D. R. (1980) *J. Neurol. Sci.* **46**, 393–404.
Oppenheimer, D. (1988) In *Autonomic Failure* (edited by R. Bannister), pp. 451–463. Oxford University Press, Oxford.
Popovitch, E. R., Wisniewski, H. M., Kaufman, M. A., Grundke-Iqbal, I. and Wen, G. Y. (1987) *Acta Neuropathol.* **74**, 97–104.
Probst, A., Anderton, B. H., Ulrich, J., Kohler, R., Kahn, J. and Heitz, P. U. (1983)

Acta Neuropathol. **60**, 175–182.
Qualman, S. J., Haupt, H. M., Young, P. and Hamilton, S. R. (1984) *Gastroenterology* **87**, 848–856.
Rafal, R. D. and Friedman, J. H. (1987) *Neurology* **37**, 1546–1549.
Reibeiz, J. J., Kolodny, E. H. and Richardson, E. P. (1967) *Trans. Am. Neurol. Assoc.* **92**, 23–26.
Reibeiz, J. J., Kolodny, E. H. and Richardson, E. P. (1968) *Arch. Neurol.* **18**, 20–33.
Richardson, J. C., Steele, J. and Olszewski, J. (1963) *Trans. Am. Neurol. Assoc.* **88**, 25–29.
Schmitt, H. P., Emser, W. and Heimes, C. (1984) *Ann. Neurol.* **16**, 642–648.
Scully, R. E., Mark, E. J. and McNeely, B. U. (1985) *N. Engl. J. Med.* **313**, 739–748.
Shy, G. M. and Drager, G. A. (1960) *Arch. Neurol.* **2**, 511–527.
Stadlan, E. M., Duvoisin, R. and Yahr, M. (1965) In *Proceedings of the Fifth International Congress of Neuropathologists, Zurich.* Excerpta Medica International Congress Series, Vol. 100, pp. 569–571.
Steele, J. C., Richardson, J. C. and Olszewski, J. (1964) *Arch. Neurol.* **10**, 333–359.
Tretiakoff, C. (1919) Thesis, University of Paris.
Vogt, C. and Vogt, O. (1920) *J. Psychol. Neurol.* **25**, 631–846.
Von Economo, E. (1931). *Encephalitis Lethargica. Its Sequelae and Treatment* (translated by K. O. Newman). Oxford University Press, London.
Woodard, J. S. (1962). *J. Neuropathol. Exp. Neurol.* **21**, 442–449.
Yoshimura, M., Shimada, H., Nakura, H. and Tomonaga, M. (1980) *Trans. Soc. Pathol. Jap.* **69**, 432.

4

The Pathophysiology of Parkinson's Disease

R. Benecke

It could be argued that it is unreasonable to ask what the general "functions" of the basal ganglia are or to try and search for the best model of basal ganglia disease to demonstrate its dysfunction. The complex anatomical and biochemical organization of the basal ganglia and its many connections with other brain regions including the supplementary motor area make it difficult to interpret motor abnormalities occurring after lesions within the basal ganglia. The question always arises whether a certain motor abnormality is an expression of primary dysfunction of the basal ganglia or of a secondary functional and/or morphological change in areas influenced by the basal ganglia. From a theoretical point of view, it is likely that both mechanisms work in parallel.

Various approaches have been used to try and elucidate the role of the basal ganglia in motor control. Experiments have been performed in animals using lesions produced pharmacologically or by the techniques of focused electric coagulation and reversible cooling. Single cell recordings from the basal ganglia nuclei have also been made in free moving animals. In addition, the clinical, pharmacological and electrophysiological examination of patients with various basal ganglia diseases has further expanded our knowledge of the role of the basal ganglia. However, in the discussion of the findings in patients one is confronted with the mystery as to why different diseases of

DISORDERS OF MOVEMENT: CLINICAL, PHARMACOLOGICAL
AND PHYSIOLOGICAL ASPECTS   ISBN 0-12-569685-X

the basal ganglia may produce opposite effects on movement. Thus, is it the hypokinetic syndrome in Parkinson's disease or the hyperkinetic syndrome in Huntington's disease which indicates the true role of the basal ganglia? According to the revised hypothesis of Penney and Young (1986) both syndromes reflect a disturbance of the basal ganglia network. They suggest that the development of either a hyper- or hypo-kinetic syndrome can occur in association with basal ganglia disease because of the inhomogeneities of the synaptic organization within the striatum and the variability of the exact site of the lesion within and between various diseases. Although the spectrum of symptoms is different between the two prototypes of basal ganglia disease mentioned above they share one common motor abnormality: the disturbance of voluntary movement described as poverty of movements, akinesia or bradykinesia. In Huntington's disease this abnormality is overshadowed by striking hyperkinesias and it is therefore often neglected (Thompson et al., 1988) whereas in Parkinson's disease it represents a leading symptom. As clinicians we may observe the inability of patients to execute complex movements like walking, dressing, writing, etc. or at least an abnormal slowness of these movements. It is not known, however, why they have difficulties in the execution of voluntary movements, and whether there is any common denominator which could explain this motor disturbance. The best way to get more information about the nature of akinesia/hypokinesia at present is through detailed physiological analysis of motor performance. Up to now such studies have concentrated on rapid simple movements. It will be shown in the present chapter that it is more the performance of complex movements which is disturbed rather than single separate movements. The underlying mechanisms will be discussed in the light of motor programming. On the basis of the presented findings there is good evidence that the basal ganglia are engaged with the execution of motor plans and the appropriate scaling of variant characteristics of motor programmes working in a predictive feed-forward control mode.

DESIGN OF SIMPLE AND COMPLEX MOVEMENTS

The only previous attempts to analyse the deficits in complex movements physiologically were made over 20 years ago. Schwab et al. (1954) asked patients to trace the outline of a triangle and to draw perpendicular lines with their dominant hand while repetitively squeezing the rubber bulb of a sphygmomanometer with the other hand. Talland and Schwab (1964) used a similar task in which patients had to press down a tally counter with the non-dominant hand and simultaneously pick up beads with a pair of tweezers with the other hand. The results of Schwab and his colleagues reflect the motor problems of patients with Parkinson's disease, but are difficult to interpret in detail. In order to investigate the deficit in complex movements in more detail, we devised a much simpler task than that of Schwab *et al.* (1954). The movements we examined were isotonic flexion at the elbow ("flex"), isometric squeezing of a force transducer between the fingers and thumb ("squeeze") and isotonic flexion of the index finger and thumb ("cut") (Fig. 4.1). Each movement could be made separately or in combination with a second movement. Within the combinations the two movements were performed either simultaneously or in a sequence and with either the same arm (unilateral arrangement) or with opposite arms (bilateral arrangement).

The initial start and the final target position, as well as the movements themselves, were displayed as vertical bars on an oscilloscope screen 60 cm in front of the subject. The general instruction for the separate and simultaneous movements was to execute each movement as rapidly as possible; in the simultaneous task both movements had to be initiated at the same time. The instruction for the sequential task was: first, execute each

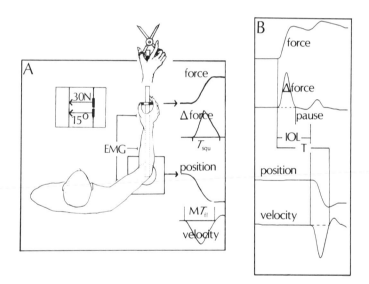

Fig. 4.1 Illustration of the experimental arrangements and measurements of movement performance. (A) The subjects were asked to perform an isometric opposition of thumb to fingers to exert a force of 30 N (squeeze), to perform an isotonic opposition of thumb to fingers (cut) from a starting angle of 135° (between the scissor arms) to an angle of 45°, and to perform an elbow flexion over an angle of 15° (flex). All movements had to be performed as fast as possible. (B) The measurements of timing in a sequential movement (e.g. squeeze then flex task). Total movement time (T) was measured from onset of the first movement to end of the second movement. The interonset latency (IOL) was the time between the onsets of both movements. The pause was the time between termination of the first to onset of the second movement.

movement of a sequence as rapidly as possible; second, initiate the second movement immediately after termination of the first. Subjects performed the separate, simultaneous and sequential movements in a cyclical order. They were given five practice trials of each movement type before 10 trials were recorded. Electromyographic activity was recorded with surface electrodes over the right biceps brachii, triceps brachii and either the right or left opponens pollicis. Joint positions, joint velocities (electronically derived from the joint position signals), force and its first derivative (Δ-force) as well as the rectified EMG signals were recorded by a PDP12 computer with a sampling rate of 250 Hz per channel. Measurements were made on each trial using the computer display unit. The duration of elbow flexion and isotonic finger flexion were measured by visual inspection of the velocity signals (from onset to the first zero crossing). Force rise time during the isometric squeezing was measured by means of Δ-force. In the sequential movements the interonset latency was measured from velocity or Δ-force onset of the first movement to the onset of the second movement. Total movement time for the complex movements (simultaneous, sequential) was measured from onset of the movements to termination of the longer lasting movement in the simultaneous task and from onset of the first movement to termination of the second movement in the sequential task. For further details of movement analysis see Benecke et al. (1986a,b, 1987a).

DISTURBANCE OF SIMPLE SEPARATE MOVEMENTS IN PARKINSON'S DISEASE

It is well known that patients with Parkinson's disease move more slowly than normal subjects in tasks involving movement at a single joint (Draper and Johns, 1964; Hallett et al., 1977). Kinesiologic analyses of ballistic goal-directed movements of different distances have shown that patients with Parkinson's disease perform these movements with inadequate speeds, so that movements of growing amplitude show increasing durations (Flowers, 1976). This is in contrast to the behaviour in normal subjects which is characterized by the use of faster speed for longer distances, keeping movement duration relatively constant (Benecke et al., 1985). EMG studies of rapid elbow movements in patients with Parkinson's disease show the patterning of muscle bursts to be abnormal (Hallett et al., 1977). In normal subjects such movements are made with a single triphasic pattern of EMG activity in agonist and antagonist muscles. The first agonist burst provides the impulsive force for the movement, the antagonist activity provides a braking force to hold the limb (Benecke et al., 1985). In patients with Parkinson's disease showing bradykinesia the first agonist burst appears not to be large enough to move the limb through the required angle. When the target zone is not visible for the patient or not displayed on an oscilloscope, this deficit results in an undershoot of the movement. However, when the target is constantly presented, additional bursts of EMG activity occur in the agonist muscle in order to reach the target (Hallett and Khoshbin, 1980). Although the movements made by parkinsonian patients are slower than those of normals, the size and duration of the first agonist EMG burst nevertheless change with movement size and added load. There is, however, a failure to appropriately match the EMG activities in the agonist and antagonist muscles to the size of movement required (Berardelli *et al.*, 1986).

A slowness of simple separate movements and abnormal EMG patterns have been observed at proximal and distal joints of the arm (Draper and Jones, 1964; Hallett and Marsden, 1979; Hallett and Koshbin, 1980; Berardelli *et al.*, 1984, 1986). Furthermore Berardelli *et al.* (1984) have stressed the point that distal movements in Parkinson's disease remain abnormal, even when activation of proximal muscles is no longer required for postural fixation. Slowness of thumb movements was the same, irrespective of whether the thumb phalanx was supported or not. However, these investigations have only studied rapid goal-directed movements at one joint. In each case it remained unclear whether all muscle groups are equally affected or whether, possibly, proximal muscle groups are more affected than distal ones. Furthermore, isometric movements in patients with Parkinson's disease have not hitherto been studied.

In our study a slowness in the separate movements of "flex" and "squeeze" could be compared in 10 patients. In five patients with Parkinson's disease all three types of movements ("flex", "squeeze" and "cut") could be compared (Fig. 4.2). The slowness of the movements was expressed as percentage increase in movement times as compared with the values in normal subjects. The mean movement durations for the three movement types obtained in an age-matched control group of 10 subjects were quite similar. The mean movement time for the rapid elbow flexion through an angle of 15° from a starting angle of 135° ("flex" task) amounted to 177 ± 37 (SD) ms, the movement time of an isotonic finger flexion ("cut" task) amounted to 160 ± 8 (SD) ms, and the movement time for an isometric finger flexion ("squeeze" task) up to a target force of 30 N amounted to 156 ± 16 (SD) ms. The percentage increase in mean movement time in patients with Parkinson's disease was 23% in the "cut", 34% in the "squeeze" and 75% in the "flex" task (see also Fig. 4.5). Thus a proximal isotonic movement ("flex") is more affected than a distal isotonic movement ("cut"), and a distal isometric movement ("squeeze") is more affected than a distal isotonic movement ("cut").

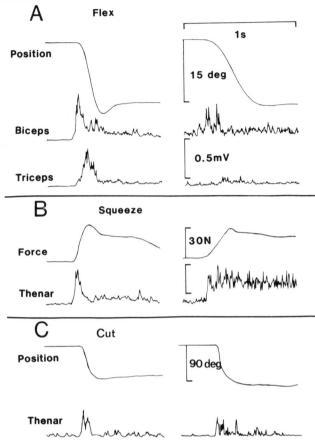

Fig. 4.2 Performance of simple separate movements (A, flex; B, squeeze; C, cut) in a normal subject (left part) and in a patient with Parkinson's disease (right part).

DISTURBANCE OF SEQUENTIAL MOVEMENTS IN PARKINSON'S DISEASE

Sequential movements play an important role in normal motor behaviour. For example, we change rapidly from sitting to standing and to walking, or we grasp an object first and then place it in a new position. In an investigation of normal subjects (Benecke *et al.*, 1986a) we found that, irrespective of whether the sequential movements of "squeeze" and "flex" were carried out with the same arm or with opposite arms, they could only be performed at maximal speed when there was a delay of about 240 ms between their onsets. At shorter interonset latencies, the speed of the second movement was progressively reduced.

From a clinical point of view it is well known that patients with Parkinson's disease have difficulties in the execution of sequential movements, but it is not known why they are slow. Is it simply because each single movement is slow? Is the timing between sequential movements disturbed? Are sequential movements slower than separate simple movements? In fact, performance of motor sequences is disturbed in all three of these ways in patients with Parkinson's disease.

Figure 4.3 (lower part) shows the performance of a unilateral sequential "squeeze then flex" task in a representative patient with Parkinson's disease in the OFF and ON condition. The movement times in the sequential task were prolonged, and their timing was disturbed. Comparison of the OFF and ON condition revealed the beneficial effect of drug administration on performance; all patients ON, however, were still impaired compared with normal subjects. Figure 4.3 (upper part) summarizes the data from all patients in the OFF condition. In the sequential movement mean squeeze time (T_{squ}) was longer in patients than in normals (260 vs 150 ms, $P < 0.001$), as was mean flex time (MT_{fl}; 490 vs 244 ms; $P < 0.001$). Also the mean interonset latency (IOL, measured from onset of "squeeze" to onset of "flex") was prolonged when compared with that of a normal subject (441 vs 244 ms; $P < 0.001$). As a result of the prolongation in interonset latency and movement times the time to complete the entire sequence (T) was strikingly prolonged in the patients as compared to normals (932 vs 488 ms; $P < 0.001$). Similar disturbances in the execution of sequential movements could be observed in the bilateral "squeeze then flex" and the unilateral "cut then flex" tasks, although the amount of the disturbance was somewhat different. The bilateral sequential tasks were less affected than the unilateral sequential tasks (see also Fig. 4.5).

DISTURBANCE OF SIMULTANEOUS MOVEMENTS IN PARKINSON'S DISEASE

As already described in previous investigations (Benecke *et al.*, 1986b) movement times increased still further in patients with Parkinson's disease when they were asked to execute two movements simultaneously rather than separately. The pronounced slowness of both movements in an unilateral "squeeze and flex" task is demonstrated for a representative patient with Parkinson's disease in the OFF and ON condition in Fig. 4.4 (lower part). When movement performance was compared in OFF and ON conditions, a pronounced improvement after drug administration was observed. Movement performance in the ON condition, however, was still slower than that in a normal subject. The upper part of Fig. 4.4 summarizes the data from all

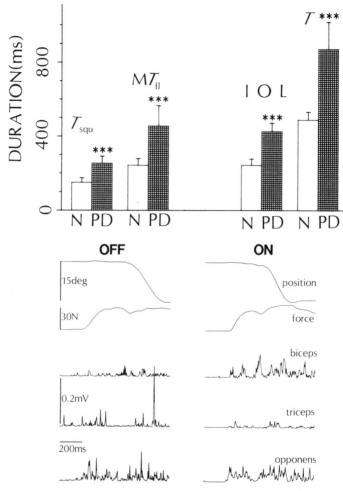

Fig. 4.3 Performance of the sequential "squeeze then flex" task in normal subjects and patients with Parkinson's disease. Upper part shows the mean values (± 1 SD) of T_{squ}, MT_{fl}, IOL, and T in nine normal subjects (open columns) and five patients (shaded columns). There were significant differences for all parameters (***$P < 0.001$). Lower part shows representative single trials in a patient with Parkinson's disease in OFF and ON conditions.

patients in the OFF condition. The mean movement time for elbow flexion was 379 ms when performed separately, but was 549 ms when performed at the same time as a "squeeze". Similarly the time taken for "squeeze" rose from 229 (separate performance) to 330 ms (simultaneous performance) ($P < 0.001$ for both movements; paired t-test). In normal subjects the

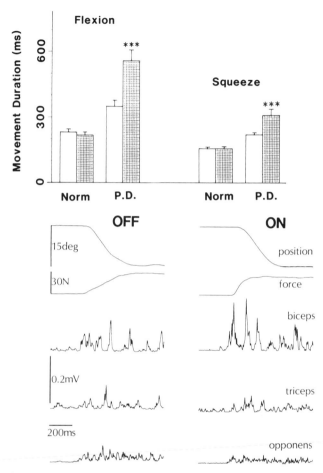

Fig. 4.4 Performance of simultaneous "squeeze and flex" task in normal subjects and patients with Parkinson's disease. Upper part shows the mean values (± 1 SD) of movement time for elbow flexion (left) and isometric finger flexion (squeeze; right) when performed separately (open columns) and simultaneously (shaded columns). In patients, but not in normal subjects, there was a significant prolongation of both movements in the simultaneous execution ($***P < 0.001$). Lower part shows representative single trials in a patient with Parkinson's disease in OFF and ON conditions.

movement times for "flex and squeeze" were the same, irrespective of whether the movements were performed separately or simultaneously. The EMG pattern in the "both together" task was not that expected from super-imposition of the behaviour in the separate tasks. If a patient had a preserved three-burst EMG pattern in biceps and triceps during a separate "flex", this changed to a multiple burst pattern or was replaced by tonic activity of biceps and triceps. Patients who already exhibited multiple bursts in the separate "flex" task changed to an exclusive tonic pattern in the "both together" task. The finger muscles recruited during "squeeze" showed a further decrease of the peak, or even a slow rise of activity rather than a step-like increase in EMG activity during the "both together" task.

A similar surplus slowness of the movements was also observed when the "squeeze" was performed with the left and the "flex" with the right hand. The amount of extra slowness was, however, much more pronounced in the unilateral than in the bilateral task (see also Fig. 4.5).

SPECIFICITY OF MOTOR ABNORMALITY AND RELATION TO CLINICALLY ESTABLISHED AKINESIA

The simple separate movements of "flex", "squeeze" and "cut" were slower in patients with Parkinson's disease than in normal subjects. This confirms

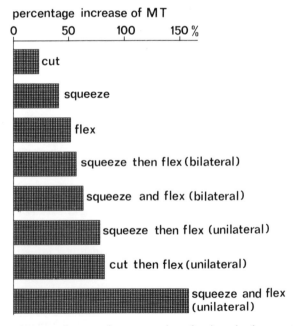

Fig. 4.5 Percentage mean increase of movement time of various simple separate and complex simultaneous or sequential tasks in patients with Parkinson's disease ($n = 5$) as compared to age-matched controls ($n = 9$) performing the same tasks. Mean values obtained in normal subjects were set to 100%.

the results of previous workers who used isotonic movements at the wrist, elbow and shoulder, but it was the first demonstration that isometric movements may be equally affected by the disease. Although movement time in simple isotonic movements is slow in Parkinson's disease, the degree of slowing is often not closely related to the degree of clinical akinesia. Furthermore, examination of the effect of L-dopa on the execution of single separate movements has revealed only slight improvements, in spite of a striking difference in the mobility of patients in the ON versus the OFF condition (Baroni *et al.*, 1984; Berardelli *et al.*, 1986; Benecke *et al.*, 1987b).

Slowness of simple separate movements is not a phenomenon specific to the akinesia/bradykinesia in Parkinson's disease. Prolongations of the three-burst pattern have also been observed in patients with cerebellar deficits (Hallett *et al.*, 1975). This slowness of simple separate movements in cerebellar patients and patients with an upper motor neuron syndrome has been confirmed in our own study (Benecke *et al.*, in preparation). When the motor deficits of these patient groups in single separate movements were compared with those in complex (simultaneous and sequential) performances, the two defects in complex movements observable in Parkinson's disease could not be seen: (1) the extra slowness of each component movement in simultaneous execution over and above that evident when the same movements were executed individually; and (2) the extra delay and slowing of movements in sequential executions. In summary, the study of complex movements in Parkinson's disease brings out specific deficits not evident in simple single movements or in other diseases causing slowness of movements, and is therefore more likely to reflect general bradykinesia and akinesia.

AKINESIA/BRADYKINESIA AS A CONSEQUENCE OF DISTURBED MOTOR PROGRAMMING

Motor programmes may be defined as a set of rules for executing a movement in which the muscles to be used (i.e. agonists, antagonists, synergists and postural fixators) and the timing of their actions are predictably related. For a movement of given amplitude and speed, the rules of the programme specify the relative activity for each muscle and the exact timing between them. A single rapid movement at one joint employs a single programme, evident in the EMG as the well-known triphasic burst pattern of the agonist, antagonist and agonist muscle. Some more complex tasks involving a series of movements at several joints can be controlled by a single motor programme or by a number of separate programmes (Benecke *et al.*, 1986a, 1987a). Motor programmes can be delivered without reference to external feedback. That humans are capable of calling up such motor programmes is well illustrated by the extensive range of both simple and complex motor acts that can be executed by a deafferented man (Rothwell *et al.*, 1982). Combinations of two movements, such as the "squeeze" and "flex" or "cut" and "flex" tasks in the present experiments, are executed by running two separate motor programmes one after the other (Benecke *et al.*, 1986a). The reason for saying this is that there is no correlation between the times taken to perform the two components of the task, as would be the case were the movements governed by single complex motor programme (Carter and Shapiro, 1984). Thus it is not possible to predict the timing of the second movement from inspection of the first. Hence there must be one set of rules controlling the execution of the first movement and another set for the second. It is these sets of rules which are referred to as motor programmes. The combination of motor programmes required to carry out a complex motor act can be termed the motor plan. Part of the function of the motor plan would be to time the onsets of the individual programmes which make up the total movement.

How is this concept of motor control disturbed in Parkinson's disease? The first conclusion to be made is that the formal aspects of motor programmes are preserved in Parkinson's disease, in that the selection of muscles and the correct sequence of muscle activation is intact. From the results of the present paper it becomes quite clear that the number and frequency of motoneurones activated is inadequate according to the aim of the patient or the instructions given. This is especially the case when complex simultaneous or sequential movements rather than separate single movements are examined. In terms of motor programming it can be concluded that the invariant characteristics of a motor programme are intact whereas the variant characteristics (speed and amplitude) are inadequately scaled. These disturbances in the organization of complex movements in patients with Parkinson's disease imply a difficulty in switching from one motor programme to the next and in running two motor programmes at the same time.

REFERENCES

Baroni, A., Benuenuti, F., Fantini, L., Pantaleo, T. and Urbani, F. (1984) *Neurology* **34**, 868–876.
Benecke, R., Meinck, H. M. and Conrad, B. (1985) *Exp. Brain Res.* **59**, 470–477.
Benecke, R., Rothwell, J. C., Day, B. L., Dick, J. P. R. and Marsden, C. D. (1986a) *Exp. Brain Res.* **63**, 585–595.
Benecke, R., Rothwell, J. C., Day, B. L. and Marsden, C. D. (1986b) *Brain* **109**, 739–757.
Benecke, R., Rothwell, J. C., Dick, J. P. R., Day, B. L. and Marsden, C. D. (1987a) *Brain* **110**, 361–379.

Benecke, R., Rothwell, J. C., Dick, J. P. R., Day, B. L. and Marsden, C. D. (1987b) *J. Neurol. Neurosurg. Psychiat.* **50**, 296–303.

Berardelli, A., Rothwell, J. C., Day, B. L. and Marsden, C. D. (1984) *Neurosci. Lett.* **47**, 47–50.

Berardelli, A., Dick, J. P. R., Rothwell, J. C., Day, B. L. and Marsden, C. D. (1986) *J. Neurol. Neurosurg. Psychiat.* **49**, 1273–1279.

Carter, M. C. and Shapiro, D. C. (1984) *J. Neurophysiol.* **52**, 787–796.

Draper, I. T. and Johns, R. J. (1964) *Bull. Johns Hopkins Hosp.* **115**, 465–489.

Flowers, K. A. (1976) *Brain* **99**, 269–310.

Hallett, M. and Koshbin, S. (1980) *Brain* **103**, 301–314.

Hallett, M. and Marsden, C. D. (1979) *J. Physiol. (Lond.)* **294**, 33–50.

Hallett, M., Shahani, B. T. and Young, R. R. (1975) *J. Neurol. Neurosurg. Psychiat.* **39**, 1163–1169.

Hallett, M., Shahani, B. T. and Young, R. R. (1977) *J. Neurol. Neurosurg. Psychiat.* **40**, 1129–1135.

Penney, J. and Young, A. B. (1986) *Movement Disorders* **1**, 3–15

Rothwell, J. C., Traub, M. M., Day, B. L., Obeso, J. A., Thomas, P. K. and Marsden, C. D. (1982) *Brain* **105**, 515–542.

Schwab, R. S. Chafetz, M. E. and Walker, S. (1954). *Arch. Neurol. Psychiat.* **72**, 591–598.

Talland, G. A. and Schwab, R. S. (1964) *Neuropsychologia* **2**, 45–53.

Thompson, P. D., Berardelli, A., Rothwell, J. C., Day, B. L., Dick, J. P. R., Benecke, R. and Marsden, C. D. (1988) *Brain* **111**, 223–244.

5

Models of Cognitive Dysfunction in Parkinson's Disease

Richard G. Brown

INTRODUCTION

Over the past three decades, an extensive body of evidence on cognitive function in Parkinson's disease has accumulated. It is, perhaps, inevitable that much of the early research adopted a "look and see" approach. In recent years, neuropsychologists have been less content with merely demonstrating the presence or absence of a deficit on a particular test or group of tests. Rather, the aim has been to define the precise nature of the deficit, and, ultimately, to relate this to brain pathology or neurochemical disturbance.

The aim of this chapter is to consider some of the models which have been proposed. These will be divided into three types, although there is clearly overlap in some cases. The *descriptive* models seek to define the characteristics of the tests on which the patients perform poorly, and those on which they are normal. The *process* models attempt to define the cognitive processes involved in producing the abnormal behavioural responses detected by the neuropsychological tests. The tradition underlying these first two types of models can be found in human cognitive psychology. Finally, the *structural* models seek to link neuropsychological disturbance with disruption of particular areas of the brain or neuronal systems. These latter models derive, largely, from a tradition of animal experimental psychology. The coming together of the two traditions promises to provide a greater understanding of cognitive function in Parkinson's disease, and of brain–behaviour relationships in general.

THE DESCRIPTIVE MODELS

A framework has evolved for describing human cognitive function in which we speak, for instance, of visuospatial function, attention, language and memory. These categories have themselves become subdivided. For example, memory is divided into remote, long term and short term, and sensory, semantic, episodic, procedural and declarative. Neuropsychology has sought dissociations between these areas following damage to different parts of the brain. One result is a vast array of measures, each designed to assess a particular aspect of cognition. In this way, tests gain labels such as "procedural learning tests" or "visuospatial tests". While this provides a convenient descriptive label, it can lead us into committing a dangerous logical error; namely, to conclude that if a patient performs poorly on a "visuospatial test", for example, then he or she has a visuospatial deficit. All tests will depend upon many aspects of cognitive function. This principle is known as "multiple determination of behaviour" and means, simply, that there are many reasons why an individual should perform poorly on any test of cognitive function. Patients with Parkinson's disease, as a group, perform poorly on many types of neuropsychological tests (see Brown and Marsden, 1987, for a review). Do these tests have some common feature?

Behavioural regulation: set and shifting

In the animal literature, the concept of "set" has been associated for some time with the functioning of the basal ganglia (Mishkin, 1964), where "set" is defined as a "relatively persisting predisposition to behave in a particular way on the occurrence of a particular stimulus" (Buchwald et al., 1975). The concept is also applied to cognitive changes in patients with Parkinson's disease. Bowen et al. (1975) assessed a group on the Wisconsin Card Sorting Test (WCST) (Berg, 1948), pre- and post-L-dopa treatment. Under both conditions, they made more errors than controls, although the number decreased after L-dopa. Rather than treating the poor WCST performance as an isolated deficit, Bowen and colleages proposed that difficulty in switching set was a more fundamental impairment in Parkinson's disease which could account for poor performance on other tests (Bowen et al., 1972,

1976). Subsequent studies confirmed that WCST performance is impaired in Parkinson's disease (e.g. Flowers, 1982; Lees and Smith, 1983; Brown and Marsden, 1986, 1988a,b), although Gotham et al. (1988) failed to find any difference in performance between patients on and off L-dopa treatment. The generality of the switching deficit was assessed by Cools et al. (1984), who studied a range of tests, all designed with a baseline and a shift phase. For instance, on a word fluency task the subjects had to generate animal names for 1 min (baseline), followed by the names of professions (shift phase). The crucial measure was performance on the shift phase relative to baseline. On all of the tests the Parkinson's disease group showed a differential impairment on this index of performance.

Requiring subjects to shift from one task to another, or to process the same stimulus in different ways, however, is not sufficient to demonstrate the deficit. Brown and Marsden (1988a) required the subjects to make left–right discriminations under two conditions, one where the stimulus was oriented in the same direction as the subject, and the other when it was rotated by 180°. When required to shift between these two conditions, both patients and controls showed increased reaction time and error rate, but this "shift effect" was the same in both groups. Despite their normal performance on this spatial task, the Parkinson's disease group were impaired on the WCST. The WCST differs from the spatial test in that the stimuli are ambiguous with no visual cues as to which response is required. This ambiguity may be crucial. Although Cools et al. (1984) described a general "shifting aptitude disorder" in Parkinson's disease, they speak in their introduction of tasks in which "behaviour is not directed by currently available sensory information".

Temporal ordering and sequencing

The concepts of set and sequencing are sometimes linked. Cools et al. (1984) spoke of the ability to "rearrange arbitrarily the serial order of the components of behavioural programmes". There is some indication, however, that even when the order is not arbitrary, a deficit may be found in Parkinson's disease. Sullivan et al. (1985) used the picture arrangement subtest of the WAIS (Weschler, 1955), as an index of sequencing ability. Patients with Parkinson's disease showed a differential impairment on this test relative to their level of vocabulary, compared to both normal controls and patients with Alzheimer's disease. Sagar et al. (1988) employed a memory task in which the subjects had to identify the content of photographs of famous events, and to date their occurrence. While the patients with Parkinson's disease were able to recognize the *content* of the scenes, they were impaired in *dating* them. This was linked to problems of recency discrimination also demonstrated in the same study. Together, these results suggest an impairment in temporal ordering and sequencing in Parkinson's disease. Whether this deficit is linked to the shifting deficit described by Cools and colleagues remains to be demonstrated.

Cognitive impairment and effort

The models described so far, while able to account for a range of cognitive impairments in Parkinson's disease, cannot be considered as general models. For instance, patients with Parkinson's disease are impaired in the free recall of information where neither set, shifting or sequencing plays any obvious role (e.g. Pirozzolo *et al.*, 1982; Tweedy *et al.*, 1982). Recall tasks, however, can require sustained effort. Recognition paradigms, which involve only a decision of familiarity, can be seen as less effort demanding, and patients with Parkinson's disease are typically unimpaired on such tasks (e.g. Lees and Smith, 1983; Flowers *et al.*, 1984). Certain other tasks such as remembering whether an item has been presented once or many times seem to be even less effort demanding and can be thought of as automatic. This distinction between effort-demanding and automatic tasks has been suggested by Weingartner *et al.* (1984) as a basis for explaining learning and memory impairment in Parkinson's disease. Can the same distinction be applied to a wide range of tests? A general model would predict that the more effort demanding the task, the greater the impairment shown by the patients with Parkinson's disease. However, studies which have presented tasks with increasing levels of difficulty have failed to show any consistent differential impairment (e.g. Wilson *et al.*, 1980; Brown and Marsden, 1986). It may be that effort and difficulty are not directly linked in a simple linear fashion. If so, then it may be difficult to predict the effort requirements of a specific task, thus limiting the heuristic value of the general model.

Locus of cognitive control

As noted above, Cools *et al.* (1984) discussed the proposed shifting deficit in Parkinson's disease with regard to tasks in which "behaviour is not directed by currently available sensory information". Although the shifting model itself may fit only a limited number of observed behavioural deficits, it is possible that the types of tasks in which it is observed can form the basis for a more general descriptive model. Brown and Marsden (1988a) suggested that the dissociation observed between the WCST and the spatial task in that study could be explained by the fact that in the latter task an explicit cue was present at each trial, which removed the ambiguity from the stimulus configuration. In the WCST, however, the subject needed to rely on "internal" cues to focus attention on the relevant stimulus attribute. A distinction was proposed between tasks in which *internal* cognitive control is needed, and those in which some degree of *external* or stimulus control is provided. Recognition paradigms, for example, include external cues, whereas in recall paradigms, the subject must rely upon internal strategies for the retrieval of the information.

Essentially the same distinction was proposed by Taylor *et al.* (1986). They administered a range of tests and noted that impairment was found on tasks which "maximized the need for self-directed planning in forming context-dependent associations or strategies", while "the execution of even highly challenging executive abilities was not impaired where rules were explicit". While Taylor *et al.* (1986) and Brown and Marsden (1988a) described similar models, both were *post hoc* explanations of results. The value of any model is judged not only by how well it explains previous results, but also by its ability to generate specific, testable hypotheses. Brown and Marsden (1988b) tested their model using a version of the Stroop test (Stroop, 1935) in which colour words were presented in the complementary colour, i.e. "RED" written in green, and "GREEN" written in red. Subjects had to identify either the meaning of the word or the colour of the "ink" in which it was written. Every 10 trials the relevant attribute changed. In one condition, subjects received an explicit (external) cue prior to each trial, reminding them of the relevant attribute. In a second condition, they received only a warning signal prior to the stimulus, and had to rely on their own internal strategies for focusing attention. The Parkinson's disease group was impaired only on the second, non-cued version of the task. While the presence or absence of external cues seemed to be a reasonable predictor of behavioural deficit in Parkinson's disease in this study, its generality requires further testing.

THE PROCESS MODELS

The various models just discussed, whatever their power in explaining previous results or in generating testable hypotheses, provide only descriptions of the factors which determine whether or not cognitive dysfunction will be observed. To state that patients with Parkinson's disease are impaired on tasks which require shifting set, or internal cognitive control, says little about the underlying processes. It is tautological to define the processes themselves as "shifting" or "internal control".

Cognitive slowing—bradyphrenia

Naville (1922) described a mental syndrome, bradyphrenia, characteristic of patients with post-encephalic parkinsonism. It consisted of a diminution of voluntary attention, spontaneous interest, initiative and the capacity for effort and work, with subjective fatiguability and a slight diminution of memory. One of the most marked features was the apparent slowing of the processes of thought. Albert *et al.* (1974) described the syndrome of "subcortical dementia", which shared many of the features described by Naville (1922). Although the nosological status of subcortical dementia has been questioned (Whitehouse, 1986; Brown and Marsden, 1988c), individual symptoms may still apply to individual disorders. Perhaps the most interesting is the proposal that cognitive processes are slowed in Parkinson's disease, in a manner analogous to the slowing of motor function. The typical paradigm used has been the two-part test in which each part has the same motor component, but different levels of cognitive complexity. Subtracting response times for the two parts gives an index of the additional cognitive processing time required for the more demanding task. If bradyphrenia is typical of Parkinson's disease, then the patients should show a slowing, relative to a control group, on this difference score. Typically, slowing has been found only on some tasks, and then only in some patient subgroups. It is difficult, from this evidence, to consider that cognitive slowing is a central feature in Parkinson's disease. Even the positive results, however, need not imply that cognitive slowing was the underlying mechanism. The time measured is that to produce the final behavioural response. If this is slowed then it can be for reasons other than slowing of the normal processes of cognition. The subject may be performing the same processes at the same speed as normal subjects, but less efficiently. If mistakes are made which need to be corrected, then this will result in a slowing of the time to produce the final behavioural response. Alternatively, the subject may use strategies which are less efficient and therefore slowing. In both cases it is inefficiency which is causing the behavioural slowing, not a slowing of normal information processing.

Processing resources

A central feature of models of human information processing such as working memory (Baddeley, 1986), is that "resources" are limited. For example, there is a limit to the number of items we can hold in short-term memory at any one time, or the number of simultaneous tasks that we can perform. Within an information processing model, "resources" can refer to a range of factors such as the capacity of the system or the number of channels available for transmitting information.

Brown and Marsden (1988a) proposed that the cues used to control cognition, either internal or external, could serve as a descriptive model of cognitive impairment in Parkinson's disease. Why should tasks which require internal cognitive control lead to impairment? One possibility is that they are more resource demanding, and that patients with Parkinson's disease have fewer resources available. In the study of Brown and Marsden (1988b), patients were impaired only on the non-cued version of the Stroop task. Control subjects showed no difference in performance on the cued and non-cued version, suggesting not that the two tasks made equal resource demands but only that both were within their level of processing resources. In an unpublished study, the authors attempted to produce the same impairment in normal subjects as seen in patients with Parkinson's disease, by reducing the amount of processing resources available. They did this by getting the subjects to perform one of three secondary tasks simultaneously with the Stroop in both the cued and non-cued conditions. The tasks were either foot tapping, repetition of a nonsense word or generating random numbers. The secondary tasks were performed during blocks 2 and 4 of the Stroop task. The random number condition was chosen as the resource-demanding task, the nonsense word condition to control for any effect of random number generation on verbal rehearsal strategies, and the foot tapping condition to control for the effect of a simple repetitive motor action. Neither foot tapping nor repetition of a nonsense syllable had any influence in either the cued or non-cued condition. In contrast, generating random numbers had a clear effect on reaction time, with a differential influence on the non-cued condition. Thus, reducing the subjects' level of processing resources produced a similar pattern of performance to that seen in Parkinson's disease. This is not proof that patients have resource limitations, nor that those limitations are responsible for a wide range of cognitive impairment. Use of working memory paradigms, however, such as that described above, provide a method of exploring the resource limitation model in a wide range of tests.

"STRUCTURAL" MODELS

The models considered so far have been concerned with behaviour and its underlying cognitive processes. It is necessary, next, to consider the neuronal and neurochemical substrate of the cognitive dysfunction.

The neurochemical models

The pathological changes of Parkinson's disease lead, directly and indirectly, to disruption of many of the brain's neurochemical systems. Because of the key role played by dopamine in the disease, much attention has focused on its possible significance in cognitive function. Several studies have assessed groups of patients before commencing L-dopa treatment, and at various stages after treatment has commenced (e.g. Loranger *et al.*, 1973; Riklan *et al.*, 1973, 1976; Portin and Rinne, 1980). In general, the studies revealed moderate initial improvement on a range of cognitive tests, but with a gradual deterioration after a few years.

An alternative approach has been to assess patients after withdrawal of treatment, or during an episode of minimal response ("off" period). This strategy allows more control over the order of testing in the two treatment conditions. Brown *et al.* (1984) assessed a group of patients on a test of verbal, numerical and spatial ability. While they were more impaired off than on L-dopa, the differences were small, and related more to the patients' self-reported level of mood and arousal than to their motor disability. Delis *et al.* (1982) described a single case in which the cognitive changes were described as "modest" in relation to the profound changes in motor status. Mohr *et al.* (1987) administered a range of tests and found on–off differences only on delayed verbal memory, while neither Girotti *et al.* (1986) nor Rafal *et al.* (1984) were able to find any changes in performance between the two states. Together, this evidence suggests that levels of dopamine (as indicated by the patient's motor status) are a poor predictor of cognitive function in Parkinson's disease: dramatic changes in motor function are accompanied by modest or no change in cognition. One further study, however, deserves mention. Gotham *et al.* (1988) assessed a range of tests thought to be sensitive to disruption of the prefrontal cortex. No consistent pattern was found between the treated and untreated states. On the WCST, the patient group was impaired in both conditions, while on a verbal fluency test they were impaired only when off L-dopa. Of some interest was the finding that performance on tests of subject-ordered pointing and conditional learning were impaired only in the medicated state. This suggests that the latter tests are sensitive to dopamine, but that the mechanism of dysfunction may be *overstimulation* rather than dopamine depletion.

Anticholinergic drugs are another common treatment in Parkinson's disease. Considerable interest has focused on the influence that they might have on memory. Miller *et al.* (1987) demonstrated that the severity of the memory impairment in two unselected samples of patients with Parkinson's disease was significantly associated with the patients' therapeutic dose of benzhexol, but not L-dopa. The link between anticholinergic medication and memory function in Parkinson's disease had been noted previously (Sadeh *et al.*, 1982; Syndulko and Tourtellotte, 1983; Koller, 1984). However, anticholinergics are known to influence learning and memory in normal healthy individuals (Collerton, 1986). Are patients with Parkinson's disease any more sensitive than age-matched controls? In a double-blind cross-over study, Dubois *et al.* (1987) demonstrated that a dose of scopolamine, insufficient to produce any effect in normal subjects, led to a significant deterioration in memory function in Parkinson's disease, suggesting that the patients had a lower level of acetylcholine and therefore a reduced threshold, a conclusion confirmed by post-mortem investigations (Agid *et al.*, 1987). Cholinergic system dysfunction (pathological, iatrogenic or in combination) may provide a useful model for memory dysfunction in Parkinson's disease, but its relationship to other areas of cognitive impairment has yet to be examined.

While attention has focused on dopamine and acetylcholine, other neurochemical systems are disrupted in Parkinson's disease, including GABA, noradrenaline and 5-HT (Agid *et al.*, 1987). The roles of these systems in cognitive function in man are still unclear, and their impact on the pattern of cognitive impairment in Parkinson's disease has yet to be examined. One study, however, suggests that altered noradrenaline metabolism may be associated with a deficit in sustained attention in Parkinson's disease (Stern *et al.*, 1984).

An anatomical model

The localizationist tradition in neuropsychology has resulted in certain areas of the brain, primarily cortical, being associated with certain broad classes of cognitive function. If parallels are sought between the pattern of cognitive impairment found in Parkinson's disease and that found following damage to areas of cerebral cortex, then the frontal region provides the closest match. "Frontal" deficits observed in Parkinson's disease include poor performance on the WCST and the impairment in shifting set and sequencing as mentioned above. Other deficits are also found: impaired setting of visual and postural vertical (Proctor *et al.*, 1964; Danta and Hilton, 1975); impaired delayed matching to sample (Horne, 1971); increased sensitivity to proactive interference in short-term memory tasks (Tweedy *et al.*, 1982); difficulty in control of the Necker cube (Talland, 1962); difficulty in performing simultaneous motor acts (Talland and Schwab, 1964; Horne, 1973); anomia (Matison *et al.*, 1982); impaired prism adaptation (Canavan and Passingham, 1985; Stern *et al.*, 1986); impaired performance on conditional learning tests (Canavan and Passingham, 1985; Gotham *et al.*, 1988) and on subject-ordered pointing tests (Gotham *et al.*, 1988); and impaired ability to reproduce a sequence of gestures (Morel-Maroger, 1977; Goldenburg *et al.*, 1986), and to respond to a gesture with a predetermined different gesture (Morel-Maroger, 1977; Flowers, 1982).

There is little evidence for structural damage to the frontal cortex in Parkinson's disease, and attention has focused instead on the neuronal connections between this region and the basal ganglia. Alexander *et al.* (1986) describe a series of "loops" connecting the striatum with the prefrontal cortex. Evidence from animal lesion studies suggests that some of these loops are involved in cognitive function and that similar behavioural changes can be obtained following lesions to different structures within the loop. In Parkinson's disease, the striatofrontal systems may be compromised through disruption of basal ganglia activity. The behavioural consequence of that disruption, however, would occur via the frontal cortex, the main cortical focus of striatal "outflow" (Taylor *et al.* 1986). This remains a fruitful avenue for future research.

CONCLUSIONS

This chapter has reviewed some of the models which seek to explain cognitive changes found in Parkinson's disease. None of them can be considered definitive, and most are limited in their applicability. However, the integration of the three types of model (descriptive, process and structural) promises to provide a more powerful framework for understanding and explaining cognitive function in Parkinson's disease and brain–behaviour relationships in general.

REFERENCES

Agid, Y., Javoy-Agid, F. and Ruberg, M. (1987) In *Movement Disorders* 2 (edited by C. D. Marsden and S. Fahn), pp. 166–230. Butterworths, London.
Albert, M. L., Feldman, R. G. and Willis, A. L. (1974) *J. Neurol. Neurosurg. Psychiat.* **37**, 121–130.
Alexander, G. E., DeLong, M. R. and Strick, P. L. (1986) *Ann. Rev. Neurosci.* **9**, 357–381.
Baddeley, A. (1986) Working Memory. Oxford University Press, Oxford.
Berg, E. A. (1948) *J. Gen. Psychol.* **39**, 15–22.
Bowen, F. P., Hoehn, M. M. and Yahr, M. D. (1972) *Neuropsychologia* **10**, 355–361.
Bowen, F. P., Kamienny, R. S., Burns, M. M. and Yahr, M. D. (1975) *Neurology* **25**, 701–704.
Bowen, F. P., Burns, M. M., Brady, E. M. and Yahr, M. D. (1976) *Neuropsychologia* **14**, 425–429.
Brown, R. G. and Marsden, C. D. (1986) *Brain* **109**, 987–1002.
Brown, R. G. and Marsden, C. D. (1987) In *Movement Disorders* 2 (edited by C. D. Marsden and S. Fahn), pp. 99–123. Butterworths, London.
Brown, R. G. and Marsden, C. D. (1988a) *Movement Disorders* **3**, 152–161.
Brown, R. G. and Marsden, C. D. (1988b). *Brain* **111**, 323–345.

Brown, R. G. and Marsden, C. D. (1988c) *Neuroscience* **25**, 363–387.

Brown, R. G., Marsden, C. D., Quinn, N. and Wyke, M. (1984) *J. Neurol. Neurosurg. Psychiat.* **47**, 454–465.

Buchwald, N. A., Hull, C. D., Levine, M. S. and Villablanca, J. (1975) In IBRO Monograph Series 1, *Growth and Development of the Brain: Nutritional Genetic and Environmental Factors* (edited by M. A. Brazier), pp. 171–189. Raven Press, New York.

Canavan, A. G. M. and Passingham, R. E. (1985) *Neurosci. Lett.* suppl. 22, S587.

Collerton, D. (1986) *Neuroscience* **19**, 1–28.

Cools, A. R., Van Den Bercken, J. H. L., Horstink, M. W. I., Van Spaendonck, K. P. M. and Berger, H. J. C. (1984) *J. Neurol. Neurosurg. Psychiat.* **47**, 443–453.

Danta, G. and Hilton, R. C. (1975) *Neurology* **25**, 43–47.

Delis, D., Direnfield, L., Alexander, M. P. and Kaplan, E. (1982) *Neurology* **32**, 1049–1052.

Dubois, B., Danze, F., Pillon, B., Cusimano, G., Lhermitte, F. and Agid, Y. (1987) *Ann. Neurol.* **22**, 26–30.

Flowers, K. A. (1982) *Behav. Brain Res.* **5**, 100.

Flowers, K. A., Pearce, I. and Pearce, J. M. S. (1984) *J. Neurol. Neurosurg. Psychiat.* **47**, 1174–1181.

Girotti, F., Carella, F., Grassi, M. P., Soliveri, P., Marano, R. and Caraceni, T. (1986) *J. Neurol. Neurosurg. Psychiat.* **49**, 657–660

Goldenberg, G., Wimmer, A., Auff, E. and Schnaberth, G. (1986) *J. Neurol. Neurosurg. Psychiat.* **49**, 1266–1272.

Gotham, A.-M., Brown, R. G. and Marsden, C. D. (1988) *Brain* **111**, 299–321.

Horne, D. J. de Lancey (1971) *J. Neurol. Neurosurg. Psychiat.* **34**, 192–194.

Horne, D. J. De Lancey (1973) *J. Neurol. Neurosurg. Psychiat.* **36**, 742–746.

Koller, W. C. (1984) *Cortex* **20**, 307–311.

Lees, A. R. and Smith, E. (1983) *Brain* **106**, 257–270.

Loranger, A. W., Goodell, H., McDowell, F. H., Lee, J. E. and Sweet, R. D. (1973) *Am. J. Psychiat.* **130**, 1386–1389.

Matison, R., Mayeux, R., Rosen, J. and Fahn, S. (1982) *Neurology* **32**, 567–570.

Miller, E., Berrios, G. E. and Politynska, B. (1987) *Acta Neurol. Scand.* **76**, 278–282.

Mishkin, M. (1964) In *Frontal Granular Cortex and Behaviour* (edited by J. M. Warren and K. Akert), pp. 219–241. McGraw-Hill, New York.

Mohr, E., Fabbrini, G., Ruggieri, S., Fedio, P. and Chase, T. N. (1987) *J. Neurol. Neurosurg. Psychiat.* **50**, 1192–1196.

Morel-Maroger, A. (1977) *Br. Med. J.* **2**, 1543–1544.

Naville, F. (1922) *Encephale* **17**, 369–375.

Pirozzolo, F. J., Hansch, E. C., Mortimer, J. A., Webster, D.D. and Kuskowski, M. A. (1982) *Brain Cognition* **1**, 71–83.

Portin, R. and Rinne, U. K. (1980) *Adv. Neurol.* **40**, 219–227.

Proctor, F., Riklan, M., Cooper, I. S. and Teuber, H.-L. (1964) *Neurology* **14**, 287–293.

Rafal, R. D., Posner, M. I., Walker, J. A. and Friedrich, F. J. (1984) *Brain* **107**, 1083–1094.

Riklan, M., Halgin, R., Maskin, M. and Weissman, D. (1973) *J. Nervous Mental Disorders* **157**, 452–464.

Riklan, M., Whelihan, W. and Cullinan, T. (1976) *Neurology* **26**, 173–179.

Sadeh, M., Braham, J. and Modan, M. (1982) *Arch. Neurol.* **39**, 666–667.

Sagar, H. J., Cohen, N. J., Sullivan, E. V., Corkin, S. and Growdon, J. H. (1988)

Brain **111**, 185–206.

Stern, Y., Mayeux, R. and Cote, L. (1984) *Arch. Neurol.* **41**, 1086–1089.

Stern, Y., Mayeux, R., Hermann, A. and Rosen, J. (1986) *Neurology* **36** (suppl. 1), 89.

Sternberg, S. (1966) *Science* **153**, 652–654.

Stroop, J. R. (1935) *J. Exp. Psychol.* **18**, 643–662.

Sullivan, E. V., Sagar, H. J., Gabrieli, J. D. E., Corkin, S. and Growdon, J. H. (1985) *J. Clin. Exp. Neuropsychol.* **7**, 160.

Syndulko, K. and Tourtelotte, W. W. (1983) *Arch. Neurol.* **40**, 390.

Talland, G. A. (1962) *J. Nervous Mental Disorders* **135**, 196–205.

Talland, G. A. and Schwab, R. S. (1964) *Neuropsychologia* **2**, 45–53.

Taylor, A. E., Saint-Cyr, J. A. and Lang, A. E. (1986) *Brain* **109**, 845–883.

Tweedy, J. R., Langer, K. G. and McDowell, F. H. (1982) *J. Clin. Neuropsychol.* **4**, 235–247.

Wechsler, D. (1955) Wechsler Adult Intelligence Scale. Psychological Corporation, New York.

Weingartner, H., Burns, S., Diebel, R. and LeWitt, P. A. (1984) *Psychiat. Res.* **11**, 223–235.

Whitehouse, P. J. (1986) *Ann. Neurol.* **19**, 1–6.

Wilson, R. S., Kaszniak, A. W., Klawans, H. L. and Garron, D. C. (1980) *Cortex* **16**, 67–72.

6

Dopaminergic Systems in Parkinson's Disease

Yves Agid, France Javoy-Agid, Merle Ruberg
and Etienne Hirsch

Dopamine-containing neurons in human brain are organized into ascending (principally nigrostriatal, mesocorticolimbic and hypothalamic) and descending (to the spinal cord) systems which are similar to those that have been identified in studies on rat brain (Lindvall and Bjorklund, 1978). Destruction of the nigrostriatal neurons and the consequent decrease in striatal dopamine (Ehringer and Hornyckiewicz, 1960) is the most distinctive biochemical characteristic of Parkinson's disease and parkinsonian-like degenerative syndromes, such as post-encephalitic parkinsonism and progressive supranuclear palsy (Hornykiewicz, 1966, 1972, 1975, 1979a, 1982; Rinne et al., 1979; Birkmayer and Rierderer, 1980; Javoy-Agid et al., 1984b, 1986). This very important observation has given rise to a cascade of questions. Is the striatal dopamine deficiency specific to parkinsonian syndromes? Can it be regarded, along with the motor symptoms, as an unequivocal hallmark of Parkinson's disease? Are the nigrostriatal dopamine neurons damaged selectively, or are the other dopamine systems also affected? Is cell loss restricted to central dopaminergic neurons, or are peripheral systems affected as well? Does the physiological state of the surviving dopaminergic neurons change in response

DISORDERS OF MOVEMENT: CLINICAL, PHARMACOLOGICAL
AND PHYSIOLOGICAL ASPECTS   ISBN 0-12-569685-X

to the decrease in striatal dopamine release? What neurochemical mechanisms underlie the beneficial effects of L-dopa and the subsequent side-effects often seen in patients? Do the various clinical signs of Parkinson's disease result uniquely from loss of dopaminergic neurons? How much cell loss can be tolerated before symptoms appear? Biochemical assays of dopaminergic markers have provided responses to some of these questions.

BIOCHEMICAL ABNORMALITIES AFFECTING DOPAMINE-CONTAINING NEURONS

Nigrostriatal system

The diagnosis of Parkinson's disease is classically confirmed by neuro-pathological evidence of cell loss in the substantia nigra. The observation of a dramatic decrease in striatal and nigral dopamine concentrations indicate that these cells correspond to the nigrostriatal dopaminergic neurons (Fig. 6.1). Loss of these neurons follows an identifiable topographic pattern. The decrease in dopamine concentrations is greater in the rostral striatum (Fahn *et al.*, 1971; Kish *et al.*, 1986). The putamen is more severely affected than the caudate nucleus (Ehringer and Hornykiewicz, 1960; Bernheimer *et al.*, 1973; Nyberg *et al.*, 1983), perhaps because cell loss in the substantia nigra is more severe in the caudal and internal portions of the structure which project preferentially to the putamen. In the substantia nigra, the dopamine deficit is more pronounced in the pars reticulata (which contains the dendrites of the nigrostriatal neurons) than in the pars compacta (which contains the cell bodies) (Ehringer and Hornykiewicz, 1960; Javoy-Agid *et al.*, 1981). In addition to the dopamine deficiency, a number of other biochemical observations converge in support of the hypothesis that the dopaminergic projections from the substantia nigra to the striatum degenerate in Parkinson's disease. (i) The activities of tyrosine hydroxylase and DOPA-decarboxylase, enzymes involved in dopamine synthesis, decrease in the substantia nigra and striatum (Table 6.1). (ii) Specific markers of aminergic nerve endings, such as the binding of ^{3}H-tetrabenazine (a ligand with high affinity for monoaminergic synaptic vesicules) (Pierot *et al.*, 1988) and ^{3}H-cocaine (which binds to the dopamine transport complex in the striatum) (Schoemaker *et al.*, 1985) decrease in the striatum of parkinsonians.

The metabolites of dopamine, dihydroxyphenylacetic acid (DOPAC) and homovanillic acid (HVA) also decrease in the substantia nigra and the striatum (Bernheimer *et al.*, 1973; Lloyd *et al.*, 1975; Rinne *et al.*, 1979; Bokobza *et al.*, 1984). This might be reflected by the reduced levels of HVA found in urine (Barbeau *et al.*, 1961; Weil-Malherbe and Van Bueren, 1969)

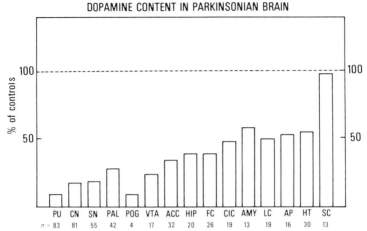

Fig. 6.1 Brain dopamine content in Parkinson's disease. Data expressed as percentage of controls are the mean of data obtained from references: Erhinger and Hornykiewicz (1960); Hornykiewicz (1963); Fahn *et al.* (1971); Bernheimer *et al.* (1973); Rinne and Sonninen (1973); Lloyd *et al.* (1975); Rinne *et al.* (1977); Price *et al.* (1978); Javoy-Agid *et al.* (1981); Ploska *et al.* (1982); Bokobza *et al.* (1984); Javoy-Agid *et al.* (1984a); Cash *et al.* (1987a); Scatton *et al.* (1982, 1983, 1986). n = number of brains. The dopamine content is statistically different from controls in all areas except spinal cord. ACC, nucleus accumbens; AMY, amygdala; AP, area postrema; Cer, cerebellum; CIC, cingular cortex (Brodmann area 24); CN, caudate nucleus; EC, entorhinal cortex; FC, frontal cortex (Brodmann area 9); HIP, hippocampus; HT, hypothalamus; LC, locus coeruleus; PAL, pallidum; POG, paraolfactory gyrus (Brodmann area 25); PU, putamen; RN, raphé nuclei; SC, spinal cord; SI, substantia innominata; SN, substantia nigra; SNC, substantia nigra, pars compacta; SNR, substantia nigra, pars reticulata; ST, subthalamic nucleus; VL, thalamus ventro-lateral nucleus; VTA, ventral tegmental area of the mesencephalon.

and CSF (Olson and Roos, 1968; Gottfries *et al.*, 1969) of patients. Interestingly, the dopamine metabolites are less reduced than the amine itself. This may in part be explained by post-mortem diffusion of dopamine from its storage sites and subsequent degradation by enzymes. Indeed, monoamine oxidase A and B (the former more specific for serotonin, the latter, 35% of brain MAO activity, more specific for dopamine) and catechol-*O*-methyltransferase (COMT), are not affected by the disease (they are primarily located outside of the dopamine neurons) (Table 6.1), or by the post-mortem conditions to which human brain samples are subject before assay. The difference between the dopamine deficiency and that of its metabolites may not be exclusively due to non-specific transformation post-mortem (Sloviter and Connor, 1977) but may also result from an increased activity of the surviving neurons (see below).

    Low striatal dopamine content, strikingly correlated with cell loss in the substantia nigra (Bernheimer *et al.*, 1973), is a common feature of parkinsonian

Table 6.1   Enzyme activities in brains from parkinsonian patients.

Enzyme* activity		Putamen	Caudate nucleus	Pallidum	Substantia nigra	Ventral tegmental area	References
Tyrosine hydroxylase		18 (75)	29 (78)	41 (59)	28 (67)	43 (7)	a,b,c,d,e,f
DOPA-decarboxylase		25 (66)	45 (66)	69 (66)	31 (41)		a,d,f
Monoamine oxidase	Total	108 (10)	101 (10)	138 (5)	81 (5)		a
	A	95 (7)	140 (7)		94 (7)		g
	B	107 (18)	125 (7)		133 (15)		a,h
Catechol-O-methyl transferase		82 (9)	70 (9)		82 (5)		a

Values are expressed as percentage of respective controls.
Numbers in parentheses represent the number of patients.
Data obtained from: (a), Lloyd et al., (1975); (b), McGeer and McGeer, 1976; (c), Riederer et al., 1977; (d), Rinne et al., 1979; (e), Javoy-Agid and Agid, 1980; (f), Nagatsu et al., 1982; (g), Jellinger and Riederer, 1984; (h), Yong and Perry, 1986.

syndromes with different etiologies: idiopathic Parkinson's disease, progressive supranuclear palsy (Bokobza et al., 1984; Ruberg et al., 1985; Kish et al., 1985), striatonigral degeneration (Sharpe et al., 1973) and post-encephalitic or manganese induced parkinsonism (Bernheimer et al., 1973). In post-encephalitic parkinsonism and progressive supranuclear palsy, however, unlike the idiopathic disease, the dopamine deficiency is particularly severe and affects the caudate nucleus and putamen to the same extent (Bernheimer et al., 1973; Ruberg et al., 1985). The biochemical observation is compatible with the histopathological evidence for a severe and uniform neuronal loss in the substantia nigra (Greenfield and Bosanquet, 1953; Hassler, 1953; Escourolle et al., 1970).

Mesocorticolimbic system

The mesolimbic and mesocortical dopaminergic systems seem to exist in human brain as well as in animals (Lindvall and Bjorklund, 1978). (i) In the mesencephalon, high levels of dopamine in the ventral tegmental area (Javoy-Agid et al., 1981) indicate the presence of dopaminergic perikarya, which have been visualized by histofluorescence in human fetuses (Nobin and

Bjorklund, 1973) and by immunohistochemical staining of tyrosine hydroxylase in adult brain (Gaspar et al., 1983). (ii) In cortical and limbic regions substantial amounts of dopamine can be detected (Price et al., 1978; Scatton et al., 1983), indicating that the structures are innervated by dopaminergic neurons, although it cannot be ruled out that the dopamine may in part be located in noradrenergic fibers and/or in blood vessels. Dopaminergic fibers have been visualized, however, by histofluorescence (Berger et al., 1980). Dopaminergic innervation of the cortical areas is sparse compared to that of the striatum, on the order of 1/500 (Scatton et al., 1983).

There is evidence that the mesocortico-limbic dopamine neurons degenerate in Parkinson's disease. Dopamine, DOPAC and HVA levels are subnormal (40–60%) in limbic (nucleus accumbens, parolfactory gyrus, cingulate and entorhinal cortex, hippocampus, olfactory tubercles) and neocortical (frontal, temporal, occipital) areas, indicating that dopamine afferents are damaged (Price et al., 1978; Scatton et al., 1983). Furthermore, morphological and biochemical changes are detected in the ventral tegmental area (VTA) suggesting loss of dopamine-containing cell bodies: (i) cell counts on serial sections of the VTA from control and parkinsonian brains provide evidence of neuronal loss (Javoy-Agid et al., 1984b; Uhl et al., 1985); (ii) the density of neurons containing tyrosine hydroxylase-like immunoreactivity is reduced (Javoy-Agid et al., 1984b) (Fig. 6.2); (iii) the activity of tyrosine hydroxylase (Javoy-Agid et al., 1982) and dopamine concentrations (Javoy-Agid et al., 1981) are decreased. The decrease in the striatum is not uniform, the putamen and the dorso-rostral caudate nucleus being most severely affected (Kish et al., 1988).

Hypothalamic systems

Dopamine concentrations decrease by 60% in the hypothalamus (disregarding intrastructural variations) of patients with Parkinson's disease compared to controls (Fig. 6.1) indicating that the dopaminergic innervation of the hypothalamus is damaged. (i) Dopaminergic systems seem to be selectively altered since the levels of other neuromediators (adrenaline, serotonin), enzymes (choline acetyltransferase, glutamic acid decarboxylase) and neuropeptides (β-endorphin, γ-lipotropin, somatostatin, methionine- and leucine-enkephalin, cholecystokinin-8) are not affected (Javoy-Agid et al., 1984a; Conte-Devolx et al., 1985; Pique et al., 1985). (ii) Discrete biochemical deficiencies restricted to specific hypothalamic nuclei, undetectable in studies on the whole structure, cannot be excluded however. Intrinsic dopaminergic neurons (Hökfelt et al., 1978) may be damaged, as suggested by the presence of Lewy bodies in the hypothalamus (Langston and Forno, 1978), but the decrease may also be due to degeneration of nigrohypothalamic afferents,

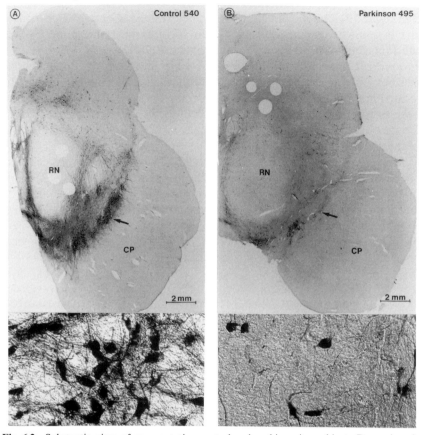

Fig. 6.2 Substantia nigra of representative control and parkinsonian subjects. Dopaminergic neurons stained by immunocytochemistry with an antibody against tyrosine hydroxylase. Lower images are enlargements of the areas indicated by the arrows. RN, red nucleus; CP, cerebral peduncle.

since neuropathological data suggest that the melanin-containing neurons in the structure are intact (Matzuk and Saper, 1985).

Other dopaminergic systems

Brainstem

The concentrations of dopamine are subnormal in a number of brain areas innervated by dopaminergic neurons of the mesencephalon (Fig. 6.1), i.e. the internal and external parts of the pallidum and the subthalamic nucleus, basal ganglia nuclei connected to the striatum and involved in motor control. Dopamine levels also decrease in nuclei such as the locus coeruleus and the area postrema (Fig. 6.1). The dopamine deficiency in the locus coeruleus suggests that the putative mesocoeruleal dopaminergic system degenerates. Decreased dopaminergic transmission in the locus coeruleus may alter the activity of the main ascending noradrenergic pathway which originates in this structure. Similarly, the dopamine deficit in the area postrema, the chemoreceptor trigger area implicated in emesis, may be of physiological importance.

Spinal cord

According to studies in animals, dopamine neurons in the periventricular grey matter of the caudal thalamus and posterior and dorsal hypothalamus, project to the spinal cord (Skagerberg and Lindvall, 1985). Biochemical evidence suggests that the human spinal cord also receives dopaminergic afferents although these are only a small fraction of catecholamine input to the structure. Less than one-tenth of the concentration of noradrenaline, spinal dopamine levels are comparable to those found in the cerebral cortex (Scatton *et al.*, 1984). Yet, the high HVA/dopamine ratio (an indirect estimation of dopamine turnover), compared to that of the striatum, indicates that despite their low density the spinal cord dopaminergic neurons may have an important functional role.

In parkinsonian patients, dopamine (Fig. 6.1) and HVA concentrations in the ventral and dorsal horns of the lumbar spinal cord are not significantly different from those of controls (Scatton *et al.*, 1986), suggesting that the descending dopaminergic system is spared in this disease. At present, these would seem to be the only dopaminergic neurons that are spared. The data must be considered with caution, however, until other segments of the spinal cord are examined.

Conclusions

The main biochemical feature of Parkinson's disease is the decrease in brain dopamine concentrations due to degeneration of central dopaminergic neurons.

(a) The neuronal systems do not all degenerate to the same degree. *The nigrostriatal pathway is more severely affected than the mesocortical or the mesolimbic systems.* Within the striatum, dopaminergic denervation is not uniform: the anterior part of the striatum, and particularly of the putamen, is preferentially affected. Given the role of the putamen in the control of motor function, this might explain why motor disability is, in most cases, the first and then predominate symptom of the disease.

(b) *Not all dopaminergic systems degenerate.* The observation that the descending dopaminergic projection to the lumbar spinal cord seems to be spared suggests that the disease does not strike selectively dopamine-containing neurons only because they are dopaminergic, although it cannot be excluded that the survival of some dopaminergic neurons is related to a particular local resistance to the pathogenic process. The assumption that Parkinson's disease is not associated with a generalized dopamine abnormality (as suggested by Barbeau, 1969) must be confirmed by pursuing the investigation of all (central and peripheral) dopaminergic systems.

(c) The mean striatal dopamine deficit in Parkinson's disease is a general estimation determined with respect to groups of subjects of variable size. Striatal dopamine levels differ from one patient to another, however. Differences in the degree of dopamine deficiency might explain, at least partly, *the variability in the clinical pictures in patients.*

(d) *Massive lesion of the nigrostriatal dopaminergic neurons is the characteristic abnormality of Parkinson's disease.* The resulting "striatal dopamine deficiency" (Hornykiewicz, 1972) is essentially, if not exclusively, the cause of the major parkinsonian symptoms. This is suggested by the drug-induced degeneration of the nigrostriatal dopamine neurons in patients intoxicated with MPTP (*N*-methyl-4-phenyl-1,2,3,6-tetrahydropyridine) which results in an unequivocal parkinsonian syndrome (Langston *et al.*, 1983; Langston, 1985a) (see below).

(e) *It is generally thought that the first parkinsonian signs appear only when at least 70–80% of the nigrostriatal dopaminergic system is damaged.* This hypothesis is based on two observations. (i) In a case of hemi-parkinsonism, a more than 80% dopamine deficit has been reported in the contralateral striatum, whereas in the homolateral striatum (contralateral to the side with no clinical signs), the dopamine deficit was less than 75% of normal (Barolin *et al.*, 1964). (ii) From an investigation performed in 39 patients, a dopamine deficiency in the caudate nucleus of about 70% seems to be necessary before parkinsonian symptoms become apparent; the dopamine deficit then increases in relation to the evolution of the disease, not in a linear manner, however (Riederer and Wuketich, 1976). It can therefore be assumed that central dopamine denervation is progressive and begins long before the consequences become clinically evident (see below).

(f) *The biochemical abnormality is not necessarily restricted to the dopamine deficiency.* It cannot be ruled out that the dopamine deficiency is the exclusive characteristic of some forms of parkinsonism, especially in young patients, at the onset of the disease. In most cases, however, other neurotransmitter-containing systems are impaired as well (i.e. in idiopathic Parkinson's disease and in complex parkinsonian syndromes such as progressive supranuclear palsy) (see below).

MECHANISMS OF COMPENSATION AFTER LESION OF DOPAMINERGIC NEURONS

Pre- and post-synaptic adjustments of dopaminergic neurotransmission are observed after denervation. (i) Presynaptically, the activity of the surviving dopaminergic neurons may be estimated indirectly from the HVA/dopamine ratio, HVA considered as an indication of the amount of dopamine released with respect to the degree of innervation represented by the dopamine concentrations. Although the HVA/dopamine ratio may increase partly because of the post-mortem degradation of dopamine, if the biochemical data are obtained from groups of carefully matched control and pathological patients, this ratio will also provide an indication of the rate of amine turnover in the remaining dopaminergic nerve terminals. (ii) Postsynaptically, changes may be estimated by determining the characteristics (density and affinity) of D1 and D2 receptors (Kebabian and Calne, 1979). The reliability of such investigations has been assessed: receptors are relatively resistant to post-mortem conditions.

Increased activity of surviving dopaminergic neurons

In most dopamine-containing areas of the parkinsonian brain, the decrease in DOPAC and HVA content is consistently less severe than the reduction in dopamine concentrations, probably because metabolite formation is increased. This accounts for the increased HVA/dopamine ratio (Fig. 6.3) and is interpreted as the consequence of a functional overactivity of the surviving nigrostriatal dopamine neurons (Hornyckiewicz, 1966, 1979b). The hypothesis has been confirmed by experimental lesions in animals. Indeed, an increase in the synthesis and turnover of striatal dopamine is observed after partial destruction of the nigrostriatal dopaminergic pathway in the rat (Agid et al., 1973; Hefti et al., 1980; Zigmond et al., 1984). The amplitude of presynaptic compensation varies as a function of the severity of the lesion (Fig. 6.4). (i) Small lesions of the substantia nigra causing limited destruction of the nigrostriatal neurons do not change the rates of dopamine synthesis or release in the striatum, suggesting that the amount of neurotransmitter released by the remaining striatal dopaminergic nerve terminals is sufficient to maintain adequate neurotransmission. (ii) When nigral lesions affect 50–60% of the dopaminergic neurons, dopamine turnover in the remaining striatal terminals increases. (iii) When denervation reaches 70–90%, despite a marked increase in the amine turnover, dopamine release from the surviving dopamine terminals is no longer sufficient to counterbalance the deficiency and to restore normal striatal dopaminergic transmission. (iv) When lesions

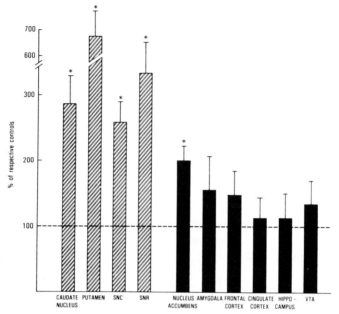

Fig. 6.3 HVA/dopamine ratio in brain areas of parkinsonian patients. Results are the mean ±SEM of data obtained on 17–21 control and 10–22 parkinsonian brains, and are expressed as % of respective control values (= 100). *Statistically significant when compared to controls. (Figure reproduced from Scatton *et al.*, 1984, with permission.)

exceed 90%, and only for such severe lesions, postsynaptic dopamine receptors become hypersensitive (Creese and Snyder, 1979), and important behavioral changes are observed (Melamed *et al.*, 1982).

In Parkinson's disease, the magnitude of the change in activity of surviving dopaminergic neurons differs among the dopaminergic systems, and between areas containing nerve terminals or cell bodies.

Compensatory hyperactivity of striatal but not cortical dopaminergic nerve terminals

The HVA/dopamine ratio is significantly increased in the putamen, caudate nucleus and the nucleus accumbens but not in the hippocampus and the frontal cortex of parkinsonian brains (Fig. 6.3). The hyperactivity observed in the striatum but not in the remaining cortical and hippocampal neurons may indicate that the lesion of the mesocortical pathway is not severe enough to induce compensatory metabolic changes. Alternatively, these neurons may not have the capacity to increase their metabolic rate. The high

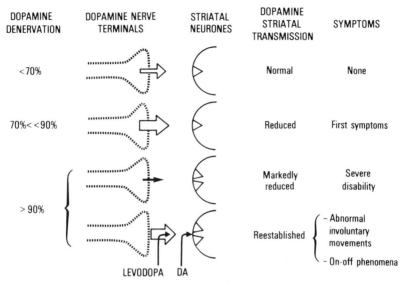

Fig. 6.4 Hypothetical role of the nigrostriatal dopamine denervation on the genesis of motor symptoms in Parkinson's disease. Degeneration of dopamine nerve terminals (broken lines), presynaptic dopaminergic activity (arrow) in the surviving neurons, dopamine receptors (triangles) on the striatal dopaminoreceptive neurons. Three phases related to the severity of the dopamine denervation (estimated by the per cent decrease in striatal dopamine concentrations compared to control levels) can be considered. (1) Phase of compensation: the dopamine deficit affects less than 70% of the neurons. The increased activity in the surviving neurons maintains a normal dopamine transmission. No symptoms expressed. (2) Phase of partial decompensation: 70–90% of the dopamine neurons degenerate. The presynaptic hyperactivity is not sufficient to support normal transmission, the first symptoms appear. (3) Phase of total decompensation: more than 90% of the dopamine innervation is destroyed. Dopamine transmission is dramatically reduced. As a consequence, postsynaptic dopamine receptors become hypersensitive. Patients are disabled. Dopamine transmission is re-established by L-dopa or dopamine agonists (DA). The clinical improvement may be associated with L-dopa-induced fluctuations of performance or abnormal involuntary movements.

HVA/dopamine ratio in the hippocampus and cerebral cortex compared to the striatum in the control human brain supports the latter hypothesis. In addition, it suggests that the cortical neurons cannot increase their metabolic rate, which may be maximal under normal conditions (Javoy-Agid *et al.*, 1984b).

Hyperactivity of the remaining dopaminergic neurons in the substantia nigra

The HVA/dopamine ratio is also increased to a greater extent in the substantia nigra pars reticulata, where the dendrites of the nigrostriatal are concentrated, than in the pars compacta, where the cell bodies are located (Fig. 6.3) (Bokobza *et al.*, 1984). The marked reduction in HVA content compared to controls indicates, however, that nigral dopaminergic transmission is not restored to normal and may thus be responsible for some disorders. (i) The metabolism of striatal dopamine may be modified. In the rat, dendritic release of dopamine inhibits dopamine release from ipsilateral striatal dopamine terminals and striatonigral afferents, which in turn modulate the activity of the nigrostriatal dopamine system (Cheramy *et al.*, 1981). (ii) A change in nigral dopamine transmission may influence the function of other systems such as those in the thalamus and the tectum, since the dopamine neurones of the substantia nigra also project to these areas (Rinvik *et al.*, 1976).

Hypersensitivity of dopamine receptors

Experiments in animals suggest that, in the striatum, D1 and D2 receptors are located on intrinsic neurons and on afferents, in particular from the cerebral cortex. In the substantia nigra, D2 receptors are present on dopamine neurons, and D1 receptors on terminals of fibers originating in the striatum (see review, Kebabian and Calne, 1979). Pharmacological (Ungerstedt, 1971), electrophysiological (Ohye *et al.*, 1970) and biochemical (Creese and Snyder, 1979; Guerin *et al.*, 1985) studies suggest that after prolonged denervation dopamine receptors may become hypersensitive.

Modifications of the characteristics of dopamine receptors may be a key element for understanding the pathophysiology of Parkinson's disease, since antiparkinsonian therapy is based essentially on the administration of dopamine agonists. However, there is no clear consensus as to the state of dopamine receptors in this disease, in particular in the striatum.

Dopamine receptors in the nigrostriatal system (Fig. 6.5)

The density of D2 receptors in the striatum of parkinsonians has been reported to increase (Lee *et al.*, 1978), or to decrease (Reisine *et al.*, 1977). No significant modification in the density (or in the affinity constant) of the receptors was observed by Rinne *et al.* (1981), although a group of patients presented a hypersensitivity and another group a hyposensitivity of striatal D2 receptors. Hypersensitivity of D2 dopamine receptors may develop in relation to the severity of the dopamine denervation. Indeed, it is detected in the putamen (where the dopamine deficiency usually exceeds 85%) and not in the caudate nucleus and nucleus accumbens where the deficit ranges from 60 to 80% (Bokobza *et al.*, 1984). Pooled together, the data suggest that the density of D2 receptors increases in the putamen but not in the

DOPAMINE RECEPTORS IN PARKINSON'S DISEASE

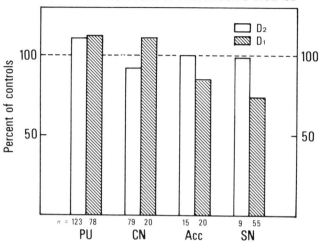

Fig. 6.5 Dopamine receptors in Parkinson's disease. The density of D1 (labelled with ³H-flupenthixol or ³H-SCH 23390) and D2 (labelled with ³H-spiperone or ³H-haloperidol) dopamine receptors in parkinsonian brains is expressed as percentage of respective control. Values are the mean data obtained from: Reisine *et al.* (1977); Lee *et al.* (1978); Quik *et al.* (1979); Rinne *et al.* (1981); Bokobza *et al.* (1984); Pimoule *et al.* (1985); Raisman *et al.* (1985); Rinne *et al.* (1985); Cash *et al.* (1987b); Pierot *et al.* (1988). The density in receptor is statistically different from controls for D2 receptors in the putamen from controls (Bokobza *et al.*, 1984) and for D1 receptors in the substantia nigra (Rinne *et al.*, 1985; Cash *et al.*, 1987b; Pierot *et al.*, 1988) and putamen (Raisman *et al.*, 1985; Rinne *et al.*, 1985). *n* = number of brains. Abbreviations, see legend to Fig. 6.1.

nucleus accumbens or in the caudate nucleus. In the substantia nigra, the absence of modification of the density of D2 receptors should be regarded with caution since the precision of the biochemical data is limited by the low density of receptor sites.

The data concerning D1 receptors are also controversial. When measured by the activity of the dopamine-sensitive adenylate cyclase, both decreased (Riederer *et al.*, 1978; Shibuya, 1979) and increased (Nagatsu *et al.*, 1978) activity of the enzyme have been reported. The data obtained by binding studies with radioactive ligands such as ³H-Sch 23390 (Raisman *et al.*, 1985; Pimoule *et al.*, 1985; Pierot *et al.*, 1988) or ³H-flupenthixol (Rinne *et al.*, 1985), specific for the D1 receptor, are equally contradictory. D1 receptors in the striatum of parkinsonian patients are reported unchanged (Pimoule *et al.*, 1985; Pierot *et al.*, 1988) or increased in number (Raisman *et al.*, 1985; Rinne *et al.*, 1985). The increase was reported to be both reversed (Raisman *et al.*, 1985) or potentiated (Rinne *et al.*, 1985) by L-dopa treatment. In the substantia nigra, the density of D1 receptors decreases (Rinne *et al.*, 1985), as do the concentrations of DARPP-32 (a protein phosphorylated in response to stimulation of the D1 receptor) (Cash *et al.*, 1987b).

Clinical consequences

(i) The number of D1 and D2 receptor sites is high in the striatum of parkinsonians considering the marked dopamine denervation. Preservation of a normal density of dopamine receptors may not be surprising since the striatal neurons, and thus the dopaminoceptive cells, seem to be intact (Greenfield and Bosanquet, 1953). Taking into account the suspected presence of D2 autoreceptor sites, i.e. located presynaptically on degenerated dopaminergic nerve terminals, the presence of normal or slightly increased dopamine receptor density in the striatum may indicate that D2 receptors are in fact markedly increased postsynaptically. This may explain the sustained efficacy of L-dopa and/or dopaminergic agonists observed in most patients after several years of evolution of the disease (Birkmayer, 1974; Barbeau, 1980).

(ii) The decrease in D1 receptors in the substantia nigra would be compatible with loss of striatonigral neurons, but is discordant if the striatum is preserved. It is evident that the localization, regulation and physiological role of the dopamine receptors is still not understood in man. Their role should not be underestimated since activation of both D1 and D2 receptors seems to be necessary to restore normal locomotor behaviour in rats treated with reserpine (Gershanik *et al.*, 1983).

(iii) An increased density of D2 receptors is observed in the most severely disabled patients, those who respond well to dopamine replacement therapy, and who tend to develop dyskinesias and fluctuations in response (on–off effects) (Rinne *et al.*, 1981). The increased density is observed preferentially in untreated patients (Lee *et al.*, 1978) or after long evolution of the disease (Bokobza *et al.*, 1984). It seems to be reversed by long-term treatment with L-dopa (Lee *et al.*, 1978; Rinne *et al.*, 1981) or dopamine agonists (Riederer *et al.*, 1983), although this is contested (Bokobza *et al.*, 1984).

(iv) It would obviously be useful for the clinician to be able to measure a marker in the periphery which reflects the state of central dopamine receptors. The binding of ^{3}H-spiperone to lymphocytes (Le Fur *et al.*, 1980) was found to be below normal in patients with Parkinson's disease, and to return to normal after L-dopa treatment (Le Fur *et al.*, 1981). The characteristics of ^{3}H-spiperone binding to human lymphocytes differ, however, from those obtained with membrane preparations from the striatum (Bloxham *et al.*, 1981; Maloteaux *et al.*, 1981; Fleminger *et al.*, 1982), and may reflect labelled ligand that is trapped in the lymphocytes, presumably in lysozomes, rather than ligand which is bound to membrane receptors. Moreover, the increase in ^{3}H-spiperone binding induced by L-dopa can be produced by other types of drugs and does not correlate with the improvement of any particular symptom (Maloteaux *et al.*, 1983). Although ^{3}H-spiperone binding to lymphocytes may not identify a dopamine receptor, the changes observed in the parkinsonian subjects may point to an abnormality of cell membranes or lysozomes.

(v) The increase in the number of striatal D2 receptor sites observed in patients with Parkinson's disease is not observed in all parkinsonian syndromes. For example, in progressive supranuclear palsy, the density of D2 receptors is reduced by 40% in the caudate nucleus and putamen both post-mortem (Bokobza *et al.*, 1984) (Fig. 6.6) and *in vivo* (Baron *et al.*, 1985). This reduction is most likely the consequence of the degeneration of dopaminoceptive cholinergic neurons in the striatum (Ruberg *et al.*, 1985). This might also explain why most, if not all, patients with progressive supranuclear palsy are insensitive to L-dopa treatment. Interestingly, striatal D1 receptor sites seem to be intact in this disease (Pierot *et al.*, 1988). The preservation of the D1 receptors may, in the future, provide the basis for effective treatment.

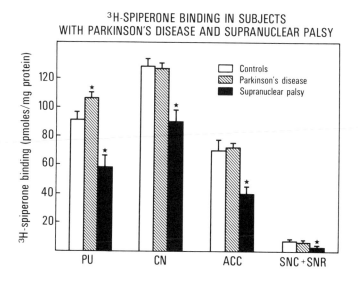

Fig. 6.6 D2 receptors labelled with ^{3}H-spiperone binding in subjects with Parkinson's disease and progressive supranuclear palsy. Results are the mean ±SEM of data obtained on 20 brains with Parkinson's disease (Bokobza *et al.*, 1984) and nine with progressive supranuclear palsy (Ruberg *et al.*, 1985). *Significant difference compared to controls. Abbreviations, see legend to Fig. 6.1.

Conclusion

Since the density of D1 and D2 receptors is not markedly affected in the striatum of patients with Parkinson's disease (Fig. 6.5), the lack of responsiveness to L-dopa in some parkinsonians does not result from receptor loss. In most patients improved by L-dopa at the onset of Parkinson's disease, the drug remains efficacious after years of evolution of the disease (Bonnet *et al.*, 1987).

In some patients, however, the efficacy of dopamine therapy obviously decreases in the course of the disease. This might be related to a reduction in the density of dopamine receptors as, for example, in patients with progressive supranuclear palsy. Alternatively, efferent projections from the basal ganglia may become damaged in the course of the disease.

Functional relevance of dopaminergic compensation

In the striatum, pre- and post-synaptic compensations seem to develop in sequence as a function of the severity of the lesion: receptors become hypersensitive when the surviving neurons can no longer compensate for cell loss by an increase in activity. These mechanisms may underlie the evolution of the symptoms of the disease through phases of functional compensation and decompensation (Hornyckiewicz, 1966, 1979b; Agid *et al.*, 1973) (Fig. 6.4).

Compensation

Less than 70% of the dopaminergic neurons are lost. The increased activity of the surviving dopamine neurons probably suffices to maintain normal dopaminergic transmission in the striatum. Parkinsonian symptoms are latent, but not manifest. Dopamine-dependent limbic and hypothalamic functions may be effectively protected by this compensatory mechanism since the dopaminergic neurons innervating these structures are usually only moderately damaged even at advanced stages of the disease. During the phase of physiological adaptation, dopamine-related symptoms may emerge, probably when dopaminergic transmission is transiently decompensated, for example after stress (Snyder *et al.*, 1985) or by short-term administration of neuroleptic drugs in otherwise asymptomatic patients (Rajput *et al.*, 1982).

Partial decompensation

The pathological process affects 70–90% of the dopamine neurons. The first symptoms appear. Though the surviving neurons are hyperactive, normal dopaminergic transmission is not maintained, but dopamine receptors are not hypersensitive. This is commonly observed post-mortem in the caudate nucleus of patients. Soon after the onset of L-dopa therapy, smooth fluctuations of performance will occur. These reflect the pharmacokinetics of the drug. However, at this stage of the disease, no L-dopa-related abnormal involuntary movements are observed, probably because receptors are not yet hypersensitive (Agid *et al.*, 1985).

Total decompensation

More than 90% of the dopaminergic innervation of the striatum is destroyed. Dopaminergic transmission is not restored to normal despite presynaptic hyperactivity of the spared neurons. As a consequence, postsynaptic D2 receptors become hypersensitive. At this stage, parkinsonian symptoms are fully expressed. They can be attenuated if striatal dopamine transmission is re-established by administration of L-dopa or dopamine agonists. The marked clinical disability is associated with severe fluctuations of performance and abnormal involuntary movements induced by L-dopa (see below).

ROLE OF THE CENTRAL DOPAMINERGIC DEFICIENCY IN PARKINSONIAN SYMPTOMS

In Parkinson's disease, the clinical expression of the central dopaminergic deficiency is different in untreated patients (symptoms) and in those receiving L-dopa or dopamine agonists (adverse reactions).

Dopaminergic lesions and parkinsonian symptoms

Motor symptoms

It is unanimously admitted that the massive dopamine deficiency in the nigrostriatal system plays a major role in the genesis of motor symptoms (akinesia, tremor and rigidity). It is less clear, however, whether it alone is responsible for all the symptoms. The notion that the nigrostriatal lesion is directly responsible for akinesia derives from two observations: in monkeys, electrolytic lesions of the substantia nigra produce a severe contralateral akinesia (Poirier and Sourkes, 1965); in parkinsonian patients, a striking correlation is observed between the severity of akinesia, the decrease in striatal dopamine content and neuronal loss in the substantia nigra (Bernheimer *et al.*, 1973). Interestingly, akinesia is primarily observed on the side contralateral to the most severely affected substantia nigra (Bernheimer *et al.*, 1973). Whether the striatal dopamine deficiency is at origin of the other symptoms is less clear. It may contribute to tremor, which can be produced in monkeys by lesioning simultaneously the rubro-olivo-cerebello-rubral loop and the nigrostriatal dopaminergic system (Larochelle *et al.*, 1970; Jenner and

Marsden, 1984). Finally, whether rigidity, thought to result from hyperactive α-motoneurons of supraspinal origin, is directly and exclusively related to the striatal dopamine deficiency is debated (Marsden *et al.*, 1982; Ellensbroek *et al.*, 1985).

Examination of cases of parkinsonism produced in man by MPTP intoxication (Langston *et al.*, 1983) should help to understand the pathophysiology of this triad of symptoms. The clinical picture of this drug-induced parkinsonian syndrome is dominated by severe akinesia, but rigidity and parkinsonian-type tremor also develop. The clinical picture in these patients is all the more interesting, since MPTP is thought to selectively destroy the nigrostriatal dopaminergic neurons in man (Davis *et al.*, 1979) and monkeys (Burns *et al.*, 1983). The selectivity of the lesion, however, is not unequivocally established. Depending on the dose of MPTP absorbed and the age of the patient when intoxicated, other dopaminergic systems (mesocortical, mesolimbic) may also be damaged (Schneider *et al.*, 1987; Elsworth *et al.*, 1987) and, therefore, be implicated in the symptomatology (Langston, 1985b). It is noteworthy that other classical neurotransmitter systems (cholinergic (Garvey *et al.*, 1986) and GABAergic in particular) seem to be unaffected by the toxin.

In summary, these observations indicate that the dopaminergic nigrostriatal lesion plays a major, if not exclusive, role in the genesis of motor symptoms observed to varying degrees in patients with Parkinson's disease. The variability of the clinical picture of this disease may reflect differences in the severity of the lesions affecting the dopaminergic systems, as well as dysfunctions of other neurotransmitter systems implicated in the expression of each of the symptoms.

Cognitive disorders

Does the central dopaminergic deficiency contribute to the development of cognitive disorders in parkinsonian patients? This hypothesis is controversial. A correlation between akinesia and impaired performance on tests of visuospatial reasoning and psychomotor speed has been reported (Mortimer *et al.*, 1982). Furthermore, the neuropsychological disabilities of parkinsonian patients may be improved by L-dopa (Loranger *et al.*, 1972), especially at early stages of the disease (Beardsley and Puletti, 1971). Though open to discussion (Marsden, 1982), the notion that the decrease in nigrostriatal dopaminergic transmission might be responsible for some intellectual changes in parkinsonian patients has been put forward: even if moderate, the pattern of mental deficits in MPTP-induced parkinsonism is very similar to that observed in the idiopathic disease (Stern and Langston, 1985). The contention that lesions of the mesocortical and limbic dopaminergic neurons are involved in the neuropsychological alterations is supported by the effects of experimental lesions in animals: for example, delayed alternation in the rat is disrupted by selective lesions of the dopaminergic mesocorticolimbic pathway (Le Moal *et al.*, 1977) and in monkeys by injections of 6-hydroxydopamine in the prefrontal cortex (Brozowski *et al.*, 1979). However, a number of neuropsychological studies of parkinsonian patients have failed to demonstrate the existence of dopamine-dependent cognitive symptoms (Brown *et al.*, 1984; Pillon *et al.*, 1989; see Ruberg and Agid, 1980). It may be that the cognitive disorders appear only when the dopaminergic systems are massively destroyed, in particular when nigrostriatal and mesocortico-limbic lesions are associated (Koob *et al.*, 1984). In summary, there is at present no definite evidence that reduced dopaminergic transmission plays a specific role in the genesis of intellectual impairment in patients. The beneficial effect of L-dopa in some patients suggests that this brain dopamine deficit might however contribute to the reduction of alertness (Brown *et al.*, 1984) and to some aspects of bradyphrenia (see below) (Lees and Smith, 1983; see Agid *et al.*, 1984) and depression (Celesia and Wanamaker, 1972; see Mayeux *et al.*, 1984), particularly at the onset of the disease, although this assumption is debatable (Gotham *et al.*, 1986).

Endocrine disorders

Endocrine abnormalities are observed in some parkinsonian patients and may result from the hypothalamic dopamine deficiency (Javoy-Agid *et al.*, 1984a). At the present time, it is not known whether all three sources of dopamine in the hypothalamus are affected (tuberoinfundibular, incerto-hypothalamic, mesolimbic). Tuberoinfundibular dopamine is itself the major prolactin-inhibiting factor. Prolactin levels have been measured in parkinsonian patients, but the data remain controversial: basal levels have been reported to be reduced (Murri *et al.*, 1980), normal (Hyyppa *et al.*, 1978; Eisler *et al.*, 1981; Laihinen and Rinne, 1986), or elevated in severely affected patients (Agnoli *et al.*, 1980), probably those with the greatest dopamine deficiency. In spite of a moderate decrease in hypothalamic dopamine (60%), hyperactivity of unlesioned neurons may maintain normal dopaminergic transmission in the hypothalamus.

Autonomic abnormalities that are often observed in patients, such as sweating, can be improved by L-dopa treatment (Goetz *et al.*, 1986), suggesting that they too may be due to a central dopamine deficiency.

Sensory disorders

Disabling sensory symptoms are sometimes observed in patients (Snider *et al.*, 1976). They have been assumed to result from a dopamine deficiency in the spinal cord since they are improved by L-dopa treatment (Nutt and Carter, 1984). However, dopamine concentrations in the spinal cord of parkinsonian patients have been found to be normal (Scatton *et al.*, 1986). The beneficial effects of L-dopa may, therefore, result from a more subtle mechanism, conceivably at the level of the basal ganglia.

Dopaminergic lesions and adverse reactions to L-dopa

Nausea and vomiting, often caused by L-dopa at the onset of treatment, may result from a sudden overstimulation of dopamine receptors in the area postrema, a region rich in D2 receptors, located outside the blood–brain barrier (Schwartz *et al.*, 1986). It is now possible to avoid this side-effect by associating L-dopa with a peripheral dopa–decarboxylase inhibitor, which prevents the production of dopamine outside the blood–brain barrier (Bartholini and Pletscher, 1969), or by administrating dopamine agonists with domperidone, a dopamine receptor antagonist which does not cross the blood–brain barrier (Agid *et al.*, 1979a).

L-Dopa-induced abnormal involuntary movements which develop in the course of the disease seem to occur only in patients with marked degeneration of the nigrostriatal dopaminergic system (Barbeau *et al.*, 1971; Agid *et al.*, 1979b; Marsden *et al.*, 1982). (i) They are provoked by medications which stimulate dopamine transmission and are reduced by dopamine D2 receptor antagonists (Klawans and Weiner, 1974). Drugs which interact with other neurotransmitters are more or less ineffective, although benzodiazepines have sometimes proved useful. (ii) They develop at critical plasma dopa concentrations, i.e. beyond a certain threshold of stimulation of central dopamine receptors (Peaston and Bianchine, 1970; Muenter *et al.*, 1977; Lhermitte *et al.*, 1977b). (iii) They are observed in patients with severe akinesia who respond dramatically to L-dopa, i.e. those with a severe central dopamine deficiency (Agid *et al.*, 1985). The topography of L-dopa-induced abnormal movements is related to impairment of dopaminergic transmission in brain structures controlling movement in the corresponding part of the body. This is probably why parkinsonian patients with predominantly unilateral signs develop dyskinesias on the more severely affected side (Mones *et al.*, 1971). The reason why the clinical picture of the movements differs from patient to patient is still debated. A choreic, dystonic, or ballic pattern may result from selective dysfunction in the caudate nucleus, the putamino-pallidal complex and the subthalamic area, respectively. The timing of these dyskinesias varies. "Mid-dose", also called "peak-dose", dyskinesias, the most frequently observed, occur during the period of maximum relief of parkinsonian disability, i.e. when dopa concentrations in plasma are the highest (Muenter and Tyce, 1971). "Onset and end-of-dose" dyskinesias, less frequent but more disabling, are observed at the beginning and at the end of the period when L-dopa is effective, i.e. when plasma L-dopa concentrations are increasing or decreasing (Muenter *et al.*, 1977; Lhermitte *et al.*, 1977a). The former most likely result from overstimulation of striatal dopamine receptors (Fig. 6.4), whereas the latter result from inadequate stimulation of the dopamine receptors or from selective stimulation of a subgroup of dopamine receptors (Agid *et al.*, 1985). "Mid-dose" dyskinesias (monophastic) can be improved through diminution and fractionation of the daily dose of L-dopa. "Onset and end-of" dose dyskinesias (biphasic) can be suppressed in some cases by increasing and fractionating the doses of L-dopa (Lhermitte *et al.*, 1978). This therapeutical approach is difficult to manage, however, because of the risk of exacerbating mid-dose dyskinesias and triggering intellectual deterioration.

In the course of long-term L-dopa treatment fluctuations of performance develop (Fahn, 1974; McDowell and Sweet, 1976; Marsden and Parkes, 1976). These fluctuations take various forms, two of which are particularly important: "end-of-dose" or "wearing-off" phenomena, characterized by smooth disappearance of clinical benefit within 3–5 h after administration of a single dose of L-dopa, observed in most patients after 3 years of L-dopa, treatment; disabling "on–off" phenomena, with abrupt onset and disappearance, not always evidently related to L-dopa intake, observed after several years of high-dose L-dopa therapy. Most, if not all, types of response swings are related to the timing of the dose of the drug (Quinn *et al.*, 1982). To prevent these fluctuations it is necessary to control the amount of L-dopa reaching its site of action, the striatal dopaminergic receptors (Fig. 6.7). Delivery of the drug may be hampered by absorption deficits or dilution of the alimentary bolus, peripheral abnormalities such as competition of the drug with other amino acids for uptake by the gut, competition between L-dopa and its metabolites, in particular methoxylated metabolites with long half-lives, or with other amino acids, at the level of the blood–brain barrier (see review in Kurlan, 1987). Within the brain, fluctuations of performance most likely result from a decreased capacity for storing the newly formed dopamine because of the progressive degeneration of dopaminergic nerve terminals (Marsden and Parkes, 1976). Reduced accumulation of 18-fluoro-dopa in the striatum, measured by PET scan in brains of parkinsonian patients with "on–off" phenomena compared to other patients (Leenders *et al.*, 1986), suggests, indeed, that the capacity to restore dopamine transmission is diminished in these parkinsonians. The hypothesis that "on–off" phenomena occur because striatal dopamine receptors are desensitized by prolonged L-dopa treatment (Fahn, 1982) seems unlikely since they disappear after intravenous infusion of L-dopa (Shoulson *et al.*, 1975; Hardie *et al.*,

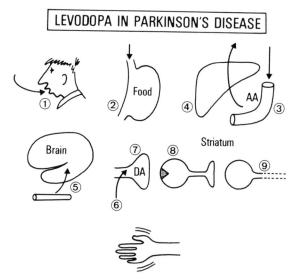

Fig. 6.7 L-Dopa in Parkinson's disease. The arrows schematize the course of L-dopa: (1) oral ingestion; (2) the drug mixes with the alimentary bolus; (3) from the gut to (4) the liver; the drug is transported from the blood (5) to the brain through the blood–brain barrier; (6) uptake of L-dopa in dopaminergic nerve terminals; the drug is metabolized to dopamine (7); the newly synthesized dopamine stimulates the dopamine receptors located (8) on the dopaminoceptive cells; restoration of the dopaminergic transmission and reactivation of the striatal output (9). Finally, parkinsonian symptoms are improved. Fluctuations of performance can result from anomalies occurring at any one of these different steps during long-term administration of L-dopa.

1984; Quinn *et al.*, 1984; Nutt *et al.*, 1984). In most patients "end-of-dose" deteriorations shift with time to "on–off" phenomena. The reason for this is not fully understood. Assuming that reduced storage of dopamine synthesized from exogenous dopa is the essential cause for fluctuations of performance, the "on–off" phenomena may result from a loss in the "buffer capacity" of the dopaminergic neurons. Consequently, these "on–off" phenomena (and the severe dyskinesias present during the "on" period) should appear when the subtle pre- and post-synaptic mechanisms of compensation are overcome, i.e. when the severity of the dopamine denervation produces a marked increase in the sensitivity of postsynaptic dopaminergic receptors (Fig. 6.4).

Mental disturbances, such as delirium and hallucinations, are the most disabling side-effects. They are most frequently observed in older patients after prolonged treatment with high daily doses of dopamine agonists (Barbeau, 1976; Presthus, 1980; Grimes and Hassan, 1983; Rondot *et al.*, 1984). The cause of these iatrogenic psychic disorders is unclear. It has been postulated that "bradyphrenia" results from deficient striatal and corticolimbic dopaminergic transmission. If the dopamine deficiencies induce hypersensitivity of the dopamine receptors in the striatum and regions of the cortex and limbic system, stimulation of the hypersensitive receptors in overtreated patients may cause the observed psychiatric reactions, in the same way that abnormal involuntary movements observed in patients treated with dopamine agonists are thought to result from stimulation of hypersensitive receptors in the striatum. The hypothesis is not unreasonable, but is an oversimplification since dysfunction of other neuronal pathways seem also to be involved, in particular the noradrenergic, serotoninergic, cholinergic systems and probably many others yet unknown.

REFERENCES

Agid, Y., Javoy, F. and Glowinski, J. (1973) *Nature New Biol.* **245**, 150–151.

Agid, Y., Pollak, P., Bonnet, A. M., Signoret, J. L. and Lhermitte, F. (1979a) *Lancet* **i**, 570–572.

Agid, Y., Bonnet, A. M., Signoret, J. L. and Lhermitte, F. (1979b) In *The Extrapyramidal System and its Disorders*. Advances in Neurology, Vol. 24 (edited by L. J. Poirier, T. L. Sourkes and P. J. Bedard), pp. 401–409. Raven Press, New York.

Agid, Y., Ruberg, M., Dubois, B. and Javoy-Agid, F. (1984) In *Parkinson-specific Motor and Mental Disorders*. Advances in Neurology, Vol. 40 (edited by R. G. Hassler and J. F. Christ), pp. 211–218. Raven Press, New York.

Agid, Y., Bonnet, A. M., Ruberg, M. and Javoy-Agid, F. (1985) In *Dyskinesia, Research and Treatment*. Psychopharmacology, Suppl. 2 (edited by D. Casey, T. N. Chase, V. Christensen and J. Gerlach), pp. 145–159. Springer-Verlag, Berlin.

Agnoli, A., Ruggieri, S., Falaschi, P. *et al.* (1980) *Adv. Biochem. Psychopharmacol.* **24**, 551–557.

Barbeau, A. (1969) In *Third Symposium on Parkinson's Disease* (edited by F. J. Gillingham and I. M. L. Donaldson), pp. 66–73. E. and S. Livingstone, London,

Barbeau, A. (1976) *Pharmacol. Therap.* **1**, 475–494.

Barbeau, A. (1980) In *Parkinson's Disease: Current Progress, Problems and Management* (edited by U. K. Rinne, M. Klinger and G. Stamm), pp. 229–239. Elsevier, North Holland Press, Amsterdam.

Barbeau, A., Murphy, G. F. and Sourkes, T. L. (1961) *Science* **133**, 1706–1707.

Barbeau, A., Mars, H. and Gillio-Joffroy, L. (1971) In *Parkinson's Disease* (edited by F. H. McDowell and C. H. Markham) pp. 203–237. F. A. Davis Co., Philadelphia.

Barolin, G. S., Bernheimer, H. and Hornyckiewicz, O. (1964) *Schw. Arch. Neurol. Psychiat.* **94**, 241–248.

Baron, J. C., Mazière, B., Loc'h, C., Sgouropoulos, P., Bonnet, A. M. and Agid, Y. (1985) *Lancet* **i**, 1163–1164.

Bartholini, G. and Pletscher, H. (1969) *J. Pharm. Pharmacol.* **21**, 323–324.

Beardsley, J. and Puletti, F. (1971) *Arch. Neurol.* **25**, 145–150.

Berger, B., Tassin, J. P., Rancurel, G. and Blanc, G. (1980) In *Enzymes and Neurotransmitters in Mental Disease* (edited by E. Usdin, T. L. Sourkes and B. H. Youdim) pp. 317–328. Wiley, London.

Bernheimer, H., Birkmayer, W., Hornykiewicz, O., Jellinger, K. and Seitelberger, F.

(1973) *J. Neurol. Sci.* **20**, 415–455.

Birkmayer, W. (1974) In *Current Concepts in the Treatment of Parkinsonism* (edited by M. Yahr), p. 141. Raven Press, New York.

Birkmayer, W. and Riederer, P. (1980) *Parkinson's Disease—Biochemistry, Clinical Pathology, and Treatment.* Springer-Verlag, Berlin.

Bloxham, C. A., Cross, A. J., Crow, T. J. and Owen, F. (1981) *Br. J. Pharmacol.* **74**, 233.

Bokobza, B., Ruberg, M., Scatton, B., Javoy-Agid, F. and Agid, Y. (1984) *Eur. J. Pharmacol.* **99**, 167–175.

Bonnet, A. M., Loria, Y., Saint-Hilaire, M. H., Lhermitte, F. and Agid, Y. (1987) *Neurology* **37**, 1539–1542.

Brown, R. G., Marsden, C. D., Quinn, N. and Wyke, M. A. (1984) *J. Neurol. Neurosurg. Psychiat.* **47**, 454–465.

Brozowski, T. J., Brown, R. M. and Goldman, P. (1979) *Science* **205**, 929–931.

Burns, R. S., Chiueh, C. C., Markey, S. P., Ebert, M. H., Jacobowitz, D. M. and Kopin, I. J. (1983) *Proc. Natl. Acad. Sci.* **80**, 4546–4560'

Cash, R., Dennis, T., L'Heureux, R., Raisman, R., Javoy-Agid, F. and Seatton, B. (1987a) *Neurology* **37**, 42–46.

Cash, R., Raisman, R., Ploska, A. and Agid, Y. (1987b) *J. Neurochem.* **49**, 1075–1083.

Celesia, G. G. and Wanamaker, W. M. (1972) *Dis. Nerv. Syst.* **33**, 577–583.

Cheramy, A., Leviel, V. and Glowinski, J. (1981) *Nature* **289**, 537–542.

Conte-Devolx, B., Grino, M., Nieoullon, A., Javoy-Agid, F., Castanas, E., Guillaume, V., Tonon, M. C., Vaudry, H. and Oliver, C. (1985) *Neurosci. Lett.* **56**, 217–222.

Creese, I. and Snyder, S. H. (1979) *Eur. J. Pharmacol.* **56**, 277–281.

Davis, G. C., Williams, A. C., Markey, S. P., Ebert, M. H., Caine, E. D., Reichert, C. M. and Kopin, I. J. (1979) *Psychiat. Res.* **1**, 249–254.

Ehringer, H. and Hornykiewicz, O. (1960) *Wien Klin. Wschr.* **38**, 1236–1239.

Eisler, T., Thorner, M. O., McLeod, R. M., Kaiser, D. L. and Calne, D. B. (1981) *Neurology* **31**, 1356–1359.

Ellenbroek, B., Schwarz, M., Sontag, K. H., Jaspers, R. and Cools, A. (1985) *Brain Res.* **345**, 132–140.

Elsworth, J. D., Deutch, A. Y., Redmond, Jr., D. E., Sladel, Jr., J. R. and Roth, R. H. (1987) *Brain Res.* **415**, 293–299.

Escourolle, R., De Recondo, J. and Gray, F. (1970) In *Monoamines et Noyaux Gris Centraux* (edited by J. de Ajuriaguerra and G. Gautier), pp. 173–229. IV Symposium de Bel Air, Genève.

Fahn, S. (1974) *Neurology* **24**, 431–441.

Fahn, S. (1982) In *Movement Disorders. Neurology*, Vol. 2 (edited by C. D. Marsden and S. Fahn), pp. 123–145. Butterworth, London.

Fahn, S., Libsch, L. R. and Cutler, R. W. (1971) *J. Neurol. Sci.* **14**, 427–455.

Fleminger, S., Jenner, F. and Marsden, C. D. (1982) *J. Pharm. Pharmacol.* **34**, 658–663.

Garvey, J., Petersen, M., Waters, C. M., Rose, S. P., Hunt, S., Briggs, R., Jenner, P. and Marsden, C. D. (1986) *Movement Disorders* **1**, 129–134.

Gaspar, P., Berger, B., Gay, M., Hamon, M., Cesselin, F., Vigny, A., Javoy-Agid, F. and Agid, Y. (1983) *J. Neurol. Sci.* **58**, 247–267.

Gershanik, O., Heikkila, R. E. and Duvoisin, R. C. (1983) *Neurology* **33**, 1489–1492.

Goetz, C. G., Lutge, W. and Tanner, C. M. (1986) *Neurology* **36**, 73–75.

Gotham, A. M., Brown, R. G. and Marsden, C. D. (1986) *J. Neurol. Neurosurg., Psychiat.* **49**, 381–389.

Gottfries, C. G., Gottfries, I. and Roos, B. F. (1969) *J. Neurochem.* **16**, 1341–1345.
Greenfield, J. G. and Bosanquet, F. D. (1953) *J. Neurol. Neurosurg. Psychiat.* **16**, 213–226.
Grimes, J. D. and Hassan, M. N. (1983) *Can. J. Neurol. Sci.* **10**, 86–90.
Guerin, B., Silice, C., Mouchet, P., Feuerstein, C. and Demenge, P. (1985) *Life Sci.* **37**, 953–961.
Hardie, R. J., Lees, A. J. and Stern, G. M. (1984) *Brain* **107**, 487–506.
Hassler, R. (1953) In *Handbuch der Innerin Medezin*, Vol. 3, pp. 676–904. Springer Verlag, Berlin.
Hefti, F., Melamed, F. and Wurtman, R. J. (1980) *Brain Res.* **195**, 95–101.
Hökfelt, T., Elde, R., Fuxe, K. *et al.* (1978) In *The Hypothalamus. Research Publications: Association for Research in Nervous and Mental Disease*, Vol. 56 (edited by S. Reichlin, R. J. Baldessarini and J. B. Martin), pp. 69–135. Raven Press, New York.
Hornykiewicz, O. (1963) *Wien Klin. Wschr.* **75**, 309–312.
Hornykiewicz, O. (1966) *Pharmacol. Rev.* **18**, 925–964.
Hornykiewicz, O. (1972) In *Neurotransmitters, Res. Publ. A.R.N.M.D.*, Vol. 50, pp. 390–415. Raven Press, New York.
Hornykiewicz, O. (1975) *Biochem. Pharmacol.* **24**, 1061–1065.
Hornykiewicz, O. (1979a) In *Neurobiology of Dopamine* (edited by A. S. Horn, J. Korf and B. H. C. Westerink), pp. 633–654. Academic Press, London.
Hornykiewicz, O. (1979b) *In The Extrapyramidal Systems and its Disorders.* Advances in Neurology, Vol. 24 (edited by L. J. Poirier, T. L. Sourkes and P. J. Bédard) pp. 275–281. Raven Press, New York.
Hornykiewicz, O. (1982) In *Movement Disorders* (edited by C. D. Marsden and S. Fahn), pp. 41–58. Butterworth, London.
Hyyppa, M. T., Langvik, V. and Rinne, U. K. (1978) *J. Neural Transm.* **42**, 151–157.
Javoy-Agid, F. and Agid, Y. (1980) *Neurology* **30**, 1326–1330.
Javoy-Agid, F., Taquet, H., Ploska, A., Cherif-Zahar, C., Ruberg, M. and Agid, Y. (1981) *J. Neurochem.* **36**, 2101–2105.
Javoy-Agid, F., Ploska, A. and Agid, Y. (1982) *J. Neurochem.* **37**, 1218–1227.
Javoy-Agid, F., Ruberg, M., Pique, L., Bertagna, X., Taquet, H., Studler, J. M., Cesselin, F., Epelbaum, J. and Agid, Y. (1984a) *Neurology* **34**, 672–676.
Javoy-Agid, F., Ruberg, M., Taquet, H., Bokobza, B., Agid, Y., Gaspar, P., Berger, B., N'Guyen-Legros, J., Alvarez, C., Gray, F., Hauw, J. J., Scatton, B. and Rouquier, L. (1984b) In *Parkinson-specific Motor and Mental Disorders.* Advances in Neurology, Vol. 40 (edited by R. G. Hassler and J. F. Christ), pp. 189–198. Raven Press, New York.
Javoy-Agid, F., Ruberg, M., Hirsch, E., Cash, R., Raisman, R., Taquet, H., Epelbaum, J., Scatton, B., Hauw, J. J. and Agid, Y. (1986) In *Recent Developments in Parkinson's Disease* (edited by S. Fahn, C. D. Marsden, P. Jenner and P. Teychenne), pp. 67–83. Raven Press, New York.
Jellinger, K. and Riederer, P. (1984) In *Parkinson-specific Motor and Mental Disorders.* Advances in Neurology, Vol. 40 (edited by R. G. Hassler and J. F. Christ), pp. 199–210. Raven Press, New York.
Jenner, P. and Marsden, C. D. (1984) In *Movement Disorders: Tremor* (edited by L. J. Findley and R. Capildeo), pp. 305–319. McMillan Press, London.
Kebabian, J. W. and Calne, D. B. (1979) *Nature* **277**, 93–96.
Kish, S. J., Chang, L. J., Mirchandani, L., Shannak, K. and Hornykiewicz, O. (1985) *Ann. Neurol.* **18**, 530–536.
Kish, S. J., Rajput, A., Gilbert, J., Rozdilsky, B., Chang, L. J., Shannak, K. and

Hornyiewicz, O. (1986) *Ann. Neurol.* **20**, 26–31.

Kish, S. J., Shannak, K. and Hornykiewicz, O. (1988) *N. Engl. J. Med.* **318**, 876–881.

Klawans, H. L. and Weiner, W. J. (1974) *J. Neurol. Neurosurg. Psychiat.* **37**, 427–430.

Koob, G. F., Simon, H., Herman, J. P. and Le Moal, M. (1984) *Brain Res.* **303**, 319–329.

Kurlan, R. (1987) *Arch. Neurol.* **44**, 1119–1121.

Laihinen, A. and Rinne, U. K. (1986) *Neurology* **36**, 393–395.

Langston, J. W. (1985a) *TINS* **8**, 79–83.

Langston, J. W. (1985b) *TIPS* **6**, 375–378.

Langston, J. W. and Forno, L. S. (1978) *Ann. Neurol.* **3**, 129–133.

Langston, J. W., Ballard, P., Tetrud, J. W. and Irwin, I. (1983) *Science* **219**, 979–980.

Larochelle, L., Bédard, P., Boucher, R. and Poirier, L. J. (1970) *J. Neurol. Sci.* **11**, 53–64.

Lee, T., Seeman, P., Rajput, A., Farley, I. J. and Hornykiewicz, O. (1978) *Nature* **273**, 59–60.

Leenders, K., Palmer, A., Turton, D., Quinn, N., Firnau, G., Garnett, S., Nahmias, C., Jones, T. and Marsden, C. D. (1986) In *Recent Developments in Parkinson's Disease* (edited by S. Fahn, C. D. Marsden, P. Jenner and P. Teychenne) pp. 103–113. Raven Press, New York.

Lees, A. J. and Smith, E. (1983) *Brain* **106**, 257–270.

Le Fur, G., Meininger, V., Phan, T., Gerard, D., Baulac, M. and Uzan, A. (1980) *Life Sci.* **27**, 1587–1591.

Le Fur, G., Meininger, V., Baulac, M., Phan, T. and Uzan, A. (1981) *Rev. Neurol.* **137**, 89–96.

Le Moal, M., Stinus, L., Simon, H., Tassin, J. P., Thierry, A. M., Blanc, G., Glowinski, J. and Cardo, B. (1977) *Advances in Biochemical Psychopharmacology, Vol. 16, Nonstriatal Dopaminergic Neurons* (edited by E. Costa and G. L. Gessa), pp. 237–245. Raven Press, New York.

Lhermitte, F., Agid, Y., Feuerstein, C., Serre, F., Signoret, J. L., Studler, J. M. and Bonnet, A. M. (1977a) *Rev. Neurol.* **133**, 445–454.

Lhermitte, F., Agid, Y., Signoret, J. L. and Studler, J. M. (1977b) *Rev. Neurol.* **133**, 297–308.

Lhermitte, F., Agid, Y. and Signoret, J. L. (1978) *Arch. Neurol.* **35**, 261–263.

Lindvall, O. and Bjorklund, A. (1978) In *Handbook of Psychopharmacology*, Vol. 9 (edited by L. L. Iversen, S. D. Iversen and S. H. Snyder), pp. 139–231. Plenum Press, New York.

Lloyd, K. G., Davidson, L. and Hornykiewicz, O. (1975) *J. Pharmacol. Exp. Therap.* **195**, 453–464.

Loranger, A. W., Goodell, H., McDowell, F. H., Lee, J. E. and Sweet, R. D. (1972) *Brain* **95**, 405–412.

Maloteaux, J. M., Waterkeyn, C. and Laduron, P. M. (1981) *Br. J. Pharmacol.* **74**, 233P.

Maloteaux, J. M., Laterre, C. E., Hens, L. and Laduron, P. M. (1983) *J. Neurol. Neurosurg. Psychiat.* **46**, 1146–1148.

Marsden, C. D. (1982) *Neurology* **32**, 514–539.

Marsden, C. D. and Parkes, J. D. (1976) *Lancet* **i**, 292–296.

Marsden, C. D., Parkes, J. D. and Quinn, N. (1982) In *Movement Disorders* (edited by C. D. Marsden and S. Fahn), pp. 96–122. Butterworth, London.

Matzuk, M. M. and Saper, C. B. (1985) *Ann. Neurol.* **18**, 552–555.

Mayeux, R., Williams, J. B. W., Stern, Y. and Cote, L. (1984) In *Parkinson-specific Motor and Mental Disorders.* Advances in Neurology, Vol. 40 (edited by R. G. Hassler and J. F. Christ), pp. 241–250. Raven Press, New York.

McDowell, F. H. and Sweet, R. D. (1976) In *Advances in Parkinsonism* (edited by W. Birkmayer and O. Hornykiewicz), pp. 603–612. Editions Roche, Basle.

McGeer, P. L. and McGeer, E. G. (1976) *J. Neurochem.* **26**, 65–76.

Melamed, E., Hefti, F. and Wurthman, R. J. (1982) *Israel J. Med. Sci.* **18**, 159–163.

Mones, R. J., Elizan, T. S. and Seigel, G. (1971) *J. Neurol. Neurosurg. Psychiat.* **34**, 668–673.

Mortimer, J. A., Pirozzolo, F. J., Hansch, E. C. and Webster, D. D. (1982) *Neurology* **32**, 133–137.

Muenter, M. D. and Tyce, G. M. (1971) *Mayo Clin. Proc.* **46**, 231–239.

Muenter, M. D., Sharpless, N. S., Tyce, G. M. and Darley, F. L. (1977) *Mayo Clin. Proc.* **52**, 163–174.

Murri, L., Indice, A., Muratorio, A., Pollerre, A., Barreca, T. and Murialdo, G. (1980) *Eur. Neurol.* **19**, 198–206.

Nagatsu, T., Kanamori, T., Kato, T., Iikuza, R. and Narabayashi, H. (1978) *Biochem. Med.* **19**, 360–365.

Nagatsu, T., Wakui, Y., Kato, T., Fujita, K., Kondo, T., Yokochi, F. and Narabayashi, H. (1982) *Biomed. Res.* 395–398.

Nobin, A. and Bjorklund, A. (1973) *Acta Physiol. Scand.* Suppl. 388, 1–40.

Nutt, J. G. and Carter, J. H. (1984) *Lancet* ii, 456–457.

Nutt, J. G., Woodward, W. R., Hammerstad, J. P., Carter, J. H. and Anderson, J. L. (1984) *N. Engl. J. Med.* **310**, 483–488.

Nyberg, P., Nordberg, A., Wester, P. and Winblad, B. (1983) *Neurochem. Pathol.* **1**, 193–202.

Ohye, C., Bouchard, R., Boucher, R. and Poirier, L. J. (1970) *J. Pharm. Exp. Therap.* **175**, 700–708.

Olson, R. and Roos, B. F. (1968) *Nature* **219**, 502–503.

Peaston, M. J. T. and Bianchine, J. R. (1970) *Br. Med. J.* **1**, 400–403.

Pierot, L., Desnos, C., Blin, J., Raisman, R., Scherman, D., Javoy-Agid, F., Ruberg, M. and Agid, Y. (1988) *J. Neurol. Sci.* **86**, 291–306.

Pillon, B., Dubois, B., Cusimano, G., Bonnet, A. M., Lhermitte, F. and Agid, Y. (1989) *J. Neurol. Neurosurg. Psychiat.* **52**, 201–206.

Pimoule, C., Schoemaker, H., Reynolds, G. P. and Langer, S. Z. (1985) *Eur. J. Pharmacol.* **114**, 235–237.

Pique, L., Jegou, S., Bertagna, X., Javoy-Agid, F., Seurin, D., Proeschel, M. F., Girard, F., Agid, Y., Vaudry, H. and Luton, J. P. (1985) *Neurosci. Lett.* **54**, 141–146.

Ploska, A., Taquet, H., Javoy-Agid, F., Gaspar, P., Cesselin, F., Berger, B., Hamon, M., Legrand, J. C. and Agid, Y. (1982) *Neurosci. Lett.* **33**, 191–196.

Poirier, L. J. and Sourkes, T. L. (1965) *Brain* **88**, 181–192.

Presthus, J. (1980) In *Parkinson's Disease: Current Progress, Problems and Management* (edited by U. K. Rinne, M. Klinger and G. Stamm), pp. 255–270. Elsevier North Holland Press, Amsterdam.

Price, K. S., Farley, I. J. and Hornykiewicz, O. (1978) In *Advances in Biochemical Psychopharmacology*, Vol. 19 (edited by P. J. Roberts and G. N. Woodruff), pp. 293–300. Raven Press, New York.

Quik, M., Spokes, E. G., Mackay, A. V. P. and Bannister, R. (1979) *J. Neurol. Sci.* **43**, 429–437.

Quinn, N., Marsden, C. D. and Parkes, J. D. (1982) *Lancet* ii, 412–415.

Quinn, N., Parkes, J. D. and Marsden, C. D. (1984) *Neurology* 34, 1131–1136.
Raisman, R., Cash, R., Ruberg, M., Javoy-Agid, F. and Agid, Y. (1985) *Eur. J. Pharmacol.* 113, 467–468.
Rajput, A. H., Rozdilsky, B., Hornykiewicz, O., Shannak, K., Lee, T. and Seeman, P. (1982) *Arch. Neurol.* 39, 644–646.
Reisine, T. D., Fields, J. Z., Yamamura, H. I., Bird, E. D., Spokes, E., Schreiner, P. S. and Enna, S. J. (1977) *Life Sci.* 21, 335–344.
Riederer, P. and Wuketich, S. (1976) *J. Neural. Transm.* 38, 277–301.
Riederer, P., Birkmayer, W., Seeman, D. and Wuketich, S. (1977) *J. Neural. Transm.* 41, 241–251.
Riederer, P., Rausch, W. D., Birkmayer, W., Jellinger, K. and Danielczyk, W. (1978) *J. Neural Transm.*, suppl. 14, 153.
Riederer, P., Reynolds, G. P., Danielczyk, W., Jellinger, K. and Seemann, D. (1983) In *Lisuride and Other Dopamine Agonists* (edited by D. B. Calne *et al.*), pp. 375–381. Raven Press, New York.
Rinne, J. O., Rinne, J. K., Laakso, K., Lönnberg, P. and Rinne, U. K. (1985) *Brain Res.* 359, 306–310.
Rinne, U. K. and Sonninen, V. (1973) *Arch. Neurol.* 28, 107–110.
Rinne, U. K., Lönnberg, P. and Koskinen, V. (1981) *J. Neural Transm.* 5:, 97–106.
Rinne, U. K., Sonninen, V. and Marttila, R. (1977) In *Parkinson's Disease—Concepts and Prospects* (edited by J. P. W. F. Lakke, J. Korf and H. Wesseling), pp. 73–83. Excerpta Medica, Amsterdam.
Rinne, U. K., Sonninen, V. and Laaksonen, H. (1979) In *The Extrapyramidal System and its Disorders.* Advances in Neurology, Vol. 24 (edited by L. J. Poirier, T. L. Sourkes and P. J. Bédard), pp. 259–274. Raven Press, New York.
Rinvik, E., Grofova, I. and Ottersen, O. P. (1976) *Brain Res.* 112, 388–394.
Rondot, P., De Recondo, J., Coignet, A. and Ziegler, M. (1984) In *Parkinson-specific Motor and Mental Disorders*, Advances in Neurology, Vol. 40 (edited by R. G. Hassler and J. F. Christ), pp. 259–269. Raven Press, New York.
Ruberg, M. and Agid, Y. (1988) In *Handbook of Psychopharmacology, Vol. 20: Psychopharmacology of the Ageing Nervous System* (edited by L. Iversen, S. D. Iversen and S. H. Snyder). Plenum Press, New York pp. 157–206.
Ruberg, M., Javoy-Agid, F., Hirsch, E., Scatton, B., Lheureux, R., Hauw, J. J., Duyckaerts, C., Gray, F., Morel-Maroger, A., Rascol, A., Serdaru, M. and Agid, Y. (1985) *Ann. Neurol.* 18, 523–529.
Scatton, B., Rouquier, L., Javoy-Agid, F. and Agid, Y. (1982) *Neurology* 32, 1039–1040.
Scatton, B., Javoy-Agid, F., Rouquier, L., Dubois, B. and Agid, Y. (1983) *Brain Res.* 275, 321–328.
Scatton, B., Monfort, J. C., Javoy-Agid, F. and Agid, Y. (1984) In *Catecholamines: Neuropharmacology and Central Nervous System. Therapeutic Aspects*, pp. 43–52. Alan R. Liss, New York.
Scatton, B., Dennis, T., Lheureux, R., Monfort, J. C., Duyckaerts, C. and Javoy-Agid, F. (1986) *Brain Res.* 380, 181–185.
Schneider, J. S., Yuweiler, A. and Markham, C. H. (1987) *Brain Res.* 411, 144–150.
Schoemaker, H., Pimoule, C., Arbilla, S., Scatton, B., Javoy-Agid, F. and Langer, S. Z. (1985) *Naunyn-Schmideberg's Arch. Pharmacol.* 329, 227–235.
Schwartz, J. C., Agid, Y., Javoy-Agid, F., Martres, M. P., Pollard, H., Bouthenet, M. L., Llorens-Cortes, C., Sales, N. and Taquet, H. (1986) In *Nausea and Vomiting: Mechanisms and Treatment* (edited by D. Grahame-Smith, C. J. Davis and C. Lake-

Bakaaz), pp. 18–30. Springer Verlag, Berlin.

Sharpe, J. A., Rewcastle, N. B., Lloyd, K. G., Hornykiewicz, O., Hill, M. and Tasker, R. R. (1973) *J. Neurol. Sci.* **19**, 275–286.

Shibuya, M. (1979) *J. Neural Transm.* **44**, 287.

Shoulson, I., Glaubiger, G. A. and Chase, T. N. (1975) *Neurology* **25**, 1144–1148.

Skagerberg, G. and Lindvall, O. (1985) *Brain Res.* **342**, 340–351.

Sloviter, R. S. and Connor, J. D. (1977) *J. Neurochem.* **28**, 1129–1131.

Snider, S. R., Fahn, S., Isgreen, W. P. and Cote, L. J. (1976) *Neurology* **26**, 423–429.

Snyder, A. M., Stricker, E. M. and Zigmond, M. J. (1985) *Ann. Neurol.* **18**, 544–551.

Stern, Y. and Langston, J. W. (1985) *Neurology* **35**, 1506–1509.

Uhl, G. R., Hedreen, J. C. and Price, D. L. (1985) *Neurology* **35**, 1215–1218.

Ungerstedt, U. (1971) *Acta Physiol. Scand.* suppl. 367, 69–93.

Weil-Malherbe, H. and Van Buren, J. M. (1969) *J. Lab. Clin. Med.* **74**, 305–318.

Yong, V. W. and Perry, T. L. (1986) *J. Neurol. Sci.* **72**, 265–272.

Zigmond, M. J., Acheson, A. L., Stachowiak, M. K. and Stricker, E. M. (1984) *Arch. Neurol.* **41**, 856–861.

7

The Neurochemistry of Parkinson's Disease as Revealed by PET Scanning

K. L. Leenders

INTRODUCTION

In recent years it has become possible to measure *in vivo* certain aspects of human brain tissue function using radiolabelled tracers and positron emission tomography (PET). Although to date measurement of regional cerebral energy metabolism and certain aspects of striatal dopaminergic function has reached an advanced stage of development, other neurochemical systems are now being addressed as well. This holds the promise that through PET studies it may now be possible to relate *in vivo* changes in cerebral neurotransmitter function to clinical features. This should be useful to study pathophysiology of movement disorders in general and in particular Parkinson's disease. In many of these conditions often no, or only mild, structural abnormalities are demonstrable in the brain while severely disturbed basal ganglia function, either in isolation or in association with other brain structures, is often evident.

In particular, longitudinal studies starting in an early phase of the disease and using various types of tracers seem to be promising. Cross-sectional cohort studies are less suitable due to the rather large variance of normal values combined with the small number of patients who can usually be

DISORDERS OF MOVEMENT: CLINICAL, PHARMACOLOGICAL
AND PHYSIOLOGICAL ASPECTS   ISBN 0-12-569685-X

scanned. This is not just because of the relatively long duration of the scanning procedures. The radiochemistry is also often complicated, and needs to be performed immediately before a scan measurement because of the short radioactive half-life of the radionuclides incorporated in the tracer molecules. However, data handling and analysis is the most time consuming aspect of measuring tissue function with PET. Reduction of count measurements into manageable units and conversion of time–activity curves into meaningful pharmacological or biochemical entities is a formidable task. It seems that in this field developments are still in an early stage. The inevitably low patient throughput per scan laboratory, in combination with the small number of PET centres worldwide, will make it understandable why accumulation of biological or clinical results with PET is a slow process.

POSITRON EMISSION TOMOGRAPHY (PET)

General remarks

For more detailed technical and methodological information the reader is referred to the appropriate handbooks (e.g. Phelps *et al.*, 1986) and Chapter 19 in this volume. The key feature of PET is the ability to quantitatively measure in absolute units special radionuclides (e.g. ^{11}C, ^{18}F or ^{15}O) independently of depth. Since these radionuclides are incorporated into molecules of physiological character or related to known drugs, quantitative and regional *in vivo* neurobiochemistry can now be pursued. Transformation of the measured radioactivity into biochemical entities is however a formidable task which has been solved for only a few tracer substances so far. The field of tracer kinetics is too complex to be discussed in this chapter.

Often PET scans are referred to as "functional scans" as opposed to CT and MRI scans which are termed "anatomical". This has resulted in attempts to design "anatomical localization methods" to be used to analyse PET scans. Although sometimes useful this concept may nevertheless be misleading, since PET measurements are not just "function" measured in isolation, but represent functional aspects of anatomical structures. They are both intimately related. When a sample (= region of interest = ROI) is required from a structure (e.g. cerebellum), this can be obtained correctly in most tracer studies by placing the ROI on the basis of shape and position of the measured radioactivity. Obviously the question is one of precision. How detailed can such a "functionally" determined ROI represent a certain brain structure? The older PET tomographs did not allow sampling of small structures, which created a need for a parallel anatomical reference. However, the more recent "state of the art" machines with multiple plane facilities and higher intrinsic spatial resolution should result in independent functional localization when processed with high-power three-dimensional image analysis techniques. That so-called "anatomical scans" can provide a false sense of precision of anatomical localization (particularly when there is a large discrepancy between spatial resolutions of CT scans and PET scans) is discussed in an editorial by Duara (1985): "PET scan images of glucose or oxygen consumption can highlight an aspect of structure of the brain more effectively than other so-called structural imaging devices".

PET tracer studies in Parkinson's disease

Several movement disorders have been studied using PET and several tracers, either alone or in combination. Table 7.1 summarizes the results as they appear in the literature. In this chapter studies related to Parkinson's disease will briefly be reviewed.

Table 7.1   Overview of results obtained with PET scans in the study of movement disorders.

	Energy metabolism	Presynaptic dopaminergic system	Postsynaptic dopaminergic system
	CBF, oxygen and glucose utilization	Fluorodopa Nomifensine	Spiperones Raclopride
Parkinson's disease	Decreased (globally) Increased (focally?)	Decreased	Normal, (sometimes increased?)
Huntington's disease	Decreased (globally, focally)	Normal	Decreased
Multiple system atrophy	Decreased? (probably)	Decreased	Decreased (?)
Progressive supranuclear palsy	Decreased (globally, focally)	Decreased	Decreased
Dystonia	Normal	Normal or decreased	Normal (sometimes increased?)

Energy metabolism

A global decrease of cerebral blood flow (CBF), oxygen and glucose utilization in patients with Parkinson's disease has been reported using PET (Kuhl *et al.*, 1984; Leenders *et al.*, 1984, 1985b; Wolfson *et al.*, 1985; Peppard *et al.*, 1989). These decreases of energy metabolism are rather modest in general, particularly if no cognitive disturbances are present. No decrease of glucose metabolism was found by Rougemont *et al.* (1984) who compared four patients with seven healthy volunteers. Perlmutter *et al.* (1985) measured CBF using PET in 11 patients and did not find significant differences compared with normal values. However, their patients were selected for unilaterality and presented with only mild symptoms and signs.

Regional differences in oxygen utilization and CBF in parkinsonian brain have also been described using PET (Wolfson *et al.*, 1985). Basal ganglia values, presumably dominated by activity in the pallidum, on the side contralateral to the affected limbs of predominantly or solely unilateral parkinsonian patients, were higher (range 12–56%) than the opposite side. This asymmetry of the globus pallidus region has also been found in two patients using 18-FDG and PET (Martin *et al.*, 1984). Glucose metabolism was 54% higher on the contralateral side in one severely affected unilateral patient and 18% in a moderately affected case. A third patient with mild signs and symptoms showed no side to side differences. On the other hand, Perlmutter *et al.* (1985) found abnormal basal ganglia CBF asymmetries in only 6 of 11 hemiparkinsonian patients and the higher value was not always found on the side contralateral to the affected limbs.

Animal experiments support the human PET asymmetry findings in

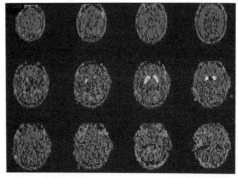

Fig. 7.1 L-[^{18}F]-6-Fluorodopa uptake in a healthy volunteer's brain. The images are contiguous transaxial cross-sections of the radioactivity distribution from the top of the brain (left upper image) to the level cutting through the cerebellum (right lower image). The top of the image is frontal and the head is seen from above. The activity shown is the cumulative activity from 30 to 90 min after administration of the tracer. (This figure appears in colour in the plate section facing page 294.)

Parkinson's disease. Wooten and Collins (1981) determined regional glucose utilization in rat brain using $[^{14}C]$-2-deoxyglucose autoradiography following unilateral lesioning of substantia nigra with 6-OHDA. These authors found an increased glucose utilization in ipsilateral globus pallidus which was maximal at 21 days after the lesion (40% higher than controls) and decreased thereafter (15% higher than controls at 104 days). Crossman *et al.* (1985) performed 2-deoxyglucose autoradiography in a monkey rendered severely parkinsonian by daily intravenous administration of MPTP. A marked increase in glucose metabolism was seen in the globus pallidus, in ventral anterior and ventral lateral thalamic nuclei.

Oxygen and glucose utilization are an expression of energy expenditure in brain tissue and most of this is used for ion transport involved in the generation of transmembrane potentials. It is suggested that nerve endings and dendrites have high surface-to-volume ratios and thus a high rate of energy expenditure to pump ions across the neural membranes. The nigrostriatal dopaminergic pathway is believed to have an inhibitory function. Hence a severe defect of this system as in Parkinson's disease should lead to disinhibition of striatal activity. The main efferent pathway of striatum is to the globus pallidus. This may explain why disinhibition of striatum results in higher oxygen and glucose utilization in the globus pallidus. However, further evidence is still needed to validate this hypothesis *in vivo* in man.

Could it be the tremor which leads to increased contralateral globus pallidus activity? Voluntary hand or finger movements in normal subjects give rise to increased CBF here and in several cortical regions. However, none of the patients with unilateral tremor had cortical increases of CBF (Leenders *et al.*, 1983; Perlmutter *et al.*, 1985). Also, PET scans repeated 1 h after L-dopa treatment still showed identical basal ganglia asymmetries even though the tremor had completely disappeared (Leenders *et al.*, 1985b).

The question has been raised whether dopamine agonist treatment in dosages which are clinically effective would influence tissue function as measured by CBF or energy metabolism. Moderate L-dopa doses resulting in adequate clinical responses do not significantly alter CBF (Leenders *et al.*, 1983, 1984, 1985b; Perlmutter *et al.*, 1985). However, high doses of L-dopa produce a global increase of CBF ranging between 10 and 80% above control values. No change of oxygen utilization was found corresponding to the CBF increase. Premedication with domperidone (a peripheral dopamine receptor antagonist) prevented the rise in CBF (Leenders, unpublished). Taken together, these findings suggest that high doses of L-dopa have a vasodilator effect unrelated to any effect on the parenchymal dopaminergic nervous tissue. This vasoactive component probably does not play a significant role under normal therapeutic regimes. The lack of effect of L-dopa on cerebral energy metabolism was corroborated by the results of

Rougemont *et al.* (1984), who found no change in glucose utilization in four patients studied with PET and 18-FDG before and 1 day after L-dopa treatment.

Dopaminergic tracers

A range of PET tracers each being trapped or bound by a particular part of the monaminergic neurotransmitter system at the level of the nerve terminal is available. The specificity and accuracy with which these tracers are able to indicate the respective functions differ, and many methodological questions are still open.

L-[^{18}F]-*Fluorodopa.* The tracer L-[^{18}F]-fluorodopa was shown to accumulate *in vivo* in man specifically in the brain region with highest intrinsic dopaminergic activity, namely striatum (Garnett *et al.*, 1983). Since then several PET groups have reported similar results (Leenders *et al.*, 1984, 1985a, 1986a,b; Martin *et al.*, 1985). Radioactivity distribution after administration of this tracer can be seen in Fig. 7.1. The tracer, like L-dopa itself, is taken up by the brain only to a small extent because the transport across the blood–brain barrier is an active, energy-dependent and strictly stereo-selective process in competition with other large neutral amino acids (Leenders *et al.*, 1985c). L-[^{18}F]-Fluorodopa is decarboxylated to L-[^{18}F]-fluorodopamine in the endothelial cells of brain capillaries and in the brain tissue itself in decarboxylase-rich regions like striatum. Firnau *et al.* (1987) showed that in monkey brain during the first 1–1.5 h after L-[^{18}F]-fluorodopa administration the radioactivity in striatum is predominantly L-[^{18}F]-fluorodopamine, although ^{18}F-HVA and ^{18}F-DOPAC also gradually increase. Still, the sum of L-[^{18}F]-fluorodopamine, ^{18}F-HVA and ^{18}F-DOPAC formation is determined by the regional decarboxylation rate. In cerebral tissues other than the striatum the O-methylated derivative was the principal labelled compound. The fact that no O-methylated derivative was found in striatum itself suggests that at least during the scan period no major transport of O-methyldopa from blood to brain took place, and thus that methylated products in non-dopaminergic tissue were formed locally from L-[^{18}F]-fluorodopa. In arterial plasma of carbidopa-pretreated subjects the main metabolite after L-[^{18}F]-fluorodopa administration was found to be the O-methylated derivative (Boyes *et al.*, 1986).

It is envisaged that more accurate L-[^{18}F]-fluorodopa uptake measurements can be made when a C-O-methyltransferase inhibitor suitable for human use can be given as premedication. By blocking methylation in the periphery the arterial radioactivity in the plasma might possibly become equivalent to L-[^{18}F]- fluorodopa itself, thus providing a purer input curve.

On the other hand blocking of tissue methylation would result in all the activity being derived from L-[^{18}F]-fluorodopa or a metabolite beyond the decarboxylation step. The kinetic modelling of cerebral L-[^{18}F]-fluorodopa uptake to estimate regional dopamine formation could then be simplified. The positive effect of a COMT inhibitor on L-[^{18}F]-fluorodopa uptake in rats has been demonstrated by Cumming *et al.* (1987).

In parkinsonian patients who clinically were grossly asymmetrically affected, L-[^{18}F]-fluorodopa uptake was found to be less in the striatum contralateral to the more affected body side (Nahmias *et al.*, 1985; Leenders *et al.*, 1987). In general, simple ratios of striatal to non-dopaminergic brain tissue activity plotted against time illustrate how L-[^{18}F]-fluorodopa handling is affected in conditions with impaired nigrostriatal pathways. In healthy subjects this ratio gradually increases over a period of 3.5–4 h following injection. In parkinsonian patients, however, it reaches a plateau at approximately 100 min after administration. This early levelling off probably indicates a reduced striatal capacity to convert the tracer into [^{18}F]-fluorodopamine and "store" it (Leenders *et al.*, 1986a). Also, patients who were in a more advanced phase and showed the "on–off" phenomenon had a significantly lower mean striatum-to-surrounding brain ratio compared to more mildly affected patients. Calne *et al.* (1985) reported decreased striatal L-[^{18}F]-fluorodopa uptake in four drug addicts who probably had been exposed to the dopaminergic cell toxin MPTP but did not show parkinsonian signs or symptoms. The findings might be taken to support the notion that a considerable part of the nigrostriatal pathway must be lost before parkinsonian signs and symptoms will become evident. On the other hand it was not excluded that the investigated patients might possibly have taken cocaine in addition to the toxic heroin. Cocaine is known to block monamine uptake into the nerve terminal and would therefore be expected to inhibit L-[^{18}F]-fluorodopa uptake giving the false impression of loss of nigrostriatal dopaminergic cells. Follow-up of these patients will possibly provide the answer.

Overall influx constants (K_i) of L-[^{18}F]-fluorodopa into striatum were calculated in 14 parkinsonian patients and plotted against severity of semiquantitatively scored bradykinesia (Leenders, 1987). A good correlation was found between bradykinesia and striatal L-[^{18}F]-fluorodopa uptake, the most severely affected patients showing the lowest specific processing of L-[^{18}F]-fluorodopa. Thus it is possible to directly demonstrate a correlation between a clinical phenomenon and a regional biochemical brain tissue activity.

Work is in progress to see whether human foetal tissue engraftment into striata of parkinsonian patients leads to improved post-operative clinical outcome in parallel to enhanced local dopaminergic activity.

[^{11}C]-*Nomifensine*. Another "presynaptic" tracer which can be applied in PET scan investigations is [^{11}C]-nomifensine. Nomifensine binds specifically to monoaminergic uptake sites of the nerve terminals (Slater and Crossman, 1984; Scatton *et al.*, 1984). Unilateral lesions of the nigrostriatal dopaminergic pathway in rats produced a marked ($\simeq 80\%$) decrease of specific striatal binding of ^{3}H-nomifensine (Scatton *et al.*, 1984). This decrease matched comparable reductions of endogenous striatal dopamine concentrations. [^{11}C]-nomifensine was first applied in Rhesus monkey PET investigations (Aquilonius *et al.*, 1987; Leenders *et al.*, 1988a). One Rhesus monkey was scanned using [^{11}C]-nomifensine and other tracers before and several times after unilateral lesioning of the nigrostriatal pathway by MPTP (Leenders *et al.*, 1988a), the neurotoxin which selectively destroys dopaminergic neurons. MPTP was slowly infused as a solution (in total 1.2 mg) through a right internal artery catheter. Within 2 days hypokinesia on the left side developed, accompanied by dystonic postures of the left limbs. The clinical signs disappeared after L-dopa or apomorphine therapy. Apomorphine induced rotation to the left during about 45 min. [^{11}C]-Nomifensine uptake in striatum was normal at 2 days after MPTP administration. However, 9 days after the lesion, specific binding, defined as the radioactivity difference between striatum and non-dopaminergic brain tissue, had decreased by between 80 and 90% in the lesioned right striatum, but was normal on the left side. Apparently MPTP had been taken up by the nerve terminals in the right striatum, resulting in rapid functional deterioration and somewhat slower structural damage to the nigrostriatal system. At 6 weeks after the lesion the [^{11}C]-nomifensine reduction in the right striatum was still severe, but 5 months later [^{11}C]-nomifensine had recovered to about 50% of normal. At that stage only mild clinical signs were still present.

Human PET studies using [^{11}C]-nomifensine are currently being undertaken by several groups. In parkinsonian patients striatal uptake is clearly reduced, and appears to match the decreases of L-[^{18}F]-fluorodopa uptake measured in the same individuals (Leenders *et al.*, in preparation).

[^{11}C]-*Methylspiperone*. The first suitable tracer available for *in vivo* measurement of dopamine receptor density using PET was [^{11}C]-methylspiperone (Wagner *et al.*, 1983). Although this compound binds selectively to D2 dopamine receptors, it is not completely specific for the dopamine receptors since it also binds to serotonin receptors particularly in cerebral cortex (Frost *et al.*, 1987), and possibly also to other receptor types.

The application of [^{11}C]-methylspiperone PET scanning in parkinsonian patients has been on a limited scale so far (Leenders *et al.* 1985a; Hägglund *et al.*, 1987). Untreated parkinsonian patients showed striatal [^{11}C]-methylspiperone uptake no different from that in healthy subjects. L-Dopa drug treatment seemed to reduce striatal [^{11}C]-methylspiperone uptake to some extent (Leenders et al., 1985a), but the number of patients studied was small. These findings are in agreement with post-mortem results showing virtually no change of dopamine D2 receptor densities in parkinsonian patients (Bokobza et al., 1984). In Parkinson's disease, which is a chronic illness based on a severe presynaptic dopaminergic lesion, apparently an intact postsynaptic dopaminergic system (at least as far as receptor density can be taken as an indicator for receptor function) prevails. This explains why dopaminergic drugs can be so effective in patients with idiopathic Parkinson's disease. Other chronic neurodegenerative diseases associated with parkinsonian features, like Steele–Richardson–Olszewski syndrome, do not readily respond to L-dopa. In that condition impaired presynaptic dopaminergic function (Leenders et al., 1988b) is accompanied by a decrease in striatal dopamine D2 receptors (Baron et al., 1985), probably due to striatal neuronal cell loss. Combination of pre- and post-synaptic tracers and possibly energy metabolism tracers may provide a means to develop a differential diagnostic system for several neurodegenerative diseases on the basis of in vivo neurochemistry.

REFERENCES

Aquilonius, S. M., Bergström, K., Eckernäs, S. A., Hartvig, P., Leenders, K. L., Lundqvist, H., Antoni, G., Gee, A., Rimland, A., Uhlin, J. and Långström, B. (1987) Acta Neurol. Scand. 76, 283–287.

Baron, J. C., Mazière, B., Loc'h, C., Sgouropoulos, P., Bonnet, A.-M. and Agid, Y. (1985) Lancet ii, 1163–1164.

Bokobza, B., Ruberg, M., Scatton, B., Javoy-Agid, F. and Agid, Y. (1984) Eur. J. Pharmacol. 99, 167–175.

Boyes, R. E., Cumming, P., Martin, W. R. W. and McGeer, E. G. (1986) Life Sci. 39, 2243–2252.

Calne, D. B., Langston, J. W., Martin, W. R. W., Stoessl, A. J., Ruth, T. J., Adam, M. J., Pat, B. D. and Shulzer, M. (1985) Nature 317, 246–248.

Crossman, A. R., Mitchell, I. J. and Sambrook, M. A. (1985) Neuropharmacology 24, 587–591.

Cumming, P., Boyes, B. E., Martin, W. R. W., Adam, M., Ruth, T. and McGeer, E. G. (1987) Biochem. Pharmacol. 36, 2527–2531.

Duara, R. (1985) J. Cereb. Blood Flow Metabol. 5, 343–344.

Firnau, G., Sood, S., Chirakal, R., Nahmias, C. and Garnett, E. S. (1987) J. Neurochem. 48, 1077–1082.

Frost, J. J., Smith, A. C., Kuhar, M. J. Dannals, R. F. and Wagner, Jr, H. N. (1987) Life Sci. 40, 987–995.

Garnett, E. S., Firnau, G. and Nahmias, C. (1983) Nature 305, 137–138.

Hägglund, J., Aquilonius, S-M., Eckernäs, S.-Å., Hartvig, P., Lundquist, H., Gullberg, P. and Långström, B. (1987) Acta Neurol. Scand. 75, 87–94.

Kuhl, D. E., Metter, E. J. and Riege, W. H. (1984) Ann. Neurol. 15, 419–424.

Leenders, K. L. (1987) In *Recent Developments in Parkinson's Disease* (edited by S. Fahn, C. D. Marsden, D. Calne and D. M. Goldstein), pp. 105–121. Macmillan Healthcare Information, New Jersey.

Leenders, K. L., Wolfson, L., Gibbs, J., Wise, R., Jones, T. and Legg, N. (1983) *J. Cereb. Blood Flow Metabol.* **3**, S488–S489.

Leenders, K. L., Wolfson, L. and Jones, T. (1984). *Monogr. Neurol. Sci.* **11**, 180–186

Leenders, K. L., Herold, S., Palmer, A. J., Turton, D., Quinn, N., Jones, T., Frackowiak, R. S. J. and Marsden, C. D. (1985a) *J. Cereb. Blood Flow Metabol.* **5** (suppl.), S517–S518.

Leenders, K. L., Wolfson, L., Gibbs, J. M., Wise, R. J. S., Causon, R., Jones, T. and Legg, N. J. (1985b) *Brain* **108**, 171–191.

Leenders, K. L., Palmer, A. J., Quinn, N., Clark, J. C., Firnau, G., Garnett, E. S., Nahmias, C., Jones, T. and Marsden, C. D. (1986a) *J. Neurol. Neurosurg. Psychiat.* **49**, 853–856.

Leenders, K. L., Poewe, W. H., Palmer, A. J., Brenton, D. P. and Frackowiak, R. S. J. (1986b) *Ann. Neurol.* **20**, 258–262.

Leenders, K. L., Aquilonius, S. M., Bergström, K., Bjurling, P., Crossman, A. R., Eckernäs, S. A., Gee, A. G., Hartvig, P., Lundqvist, H., Långström, B., Rimland, A. and Tedroff, J. (1988a) *Brain Res.* **445**, 61–67.

Leenders, K. L., Frackowiak, R. J. S. and Lees, A. J. (1988b) *Brain* **111**, 615–630.

Martin, W. R. W., Beckman, J. H., Calne, D. B., Adam, M. J., Harrop, R., Rogers, J. G., Ruth, T. J., Sayre, Cl. and Pate, B. D. (1984) *Can. J. Neurol. Sci.* **11**, 169–173.

Martin, W. R. W., Boyes, B. E., Leenders, K. L. and Patlak, C. S. (1985) *J. Cereb. Blood Flow Metabol.* **5** (suppl.), S593–S594.

Nahmias, C., Garnett, E. S., Firnau, G. and Lang, A. (1985) *J. Neurol. Sci.* **69**, 223–230.

Peppard, R. F., Martin, W. R. W., Guttman, M., Grochowski, E., Okada, J., McGeer, P. L., Carr, G. D., Phillips, A. G., Steele, J. C., Tsui, J. K. C. and Calne, D. B. (1989) In *Neural Mechanisms in Disorders of Movement* (edited by A. Crossman). John Libbey, London (in press).

Perlmutter, J. S., Powers, W. J., Herscovitch, P., Fox, P. T. and Raichle, M. E. (1985) *J. Cereb. Blood Flow Metabol.* **5**, S641–S642.

Phelps, M. E., Mazziotta, J. C. and Schelbert, H. R. (Editors) (1986) *Positron Emission Tomography and Autoradiography. Principles and Applications for the Brain and Heart.* Raven Press, New York.

Rougement, D., Baron, J. C., Collard, P., Bustany, P., Comar, D. and Agid, Y. (1984) *J. Neurol. Neurosurg. Psychiat.* **47**, 824–830.

Scatton, B., Dubois, A., Dubocovitch, M. L., Zahniser, N. R. and Fage, D. (1984) *Life Sci.* **36**, 815–822.

Slater, P. and Crossman, A. R. (1984) In *Nomifensine. A Pharmacological and Clinical Profile* (edited by W. Linford-Rees and R. G. Priest), pp. 15–19. The Royal Society of Medicine, London.

Wagner, H. N., Burns, H. D., Dannals, R. F., Wong, D. F., Langstrom, B., Duelfer, T., Frost, J. J., Ravert, H. T., Links, J. M., Rosenbloom, S. B., Lukas, S. E., Kramer, A. V. and Kuhar, M. J. (1983) *Science* **221**, 1264–1266.

Wolfson, L. I., Leenders, K. L., Brown, L. L. and Jones, T. (1985) *Neurology* **35**, 1399–1405.

Wooten, G. F. and Collins, R. C. (1981) *J. Neurosci.* **1**, 285–291.

8

Sleep and Parkinson's Disease

David Parkes

INTRODUCTION

Several years ago we did a strange experiment. Parkinson's disease was clearly designed to answer one simple question about sleep—what is its function? We argued that dopamine receptor configuration, number or affinity changed with sleep. If sleep "restores" receptors, then a night without sleep should make Parkinson's disease worse, as well as impair the response to L-dopa and dopamine agonists. David Marsden had previously described the phenomenon of sleep benefit in parkinsonism, with the implied assumption that sleep must alter some important aspect of dopamine neurotransmission. Our experimental design was simple. It consisted of a tired doctor sitting beside a patient with Parkinson's disease, both attempting to keep awake for 24 h. This proved difficult, and the results showed our hypothesis was wrong. The response to L-dopa was enhanced, not reduced, after 24 h without sleep. However, the degree of sleep deprivation actually achieved was uncertain, and the results were not published even in a conference proceedings. The point of this history has little to do with receptors, L-dopa, sleep or international Parkinson's disease conferences. It is a tribute to David Marsden's ability to stimulate scientific research, ask fundamental questions, and challenge accepted dogma. The experiment was done successfully by

Bertolucci *et al.* (1987), who showed that stiffness, walking, akinesia, postural disturbance and functional disability in Parkinson's disease all improved for up to 2 weeks after a single night of total sleep deprivation.

What do sleep deprivation experiments in parkinsonism teach? Few researchers have personally observed parkinsonian patients over prolonged periods. As well as great difficulty in keeping parkinsonian patients awake, the surprising temporal variability in disease severity, and in particular in tremor, even in untreated subjects, is revealed. Vocalizations, grunting, muscle jerking, blepharospasm, facial tics and tremor are very conspicuous in sleeping patients with Parkinson's disease, both in light and rapid eye movement (REM) sleep, and rigidity is almost entirely abolished. The degree of night movement does not parallel the degree of waking akinesia. A few parkinsonians are almost impossible to arouse from sleep, whatever the stimulus. However, despite these observations, Parkinson's disease appears to have little obvious effect in most instances on sleeping or waking mechanisms, and any sleep disturbance present appears to be mainly due to the motor defect. Basal ganglia involvement in sleep mechanisms, with the exception of sleep atonia and, possibly, eye movements (REMs) within REM sleep may not be very important; at the physiological level, substantia nigra neuronal discharge rates show little spontaneous alteration in frequency between sleep and waking (Trulson *et al.*, 1981). Also, antiparkinsonian drugs have on the whole only minor effects on sleep and waking, modest rather than dramatic in nature.

Despite these somewhat disappointing findings, a study of sleep in parkinsonism has thrown light on a number of important anatomical and pharmacological sleep mechanisms. Pathological studies last century led Mauthner and others to conclude that lesions around the back of the third ventricle and aqueduct resulted in drowsiness, a finding amply confirmed by the studies of encephalitis lethargica by von Economo (1930). A correlation of focal brain lesions in encephalitis lethargica with different forms of sleep disturbance (described below) has given important evidence on the anatomical basis of time clock as well as sleep–wake systems in man. More recently, studies of brain stem tegmental pontine magnocellular "on" neurons in progressive supranuclear palsy have given important information about mechanisms of REM sleep. Data from human and animal pharmacological studies suggest that dopamine, acetylcholine and 5-HT are not primary neurotransmitters governing sleep onset or controlling wakefulness, although they may all have important secondary roles, particularly in the case of acetylcholine and REM sleep.

Table 8.1 Nocturnal symptoms experienced since diagnosis of Parkinson's disease ($n = 220$).

Symptoms	Experienced by (%)	Most troublesome (%)
Need to visit lavatory	79	29
Inability to turn over in bed	65	39
Painful leg cramps	55	15
Vivid dreams/nightmares	48	9
Inability to get out of bed unaided	35	15
Limb or facial dystonia	34	10
Back pain	34	9
Myoclonic jerks	33	5
Visual hallucinations	16	3

SLEEP IN PARKINSON'S DISEASE

Many studies have shown that night sleep is poor in idiopathic Parkinson's disease. Sleep is restless and disturbed, with bed discomfort, the patient taking longer to fall asleep, with more frequent and prolonged arousals than controls. The greater the disease severity, the greater the sleep disturbance. Tremor may interfere with or prevent sleep onset, turning in bed is often very difficult, and total night sleep may be reduced by as much as half in late-stage illness (Mouret, 1975; Wein *et al.*, 1979; Friedman, 1980; Laihinen *et al.*, 1987). Table 8.1 taken from Lees (1988), lists symptomatic problems in 220 patients with Parkinson's disease. Only five of these subjects did not report any difficulty at night or on waking.

Waking symptoms include difficulty in turning over and getting out of bed unaided, severe tremor and foot dystonia (see Table 8.2). In contrast, about a third of patients with symptom variability across the day experience their best motility, lasting 1–3 h, upon arising at their usual hour of awakening, i.e. sleep "benefit" (Clark and Feinstein, 1977).

Table 8.2 Symptoms on waking experienced since diagnosis of Parkinson's disease ($n = 220$).

Symptoms	Experienced by (%)	Most troublesome (%)
Difficulty on turning over on waking	64	37
Difficulty in getting out of bed unaided	50	30
Severe tremor	22	10
Foot dystonia	20	10

DAYTIME DROWSINESS

Severe daytime drowsiness is an important but neglected symptom in parkinsonism, reported by up to 20% of all subjects with idiopathic Parkinson's disease. Daytime sleepiness increases in duration and severity with progression of the illness, and with prolonged treatment (Nausieda *et al.*, 1984). The cause of daytime drowsiness is uncertain, but at least in part it may be attributable to chronic insomnia. It is not related to sleep apnoea, for this is not characteristic of Parkinson's disease (Apps *et al.*, 1985; see below). In most parkinsonians, L-dopa–decarboxylase inhibitor and anticholinergic drug treatment does not cause gross drowsiness, but the use of these drugs, with or without more sedative ergot compounds, may increase daytime difficulties resulting from chronic insomnia, and sometimes produce very considerable daytime drowsiness.

SLEEP LABORATORY FINDINGS IN PARKINSONISM

Sleep laboratory recordings in idiopathic Parkinson's disease show a number of minor abnormalities, but these are not specific to the illness, being mainly if not entirely due to the motor defect or drug treatment. Similar findings are present in many elderly subjects who do not have Parkinson's disease. Sleep is commonly fragmented, with a long sleep latency and frequent arousals. There is a decrease in both deep sleep (stages 3–4 NREM sleep) and REM sleep. However, the density of rapid eye movements within REM sleep may be increased, not decreased, in parkinsonism, either as a result of illness or L-dopa treatment (Lavie *et al.*, 1980a). Sleep spindles are unusually rare, restored by L-dopa treatment. Some parkinsonian subjects have disruption of the whole circadian sleep–wake cycle, with sleep distributed throughout the entire 24 h rather than consolidated at night. Similar findings to the above, with an increased number of awakenings, increased wake time after sleep onset, and increased time spent in stage 1 sleep at the expense of deep sleep, with REM sleep fragmentation, occur in many depressed subjects. In parkinsonism, tremor is less severe during sleep than wakefulness, but may persist into stage 1–2 NREM sleep, very occasionally in deeper sleep stages. Tremor recurs with each arousal. As well as tremor, myoclonus, periodic leg movements in sleep, muscle twitches, dystonic episodes, both drug-related and drug-independent, and blepharospasm, occur in both NREM and REM sleep. Parkinsonian patients with myoclonic jerking during the day due to L-dopa almost always have similar jerks at night (Klawans *et al.*, 1975). Painful sleep dystonic spasms may produce arousal. Nocturnal vocalizations and incomplete somnambulism (sitting up in bed rather than walking), despite akinesia, are common.

No gross asymmetry of EEG characteristics is found in parkinsonism, even in unilateral disease. Myslobodsky *et al.* (1982) showed that in seven patients with hemiparkinsonism and poor sleep there were no asymmetries of EEG sleep pattern.

Upper airway and chest wall motor activity, stiffness, tremor and drug-induced dyskinesias involving oral, facial, glossal, laryngeal and pharyngeal muscles can markedly affect breathing, both awake and asleep, in parkinsonism, and it is easy to demonstrate these abnormalities in the sleep laboratory. However, in idiopathic Parkinson's disease, these changes during sleep are usually minor, not clinically important, and do not result in upper airway obstruction, obstructive or central sleep apnoea (Apps *et al.*, 1985). Arterial oxygen saturation throughout the night is usually not compromised in uncomplicated idiopathic Parkinson's disease.

There seems little if any indication for sleep laboratory recording in routine clinical practice in Parkinson's disease in the investigation of either insomnia or daytime drowsiness, unless anatomical factors, e.g. mandibular deformity, obesity, together with a history of snoring and observed apnoeas, indicate a possible diagnosis of symptomatic obstructive sleep apnoea. This, when present, is not due to Parkinson's disease.

SLEEP IN PROGRESSIVE SUPRANUCLEAR PALSY

Sleep changes in progressive supranuclear palsy (PSNP) are usually more extensive than in idiopathic Parkinson's disease (Mouret, 1975). The constellation of sleep abnormalities found in PSNP (and also in striatonigral degeneration) is very similar to that of vascular ventral tegmental lesions, and also that characteristic of olivopontocerebellar atrophy. All these conditions affect brain stem structures important for human sleep, with degeneration of pontine and mesencephalic tegmental nuclei, damage to the locus coeruleus and raphé nuclei. At night, NREM sleep is very disorganized, and REM sleep may be absent or markedly reduced (Leygonie *et al.*, 1976; Gross *et al.*, 1978; Laffont *et al.*, 1979; Osorio and Daroff, 1980). Unlike the situation in idiopathic Parkinson's disease, lack of sleep at night in PSNP is not compensated for by sleep during the day, and the system which under normal conditions determines that loss of REM sleep is followed by a rebound of REM sleep may not be functional in PSNP. Surprisingly, chronic loss of REM sleep in PSNP has not been clearly associated with any gross behavioural abnormality, and the level of memory disturbance is not closely related to the degree of loss of REM sleep in either PSNP or striatonigral degeneration (Perret *et al.*, 1979). However, the loss of REM sleep and REMs within this sleep phase is paralleled by the disruption of saccadic eye movements in wakefulness. Osorio and Daroff (1980) have suggested that the same pontine ocular motor neuronal system generates rapid eye movements both sleeping and waking.

A dissociation between the behavioural manifestations and electrographic changes of sleep has been reported in PSNP, but may be more apparent than real. As much as half the time of behavioural sleep may be accompanied by the EEG patterns of wakefulness, and α-frequency range activity sometimes persists throughout sleep, intermixed with sleep spindles of 12–15 Hz (Mouret, 1975). This may reflect the lack of deep sleep, poor light sleep, and difficulty in sleep staging in PSNP, rather than a more fundamental dissociation between the electrical and behavioural aspects of sleep.

MULTISYSTEM ATROPHY

Loss of the autonomic chemoreceptor control of ventilation in addition to laryngeal obstruction may result in obstructive or central sleep apnoea in multisystem atrophy and related degenerative brain stem disorders. Here sleep apnoea may be a presenting feature resulting in daytime drowsiness, particularly in cases with an early autonomic neuropathy and laryngeal palsy. In these instances, sleep apnoea is usually accompanied by rapid changes in blood pressure, but without the pattern of tachycardia/bradycardia associated with other forms of sleep apnoea, in which cardiac vagal innervation remains intact. In these instances, respiratory studies during sleep, oximetry and sleep laboratory investigation, may all be necessary for diagnosis.

POST-ENCEPHALITIC PARKINSONISM

In post-encephalitic parkinsonism, in contrast to the idiopathic disease, respiratory abnormalities, hypoventilation, and central and obstructive apnoeas may all occur during sleep, at the same time as dystonic postures, hiccup, chorea, myoclonus and tics, characteristic of this disorder (Apps *et al.*, 1975; Efthimiou *et al.*, 1987). The recumbent posture of sleep, sleep dystonia and sleep apnoea occasionally result in hypoxic fits during sleep. Otherwise sleep changes in post-encephalitic parkinsonism are similar to those found in the idiopathic disease.

Post-encephalitic parkinsonism was reported in approximately 25% of all subjects with a clinical diagnosis of encephalitis lethargica. In contrast, post-encephalitic narcolepsy–cataplexy was a rarity. Why this difference in disease sequelae? Perhaps the reason was genetic, with the aetiologic agents of encephalitis lethargica precipitating narcolepsy–cataplexy only in a genetically susceptible host. Recent studies demonstrated a positive HLA DR2 status (the HLA antigen associated with idiopathic narcolepsy) in all the six subjects considered to have "post-encephalitic" narcolepsy (Andreas-Zietz et al., 1986), whereas in our experience there is no specific HLA linkage in post-encephalitic parkinsonism.

Von Economo (1930) described three major patterns of sleep disturbance in the acute phase of encephalitis lethargica. These were excessive sleepiness, sleeplessness, and reversal of the sleep rhythm. Lesions causing insomnia were mainly situated in the basal forebrain, while those associated with hypersomnia were situated in the mesencephalic tegmentum and posterior hypothalamus. Von Economo recognized the association between lesions of the grey matter of the midbrain and the caudal parts of the third ventricle and sleep disruption. With the subsequent recognition of an ascending reticular activating system (ARAS) it was considered that excessive sleepiness in encephalitis resulted from bilateral damage to rostral fibres of this system. However, kainic acid lesion experiments resulting in loss of cell bodies, but not fibres, within the ARAS, have not demonstrated behavioural or electrographic sleep. Wakefulness systems may be more rostral than von Economo considered. Somnolence in encephalitis lethargica may have been pharmacological as well as pathological in origin. Thus for example, viral infection may result in the release of immune mediators such as interleukin, a substance which causes drowsiness and enhances slow-wave activity in the sleep EEG (Tobler et al., 1984). A close parallel here is the profound daytime drowsiness, as well as fatigue, of mononucleosis (Guilleminault and Mondini, 1986).

DRUGS AND SLEEP IN PARKINSON'S DISEASE

Dopamine does not appear to have any major effect on behavioural or electrographic aspects of sleep or arousal. The reported effects of L-dopa on sleep in Parkinson's disease, normal volunteers, narcoleptic and depressed subjects are highly variable, but usually minor. In Parkinson's disease, low doses of L-dopa, and to a lesser extent the dopamine-releasing drug amantadine, improves sleep. The sleep disturbance, in parallel with motor disability, is reversed, with longer total sleep duration and occasionally an increase in stage 4 NREM sleep. That the effect of L-dopa on sleep is consequent upon improved motor performance is suggested by a finding that normal or depressed subjects given L-dopa have little or no change in total sleep time (Mendelson, 1987).

The effect of L-dopa on REM sleep is extremely variable. One review described increases in REM sleep in seven subjects, decreases in four, and no consistent change in another four (Mendelson *et al.*, 1977).

It has been suggested that the effect of L-dopa on sleep in parkinsonism is dose dependent, low dosages resulting in sleep improvement, high dosages in deterioration. This result is probably artefactual, improved sleep with low doses in early disease accompanying a general increase in well-being, high doses resulting in toxicity, hallucinosis, involuntary movements and sickness (Klinger, 1982). Whatever the mechanism, high L-dopa dosages sometimes cause total sleep suppression (Wyatt *et al.*, 1970).

REM sleep latency may be slightly altered by L-dopa and other dopamine-active drugs. Occasionally intravenous L-dopa doubles REM sleep latency (Gillin *et al.*, 1973). It is noteworthy that hydroxybutyrolactone, which increases impulse traffic in dopamine neurones, is one of the few known drugs that decreases rather than increases REM sleep latency.

L-Dopa-treated parkinsonians often report unusually vivid and detailed dreams. This may be entirely dependent on poor sleep, easy arousal and good dream recall. However, Lavie *et al.* (1980b) associated this finding with an increase in oculomotor activity within REM sleep. There is no good evidence that the density of REMs within REM sleep results in any major alteration in dream mentation.

LONG-ACTING L-DOPA FORMULATIONS

There have been several attempts to control nocturnal symptoms in parkinsonism by the use of long-acting L-dopa preparations taken in the evening. The bioavailability of these formulations of Madopar and Sinemet is usually a little less than that of the same dose of conventional formulations with slightly lower plasma peak dopa levels, although on a theoretical basis these compounds should prove successful in prolonging L-dopa responsiveness both waking and sleeping. At present there is little definite evidence in favour of this view. L-Dopa plasma pharmacokinetic profiles following oral dosage are similar waking and sleeping, by day and by night (Critchley and Jenner, personal communication), and in our experience sleep laboratory profiles, Parkinson's disease subjective ratings and sleep quality do not greatly differ between conventional and delayed-release L-dopa treatment in bio-equivalent dosages. However, more work on this important topic is needed. A study by Lees (1987) of a sustained-release formulation of L-dopa (Madopar HBS) in the treatment of nocturnal and early morning disabilities in Parkinson's disease indicated that either conventional Madopar or Madopar HBS taken at bedtime significantly reduced the severity of nocturnal and early morning symptoms of Parkinson's disease relative to no treatment values, although total severity scores for nocturnal and early morning symptoms did not differ between the two treatments.

DOPAMINE AGONISTS AND RELATED DRUGS

The MAO-B inhibitor selegiline is interconverted to amphetamine and methamphetamine, and would be expected to cause insomnia if taken in the late evening. Selegiline in high dosages, 20–30 mg per 24 h, causes marked arousal in the treatment of narcolepsy, but in the low dosages employed in parkinsonism, 5 mg, taken in the morning, does not cause insomnia (Lavie *et al.*, 1980a).

Apomorphine s.c. or i.v., in contrast to oral L-dopa, causes considerable sedation in both healthy volunteers and drug-naive parkinsonians. However, tolerance develops to this sedative effect, and apomorphine does not cause marked drowsiness in parkinsonian patients previously treated for a long period with L-dopa (Bassi *et al.*, 1979).

The ergot derivatives *bromocriptine, lisuride* and *pergolide,* by mouth, or in the case of lisuride, s.c., can all cause considerable daytime drowsiness, although in our experience tolerance rapidly develops. Clinical use indicates that all these drugs have a greater initial sedative effect than comparable doses of L-dopa in the management of parkinsonism, but controlled studies are lacking. The effect of other ergot derivatives on sleep and wakefulness shows considerable variability. For example, lysergic acid monoethylamide has a considerable sedative effect, lysergic acid diethylamide is hallucinatory but not sedative, whilst systemic administration of ergotamine and other ergopeptine derivatives leads to behavioural stimulation, not sleep, in most animal species. Meldrum and Naquet (1970) showed that in the baboon, methysergide and methylergometrine caused drowsiness and muscle atonia, whereas ergotamine, lysergic acid and dihydroergotoxine were without effect.

The sedative effect of ergot derivatives depends on the route of administration as well as the compound. In addition, results are different in different species. With all these variables, the pharmacological mechanism of sleep induction by bromocriptine, lisuride and pergolide in Parkinson's disease remains uncertain. Unfortunately these observations tell us little about the basic neurochemistry of sleep. Perhaps the occasional marked effect is due to combined dopamine and serotonin actions. As compared with barbiturates and benzodiazepines, the 5-HT precursor 1-tryptophan has at best only minor hypnotic effects.

The *anticholinergic* drugs used in parkinsonism have on the whole minor sedative effects. Atropine and scopolamine suppress REM sleep, whilst REM sleep is increased by anticholinesterases. It seems unlikely that anticholinergic drug treatment of parkinsonism has any major effect on behavioural aspects of sleep, but we still sometimes give small evening doses of benzhexol to young parkinsonians who complain of insomnia and who do not have any features of mental impairment. There is no good scientific evidence to favour this practice, and in elderly or demented parkinsonian subjects anticholinergic drugs are best avoided.

TREATMENT OF SLEEP PROBLEMS IN PARKINSONISM

(1) Commode by the bed.

(2) Nylon sheets are preferable to cotton, and help turning in bed (is this advice, which is often quoted, correct?).

(3) Patients with Parkinson's disease best sleep alone. They make impossible bedmates, and their partners in any case need a good night's sleep to help cope during the day.

(4) Evening antiparkinsonian drugs may help sleep, but dosage and timing are critical.

    (a) A low evening dose of L-dopa may reduce sleep latency and reduce awakenings. High dosages may cause insomnia.

    (b) Consider the use of a small evening dose of bromocriptine, pergolide or long-acting L-dopa preparation. There may be a role for a 2 a.m. L-dopa dose.

    (c) Sleep onset is difficult: consider short-acting benzodiazepine, e.g. triazolam. With frequent early morning arousals, use medium duration benzodiazepine, e.g. temazepam.

    (d) Avoid evening amantadine or selegiline.

(5) In late disease, with marked sleep fragmentation, daytime drowsiness and evening sub-alertness, disorientation, aggressive behaviour, hallucinosis, consider limiting antiparkinsonian drug treatment to 6 a.m.–2 p.m.

REFERENCES

Andreas-Zietz, A., Scholz, S., Roth, B., Sonka, K., Naskova, E., Schulz, H., Keller, E., Albert, E. D., Nevsimalova, S., Dolekal, P. and Geisler, P. (1986) *Lancet* **ii**, 684–685.

Apps, M. C., Sheaff, P. C., Ingram, D. A., Kennard, C. and Empey, D. W. (1985) *J. Neurol. Neurosurg. Psychiat.* **48**, 1240–1245.

Bassi, S., Albizzati, M. G., Frattola, L., Passerini, D. and Trabucchi, M. (1979) *J. Neurol. Neurosurg. Psychiat.* **42**, 458–460.

Bertolucci, P. H., Andrade, L. A., Lima, J. G. and Carlini, E. A. (1987) *Arq.*

Neuropsiquitr. **45**, 224–230.

Clark, E. C. and Feinstein, B. (1977) *Adv. Expl. Biol. Med.* **90**, 175–182.

Efthimiou, J., Ellis, S. J., Hardie, R. J. and Stern, G. M. (1987) *Adv. Neurol.* **45**, 275–276.

Friedman, A. (1980) *Acta Med. Pol.* **21**, 193–199.

Gillin, J. C., Post, R. M., Wyatt, R. J., Goodwin, F. K., Snyder, F. and Bunney, W. E. (1973) *Electroenceph. Clin. Neurophysiol.* **35**, 181–186.

Gross, R. A., Spehlmann, R. and Daniels, J. C. (1978) *Electroenceph. Clin. Neurophysiol.* **45**, 16–25.

Guilleminault, C. and Mondini, S. (1986) *Arch. Int. Med.* **146**, 1333–1335.

Klawans, H., Goetz, C. and Bergen, D. (1975) *Arch. Neurol.* **32**, 331–334.

Klinger, M. (1982) *Int. J. Clin. Pharmacol. Ther. Toxicol.* **20**, 190–193.

Laffont, F., Autret, A., Minz, M., Beillevaire, T., Gilbert, A., Cathala, H. P. and Castaigne, P. (1979) *Rev. Neurol.* **135**, 127–142.

Laihinen, A., Alihanka, J., Raitasuo, S. and Rinne, U. K. (1987) *Acta Neurol. Scand.* **76**, 64–68.

Lavie, P., Bental, E., Goshen, H. and Sharf, B. (1980a) *J. Neural. Transm.* **47**, 61–67.

Lavie, P., Wajsbort, J. and Youdim, M. B. (1980b) *Commun. Psychopharmacol.* **4**, 303–307.

Lees, A. J. (1987) *Eur. Neurol.* **27** (Suppl. 1) 126–134.

Lees, A. (1988) Poster presented at the IX International Symposium on Parkinson's Disease, 5–9 June 1988, Jerusalem, Israel.

Leygonie, F., Thomas, J., Degos, J. D., Bouchareine, A. and Barbizet, J. (1976) *Rev. Neurol.* **132**, 125–136.

Meldrum, B. S. and Naquet, R. (1970) *Br. J. Pharmacol.* **40**, 144P–145P.

Mendelson, W. B. (1987) *Human Sleep. Research and Clinical Care*, p. 47. Plenum Press, New York.

Mendelson, W. B., Gillin, J. C. and Wyatt, R. J. (1977) *Human Sleep and its Disorders*, pp. 147–210. Plenum Press, New York.

Mouret, J. (1975) *Electroenceph. Clin. Neurophysiol.* **38**, 563–567.

Myslobodsky, M., Mintz, M., Ben-Mayor, V. and Radwan, H. (1982) *Electroencephalogr. Clin. Neurophysiol.* **54**, 227–231.

Nausieda, P. A., Glantz, R., Weber, S., Baum, R. and Klawans, H. L. (1984) In *Adv. Neurol.* **40**, 271–277.

Osorio, I. and Daroff, R. B. (1980) *Ann. Neurol.* **7**, 277–280.

Perret, J. L., Tapissier, J. and Jouvet, M. (1979) *Electroenceph. Clin. Neurophysiol.* **47**, 499–502.

Tobler, I., Borbely, A. A., Schwyzer, M. and Fontana, A. (1984) *Eur. J. Pharmacol.* **10**, 191–192.

Trulson, M. E., Preussler, D. W. and Howell, G. A. (1981) *Neurosci. Lett.* **26**, 183–188.

von Economo, C. (1930) *J. Nerv. Ment. Dis.* **71**, 249–259.

Wein, A., Golubev, V. and Yakhno, N. (1979) *Waking–Sleeping* **3**, 31–40.

Wyatt, R. J., Chase, T. N., Scott, J., Snyder, F. and Engelman, K. (1970) *Nature* **228**, 999–1001.

9

Different Therapeutic Approaches to Complicated Parkinson's Disease

Fabrizio Stocchi, Stefano Ruggieri, Angelico Carta, Peter Jenner and Alessandro Agnoli

INTRODUCTION

The fact that sooner or later almost 100% of parkinsonian patients treated with L-dopa experience fluctuations in motor performance and/or abnormal involuntary movements (Fahn, 1983) calls for departure from the classical therapeutic approach to Parkinson's disease to find new drugs and strategies to improve these disabling symptoms.

It is becoming evident that many of these fluctuations are determined by the rate of L-dopa delivery to the brain, and consequently by L-dopa plasma concentrations (Nutt et al., 1986). It is very difficult to ensure the presence of adequate plasma levels of L-dopa throughout the day using standard L-dopa formulations. The excessive fractionating of therapy, in fact, often itself produces unpredictable "off" periods, so that sometimes it is more convenient for the patients to resume a thrice-daily drug regime and cram all their daily tasks into periods of predictable mobility.

Therefore, drugs or strategies able to give a sufficient and constant stimulation of dopaminergic receptors at the striatal level, or alternatively

DISORDERS OF MOVEMENT: CLINICAL, PHARMACOLOGICAL
AND PHYSIOLOGICAL ASPECTS   ISBN 0-12-569685-X

to rapidly reverse "off" periods, can be very useful in the treatment of brittle parkinsonian patients.

LISURIDE SUBCUTANEOUS INFUSION

Many authors have demonstrated that it is possible to give fluctuating parkinsonian patients constant periods of good mobility by means of a continuous intravenous infusion of L-dopa (Marion *et al.*, 1986a,b; Stocchi *et al.*, 1986). Unfortunately, this treatment is not suitable for chronic therapy because of the low solubility and high acidity of L-dopa itself.

Lisuride is a potent DA agonist acting mainly on postsynaptic D2 receptors; it is highly water soluble and does not produce severe local reactions when infused intravenously or injected subcutaneously (Obeso *et al.*, 1986; Ruggieri *et al.*, 1986).

Obeso *et al.* (1983) first reported a stable improvement in mobility in three brittle parkinsonians given a continuous infusion of 0.5–2 mg of lisuride over 8 h in addition to conventional oral L-dopa treatment. In our own experience, we have given 20 patients an intravenous lisuride infusion during three consecutive days without any other antiparkinsonian medicament. We found that lisuride alone was able to ensure constant "on" periods during the infusion in almost 35% of the patients, and to produce at least some time "on" in a further 35% of patients. However, six patients out of 20 (30%) did not respond to the drug (Stocchi *et al.*, 1986; Agnoli *et al.*, 1987).

Considering the high solubility of the drug and the encouraging results obtained from the intravenous study, we then administered lisuride subcutaneously to a group of 24 fluctuating parkinsonian patients. All these patients were selected because of their severe motor fluctuations or disabling dyskinesias. Patients older than 65 and/or having a history of psychiatric disturbance were excluded from this study. The drug was given subcutaneously using a programmable micropump, usually for 12 h during the active day, alternating with drug-free nights. However, four patients out of 24 received a 24-h round-the-clock infusion because of severe nocturnal akinesia and/or a long delay to turn on in the morning. Two of these four withdrew from the study because of psychiatric side-effects, and the remaining two were able to turn off the pump during the night after 6 months of treatment.

All 24 patients had a striking improvement of their motor oscillations during the infusion treatment. Ten of them were treated with lisuride alone, whilst the remaining 14 required the addition of small amounts of oral L-dopa. These patients have been on lisuride treatment for 1–31 months and during this period nine of them have had to abandon it. The reasons for stopping lisuride were as follows: psychiatric side-effects (*n* = 3) after 1, 7 and 12 months, respectively (the last two were on 24-h infusion regimes); concurrent myocardial ischaemia ($n = 1$); intractable vomiting ($n = 1$) or diarrhoea ($n = 1$); the remaining three withdrew because they felt uneasy about being dependent on a subcutaneous pump system.

On the other hand, considering the 15 patients who are still experiencing beneficial effects from the subcutaneous treatment, seven are receiving lisuride alone (mean dosage $65\,\text{ng}\,\text{h}^{-1}$), while eight are on lisuride (mean dosage $80\,\text{ng}\,\text{h}^{-1}$) plus oral L-dopa tablets (mean dosage $384\,\text{mg}\,\text{day}^{-1}$ plus peripheral decarboxylase inhibitor – PDI). In the group treated with lisuride alone a significant reduction in hours spent "off" (0.5 vs 3.8) was observed as well as in the group treated with lisuride plus oral L-dopa (0.7 vs 4.8).

The clinical improvement of these patients is still unchanged both in terms of Parkinson's severity score and in terms of hours spent in "off" during the infusion time. After this consistent period of subcutaneous infusion treatment (most of these patients have been treated for more than 12 months) it is possible to draw the following conclusions.

(1) There are no discernible clinical predictors which differentiate between patients requiring additional oral L-dopa and patients who do well with lisuride alone.

(2) Clinical improvement and dosage required (both of lisuride and L-dopa) remained practically unchanged.

(3) Psychiatric side-effects occur mainly in patients receiving a continuous 24-h regime.

(4) Not all patients can tolerate this treatment; in purely mechanical terms, the technique can be difficult for the patient to handle, and psychologically not easy to accept.

APOMORPHINE SUBCUTANEOUS INFUSION

Stibe *et al.* (1988) have recently reported that it is possible to control fluctuations in parkinsonian patients by giving a continuous subcutaneous infusion of apomorphine combined with oral L-dopa. These authors reported that in their experience apomorphine was more effective than lisuride and did not induce psychiatric side-effects.

We therefore gave subcutaneous apomorphine infusions to six patients for 1 week. Two of these six were previously full responders to lisuride, and these responded equally well to apomorphine. In contrast, of the other four patients, three had only a partial response to lisuride, and one had failed to respond at all. All four turned on with apomorphine, but the drug was not able to give a constant and predictable response despite the addition of L-dopa tablets. Side-effects (somnolence and hypotension) were mild. However, our results are not extensive enough to establish any clear difference between the effects of lisuride and apomorphine; controlled trials need to be performed to assess the relative merits of the two drugs given by subcutaneous infusion.

L-DOPA METHYLESTER ADMINISTRATION

L-Dopa still represents the most effective drug in the treatment of Parkinson's disease. All patients with idiopathic Parkinson's disease satisfactorily respond to this drug and this response to L-dopa is rarely, if ever, completely lost.

The esterification of the carboxylic acid moiety of the L-dopa molecule produces a highly soluble neutral prodrug. L-Dopa methylester (ME) has been shown in animal studies to be equipotent to L-dopa itself in reversing reserpine-induced akinesia in mice. The compound is metabolized *in vivo* to L-dopa and methanol (Cooper *et al.*, 1984).

At least a sizeable contribution to fluctuations in motor performance experienced by Parkinson's disease patients receiving chronic oral L-dopa therapy is due to the peculiar pharmacokinetics of the drug. Thus, L-dopa has to transit from the stomach to the small intestine, cross the gut barrier to the entero-hepatic circulation and thence to the systemic circulation and then gain access to the brain through the blood–brain barrier. Many of these factors can be circumvented by giving L-dopa intravenously, but this cannot be done as a chronic treatment because of the large amount of solution which has to be administered (Juncos *et al.*, 1987a,b). Therefore, considering the superior solubility of the ME prodrug, we infused L-dopa ME ($250 \, \text{mg ml}^{-1}$; pH 6.3) intravenously in three severely fluctuating parkinsonian patients for three consecutive days, comparing clinical effect and blood levels to those obtained after intravenous standard L-dopa previously administered to the same patients on different days. Oral PDI pretreatment was given with each preparation.

The clinical effects of the two drugs were similar. The patients remained mobile during each single infusion day, and only one of them required an increase of dosage late in the afternoon, with both drugs. However, the amount of ME required to keep the patients continuously "on" was three times higher when compared to L-dopa alone. The two drugs showed the same pharmacokinetic patterns even if the area under the curve (AUC) of ME was slightly larger. The increase of the AUC of 3-*O*-methyldopa detected during ME infusion was more evident when compared to that obtained during standard L-dopa infusion.

Unfortunately, because of the higher dosage required and local reaction observed at the site of injections, even ME continuous infusion cannot be a practical choice in the current management of fluctuating parkinsonian patients. However, considering the good solubility of the drug (250 mg ml^{-1}), we then went on to give six fluctuating parkinsonian patients 200 mg of ME solution as an oral bolus, comparing in a double-blind cross-over study clinical results and plasma levels with those obtained with a solution containing an equal weight (but not equal molarity) of standard L-dopa. These patients received on two different days 40 ml of water containing 200 mg of ME (Chiesi Farmaceutici) or conventional L-dopa. One hour before the beginning of the study, the patients received an oral dose of 25 mg of carbidopa, having discontinued all other antiparkinsonian medication 12 h previously. The clinical evaluation was performed every 10 min by means of King's College Hospital Rating Scale and "on–off" charts; at the same time blood samples were taken for the assessment of dopa and 3-O-methyldopa blood levels.

The results showed an identical degree of clinical improvement for the two drugs, but the delay to turn "on" was shorter with the ME bolus than after the L-dopa one (Table 9.1). The overall evaluation of the corresponding blood levels fits well with the clinical data. In fact, the T_{max} was 20 min for ME and 30 min for conventional L-dopa even if the AUC for ME appeared slightly lower than that for L-dopa (Fig. 9.1).

Encouraged by these interesting data, we then tested six other fluctuating parkinsonians, giving them 200 mg of ME sublingually (again with PDI pretreatment), monitoring clinical effects and related L-dopa blood levels. The drug was given in the form of a 1 ml solution, asking the patients to hold it in their mouth for at least 20 min. Using this route of administration, the patients turned "on" after a mean of 40 min, while the clinical benefit lasted for a mean of 175 min. The assessment of L-dopa blood levels showed a T_{max} of 40 min; the mean curve presented a square shape without a real peak. The AUC was 10.53, very similar to that obtained after the oral bolus (Fig. 9.2).

Table 9.1 Effect of L-dopa and L-dopa ME oral bolus (200 mg).

Patient	L-Dopa		L-Dopa ME	
	Time to turn on	Time on	Time to turn on	Time on
G.A.	50	150	30	100
J.O.	20	140	30	70
M.I.	40	150	15	170
C.O.	50	190	30	210
R.A.	30	120	35	145
D.F.	35	150	20	120
Mean	37.5	150	26.7	145.8
SD	11.7	22.8	7.53	47.5

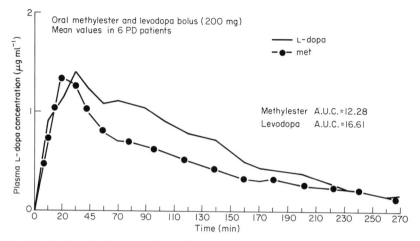

Fig. 9.1   Oral methylester vs L-dopa bolus (200 mg), both given one hour after oral PDI. Mean values of six Parkinson's disease patients. (—) L-dopa, AUC = 16.61; (●) methyl ester, AUC = 12.28.

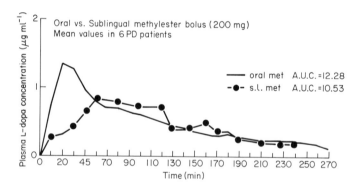

Fig. 9.2   Oral vs sublingual methylester bolus (200 mg), both given one hour after oral PDI. Mean values of six Parkinson's disease patients. (—) Oral, AUC = 12.28; (●) sublingual, AUC = 10.53.

These preliminary results show that the absorbtion of the methyl ester is faster than the solution of standard L-dopa and therefore potentially useful in the management of patients with prolonged drug-resistant "off" periods. Moreover the drug can be given sublingually with clear advantages for patients with swallowing problems.

SLOW-RELEASE PREPARATIONS

The recognition of the importance of steady-state plasma and brain L-dopa concentrations for the smooth control of parkinsonian disability has brought a revival of interest in sustained or slow-release formulations of L-dopa. Madopar hydrodynamically balanced system (HBS) is a galenic formulation of L-dopa plus benserazide with a slow gastric dissolution. In a pharmaco-kinetic study HBS showed more sustained concentrations with a smoother peak than standard Madopar although (because of reduced bioavailability) the dosage required was higher for the HBS (Marion et al., 1986a,b).

We gave Madopar HBS to a group of 10 fluctuating parkinsonian patients for a period of 6 weeks comparing the results with their previous conventional Madopar therapy. The substitution for Madopar (631 mg day^{-1}) of the HBS formulation (1050 mg day^{-1}) led to an improvement in off period severity and nocturnal akinesia, but the mean time spent in the off condition by the patients increased from 267.5 to 351 min over a 12 h day. This increase in off period duration was due to a longer latency of clinical response with HBS. Our attempts to correct this resulted in increasing HBS dosage leading to a worsening of dyskinesias and to the appearance of more unpredictable off periods. This evidence suggested we should keep the dosage of HBS unchanged and, instead, add some doses of standard Madopar. When we applied this strategy in the second part of the study, there was an improvement of wearing off with a consequent significant decrease of the time spent in off periods (141 min vs 264 at baseline). Off period severity and nocturnal akinesia also improved. The mean dosage of HBS required to achieve these clinical results was 1050 mg day^{-1} with the addition of 240 mg of standard Madopar. In contrast to these beneficial effects, diphasic dyskinesias worsened during HBS treatment and some unpredictable off periods also appeared.

Another interesting compound is the Merck CR 4, a slow dissolving matrix tablet of L-dopa/carbidopa (50 mg/200 mg) which has been found to be the most successful among a series of different CR preparations (I to V) (Goetz et al., 1987; Juncos et al., 1987a,b).

Recently, Sage and Mark (1988) compared CR 4 with the standard Sinemet formulation (25/100) in a double-blind cross-over study. The results of this study showed a significant amelioration in terms of "on" hours after CR 4 administration. The use of CR 4 also allowed most of the patients to reduce the number of daily doses. Even with this formulation a major problem was the long delay of the beginning of clinical effect of each single dose and a worsening, in some cases dramatic, of abnormal involuntary movements.

CONCLUSION

Continuous dopaminergic stimulation represents the best current strategy to control fluctuations in Parkinson's disease. This goal can be achieved by means of subcutaneous infusions of DA agonists, but this technique is not easy to handle and cannot be tolerated by all patients.

Slow-release oral preparations of L-dopa are of considerable theoretical interest, but the results obtained with the formulations available to date do not show a striking advantage in comparison with the standard formulations.

L-Dopa methylester is a promising prodrug because of its high solubility. It can be conveniently administered to patients with complicated fluctuations in order to rapidly reverse off periods. Furthermore, L-dopa ME may prove useful in patients with swallowing problems or in association with other slow release oral and/or subcutaneous treatments.

ACKNOWLEDGEMENTS

Thanks go to Prof. C. David Marsden for giving the opportunity to carry out the oral methyl ester work in his Department, and for reviewing the protocol and the data.

We gratefully acknowledge the kind gift of intravenous L-dopa and methyl ester solutions from Chiesi Farmaceutici Parma; of oral Carbidopa from Merck Sharp and Dohme Roma; of Madopar HBS from Roche Milano; and of intravenous and subcutaneous Lisuride from Schering A. G. Berlin.

REFERENCES

Agnoli, A., Stocchi, F., Carta, A., Antonini, A., Bragoni, M. and Ruggieri, S. A. (1988) *Mount Sinai J. Med.* **55**, 62–66.

Cooper, D. R., Marrel, C., Testa, B., Van de Waterbeemd, H., Quinn, N., Jenner, P. and Marsden, C. D. (1984) *Clin. Neuropharmacol.* **7**, 89–98.

Fahn, S. (1983) In *Movement Disorders* (edited by C. D. Marsden and S. Fahn), pp. 123–145. Butterworths, London.

Goetz, C. G., Tanner, C. M., Klawans, H. L., Shannon, K. M. and Carroll, V. S. (1987) *Neurology* **37**, 875–878.

Juncos, J. L., Fabbrini, G., Mouradian, M. M. and Chase, T. N. (1987a) *Arch. Neurol.* **44**, 1010–1012.

Juncos, J. L., Mouradian, M. M., Fabbrini, G., Serrati, C. and Chase, T. N. (1987b) *Neurology* **37**, 1242–1245.

Marion, M. H., Stocchi, F., Quinn, N. P., Jenner, P. and Marsden, C. D. (1986a) *Clin. Neuropharmacol.* **9**, 165–181.

Marion, M. H., Stocchi, F., Quinn, N. P., Jenner, P. and Marsden, C. D. (1986b) *Adv. Neurol.* **45**, 493–496.

Nutt, J. G., Woodward, W. R., Hammerstad, J. P., Carter, J. H. and Anderson, J. L. (1986) *N. Engl. J. Med.* **310**, 483–488.

Obeso, J. A., Luquin, M. R. and Martinez Lage, J. M. (1983) *Ann. Neurol.* **14**, 134.

Obeso, J. A., Luquin, M. R. and Martinez-Lage, J. M. (1986) *Lancet* i, 467–470.

Ruggieri, S., Stocchi, F. and Agnoli, A. (1986) *Lancet* **ii**, 860.
Sage, J. I. and Mark, M. H. (1988) *Clin. Neuropharmacol.* **11**, 174–179.
Stibe, C. M. H., Lees, A. J., Kempster, P. A. and Stern, G. M. (1988) *Lancet* **i**, 403–406.
Stocchi, F., Ruggieri, S., Brughitta, G. and Agnoli, A. (1986) *J. Neural. Transm.* suppl. **22**, 223–229.

10

The Antiparkinsonian Effects of Transdermal +PHNO

R. J. Coleman, J. A. Temlett, M. Nomoto, N. P. Quinn,
P. Jenner and S. M. Stahl

INTRODUCTION

For many patients with Parkinson's disease, the development of motor
fluctuations blights their otherwise good response to L-dopa. This "on–off"
syndrome consists of a variable and unpredictable response to regular oral
L-dopa and is now recognized as a major source of disability in patients
with Parkinson's disease on chronic L-dopa therapy. Such patients develop
a tendency to switch suddenly between "off" and "on" states in a way that
cannot readily be predicted from the timing of individual L-dopa doses or
from the levels of L-dopa in the plasma (Fahn, 1974; Marsden and Parkes,
1976).

   It is known that an intravenous infusion of L-dopa can significantly
improve these fluctuations by keeping plasma concentrations of L-dopa
constant. In attempting to adapt this model of continuous dopaminergic
stimulation for practical use, various trials have been carried out using
sustained-release L-dopa tablets and subcutaneous infusions of the agonists
lisuride and apomorphine (Juncos et al., 1987; Stibe et al., 1987).

(+)-4-Propyl-9-hydroxynaphthoxazine (+ PHNO) is a novel drug with the properties of a highly potent dopamine agonist (Martin *et al.*, 1984). It reverses the parkinsonian deficits produced by 1-methyl-4-phenyl-1,2,3,6-tetrahydropyridine (MPTP) in marmosets and is effective in patients with Parkinson's disease when given as a single oral dose (Stoessl *et al.*, 1985; Grandas-Perez *et al.*, 1986; Nomoto *et al.*, 1987). Although the effect of a single dose of + PHNO is brief, it would appear to be a good candidate for use in the control of "on–off" fluctuations if it could be administered in a continuous manner.

The behavioural changes induced in rodents by oral or intraperitoneal + PHNO have also been seen when a solution of + PHNO is applied to the skin (Koller *et al.*, 1987). This implies transdermal absorption of + PHNO, which was first shown to occur in man when pharmaceutical workers developed vomiting after handling the compound. + PHNO is soluble in both aqueous and lipid media and this is an important physical property of drugs that are capable of transdermal absorption (Wepierre and Marty, 1979). This, therefore, would be an ideal method for administration of + PHNO in view of its high potency and significant first pass metabolism, in addition to the desire for continuous delivery. The technology required to design and manufacture skin patches already exists and has been applied to nitroglycerin for angina, clonidine for hypertension, scopolamine for motion sickness and oestrogens for post-menopausal hormone replacement therapy (Shaw and Urquhart, 1981).

We have therefore studied transdermal absorption of + PHNO in marmosets with MPTP-induced parkinsonism and in patients with Parkinson's disease who have developed "on–off" fluctuations.

ANIMAL STUDIES

Topical application of + PHNO solution

A group of marmosets were rendered parkinsonian after treatment with MPTP (Jenner *et al.*, 1984). These animals were then treated by painting their abdominal skin with an alcoholic solution of + PHNO which quickly dried. Three different-sized areas of skin were treated (0.5 cm^2, 1.23 cm^2 and 2.97 cm^2) and three doses of + PHNO were applied to each (25 μg, 50 μg and 100 μg). This created nine experimental groups with three animals in each group. Direct observation of each animal determined whether the + PHNO had or had not made the movements qualitatively normal; Table 10.1 shows how many of the three animals in each group responded to + PHNO in this way. Using the smallest area (0.5 cm^2), no animal responded

Table 10.1   Results of alcoholic solution of +PHNO applied to the skin of MPTP-treated marmosets. The numbers show how many animals had a detectable response ($n = 3$ in each treatment group).

Area of skin covered (cm^2)	Quantity of +PHNO (μg)		
	25	50	100
0.5	0	0	0
1.23	0	2	3
2.97	1	3	3

to +PHNO in doses up to $100\,\mu g$; with the next area ($1.23\,cm^2$), no animals responded to $25\,\mu g$, two animals responded to $50\,\mu g$ and all three animals responded to $100\,\mu g$; and using the largest area ($2.97\,cm^2$), one animal responded to $25\,\mu g$ and all three responded to both 50 and $100\,\mu g$ doses.

In addition to this qualitative assessment, the animals were assessed by placing them in a test cage fitted with infrared light beams which counted their movements. For each area used, larger quantities of drug gave rise to higher movement counts (Fig. 10.1). In addition it can be seen that the effect of the largest dose is different when applied over $1.23\,cm^2$ and $2.97\,cm^2$. The larger area produces a large, early, brief effect (maximum 910 counts h^{-1} at 1–2 h) whereas the smaller area produces a smaller, more prolonged effect (maximum 230 counts h^{-1} at 2 days) (Fig. 10.1).

The earliest effects of +PHNO were seen during the first hour after

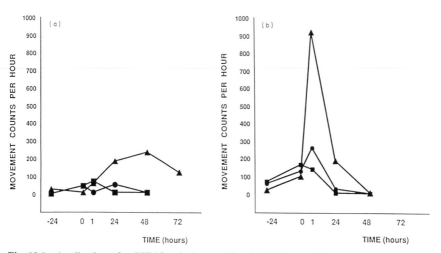

Fig. 10.1   Application of +PHNO solution to skin of MPTP-treated marmosets. (a) Area of skin covered = $1.23\,cm^2$; (b) area of skin covered = $2.97\,cm^2$. Quantity of +PHNO: (—●—) $25\,\mu g$; (—■—) $50\,\mu g$; (—▲—) $100\,\mu g$.

application and in some cases they continued throughout the 3 days of the study. The exact length of action of each dose was not determined but, in general, smaller skin areas produced more prolonged responses so long as an effective quantity of drug had been used.

These data illustrate an important principle of transdermal drug absorption: the area of skin controls the size of the response and the quantity of drug applied to that area controls the duration of response.

Application of skin patches

Four marmosets with MPTP-induced parkinsonism were treated with circular patches $1.27\,cm^2$ in area, which were capable of delivering $+$PHNO at a rate substantially greater than the rate of skin penetration. Hair covering the chest wall was clipped (not shaved) and a patch was applied. It was kept firmly in place by a strip of Elastoplast which completely encircled the animal's thorax. After 5 days the patch was removed. Behavioural testing was again carried out by placing each animal in a test cage equipped with infrared light beams to measure movement activity.

All four marmosets responded to $+$PHNO skin patches by showing increased movement counts between 4 and 7 h after application. In one animal the effect only lasted for this brief period. In the second animal the effect gradually wore off over a period of 48 h. In animals 3 and 4 the increase in movement was greater, reaching a peak on day 2 and still detectable on day 4 (Fig. 10.2). Technical difficulties in maintaining good skin contact appeared to account for the differences in the degree and duration of response between monkeys. This was supported by subsequent analysis of patches for residual $+$PHNO which indicated that the two animals with the more dramatic response had received considerably more $+$PHNO than the two animals with the less good response (see legend to Fig. 10.2).

These results confirm the antiparkinsonian actions of $+$PHNO and demonstrate that dermal application of $+$PHNO using skin patches can produce effects in the central nervous system.

HUMAN STUDIES

$+$PHNO patches (of identical composition to those used in the animal studies) have been tested in four patients with Parkinson's disease who exhibited "on–off" fluctuations in their response to regular oral L-dopa. These results have been reported in greater detail elsewhere (Coleman *et al.*, 1989a). Two sizes of patch were available ($2\,cm^2$ and $5\,cm^2$) and these were

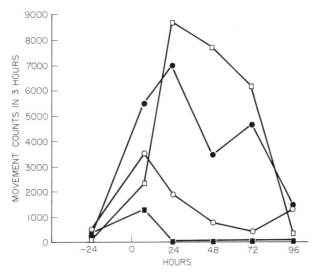

Fig. 10.2   The response of four MPTP-treated marmosets to + PHNO skin patches. The total dose of + PHNO delivered over 5 days (calculated by subsequent patch analysis) was as follows: animal 1 (—■—) 33 μg; animal 2 (—○—) 23 μg; animal 3 (—□—) 133 μg; animal 4 (—●—) 185 μg.

applied in various combinations to the anterior chest wall so that larger areas of skin were covered on each occasion. Patches were applied on alternate days so that there was at least a 24-h wash-out period between doses.

On each test day the patients took no antiparkinsonian medication after midnight so that they would expect to be "off" on waking the next morning. If no effect from + PHNO was detected (i.e. if the patient did not switch "on") then a single dose of L-dopa was given and repeated if necessary every 4 h. The patients were monitored by bedside observation which allowed their treatment response to be recorded using an "on–off" chart (Coleman *et al.*, 1989b).

The first two patients had patches applied in the early morning and these remained in place for up to 8 h. In neither case was any clear effect seen until the maximum area of 20 cm² was used. With this dose an effect was found which started between 4 and 6 h after patch application, and this is illustrated by the relevant "on–off" charts (Fig. 10.3). Without + PHNO, a standard dose of L-dopa (one and a half tablets of SINEMET 110) produced a response (i.e. the patient turned "on") which lasted for between 35 and 160 min. After application of 20 cm² + PHNO skin patches, an initial dose of L-dopa gave rise to the normal duration of "on" response. Patient 1 then switched "on" again without further L-dopa and after a number of "on–off" fluctuations stayed "on" from 6 h onwards. Patient 2 required a second dose of L-dopa,

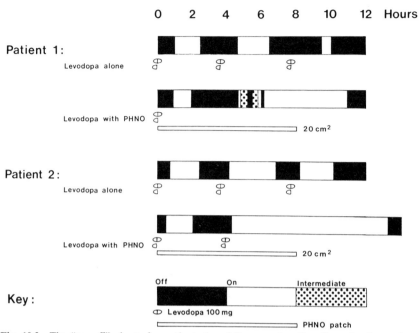

Fig. 10.3 The "on–off" charts for patients 1 and 2 comparing the effect of L-dopa alone (normal response) and L-dopa plus +PHNO (20 cm² patch). Hour 0 was approximately 8 a.m. (Reproduced from Coleman *et al.*, 1989a.)

but its effect was prolonged far beyond what would normally have been expected. After the patches were removed, the response continued in both patients without further L-dopa (in patient 1 for a further 3 h and in patient 2 for just over 5 h).

The +PHNO plasma concentrations from these two doses (shown in Fig. 10.4) correspond with the clinical effect; the levels rose 3–4 h after application and in patient 2 they continued to rise for 4 h after removal. The delay between patch application and rising plasma levels represents the time taken for +PHNO to permeate the epidermis and reach the dermal capillaries. The skin and subcutaneous tissues appear to act as an additional "drug reservoir" between the patch and the systemic circulation so that the patch must saturate this compartment before the drug can begin to enter the blood stream. +PHNO retained in the skin continues to enter the circulation after the patches have been removed.

Because +PHNO levels in patients 1 and 2 were still rising 8 h after patch application, it was decided to monitor the effects of all patch sizes over a period of 24 h in patients 3 and 4. These patients had patches applied at night and their response was assessed during the following day. None of the

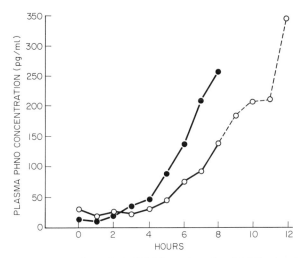

Fig. 10.4   Plasma concentrations of +PHNO in patients 1 and 2 following the application of 20 cm² patches applied at hour 0 and removed at hour 8. (—●—) Patient 1; (—○—) patient 2; (‑‑○‑‑) patient 2 after patch removal. (Reproduced from Coleman *et al.*, 1989a.)

skin patches (from 4 cm² to 20 cm²) turned the patients "on" when they awoke, so they were given L-dopa. The patches of 4 cm² size had little effect, but larger areas gave rise to progressively longer "on" periods in response to regular L-dopa doses. The plasma drug concentrations reached a steady state between 18 and 24 h after patch application (Fig. 10.5). The final +PHNO levels were governed by the size of patch used and so, given a constant rate of drug delivery, the area of skin covered represents the dose of +PHNO administered.

CONCLUSIONS

+PHNO is a potent dopamine agonist. Its antiparkinsonian actions have been demonstrated before and are confirmed by these studies. The ability of +PHNO to pass across intact animal and human skin provides an excellent means of continuous drug delivery, avoiding the hazards of unpredictable gastrointestinal absorption and first pass metabolism. The effect resembles a continuous subcutaneous infusion but without the need for needles or pumps.

In previous studies it has been shown that +PHNO can turn patients "on" when given without concurrent L-dopa. In these transdermal studies we have shown that the two drugs may be successfully used in combination because +PHNO lengthens the duration of action of individual L-dopa doses.

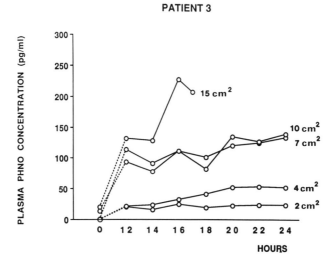

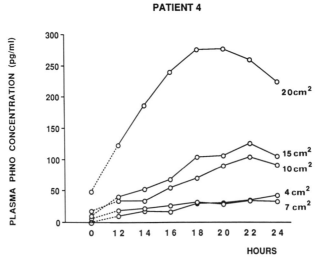

Fig. 10.5 Plasma concentrations of +PHNO achieved by increasing patch area in patients 3 and 4. (Reproduced from Coleman *et al.*, 1989a.)

ACKNOWLEDGEMENTS

JAT was a Medical Research Council (RSA) post-doctoral fellow.

We gratefully thank Drs A. E. Loper, J. V. Bondi and M. Hichens (of Merck, Sharp and Dohme Research Laboratories, West Point, PA, USA)

for their help in supplying the patches and carrying out the +PHNO assays.
Figures 3, 4 and 5 have been reproduced by kind permission of Raven
Press Ltd, the publishers of *Movement Disorders*.

REFERENCES

Coleman, R. J., Lange, K. W., Quinn, N. P. *et al.* (1989a) *Movement Disorders* **4**,
   129–138.
Coleman, R. J., Temlett, J. A., Quinn, N. P., Stahl, S. M. and Marsden, C. D. (1989)
   Clin. Neuropharmacol. **12**, 37–45.
Fahn, S. (1974) *Neurology* **24**, 431–441.
Grandas-Perez, F. J., Jenner, P. G., Nomoto, M. *et al.* (1986) *Lancet* **i**, 906.
Jenner, P., Rupniak, N. M. J., Rose, S., Kelly, E., Kilpatrick, G., Lees, A. J. and
   Marsden, C. D. (1984) *Neurosci. Lett.* **50**, 85–90.
Juncos, J. L., Fabbrini, G., Mouradian, M. M. and Chase, T. N. (1987) *Arch. Neurol.*
   44, 1010–1012.
Koller, W., Herbster, G. and Gordon, J. (1987) *Movement Disorders* **2**, 193–199.
Marsden, C. D. and Parkes, J. D. (1976) *Lancet* **i**, 292–296.
Martin, G. E., Williams, M., Pettibone, D. J., Yarbrough, G. G., Clineschmidt, B. V.
   and Jones, J. H. (1984) *J. Pharmacol. Exp. Ther.* **230**, 569–576.
Nomoto, M., Stahl, S., Jenner, P. and Marsden, C. D. (1987) *Movement Disorders* **2**,
   37–45.
Shaw, J. E. and Urquhart, J. (1981) *Br. Med. J.* **283**, 875–876.
Stibe, C. M. H., Lees, A. J. and Stern, G. M. (1987) *Lancet* **i**, 871.
Stoessl, A. J., Mak, E. and Calne, D. B. (1985) *Lancet* **ii**, 1330–1331.
Wepierre, J. and Marty, J. P. (1979) *TIPS* **1**, 23–26.

11

MPTP-induced Parkinsonism: the Relevance to Idiopathic Parkinson's Disease

P. Jenner

INTRODUCTION

Compared to many neurodegenerative diseases Parkinson's disease is well researched. Pathological changes occurring in Parkinson's disease involve degeneration of pigmented brain stem nuclei, particularly those dopamine-containing cells within the zona compacta of substantia nigra (Forno, 1982). Wherever cell loss occurs in Parkinson's disease dense eosinophilic inclusions, termed Lewy bodies, are found. Investigation of transmitter changes in Parkinson's disease has shown a primary loss of forebrain dopamine content but also changes in noradrenaline, 5HT, acetylcholine, GABA and neuropeptides (Hornykiewicz, 1982; Agid et al., 1987). Dopamine replacement therapy forms the basis of treatment involving L-dopa and synthetic dopamine agonist drugs, such as bromocriptine. This emphasizes the key role of dopamine in the symptomatology of the illness and drugs acting on other neuronal systems are in general without effect (see for example Sheehy et al., 1981; Jenner et al., 1983).

However, despite the extensive knowledge on Parkinson's disease and its treatment there are many problems which remain unresolved. There are a few clues as to the cause of the disease or to the mechanism underlying cellular degeneration (see Duvoisin, 1982). There is at present no means of stopping or slowing the progression of the disease process once established.

DISORDERS OF MOVEMENT: CLINICAL, PHARMACOLOGICAL AND PHYSIOLOGICAL ASPECTS   ISBN 0-12-569685-X

There is as yet no cure for Parkinson's disease. These are problems of increasing importance since the limitations of symptomatic drug treatment have been clearly established (Marsden *et al.*, 1982). Thus, after some years of therapy with L-dopa the effectiveness of the drug becomes shortened and unpredictable. At the same time there is frequently the onset of complex dyskinetic phenomena which presumably reflect the progression of the disease process.

One problem in obtaining answers to these critical questions relates to the difficulty of studying them experimentally since Parkinson's disease appears peculiar to man. A variety of animal models have been utilized to study the illness but none of these closely parallel the disease as it occurs in man. Some drugs may be used to reduce brain dopamine function and so diminish movement, for example, reserpine or α-methylparatyrosine (AMPT) can be used to deplete the brain dopamine content while neuroleptic drugs can be used to block postsynaptic dopamine receptors so reducing dopaminergic transmission. Rodents rendered akinetic by the administration of reserpine or AMPT respond to drugs effective in the treatment of Parkinson's disease (Carlsson *et al.*, 1957; Dolphin *et al.*, 1976). However, while reserpine or AMPT may produce the motor deficits of Parkinson's disease they do not mimic the pathology of the disorder and their effects are readily reversible. Another commonly employed approach has been to destroy ascending dopamine fibres in the brain of rodents using the neurotoxin 6-hydroxydopamine. Such lesions have been widely employed to produce the circling rodent model which has proved a useful test-bed for the development of anti-parkinsonian agents (Ungerstedt and Arbuthnott, 1970; Ungerstedt, 1971a,b; von Voigtlander and Moore, 1973; Reavill *et al.*, 1983). However, the lesions can only be produced successfully on a unilateral basis since bilateral lesioning results in the occurrence of adipsia and aphagia. So, although the major pathology of Parkinson's disease can be produced in this model the motor deficits produced in no way resemble those occurring in Parkinson's disease itself. Other types of lesions, for example electrolytic lesions or kainic acid lesions, can be employed in a similar manner but the same problems are evident. Perhaps the closest approximation to Parkinson's disease was produced by the use of electrolytic or radiofrequency lesions of the brain stem of monkeys which produced motor deficits identical to those occurring in Parkinson's disease (Goldstein *et al.*, 1973; Pechadre *et al.*, 1976). However, these lesions were difficult to place reproducibly and invariably different animals exhibited different parkinsonian signs. Also, in many animals tremor was the major component due to interruption of the cerebellorubrothalamic pathway. Again, there was the problem of producing bilateral lesions without grossly impairing the ability of the animals to feed and drink. The overall picture a few years ago, therefore, was of a lack of an effective animal model of Parkinson's disease in which to study many of the remaining problems.

THE INTRODUCTION OF MPTP

The absence of a viable model of Parkinson's disease was remedied by a chance occurrence in California in the late 1970s and early 1980s. Langston, faced by an epidemic of young drug addicts presenting with clinical parkinsonism, discovered that they were abusing a synthetic pethidine derivative (MPPP) which they used as a heroin substitute (Langston et al., 1983). Indeed, an identical occurrence involving a single patient had been reported some years previously (Davis et al., 1979). This patient had abused the pethidine derivative without ill effect for some months before exhibiting motor abnormalities so Langston suspected that MPPP itself was not responsible for the outbreak of parkinsonism. Analysis of samples of the drug being abused showed it to be contaminated with varying amounts of another substance, namely 1-methyl-4-phenyl-1,2,3,6-tetrahydropyridine (MPTP), and it was this substance which proved to be the causative agent (Langston et al., 1983).

Langston initially described four patients presenting with parkinsonism and subsequently described seven cases in detail (Ballard et al., 1985). All of these patients exhibited classical symptomatology of idiopathic Parkinson's disease and responded to antiparkinsonian therapy. All developed the long-term complications of drug treatment of Parkinson's disease shortly after commencing L-dopa therapy. This is probably a reflection of the severity of the disease at onset coupled with the young age of the patients concerned. Langston has traced some 400 people exposed to MPTP but the majority of these remain asymptomatic (see, for example Langston, 1986). Some have a partial loss of brain dopamine neurons as detected using ^{11}F-dopa uptake on PET scanning and may well develop Parkinson's disease at an early age (Calne et al., 1985). Others, however, appear totally unaffected by their exposure to MPTP.

The discovery of MPTP raised the possibility of establishing an exact experimental model of Parkinson's disease. However, the initial results were disappointing. The administration of large amounts of MPTP to rodent species failed to produce any motor deficits (Chiueh et al., 1984; Boyce et al., 1984). Indeed, it was only when MPTP was administered to primate species that effects resembling those in man were observed (Burns et al., 1983; Langston et al., 1984a; Jenner et al., 1984). The reason for this species difference became clear with the discovery of the mechanism by which MPTP induced parkinsonism. Pathological examination of brains from primates treated with MPTP showed a selective destruction of dopamine-containing cells in the zona compacta of substantia nigra (Burns *et al.*, 1983; Langston *et al.*, 1984a). Indeed, this was identical to pathological change found in the brain of the first patient reported to have been exposed to MPTP (Davis *et al.*, 1979). In most rodent species, however, MPTP appeared not to produce cell loss within the substantia nigra. So here was a toxin which mimicked the major pathological change in Parkinson's disease and which in primate species produced clinical parkinsonism. But the question remained as to how exact a model of Parkinson's disease could be produced by the administration of MPTP.

THE MPTP MODEL

The administration of MPTP to a variety of primate species produces typical parkinsonian signs within a few days of the start (Burns *et al.*, 1983; Langston *et al.*, 1984a; Jenner *et al.*, 1984; Doudet *et al.*, 1985; Eidelberg *et al.*, 1986). Animals become increasingly bradykinetic and akinetic, exhibiting rigidity of limbs and trunk, postural abnormalities, loss of vocalization and a loss of blink reflex. Animals also exhibit other features associated with Parkinson's disease such as a hyper-reflexic bladder (Albanese *et al.*, 1988). Tremor is not always a marked component of the MPTP-induced syndrome in primates. Some species appear to show a marked tremor while others show little or no parkinsonian tremor. There are other differences to Parkinson's disease in that symptoms once established do not progress but in contrast they may show some reversal over a period of months (Eidelberg *et al.*, 1986). So, in these respects MPTP did not produce a precise model of idiopathic Parkinson's disease. In addition, other neurological signs apart from deficits can be observed in MPTP-treated primates. Common marmosets treated with MPTP may show an exaggerated startle response, action myoclonus and dystonic posturing of the neck and trunk. However, all the deficits produced by MPTP are responsive to the administration of L-dopa plus a peripheral decarboxylase inhibitor or to the administration of synthetic dopamine agonist drugs.

In the young adult primates usually employed for the administration of MPTP, the pathological picture reveals MPTP to act as a selective nigral toxin. There is a selective destruction of tyrosine hydroxylase positive cells within the zona compacta of substantia nigra with a smaller and more variable damage to the adjacent cells in the ventral tegmental area (Burns *et al.*, 1983; Langston *et al.*, 1984a; Kitt *et al.*, 1986; Waters *et al.*, 1987). In these animals there is little evidence for damage to other pigmented brain stem nuclei or other cell groups within the brain stem. So far, in young adult primates there has been no evidence for the occurrence of Lewy bodies.

Overall, the pathological profile of MPTP is far more limited than that which occurs in idiopathic Parkinson's disease. There is also evidence of a selective effect of MPTP on different cell groups within zona compacta of substantia nigra (Elsworth *et al.*, 1987a; Schneider *et al.*, 1987; German *et al.*, 1988; Gibb *et al.*, 1989). There is controversy as to whether these resemble the pattern of cell loss in idiopathic Parkinson's disease or not. Recently Hornykiewicz and colleagues have presented data to suggest that MPTP produces changes more closely resembling those occurring in post-encephalitic Parkinson's disease than the idiopathic form of the illness (Pifl *et al.*, 1988). So, this again may provide a difference between the effects of MPTP and what occurs in idiopathic Parkinson's disease.

Biochemical changes observed in the brains of primates treated with MPTP reflect the limited pathology. Thus, there is a marked and persistent decrease in caudate-putamen dopamine content and a corresponding fall in ^{3}H-dopamine uptake and ^{3}H-mazindol binding consistent with the loss of nerve terminals (Burns *et al.*, 1983; Jenner *et al.*, 1984; Joyce *et al.*, 1985; Elsworth *et al.*, 1987b; Rose *et al.*, 1989a). In the nucleus accumbens a transient and reversible fall of dopamine content is observed, again consistent with the limited pathology in the ventral tegmental area (Rose *et al.*, 1989b). In contrast, there are few changes in noradrenaline, 5HT, acetylcholine, GABA or neuropeptide markers within basal ganglia or other brain regions of MPTP-treated primates (see, for example Garvey *et al.*, 1986; Jenner *et al.*, 1986a; Taquet *et al.*, 1988; but see also Zamir *et al.*, 1984; Allen *et al.*, 1986). So, in contrast to the widespread changes in neurochemical indices observed in idiopathic Parkinson's disease, administration of MPTP to primates produces only a partial model of the illness as it affects man.

AGE AS A DETERMINANT OF MPTP TOXICITY

There may be many reasons why MPTP does not produce a precise model of Parkinson's disease. For example, most studies attempt to mimic the effects of an illness which probably takes decades to develop in man by using an acute treatment regime in primate species. However, another possible explanation for the limited pathology produced by MPTP may relate to the age of the animals employed. Overall, young adult primates are used to produce a model of an illness which is a disease of increasing age in man. Indeed, there is some experimental evidence to suggest that use of MPTP in elderly primates may produce a better model of Parkinson's disease.

For example, Forno *et al.* (1986, 1988) have shown that the administration of MPTP to elderly monkeys produces not only a destruction of cells within substantia nigra but also within the locus coeruleus. This more extensive cell loss is accompanied by evidence for the occurrence of inclusions resembling Lewy bodies. Mitchell *et al.* (1985) have reported similar findings and it may have been the age of their animals which led to the spread of pathology. In our own studies we have found that adult and aged marmosets are more susceptible to the effects of MPTP than juvenile animals (unpublished observations). In elderly animals once motor deficits are established there is less recovery than occurs in either adult or juvenile animals, suggesting the limited ability of the aged brain to adapt to the toxic effects of MPTP.

Similarly, the duration of administration may be a critical fact. In a recent study we have found the once-weekly administration of MPTP to marmosets for 16 or 24 weeks leads to production of motor deficits which do not recover on cessation of MPTP administration (unpublished observations). This might suggest that chronic exposure to MPTP produces effects more consistent with those occurring during the development of Parkinson's disease. Indeed, it would seem that short-term low-dose exposure to MPTP may lead to a selective destruction of substantia nigra and to parkinsonian deficits which recover over a period of time. In contrast, it may be that high-dose long-term exposure to MPTP leads to the destruction not only of cells in substantia nigra but also within other brain regions and that this produces a more persistent model of parkinsonism.

RELEVANCE OF THE MPTP MODEL OF PARKINSON'S DISEASE

Although MPTP may not provide an exact model of Parkinson's disease, it does, however, offer possibilities of studying the major outstanding problems related to Parkinson's disease. In the following sections we will consider the relevance of the MPTP model to the cause of Parkinson's disease, the evaluation of novel treatments and the onset of motor side-effects and to the possibilities of a "cure" for Parkinson's disease based on the transplantation of foetal nigral cells. In each of these areas the use of MPTP has provided considerable impetus into research into Parkinson's disease.

MECHANISM OF ACTION OF MPTP

The mechanism of action of MPTP may provide clues to the process underlying cell death in Parkinson's disease and to the selective vulnerability of nigral dopamine-containing cells. A number of key discoveries have already revealed critical steps in the pathway leading to neurotoxicity. An important initial discovery was the finding that in the brains of primates

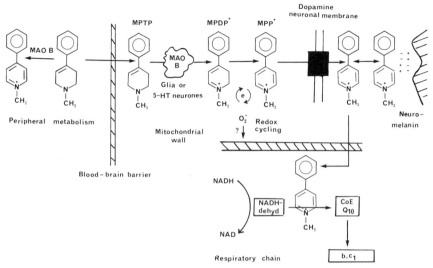

Fig. 11.1   Scheme to illustrate the proposed metabolism, mechanism of action and neurotoxic effect of MPTP (and MPP$^+$) to dopamine cells in substantia nigra.

treated with MPTP another substance, namely 1-methyl-4-phenylpyridinium species (MPP$^+$), was present and persisted over long periods (Markey *et al.*, 1984; Johannessen *et al.*, 1985). MPTP was subsequently shown to be an unexpected substrate for MAO B being converted in two steps to MPP$^+$ (Chiba *et al.*, 1984; Salach *et al.*, 1984; Heikkila *et al.*, 1985). The site of this conversion is disputed but it appears to occur either within glial cells or within 5HT neurons (see Shen *et al.*, 1985; Ransom *et al.*, 1987). MPP$^+$ appears to be at least one reactive species formed from MPTP since it is toxic to mesencephalic cells in culture and it is also toxic on direct focal injection into the substantia nigra (in contrast to the relatively low toxicity of MPTP) (Bradbury *et al.*, 1985b, 1986; Mytilineou *et al.*, 1985; Sanchez-Ramos *et al.*, 1986). Why MPTP selectively kills dopamine cells was partially resolved when it was discovered to be a substrate for dopamine re-uptake mechanisms so being accumulated within dopamine neurons (Javitch *et al.*, 1985). Once within these neurons its accumulation may be enhanced by an association with neuromelanin (D'Amato *et al.*, 1987).

These data do not, however, explain how MPP$^+$ produces its toxic actions. Initially it was thought that MPP$^+$ may redox cycle to produce toxic free radical species but this idea was discarded because of the relative stability of MPP$^+$ and its low electrochemical potential which makes its bioreduction improbable (Di Monte *et al.*, 1986a; Sayre *et al.*, 1986; Frank *et al.*, 1987; but see below). Consequently, other explanations were sought. Interest subsequently centred on the ability of MPP$^+$ to inhibit NADH-linked components of mitochondrial energy metabolism, and specifically complex I (Nicklas *et al.*, 1985; Vyas *et al.*, 1986). In the *in vitro* systems used, millimolar concentrations of MPP$^+$ were required to inhibit mitochondrial function and initially it was thought that these were unlikely to be reached *in vivo*. However, mitochondria actively accumulate MPP$^+$ and concentrations up to 50 times over the external level can be achieved (Ramsay and Singer, 1986; Ramsey *et al.*, 1986a,b). The site of action of MPP$^+$ appears to be on complex I at some point between NADH dehydrogenase and coenzyme Q10 but precisely where remains unknown (see Singer *et al.*, 1987). The effects of MPP$^+$ on mitochondria lead to a depletion of ATP, a decrease in reduced glutathione, and changes in intracellular calcium content (Di Monte *et al.*, 1986b, 1987; Kass *et al.*, 1988). The importance of interference with energy metabolism was shown recently by the ability of fructose to prevent the toxic effects of MPTP (or MPP$^+$) on isolated hepatocyte preparations (Di Monte *et al.*, 1988). Glucose was much less effective, presumably since it is a poorer substrate for glycolysis. So, all this data implicates interference with mitochondrial function as the primary mechanism of action of MPTP but as yet does not identify the exact manner in which MPP$^+$ brings about this impairment.

The mitochondrial concept of MPTP toxicity presumes it is MPP$^+$ alone which is responsible for cell death. MPTP is metabolized by MAO B in a two-step reaction to MPP$^+$ via the intermediate dihydropyridine derivative, MPDP$^+$ (Castagnoli *et al.*, 1985; Petersen *et al.*, 1985; Trevor *et al.*, 1987). Along this pathway there is considerable scope for the generation of radical species. Indeed, recent evidence has again suggested that free radical mechanisms might be operative as a part of MPTP toxicity (Rossetti *et al.*, 1988). The incorporation of MPTP into mitochondrial preparations was found to generate an ESR signal indicating active oxygen species production and which was prevented by the incorporation of an MAO B inhibitor, and by the incorporation of superoxide dismutase. This evidence suggests that the ESR signal is dependent on the metabolism of MPTP and that it is due to the generation of a superoxide radical. While MPP$^+$ alone or MPDP$^+$ alone in the absence of the tissue preparation did not generate an ESR signal, the combination of MPP$^+$ and MPDP$^+$ did produce a signal which increased with time. These data suggest that a redox reaction may occur between these two MPTP metabolites which results in the formation of toxic superoxide radicals. As a result of this data a free radical hypothesis of MPTP action cannot be dismissed. Whether the production of superoxide radicals contributes to the inhibition of mitochondrial function induced by MPTP or whether it provides a separate mechanism by which neurotoxicity can be explained remains to be resolved.

The schemes proposed for the mechanism of action of MPTP are sufficient to explain the toxicity of MPTP to nigral dopamine-containing neurones, but do not explain the lack of toxicity to other dopamine-containing cell groups within brain or the lack of toxicity in other species. One explanation for the difference between species may lie in the activity of MAO in the blood–brain barrier (Kalaria et al., 1987; Riachi et al., 1988a). Thus, rats and mice have higher MAO B activity in the blood–brain barrier than found in primates or man. So, the amount of MPTP able to enter the brain may differ markedly between species. Since MPP^+ as an ionized molecule would not penetrate into brain readily, the toxicity of MPTP can be curtailed at the level of the blood–brain barrier.

DRUG ASSESSMENT USING THE MPTP MODEL

The MPTP model allows assessment of a number of aspects of drug action in Parkinson's disease. These include ways of preventing or reversing the actions of MPTP-like toxins, the development of drugs for the symptomatic treatment of Parkinson's disease, and the study of the long-term problems associated with the use of dopamine replacement therapy.

The mechanism by which MPTP exerts its toxicity suggests a number of possibilities for preventing the onset of parkinsonism due to MPTP-like toxins. Thus, selective MAO B, but not MAO A, inhibitors prevent the formation of MPP^+ and block the toxicity of MPTP in rodent and primate species (Heikkila et al., 1984; Langston et al., 1984b; Cohen et al., 1985). Similarly, dopamine re-uptake blockers, such as nomifensine, can prevent the entry of MPP^+ into dopamine neurones. However, while the effectiveness of this drug group has been established in rodents (Mayer et al., 1986), there is doubt over the efficacy of such compounds in preventing the toxicity of MPTP in primates (see Schultz et al., 1986; Barnes et al., 1986). Antioxidants, for example α-tocophorol, might be expected to prevent the final stages of MPTP toxicity but the evidence in primates is limited (Perry et al., 1985, 1987; Yong et al., 1986; Sanchez-Ramos et al., 1988).

Whether any of these classes of compounds could be effective in preventing the onset or progression of Parkinson's disease in man remains to be determined. The prime determinant would be whether an MPTP-like toxin is indeed involved in the production of Parkinson's disease.

An important facet of the MPTP-treated primate is the ability of all established antiparkinsonian drugs to reverse the motor deficits. This would suggest the model to be useful in the development of novel antiparkinsonian therapy. Indeed, the highly potent dopamine agonist (+)-PHNO is effective in reversing motor deficits in MPTP-treated marmosets in a manner predictive of its actions in man (Nomoto *et al.*, 1987). The model may also reveal the role of D1 receptors in Parkinson's disease. For example, the D1 agonist SKF 38393 might be thought to be potentially effective in Parkinson's disease since it causes contraversive rotation in 6-OHDA lesioned rats (Setler *et al.*, 1978). However, SKF 38393 does not reverse MPTP-induced parkinsonism (Nomoto *et al.*, 1985; Close *et al.*, 1985) and it is ineffective in parkinsonian patients (Braun *et al.*, 1987). Indeed, in MPTP-treated primates SKF 38393 inhibits the antiparkinsonian activity of the D2 agonist quinpirole (Nomoto *et al.*, 1988). So, the organization of brain D1 and D2 sites may be different in primates and man compared to rodent species.

A further application of the MPTP model is in establishing novel routes of drug administration in Parkinson's disease. In this context we have shown both (+)-PHNO and another highly potent dopamine agonist, N-0437, to be effective in producing normal motor function in MPTP-treated marmosets following either application to the skin or by the use of transdermal patches (Jenner *et al.*, 1986b; Loschmann *et al.*, 1989). It may be that the prolonged period of drug action resulting from the transcutaneous route of administration will lead to new developments in the therapy of Parkinson's disease.

Finally, the MPTP-treated primate may be an effective means of discovering the cause of the dyskinesias and dystonias which accompany chronic L-dopa therapy. Thus, MPTP-treated monkeys rapidly develop peak dose dyskinesias and dystonias in response to repeated L-dopa administration (Bedard *et al.*, 1986; Crossman *et al.*, 1987; Falardeau *et al.*, 1988). Interestingly, the chronic administration of bromocriptine does not appear to produce the same pattern of involuntary movements consistent with claims made in man (Bedard *et al.*, 1986).

MPTP-TREATED PRIMATES AS A TEST-BED FOR THE DEVELOPMENT OF FOETAL CELL TRANSPLANTATION

As described elsewhere in this volume the transplantation of foetal nigral cells may offer an opportunity to reinnervate the denervated caudate-putamen in Parkinson's disease. Many features of transplantation remain to be evaluated and the MPTP-treated primate may offer an opportunity to study these factors. Redmond *et al.* (1986) showed that foetal nigral cells transplanted into MPTP-treated monkeys remained viable and caused a reversal of motor deficits. Our studies showed the survival over a period of months of foetal ventral mesencephalon cells implanted into the brain of adult MPTP-treated marmosets as judged by the presence of tyrosine hydroxylase immunoreactive cells (Fine *et al.*, 1989). However, we were unable to observe any change in spontaneous motor behaviour although the response of the animals to amphetamine administration was dramatic, presumably due to the release of dopamine from the graft terminals. So it does indeed appear feasible to use this model for assessing factors such as graft size, graft position, the need for immunosuppressants, and the extent of functional recovery using a primate species, all of which must be determined before the technique can be applied routinely to man.

RELEVANCE OF THE MPTP MODEL TO PARKINSON'S DISEASE

The administration of MPTP to primates including man produces parkinsonism which closely resembles the motor deficits occurring in the idiopathic disease (Table 11.1). The general picture is that of MPTP being a selective nigral toxin and as such, while mimicking the major pathological and biochemical changes of Parkinson's disease, not mirroring all the changes which occur in this disorder. It may be that in older primates a more appropriate model of parkinsonism will be produced. However, the mechanism of action of MPTP may provide clues as to the cause of Parkinson's disease and to the underlying mechanism of cell death. The model itself provides a most appropriate means of devising novel strategies for the treatment of Parkinson's disease. Finally, the MPTP-treated primates may hold the answer to many of the problems which surround foetal cell transplantation into man. The MPTP story as a whole has raised awareness as to the involvement of toxins as a cause of Parkinson's disease and, in particular, to questions about the environment.

Since the discovery of MPTP approximately 100 MPTP analogues have been described (see, for example Bradbury et al., 1985a; Fuller et al., 1987; Hoppel et al., 1987; Finnegan et al., 1987; Youngster et al., 1987; Heikkila et al., 1988a; Riachi et al., 1988b). A number of these are substrates for monoamine oxidase and are converted to the corresponding pyridinium species and a few are also neurotoxic. Two interesting features arise out of this series of MPTP derivatives. First, one derivative, namely 2'-methyl-MPTP, is 7–8 times more potent than MPTP itself in mice (Youngster et al., 1986). However, in primates 2'-methyl-MPTP is, if anything, slightly less potent than MPTP (unpublished data). This raises the concept that species differences in MPTP analogue toxicity can occur. Second, there is evidence that MAO A as well as MAO B may be involved in the toxicity of such derivatives. Thus 2'-methyl-MPTP appears to be metabolized in a complex manner by both MAO A and B while 2'-ethyl-MPTP is selectively metabolized by MAO A (Kindt et al., 1988; Heikkila et al., 1988b).

So, rather than just MPTP there is a family of neurotoxins, the structural limitations of which remain to be identified and which can involve both

Table 11.1 Comparison of the effects of MPTP in primates with the characteristics of idiopathic Parkinson's disease in man.

Parameter	Parkinson's disease	MPTP-induced parkinsonism
Motor deficits		
Akinesia	+ + +	+ + +
Rigidity	+ + +	+ +
Tremor	+ + +	+
Postural abnormalities	+ + +	+ +
Bladder hyper-reflexia	+ + +	+ + +
Other motor disorders	0	+
Recovery	0	+ +
Drug responsiveness		
L-Dopa	+ + +	+ + +
Bromocriptine	+ + +	+ + +
PHNO	+ + +	+ + +
SKF 38393	0	0
CY 208-243	+ +	+ +
Dyskinesias	+ +	+ +
Wearing-off	+ +	+
On–off	+ +	?
Pathology		
Substantia nigra	+ + +	+ + +
Ventral tegmental area	+ +	+
Locus coeruleus	+ + +	+
Substantia innominata	+ +	0
Lewy bodies	+ + +	+
Biochemistry		
Dopamine		
–caudate	+ + +	+ + +
–putamen	+ + +	+ + +
–nucleus accumbens	+ +	+
Noradrenaline	+ +	+
5-HT	+ +	0
Acetylcholine	+ +	0
GABA	+	0
Peptides	+ +	+
^{3}H-mazindol	+ + +	+ + +
^{3}H-spiperone	+ +	+

isoenzymes of MAO in their metabolism and subsequent toxicity. Consequently, it is necessary to determine how many related substances in the environment are potentially neurotoxic in this manner.

There is a strong structural similarity between MPP^+ and the herbicide paraquat. Indeed, MPP^+ also known as cyperquat was examined in field trials as a herbicide in the early 1950s. Initially, the structural similarity between paraquat and MPP^+ gave rise to the idea that they may have a similar mechanism of action. Although this appears unlikely it led to the idea that herbicides and related substances might contribute to the cause of Parkinson's disease (Barbeau *et al.*, 1985, 1986, 1987; but see also Poirier *et al.*, 1987). Subsequent studies have suggested early exposure to a rural environment and the drinking of well water may lead to an increased incidence of Parkinson's disease (Rajput *et al.*, 1987; Schoenberg, 1987; Tanner *et al.*, 1987). It may be that indirectly the MPTP story will lead us to examine more closely factors within the environment which contribute to onset of this disorder.

The final question raised by the MPTP story is whether there is any direct evidence of a neurotoxic process contributing to the underlying pathology of Parkinson's disease. If Parkinson's disease occurs as a result of a single toxic insult either *in utero* or during early life, then evidence of such toxicity will not be evident at subsequent post-mortem. However, if the onset of Parkinson's disease is associated with some continuing toxic process then it may be possible to detect this in post-mortem tissues. Recently, we have shown that the substantia nigra, but not other brain areas, of patients dying with Parkinson's disease contain a decreased content of substrate for lipid peroxidation namely, polyunsaturated fatty acids (Dexter *et al.*, 1986, 1989a). In addition, there are increased levels of a stable intermediate in lipid peroxidation, malondialdehyde. These data suggest that some on-going process such as free radical attack can be detected at the end stage of Parkinson's disease when most dopamine neurones have already been destroyed. What could be stimulating cellular degeneration in the parkinsonian brain? One possibility is that it is not due to an environmental toxin but to some endogenous factor. Indeed, the substantia nigra of patients dying with Parkinson's disease contains an excess of iron that is not found in other brain regions (Earle, 1968; Dexter *et al.*, 1987, 1989b; Sofic *et al.*, 1988). The ability of iron to stimulate the formation of toxic oxygen radicals may lead to an increase in lipid peroxidation (see Halliwell and Gutteridge, 1985). It is unlikely that L-dopa treatment contributes to the increased accumulation of iron since in 1968 Earle found an enhanced content of iron in the brain stem of formalin-fixed tissues from patients who died prior to the advent of the L-dopa era. It should be noted, however, that not all forms of iron are capable of exerting toxicity via the formation of free radicals. In brain an increasing iron content would normally trigger increased synthesis of the iron-binding protein ferritin. Therefore, the excess iron present in parkinsonian substantia nigra may be in a form which is inactivated. However, when we measured brain ferritin levels in patients dying with Parkinson's disease there was a generalized decrease (Dexter *et al.*, 1989c; but see Riederer *et al.*, 1989). This suggests that Parkinson's disease may be accompanied by some alteration in iron handling. While it remains to be determined whether the increased content of iron in Parkinson's disease is a cause or a consequence of this disorder or whether it is specific to this one illness, we at least appear to have some leads as to potential causes of the disease. So yet again the MPTP story has led us to examine factors which at least at first do not appear directly related to the actions of the toxin itself.

There may, however, be a common link between the underlying disease process of Parkinson's disease and the action of MPTP. Thus, very recently, we have demonstrated that there is a selective increase in mitochondrial superoxide dismutase activity (Saggu et al., 1989; but see Marttila et al., 1988) and an inhibition of mitochondrial complex I activity in the parkinsonian substantia nigra (Schapira et al., 1989).

The discovery of MPTP has provided enormous impetus for research into Parkinson's disease. It has provided the most appropriate model of the disorder so far available. It has allowed us to tackle major problems which remain to be resolved in Parkinson's disease. It may be that MPTP itself has nothing to do with the cause of Parkinson's disease. However, the interest raised by the discovery of the toxin has led to lines of research which previously would not have been envisaged. It may also be that MPTP has underlined reasons for the selective vulnerability of nigral dopamine-containing cells whose death results in the motor symptoms characterizing Parkinson's disease.

ACKNOWLEDGEMENTS

This study was supported by the Parkinson's Disease Society, the Medical Research Council and the Research Funds of the Bethlem Royal and Maudsley Hospitals, and King's College Hospital.

REFERENCES

Agid, A., Javoy-Agid, F. and Ruberg, M. (1987) In *Movement Disorders 2* (edited by C. D. Marsden and S. Fahn), pp. 166–230. Butterworths, London.

Albanese, A., Jenner, P., Marsden, C. D. and Stephenson, J. D. (1988) *Neurosci. Lett.* **87**, 46–50.

Allen, J. M., Cross, A. J., Yeats, J. C., Ghatei, M. A., McGregor, G. P., Close, S. P., Pay, S., Marriott, A. S., Tyers, M. B., Crow T. J. and Bloom, S. R. (1986) *Brain* **109**, 143–157.

Ballard, P. A., Tetrud, J. W. and Langston, J. W. (1985) *Neurology* **35**, 949–956.

Barbeau, A., Cloutier, T., Roy, M., Plasse, L., Paris, S. and Poirier, J. (1985) *Lancet* **ii**, 1213–1216.

Barbeau, A., Roy, M., Cloutier, T., Plasse, L. and Paris, S. (1986) In *Advances in Neurology*, Vol. 45 (edited by M. D. Yahr and K. J. Bergmann), pp. 299–306. Raven Press, New York.

Barbeau, A., Roy, M., Bernier, G., Campanella, G. and Paris, S. (1987) *Can. J. Neurol. Sci.* **14**, 36–41.

Barnes, N. J. G., Costall, B. and Naylor, R. J. (1986) *Br. J. Pharmacol.* **90**, 241P.

Bedard, P. J., Di Paolo, T., Falardeau, P. and Boucher, R. (1986) *Brain Res.* **379**, 294–299.

Boyce, S., Kelly, E., Reavill, C., Jenner, P. and Marsden, C. D. (1984) *Biochem. Pharmacol.* **33**, 1747–1752.

Bradbury, A. J., Costall, B., Domeney, A. M., Testa, B., Jenner, P., Marsden, C. D. and Naylor, R. J. (1985a) *Neurosci. Lett.* **61**, 121–126.

Bradbury, A. J., Costall, B., Jenner, P. G., Kelly, M. E., Marsden, C. D. and Naylor, R. J. (1985b) *Neurosci. Lett.* **58**, 177–181.

Bradbury, A. J., Costall, B., Domeney, A. M., Jenner, P., Kelly, M. E., Marsden, C. D. and Naylor, R. J. (1986) *Nature* **319**, 56–57.

Braun, A., Fabbrini, G., Mouradian, M. M., Serrati, C., Barone, P. and Chase, T. N. (1987) *J. Neural Transm.* **68**, 41.

Burns, R. S., Chiueh, C. C., Markey, S. P., Ebert, M. H., Jacobowitz, D. M. and Kopin, I. J. (1983) *Proc. Natl. Acad. Sci. USA* **80**, 4546–4550.

Calne, D. B., Langston, J. W., Martin, W. R. W., Stoessl, A. J., Ruth, T. J., Adam, M. J., Pate, B. D. and Schulzer, M. (1985) *Nature* **317**, 246–248.

Carlsson, A., Lindquist, M. and Magnusson, T. (1957) *Nature* **180**, 1200.

Castagnoli, N. Jr., Chiba, K. and Trevor, A. J. (1985) *Life Sci.* **36**, 225–230.

Chiba, K., Trevor, A. and Castagnoli, N. Jr. (1984) *Biochem. Biophys. Res. Commun.* **120**, 574–578.

Chiueh, C. C., Markey, S. P., Burns, R. S., Johannessen, J. N., Pert, A., and Kopin, I. J. (1984) *Eur. J. Pharmacol.* **100**, 189–194.

Close, S. P., Marriott, A. S. and Pay, S. (1985) *Br. J. Pharmacol.* **85**, 320–322.

Cohen, G., Pasik, P., Cohen, B., Leist, A., Mytilineou, C. and Yahr, M. D. (1985) *Eur. J. Pharmacol.* **106**, 209–210.

Crossman, A. R., Clarke, C. E., Boyce, S., Robertson, R. G. and Sambrook, M. A. (1987) *Can. J. Neurol. Sci.* **14**, 428–435.

D'Amato, R. J., Benham, D. F. and Snyder, S. H. (1987) *J. Neurochem.* **48**, 653–658.

Davis, G. C., Williams, A. C., Markey, S. P., Ebert, M. H., Caine, E. D., Reichert, C. M. and Kopin, I. J. (1979) *Psychiat. Res.* **1**, 249–254.

Dexter, D. T., Carter, C., Agid, F., Agid, Y., Lees, A. J., Jenner, P. and Marsden, C. D. (1986) *Lancet* **ii**, 639–640.

Dexter, D. T., Wells, F. R., Agid, F., Agid, Y., Lees, A. J., Jenner, P. and Marsden, C. D. (1987) *Lancet* **ii**, 1219–1220.

Dexter, D. T., Carter, C. J., Wells, F. R., Javoy-Agid, F., Agid, Y., Lees, A., Jenner, P. and Marsden, C. D. (1989a) *J. Neurochem.* **52**, 381–389.

Dexter, D. T., Wells, F. R., Lees, A. J., Agid, F., Agid, Y., Jenner, P. and Marsden, C. D. (1989b) *J. Neurochem.* **52**, 1830–1836.

Dexter, D. T., Carayon, A., Vidailhet, M., Ruberg, M., Agid, F., Agid, Y., Lees, A. J., Wells, F. R., Jenner, P. and Marsden, C. D. (1989c) *J. Neurochem*, submitted.

Di Monte, D., Sandy, M. S., Ekstrom, G. and Smith, M. T. (1986a) *Biochem. Biophys. Res. Commun.* **137**, 303–309.

Di Monte, D., Jewell, S. A., Ekstrom, G., Sandy, M. S. and Smith, M. T. (1986b) *Biochem. Biophys. Res. Commun.* **137**, 310–315.

Di Monte, D., Sandy, M. S. and Smith, M. T. (1987) *Biochem. Biophys. Res. Commun.* **148**, 153–160.

Di Monte, D., Sandy, M. S., Blank, L. and Smith, M. T. (1988) *Biochem. Biophys. Res. Commun.* **153**, 734–740.

Dolphin, A. C., Jenner, P. and Marsden, C. D. (1976) *Pharmacol. Biochem. Behav.* **4**, 441–670.

Doudet, D., Gross, C., Lebrun-Grandie, P. and Bioulac, B. (1985) *Brain Res.* **335**, 194–199.

Duvoisin, R. (1982) in *Neurology 2, Movement Disorders* (edited by C. D. Marsden and S. Fahn), pp. 8–24. Butterworth Scientific, London.

Earle, K. M. (1968) *J. Neuropathol. Exp. Neurol.* **27**, 1–14.

Eidelberg, E., Brooks, B. A., Morgan, W. W., Walden, J. G. and Kokemoor, R. H. (1986) *Neuroscience* **18**, 817–822.

Elsworth, J. D., Deutch, A. Y., Redmond, D. E. Jr., Sladek, J. R. and Roth, R. H. (1987a) *Life Sci.* **40**, 193–202.

Elsworth, J. D., Deutch, A. Y., Redmond, D. E. Jr., Sladek, J. R. and Roth, R. H. (1987b) *Brain Res.* **415**, 293–299.

Falardeau, P., Bouchard, S., Bedard, P. J., Boucher, R. and Di Paolo, T. (1988) *Eur. J. Pharmacol.* **150**, 59–66.

Fine, A., Oertel, W. H., Hunt, S. P., Nomoto, M., Chong, P. N., Bond, A., Temlett, J., Annett, L., Dunnett, S., Jenner, P. and Marsden, C. D. (1989) In *Prog. Brain Res. 74: Transplantation into the Mammalian CNS: Preclinical and Clinical Studies* (edited by D. M. Gash and J. R. Sladek Jr.) (in press).

Finnegan, K. T., Irwin, I., Delanney, L. E., Ricaurte, G. A. and Langston, J. W. (1987) *J. Pharmacol. Exp. Therap.* **242**, 1144–1151.

Forno, L.S. (1982) *In* Neurology 2, Movement Disorders (edited by C. D. Marsden and S. Fahn). pp. 25–40. Butterworth Scientific, London.

Forno, L. S., Langston, J. W., DeLanney, L. E., Irwin, I. and Ricaurte, G. A. (1986) *Ann. Neurol.* **20**, 449–455.

Forno, L. S., Langston, J. W., DeLanney, L. E. and Irwin, I. (1988) *Brain Res.* **448**, 150–157.

Frank, D. M., Arora, P. K., Blumer, J. L. and Sayre, L. M. (1987) *Biochem. Biophys. Res. Commun.* **147**, 1095–1104.

Fuller, R. W., Robertson, D. W. and Hemrick-Leucke, S. K. (1987) *J. Pharmacol. Exp. Therap.* **240**, 415–420.

Garvey, J., Petersen, M., Waters, C. M., Rose, S. P., Hunt, S., Briggs, R., Jenner, P. and Marsden, C. D. (1986) *Movement Disorders* **1**, 129–134.

German, D. C., Dubach, M., Askari, S., Speciale, S. G. and Bowden, D. M. (1988) *Neuroscience* **24**, 161–174.

Gibb, W. R. G., Terruli, M., Lees, A. J., Jenner, P. and Marsden, C. D. (1989) *Movement Disorders* in press.

Goldstein, M., Battista, A. F., Ohmoto, T., Anagnoste, B. and Fuxe, K. (1973) *Science* **179**, 816–817.

Halliwell, B. and Gutteridge, J. M. C. (1985) *Free Radicals in Biology and Medicine.* Clarendon Press, Oxford.

Heikkila, R. E., Manzino, L., Cabbat, F. S. and Duvoisin, R. C. (1984) *Nature* **311**, 467–469.

Heikkila, R. E., Manzino, L., Cabbat, F. S. and Duvoisin, R. C. (1985) *J. Neurochem.* **45**, 1049–1054.

Heikkila, R. E., Youngster, S. K., Panek, D. U., Giovanni, A. and Sonsalla, P. K.

(1988a) *Toxicology* **49**, 493–501.

Heikkila, R. E., Kindt, M. V., Sonsalla, P. K., Giovanni, A., Youngster, S. K., McKeown, K. A. and Singer, T. P. (1988b) *Proc. Natl. Acad. Sci. USA* **85**, 6172–6176.

Hoppel, C. L., Greenblatt, D., Kwok, H.-C., Arora, P. K., Singh, M. P. and Sayre, L. M. (1987) *Biochem. Biophys. Res. Commun.* **148**, 684–693.

Hornykiewicz, O. (1982) In *Neurology 2, Movement Disorders* (edited by C. D. Marsden and S. Fahn), pp. 41–58. Butterworth Scientific, London.

Javitch, J. A., D'Amato, R. J., Strittmatter, S. M. and Snyder, S. H. (1985) *Proc. Natl. Acad. Sci. USA* **82**, 2173–2177.

Jenner, P., Sheehy, M. P. and Marsden, C. D. (1983) *Br. J. Clin. Pharmacol.* **15**, 277S–289S.

Jenner, P., Rupniak, N. M. J., Rose, S., Kelly, E., Kilpatrick, G., Lees, A. and Marsden, C. D. (1984) *Neurosci. Lett.* **50**, 85–90.

Jenner, P., Taquet, H., Mauborgne, A., Benoliel, J. T., Cesselin, F., Rose, S., Javoy-Agid, F., Agid, Y. and Marsden, C. D. (1986a) *J. Neurochem.* **47**, 1548–1551.

Jenner, P., Marsden, C. D., Nomoto, M. and Stahl, S. (1986b) *Br. J. Pharmacol.* **89**, 629P.

Johannessen, J. N., Chiueh, C. C., Burns, R. S. and Markey, S. P. (1985) *Life Sci.* **36**, 219–224.

Joyce, J. N., Marshall, J. F., Bankiewicz, K. S., Kopin, I. J. and Jacobowitz, D. M. (1985) *Brain Res.* **383**, 360–364.

Kalaria, R. N., Mitchell, M. J. and Harik, S. I. (1987) *Proc. Natl. Acad. Sci. USA* **84**, 3521–3525

Kass, G. E. N., Wright, J. M., Nicotera, P. and Orrenius, S. (1988) *Arch. Biochem. Biophys.* **260**, 789–797.

Kindt, M. V., Youngster, S. K., Sonsalla, P. K., Duvoisin, R. C. and Heikkila, R. E. (1988) *Eur. J. Pharmacol.* **146**, 313–318.

Kitt, C. A., Cork, L. C., Eidelberg, F., Joh, T. H. and Price, D. L. (1986) *Neuroscience* **17**, 1089–1103.

Langston, J. W., Ballard, P., Tetrud, J. W. and Irwin, I. (1983) *Science* **219**, 979–980.

Langston, J. W., Forno, L. S., Rebert, C. S. and Irwin, I. (1984a) *Brain Res.* **292**, 390–394.

Langston, J. W., Irwin, I., Langston, E. B. and Forno, L. S. (1984b) *Science* **225**, 1480–1482.

Langston, J. W. (1986) In *Recent Development in Parkinson's Disease* (edited by S. Fahn, C. D. Marsden, P. Jenner and P. Teychenne), pp. 119–126. Raven Press, New York.

Loschmann, P.-A., Chong, P. N., Nomoto, M., Tepper, P. G., Horn, A. S., Jenner, P. and Marsden, C. D. (1989) *Eur. J. Pharmacol,* in press.

Markey, S. P., Johannessen, J. N., Chiueh, C. C., Burns, R. S. and Herkenham, M. A. (1984) *Nature* **311**, 464–467.

Marsden, C. D., Parkes, J. D. and Quinn, N. (1982) In *Neurology 2: Movement Disorders* (edited by C. D. Marsden and S. Fahn), pp. 96–122. Butterworth Scientific, London.

Marttila, R. J., Lorentz, H. and Rinne, U. K. (1988) *J. Neurol. Sci.* **86**, 321–331.

Mayer, R. A., Kindt, M. V. and Heikkila, R. E. (1986) *J. Neurochem.* **47**, 1073–1079.

Mitchell, I. J., Cross, A. J., Sambrook, M. A. and Crossman, A. R. (1985) *Neurosci. Lett.* **61**, 195–200.

Mytilineou, C., Cohen, G. and Heikkila, R. E. (1985) *Neurosci. Lett.* **57**, 19–24.

Nicklas, W. J., Vyas, I. and Heikkila, R. E. (1985) *Life Sci.* **36**, 2503–2508.

Nomoto, M., Jenner, P. and Marsden, C. D. (1985) *Neurosci. Lett.* **57**, 37–41.

Nomoto, M., Stahl, S., Jenner, P. and Marsden, C. D. (1987) *Movement Disorders* **2**, 37–45.

Nomoto, M., Jenner, P. and Marsden, C. D. (1988) *Neurosci. Lett.* **93**, 275–280.

Pechadre, J. C., Larochelle, L. and Poirier, L. J. (1976) *J. Neurol. Sci.* **28**, 147–157.

Perry, T. L., Yong, V. W., Clavier, R. M., Jones, K., Wright, J. M., Foulks, J. G. and Wall, R. A. (1985) *Neurosci. Lett.* **60**, 109–114.

Perry, T. L., Yong, V. W., Hansen, S., Jones, K., Bergeron, C., Foulks, J. G. and Wright, J. M. (1987) *J. Neurol. Sci.* **81**, 321–331.

Petersen, L. A., Caldera, P. S., Trevor, A., Chiba, K. and Castagnoli, N. Jr. (1985) *J. Med. Chem.* **28**, 1432–1436.

Pifl, Ch., Schingnitz, G. and Hornykiewicz, O. (1988) *Neurosci. Lett.* **92**, 228–233.

Poirier, J., Roy, M., Campanella, G., Cloutier, T. and Paris, S. (1987) *Lancet* **ii**, 386.

Rajput, A. H., Uitti, R. J., Stern, W., Laverty, W., O'Donnell, K., O'Donnell, D., Yuen, W. K. and Dua, A. (1987) *Can. J. Neurol. Sci.* **14**, 414–418.

Ramsay, R. R. and Singer, T. P. (1986) *J. Biol. Chem.* **17**, 7585–7587.

Ramsay, R. R., Salach, J. I. and Singer, T. P. (1986a) *Biochem. Biophys. Res. Commun.* **134**, 743–748.

Ramsay, R. R., Dadgar, J., Trevor, A. and Singer, T. P. (1986b) *Life Sci.* **39**, 581–588.

Ransom, B. R., Kunis, D. M., Irwin, I. and Langston, J. W. (1987) *Neurosci. Lett.* **75**, 323–328.

Reavill, C., Jenner, P. and Marsden, C. D. (1983) *Biochem. Pharmacol.* **32**, 865–870.

Redmond, D. E., Roth, R. H., Elsworth, J. D., Sladek, J. R. Jr., Collier, T. J., Deutch, A. Y. and Haber, S. (1986) *Lancet* **ii**, 1125–1127.

Riachi, N. J., Harik, S. I., Kalaria, R. N. and Sayre, L. M. (1988a) *J. Pharmacol. Exp. Therap.* **244**, 443–448.

Riachi, N. J., Arora, P. K., Sayre, L. M. and Harik, S. I. (1988b) *J. Neurochem.* **50**, 1319–1321.

Riederer, P., Sofic, E., Rausch, W.-D., Schmidt, B., Reynolds, G. P., Jellinger, K. and Youdim, M. B. H. (1989) *J. Neurochem.* **52**, 515–520.

Rose, S., Nomoto, M., Kelly, E., Kilpatrick, G., Jenner, P. and Marsden, C. D. (1989a) *Neurosci. Lett.* **101**, 305–310.

Rose, S., Nomoto, M., Jenner, P. and Marsden, C. D. (1989b) *Biochem. Pharmacol*, in press.

Rossetti, Z. L., Sotgiu, A., Sharp, D. E., Hadjiconstantinou, M. and Neff, N. H. (1988) *Biochem. Pharmacol.* **37**, 4573–4574.

Saggu, H., Cooksey, J., Dexter, D., Wells, F. R., Lees, A., Jenner, P. and Marsden, C. D. (1989) *J. Neurochem*, in press.

Salach, J. I., Singer, T. P., Castagnoli, N. Jr. and Trevor, A. (1984) *Biochem. Biophys. Res. Commun.* **125**, 831–835.

Sanchez-Ramos, J., Barratt, J. N., Goldstein, M., Weiner, W. J. and Hefti, F. (1986) *Neurosci. Lett.* **72**, 215–220.

Sanchez-Ramos, J. R., Michel, P., Weiner, W. J. and Hefti, F. (1988) *J. Neurochem.* **50**, 1934–1943.

Sayre, L. M., Arora, P. K., Feke, S. C. and Urbach, F. L. (1986) *J. Am. Chem. Soc.* **108**, 2464–2466.

Schapira, A. H. V., Cooper, J. M., Dexter, D., Jenner, P., Clark, J. B. and Marsden, C. D. (1989) *J. Neurochem.* submitted.

Schneider, J. S., Yuwiler, A., Markham, C. H. (1987) *Brain Res.* **411**, 144–150.

Schoenberg, B. S. (1987) *Can. J. Neurol. Sci.* **14**, 407–413.

Schultz, W., Scarnati, E., Sundstrom, E., Tsutsumi, T. and Jonsson, G. (1986) *Exp. Brain Res.* **63**, 216–220.

Setler, P. E., Sarau, H. M., Zirkle, C. L. and Saunders, H. L. (1978) *Eur. J. Pharmacol.* **50**, 419–430.

Sheehy, M. P., Schachter, M., Parkes, J. D. and Marsden, C. D. (1981) In *Research Progress in Parkinson's Disease* (edited by F. Clifford-Rose and R. Capildeo), pp. 309–317. Pitman Medical, Tunbridge Wells.

Shen, R.-S., Abell, C. W., Gessner, W. and Brossi, A. (1985) *FEBS Lett.* **189**, 225–230.

Singer, T. P., Castagnoli, N. Jr., Ramsay, R. R. and Trevor, A. J. (1987) *J. Neurochem.* **49**, 1–8.

Sofic, E., Riederer, P., Heinsen, H., Beckmann, H., Reynolds, G. P., Hebenstreit, G. and Youdim, M. B. H. (1988) *J. Neural Transm.* **74**, 199–205.

Tanner, C. M., Chen, B., Wang, W.-Z., Peng, M.-L., Liu, Z.-L., Liang, X.-L., Kao, L. C., Gilley, D. W. and Schoenberg, B. S. (1987) *Can. J. Neurol. Sci.* **14**, 419–423.

Taquet, H., Nomoto, M., Rose, S., Jenner, P., Javoy-Agid, F., Mauborgne, A., Benoliel, J. J., Marsden, C. D., Legrand, J. C., Agid, Y., Hamon, M. and Cesselin, F. (1988) *Neuropeptides* **12**, 105–110.

Trevor, A. J., Castagnoli, N. Jr., Caldera, P., Ramsay, R. R. and Singer, T. P. (1987) *Life Sci.* **40**, 713–719.

Ungerstedt, U. (1971a) *Acta Physiol. Scand. (Suppl.)* **367**, 49–68.

Ungerstedt, U. (1971b) *Acta Physiol. Scand. (Suppl.)* **367**, 69–93.

Ungerstedt, U. and Arbuthnott, G. W. (1970) *Brain Res.* **24**, 485–493.

von Voigtlander, P. F. and Moore, K. E. (1973) *Neuropharmacology* **12**, 451–462.

Vyas, I., Heikkila, R. E. and Nicklas, W. J. (1986) *J. Neurochem.* **46**, 1501–1507.

Waters, C. M., Hunt, S. P., Jenner, P. and Marsden, C. D. (1987) *Neuroscience* **23**, 1025–1039.

Yong, V. W., Perry, T. L. and Krisman, A. A. (1986) *Neurosci. Lett.* **63**, 56–60.

Youngster, S. K., Duvoisin, R. C., Hess, H., Sonsalla, P. K., Kindt, M. V. and Heikkila, R. E. (1986) *Eur. J. Pharmacol.* **122**, 283–287.

Youngster, S. K., Sonsalla, P. K. and Heikkila, R. E. (1987) *J. Neurochem.* **48**, 929–934.

Zamir, N., Skofitsch, G., Bannon, M. J., Helke, C. J., Kopin, I. J. and Jacobwitz, D. M. (1984) *Brain Res.* **322**, 356–360.

12

The Role of Dopamine in the Selectivity of the Neurotoxin MPTP for Nigrostriatal Dopaminergic Neurons

Eldad Melamed

The etiology of Parkinson's disease is unknown but one of the popular theories suggests that it may be caused by an exogenous neurotoxin (Calne and Langston, 1983). A major criticism against this hypothesis argued that it would be unlikely for such a putative toxin to so selectively destroy the nigrostriatal dopaminergic neurons while leaving other neuronal populations in the brain totally unaffected. The story of 1-methyl-*n*-phenyl-1,2,5,6-tetrahydropyridine (MPTP) indicates that such a scenario is not impossible and gave tremendous support for the neurotoxic hypothesis (Langston *et al.*, 1983; Ballard *et al.*, 1985). MPTP came on the scene as a by-product of illicit "designer's drug" synthesis that caused a syndrome indistinguishable from that of the "natural" idiopathic Parkinson's disease in a group of young drug addicts (Langston, 1985). It is now well established that MPTP damages nigrostriatal dopaminergic neurons in several mammalian species including humans, primates and mice (Burns *et al.*, 1983; Heikkilla *et al.*, 1984a; Jenner *et al.*, 1984; Kopin *et al.*, 1986). MPTP is a relatively simple molecule and it is not unlikely that our environment may be polluted with similar or other substances derived from various industrial sources with potent dopaminergic

DISORDERS OF MOVEMENT: CLINICAL, PHARMACOLOGICAL AND PHYSIOLOGICAL ASPECTS   ISBN 0-12-569685-X

neurotoxic properties (Barbeau, 1984). Although there is yet no evidence that idiopathic Parkinson's disease is caused by an MPTP-like neurotoxin, investigations of its mechanisms of action are of great interest in elucidating the pathogenesis of this common but enigmatic neurological disorder.

There are several intriguing strata of selectivity related to the dopaminergic neurotoxicity of MPTP.

SPECIES SELECTIVITY

There are marked species differences in susceptibility to systemic administration of MPTP. Humans and primates are extremely sensitive and their nigrostriatal neurons are massively and irreversibly destroyed by small doses of the toxin (Kopin et al., 1986). By contrast, rats are immune to the toxic effects of exogenously administered MPTP (Boyce et al., 1984). Mice represent an intermediate situation. This species requires relatively much larger doses of MPTP to develop depletions of dopamine in the striatum (Heikkilla et al., 1984a; Melamed et al., 1985a; Kopin et al., 1986). The latter are mainly due to lesion of the dopaminergic nerve terminals in the form of distal axonopathy, are transitory and associated with no or only partial loss of dopaminergic neuronal cell bodies in the substantia nigra pars compacta (Hallman et al., 1985; Ricaurte et al., 1986). The causes for the variable species sensitivity to the toxicity of MPTP are unknown. One possible mechanism may involve longer persistence of MPTP in the central nervous systems of humans and primates while it is more rapidly cleared from brains of rodents (Johanessen et al., 1985). Another possibility (D'Amato et al., 1986; Melamed et al., 1987) is presence of neuromelanin in human and primate substantia nigra pars compacta dopaminergic neuronal peikarya while this pigment is absent in rodents (Marsden, 1961).

DOPAMINERGIC VERSUS OTHER NEURONAL POPULATIONS SELECTIVITY

Although MPTP readily penetrates brain parenchyma following its systemic administration, it damages almost exclusively central dopaminergic neurons. It largely spares the multitude of other neuronal systems. Even other monoaminergic neurons, e.g. serotonergic and noradrenergic, generally escape the toxic effects of MPTP (Hallman et al., 1985; Melamed et al., 1985a).

NIGROSTRIATAL VERSUS OTHER CENTRAL DOPAMINERGIC NEURONS SELECTIVITY

Although MPTP is a dopaminergic neurotoxin, it does not have the same effect on the different subpopulations of central dopaminergic neurons. The most pronounced lesion produced by systemically administered MPTP affects the nigrostriatal neurons originating from the A9 dopaminergic perikaryal nuclear collection in the substantia nigra pars compacta (Kopin et al., 1986). Nigromesolimbic dopaminergic neurons ascending from the A10 region in the nigra are damaged by MPTP but to a much lesser extent (Melamed et al., 1985a). In contrast, nigromesocortical, hypothalamic and retinal-amacrine dopaminergic neurons are either completely spared or only mildly damaged by the neurotoxin (Melamed et al., 1985a).

The greatest riddle concerning MPTP is why the toxin is so selective for the nigrostriatal dopaminergic neurons. A feeling shared by investigators involved in MPTP research is that solving this puzzle may yield exceedingly important insights into the mechanisms responsible for the nigrostriatal degeneration that constitutes the pathogenic substrate of idiopathic Parkinson's disease. Extensive research was successful in unravelling some of the mechanisms responsible for the dopaminergic neurotoxicity of MPTP but the picture is still far from being complete (Langston, 1985; Kopin et al., 1986; Snyder and D'Amato, 1986).

THE ROLE OF MONOAMINE OXIDASE (MAO) TYPE B IN THE NEUROTOXICITY OF MPTP

Under normal circumstances, one of the mechanisms utilized to terminate the action of dopamine molecules released from presynaptic dopaminergic nerve endings into nigrostriatal synapses involves degradation of the neurotransmitter by MAO. It is known that there are at least two subclasses of MAO. MAO-A is more specific for degradation of serotonin and norepinephrine while MAO-B is more selective for dopamine metabolism. It was found that a crucial initial step in the mechanism of MPTP toxicity is its oxidative conversion to the toxic metabolite 1-methyl-n-pyridinium ion (MPP+) (Markey et al., 1984; Langston et al., 1984a). This reaction is catalyzed in the brain by MAO-B (Chiba et al., 1984; Heikkilla et al., 1985; Melamed et al., 1986). Furthermore, it was shown that MPTP binds to MAO-B and also inhibits the activity of this enzyme (Singer et al., 1985). In addition, animals can be protected against the toxic effects of MPTP by pretreatment with specific MAO-B inhibitors, such as pargyline or deprenyl, that prevent the formation of MPP^+ from MPTP by the enzyme (Heikkilla *et al.*, 1984b; Langston *et al.*, 1984b, Cohen *et al.*, 1984). Although it would appear at first sight that the role of MAO-B in MPTP toxicity might explain the selectivity of the toxin for dopaminergic neurons, this is definitely not the case. Despite the fact that MAO-B is the specific enzyme for breakdown of dopamine, it is not present with nigrostriatal or other dopaminergic neurons. Rather surprisingly, immunohistochemical studies have shown that MAO-B is located outside of dopaminergic neurons predominantly in glial cells but also in serotonergic neurons (Levitt *et al.*, 1982; Westlund *et al.*, 1985). Furthermore, we recently demonstrated that the oxidation of MPTP to MPP^+ takes place in extradopaminergic sites within the striatum, in glia, and not in serotonergic nerve terminals (Melamed *et al.*, 1986). All these data indicate that MAO-B is probably not the determinant factor in the selectivity of MPTP for dopaminergic neurons.

THE ROLE OF THE DOPAMINE RE-UPTAKE SYSTEM IN THE NEUROTOXICITY OF MPTP

Under physiological conditions, a large fraction of dopaminergic molecules released into the synapse re-enter the presynaptic nigrostriatal nerve terminals via the specific dopamine re-uptake channels in order to be reutilized. Many studies have shown that MPP^+, the toxic product of MPTP oxidation by MAO-B, gains entry into dopaminergic nerve endings through the specific dopamine re-uptake system (Javitch and Snyder, 1985; Kopin *et al.*, 1986). Furthermore, the dopaminergic neurotoxicity of MPTP can be prevented if animals are pretreated with drugs that block re-uptake of dopamine and concomitantly that of MPP^+, such as nomifensine, mazindol or benztropine (Sundstrom and Jonsson, 1985; Melamed *et al.*, 1985b,c). Since only dopaminergic neurons contain the specific channels for dopamine uptake, this mechanism could provide a perfect explanation for the selectivity of MPTP for dopaminergic neurons (Snyder and D'Amato, 1986). However, although this is undoubtedly an important feature of such selectivity, it cannot be the only cause. For instance, the rat brain contains MAO-B which is capable of converting MPTP to MPP^+ (Melamed *et al.*, 1985d). Dopaminergic neurons in the rat central nervous system have dopamine re-uptake channels and yet this species is resistant to the toxic effects of systemically administered MPTP. Furthermore, not only nigrostriatal nerve terminals but also those of the nigromesolimbic, nigromesocortical, hypothalamic and retinal dopaminergic neurons contain the specific dopamine re-uptake systems and are capable of taking up MPP^+ (Kopin *et al.*, 1986). Nevertheless, these neuronal projections are less or not at all damaged by the neurotoxin. In addition, MPP^+ is produced outside of dopaminergic neurons and mainly in glia but there is no evidence that these cellular elements are destroyed by the toxin. Taken together, these data suggest that penetration of MPP^+ into dopaminergic neurons through the specific dopamine re-uptake channels cannot represent the only cause for selectivity of MPTP.

IS THERE A ROLE FOR PRESENCE OF DOPAMINE ITSELF WITHIN NIGROSTRIATAL NEURONS IN THE SELECTIVE NEUROTOXICITY OF MPTP?

Another unique feature of dopaminergic neurons (in addition to dopamine re-uptake channels) is simply that they contain their neurotransmitter dopamine. It is possible (at least theoretically) that presence of dopamine itself within dopaminergic neurons makes them the specific target for the attack of MPTP. It is unknown how, once inside the dopaminergic neurons, $MPTP/MPP^+$ exert their damaging effects. One possibility is that MPP^+ poisons the mitochondrial respiratory enzymes and thus causes cellular death (Nicklas et al., 1985; Ramsay and Singer, 1986). Another mechanism may involve generation of cytotoxic free radical species (Kopin et al., 1986). There is a chemical similarity between $MPTP/MPP^+$ and the herbicide paraquat. Paraquat diverts electron flow from the water-producing cytochrome pathway to a superoxide-producing route (Hassan and Fridovich, 1978). Likewise, other specific dopaminergic neurotoxins, such as 6-hydroxydopamine and manganese, may also operate through the generation of free radicals. Pretreatment of animals with diethyldithiocarbamate, an inhibitor of superoxide dismutase which is a scavenger enzyme for superoxides, greatly potentiates the dopaminergic neurotoxicity of MPTP (Corsini et al., 1985). There is evidence for further oxidation of MPP^+ inside dopaminergic neurons which could result in intraneuronal superoxide formation (Mytilineu and Cohen, 1985). Dopamine itself may undergo autooxidation and produce free radicals (Cohen, 1984). It was even hypothesized that excessive cytotoxic free radical formation by intraneuronal dopamine oxidation may cause the progressive nigrostriatal degeneration in Parkinson's disease (Calne and Langston, 1983; Cohen, 1984). In vitro studies have shown that $MPTP/MPP^+$ catalyse and enhance the autooxidation of dopamine and the generation of superoxides (Poirier et al., 1985; Poirier and Barbeau, 1985). Hypothetically, MPP^+ may combine with dopamine to form another toxic agent. All these data suggest that interaction of $MPTP/MPP^+$ with dopamine inside dopaminergic neurons may represent an important characteristic of selectivity of the neurotoxin. We, therefore, undertook several studies in an attempt to establish or rule out this possibility. Investigations were carried out in male C57 black mice (initial weight 20–25 g). MPTP was given subcutaneously at a dose of $40 \, mg \, kg^{-1}$. Animals were usually decapitated 3 weeks after MPTP injection. We measured depletions in striatal dopamine concentrations as an index for MPTP toxicity using high performance liquid chromatography with electrochemical detection (Hefti *et al.*, 1981).

Effect of increased dopamine concentrations within nigrostriatal dopaminergic neurons on the toxicity of MPTP

If interaction of MPTP/MPP$^+$ with intraneuronal dopamine is an important factor in MPTP toxicity, enhancement of dopamine levels within nigrostriatal neurons may amplify MPTP-induced striatal dopamine depletions. In that context it is of interest that in a comparative study we found that mouse striatum contains significantly more dopamine (by about 30–80%) as compared with that in the rat, a species immune to systemic treatment with MPTP. In the biosynthesis of monoamines, dopamine is synthesized from its precursor L-dopa in a reaction catalysed by the enzyme dopa decarboxylase (aromatic L-amino acid decarboxylase). Under normal circumstances, dopa decarboxylase is predominantly localized in rodent striatum within nigrostriatal nerve terminals. Systemic administration of L-dopa produces marked increases in striatal dopamine concentrations (Hefti *et al.*, 1981). In animals with intact nigrostriatal projections, most of the dopamine formed from exogenous L-dopa in striatum is located in nigrostriatal dopaminergic nerve endings (Hefti *et al.*, 1981). We pretreated mice with a large dose of L-dopa ($50 \, mg \, kg^{-1}$ intraperitoneally) in combination with the peripheral dopa decarboxylase inhibitor carbidopa, to enhance L-dopa penetration into, and dopamine formation within, nigrostriatal neurons. One hour later the animals received MPTP and were decapitated 3 weeks later. Prior administration of L-dopa, producing acute increases in striatal dopamine levels, did not increase the striatal dopamine depletions caused by treatment with MPTP alone. In fact, in our experiment, pretreatment with L-dopa even showed some protective effect and rather attenuated the MPTP-induced dopamine decreases in the striatum. This could perhaps be due to competition of L-dopa or the formed dopamine with MPTP on MAO-B or on the dopamine re-uptake system. In another study, mice were given a single acute challenge with MPTP and then received injections of L-dopa and carbidopa, once daily, for 3 weeks. Animals were decapitated after an additional week of washout. Continuous loading of the animals with L-dopa and long-term increases of dopamine concentrations after MPTP administration did not augment effects of the neurotoxin. MPTP-induced striatal dopamine depletions were identical in animals given chronic treatment with L-dopa and in those receiving MPTP alone. In an additional experiment we injected male albino rats (Hebrew University strain, initial weight 180 g) with L-dopa and carbidopa prior to systemic administration of MPTP in an attempt to overcome resistance of this species to the neurotoxin. Neither MPTP alone nor MPTP given after an acute challenge with L-dopa (that produced large increases in striatal dopamine levels) caused any decreases in dopamine levels in rat striatum. These findings suggest that increased concentrations of dopamine within nigrostriatal nerve terminals do not enhance the toxic effects of MPTP.

Effect of decreased dopamine concentrations within nigrostriatal dopaminergic neurons on the toxicity of MPTP

If the presence of dopamine within dopaminergic nerve terminals constitutes an important feature of MPTP toxicity, depletion of dopamine prior to treatment with the neurotoxin might have a protective effect. We used two experimental designs to study this possibility. Reserpine blocks the formation of storage vesicles and produces depletions of dopamine in nigrostriatal neurons. We pretreated mice with reserpine at a dose that causes massive but reversible dopamine depletions in striatum and then administered MPTP. Dopamine falls in striatum produced by MPTP (at 3 weeks post-treatment) were not abolished or suppressed by prior administration of reserpine. In fact, in reserpinized mice MPTP caused slightly greater dopamine decreases. Another substance, α-methyl-paratyrosine, also causes dopamine depletions in nigrostriatal neurons by inhibiting the enzyme tyrosine hydroxylase that catalyses the conversion of tyrosine to L-dopa. As was the case with reserpine, pretreatment of mice with this tyrosine hydroxylase inhibitor did not prevent or alter in any way the toxic effects of MPTP. A similar finding was reported by Schmidt et al. (1985). These data indicate that decreased levels of dopamine within nigrostriatal nerve endings do not diminish the dopaminergic neuro-toxicity of MPTP.

Effect of acceleration and suppression of impulse flow and dopamine turnover within nigrostriatal dopaminergic neurons on the toxicity of MPTP

Since simple increases or decreases in dopamine concentrations did not make nigrostriatal neurons more or less vulnerable, respectively, to the toxic effects of MPTP, we wondered whether the latter may depend on the rates of firing and dopamine synthesis and release (turnover) of the dopaminergic neurons. For instance, there are differences in dopamine turnover among the various subpopulations of central dopaminergic neurons which might, theoretically, explain the differences in their susceptibility to MPTP. Neuroleptic drugs that block post- and pre-synaptic dopaminergic receptors in striatum, accelerate impulse flow and dopamine turnover in nigrostriatal neurons. We therefore pretreated mice with an acute single intraperitoneal injection of haloperidol or chlorpromazine at a dose that increased the rates of dopamine synthesis and release (manifested by augmentation of levels of the dopamine metabolites DOPAC and HVA). After 1 h, animals were given MPTP and decapitated 3 weeks later. MPTP-induced dopamine depletions were similar in neuroleptic pretreated animals and in those given MPTP alone. In another experiment, mice were administered MPTP once and then injected chronically with haloperidol, once daily, for 3 weeks and decapitated after an additional washout period. Continuous treatment of the animals with haloperidol, inducing long-term acceleration of dopamine turnover, did not alter effects of the neurotoxin.

Direct dopaminergic agonists, such as apomorphine or bromocriptine, stimulate both post- and pre-synaptic dopaminergic receptors and consequently suppress firing rates and dopamine turnover in nigrostriatal neurons. We gave mice either apomorphine or bromocriptine prior to an acute challenge with MPTP. Here again, decreases in striatal concentrations of dopamine induced by MPTP in animals pretreated with apomorphine or bromocriptine were similar to those in mice given MPTP alone.

Taken together, our findings do not support a key role for presence of dopamine within nigrostriatal neurons in the selectivity of MPTP towards this neuronal projection. Other (partly hypothetical) biochemical characteristics that are unique for dopaminergic neurons (e.g. neuromelanin, lack of neutralizing enzymes, differences in the mitochondrial respiratory apparatus) must be sought for to identify the cause for their extraordinary susceptibility to MPTP.

ACKNOWLEDGEMENTS

Supported, in part, by the Karen and Erich Segal Foundation, by the Jacob and Hilda Blaustein Foundation, Inc. and by the Adeline and Dr Benjamin Boshes Endowment Fund for Research of the Neurological Disorders of Aging and Parkinson's Disease.

REFERENCES

Ballard, P. A., Tetrud, J. W. and Langston, J. W. (1985) *Neurology* **35**, 949–956.
Barbeau, A. (1984) *Can J. Neurol. Sci.* **11**, 24–28.
Boyce, S., Kelly, E., Reavill, C., Jenner, P. and Marsden, C. D. (1984) *Biochem.*

Pharmacol. **33**, 1747–1759.
Burns, R. S., Ciueh, C. C., Markey, S., Ebert, M. H., Jacobowitz, P. and Kopin, I. J. (1983) *Proc. Natl. Acad. Sci. USA* **80**, 4546–4550.
Calne, D. B. and Langston, J. W. (1983) *Lancet* **ii**, 1457–1459.
Chiba, K., Trevor, A. J. and Castagnoli, J. N. (1984) *Biochem. Biophys. Res. Commun.* **120**, 574–578.
Cohen, G. (1984) *Neurotoxicology* **5**, 77–82.
Cohen, G., Pasik, P., Cohen, B., Leist, A., Mytilineu, C. and Yahr, M. D. (1984) *Eur. J. Pharmacol.* **106**, 209–210.
Corsini, G. U., Pintus, S., Chiueh, C. C., Weiss, J. F. and Kopin, I. J. (1985) *Eur. J. Pharmacol.* **119**, 127–128.
D'Amato, R. J., Lippman, Z. P. and Snyder, S. H. (1986) *Science* **231**, 987–989.
Hallman, H., Lange, J., Olson, L., Stromberg, I. and Johnsson, G. (1985) *J. Neurochem.* **44**, 117–127.
Hassan, H. M. and Fridovich, I. (1978) *J. Biol. Chem.* **253**, 8143–8148.
Hefti, F., Melamed, E. and Wurtman, R. J. (1981) *J. Pharmacol. Exp. Ther.* **217**, 1451–1457.
Heikkilla, R. E., Hess, H. and Duvoisin, R. C. (1984a) *Science* **224**, 1451–1453.
Heikkilla, R. E., Manzino, L., Cabbat, F. S. and Duvoisin, R. C. (1984b) *Nature* **311**, 467–469.
Heikkilla, R. E., Manzino, L., Cabbat, F. S. and Duvoisin, R. C. (1985) *J. Neurochem.* **45**, 1049–1054.
Javitch, J. A. and Snyder, S. H. (1985) *Eur. J. Pharmacol.* **106**, 455–456.
Jenner, P., Nadia, M. T., Rupniak, S. R., Kelly, E., Kilpatrick, G., Lees, A. and Marsden, C. D. (1984) *Neurosci. Lett.* **50**, 85–90.
Johanessen, J. N., Chiueh, C. C., Burns, R. S. and Markey, S. P. (1985) *Life Sci.* **36**, 219–224.
Kopin, I. J., Burns, S. R., Chiueh, C. C. and Markey, S. P. (1986) In *Alzheimer's and Parkinson's Diseases: Strategies for Research and Development* (edited by A. Fisher, I. Hanin and C. Lachman), pp. 519–530. Plenum Press, New York.
Langston, J. W. (1985) *Trends Neurosci.* **80**, 79–83.
Langston, J. W., Ballard, P. A., Tetrud, J. W. and Irwin, I. (1983) *Science* **219**, 979–980.
Langston, J. W., Irwin, I., Langston, E. B. and Forno, L. S. (1984a) *Neurosci. Lett.* **48**, 87–92.
Langston, J. W., Irwin, I., Langston, E. B. and Forno, L. S. (1984b) *Science* **225**, 1480–1482.
Levitt, P., Pintar, J. E. and Breakfield, X. U. (1982) *Proc. Natl. Acad. Sci. USA* **79**, 6385–6388.
Markey, S. P., Johanessen, J. N., Chiueh, C. C., Burns, R. S. and Herkenheim, M. A. (1984) *Nature* **311**, 464–467.
Marsden, C. D. (1961) *J. Anat.* **95**, 256–261.
Melamed, E., Rosenthal, J., Globus, M., Cohen, O. and Frucht, Y. (1985a) *Eur. J. Pharmacol.* **114**, 97–100.
Melamed, E., Rosenthal, J., Cohen, O., Globus, M. and Uzzan, A. (1985b) *Eur. J. Pharmacol.* **116**, 179–181.
Melamed, E., Rosenthal, J., Globus, M., Cohen, O. and Uzzan, A. (1985c) *Brain Res.* **342**, 401–404.
Melamed, E., Youdim, M. B. H., Rosenthal, J., Spanier, I., Uzzan, A. and Globus, M. (1985d) *Brain Res.* **359**, 360–363.

Melamed, E., Pikarsky, E., Goldberg, A., Rosenthal, J. and Uzzan, A. (1986) *Brain Res.* **399**, 178–180.

Melamed, E., Soffer, D., Rosenthal, J., Pikarsky, E. and Reches, A. (1987) *Neurosci. Lett.* **83**, 41–46.

Mytilineu, C. and Cohen, G. (1985) *Neurochemistry* **45**, 1951–1953.

Nicklas, W. J., Byas, I. and Heikkila, R. E. (1985) *Life Sci.* **36**, 2503–2508.

Poirier, J. and Barbeau, A. (1985) *Biochem. Biophys. Res. Commun.* **131**, 1248–1289.

Poirier, J., Donaldson, J. and Barbeau, A. (1985) *Biochem. Biophys. Res. Commun.* **128**, 25–33.

Ramsay, R. R. and Singer, T. P. (1986) *J. Biol. Chem.* **261**, 7585–7587.

Ricaurte, G. H., Langston, J. W., DeLanney, L. E., Irwin, I., Peroutka, S. J. and Forno, L. S. (1986) *Brain Res.* **376**, 117–124.

Schmidt, C. J., Bruckwick, E. and Lovenberg, W. (1985) *Eur. J. Pharmacol.* **113**, 149–150.

Singer, T. P., Salach, J. I. and Crabtree, D. (1985) *Biochem. Biophys. Res. Commun.* **127**, 707–712.

Snyder, S. H. and D'Amato, R. J. (1986) *Neurology* **36**, 250–258.

Sundstrom, E. and Jonsson, G. (1985) *Eur. J. Pharmacol.* **110**, 293–299.

Westland, K. N., Denney, R. M., Kochersperger, L. M., Rose, R. M. and Abell, C. W. (1985) *Science* **230**, 181–183.

13

Effect of Dopamine D1 and D2 Receptor Stimulation in MPTP Monkeys

Paul J. Bedard, René Boucher and Thérèse Di Paolo

Dyskinesia is one of the most frequent side-effects of L-dopa therapy occurring in more than 50% of patients (Mones et al., 1971; Rinne, 1987). The cause or mechanism of dyskinesia is unclear but the fact that it generally begins and remains more prominent in the first affected side suggests that denervation itself plays some role. Moreover in patients whose parkinsonian syndrome is caused by MPTP, and is very severe from the onset, dyskinesia appear very rapidly (Langston and Ballard, 1984).

On the other hand, the direct dopamine agonist bromocriptine when given *de novo* to parkinsonian patients is much less likely to induce dyskinesia even after several years (Lees and Stern, 1981; Rascol et al., 1979; Rinne, 1987). This suggests that the nature of the drug may play a role in the induction of dyskinesia.

To explain such a difference, one can think of several hypotheses: (1) L-dopa may be intrinsically a more potent stimulant of the dopamine receptors and the dyskinesia is simply an index of overstimulation; (2) the difference may lie in the fact that L-dopa via dopamine affects the two types of dopamine receptors, D1 and D2 (Kebabian and Calne, 1979), while bromocriptine affects mainly the D2 type and may even inhibit the D1 receptor.

If the latter proposition is true, this means that chronic D1 receptor activation is in some way responsible for dyskinesia. This can be viewed in several ways. (a) The D1 receptor may acquire new functions under the conditions of denervation (Arnt and Hyttel, 1984; Arnt, 1985). If so, a selective D1 agonist should be able to elicit dyskinesia by itself. (b) Dyskinesia may be the result of potentiation or synergy due to simultaneous stimulation of D1 and D2 receptors in a denervated preparation (Robertson and Robertson, 1986; Mashurano and Waddington, 1986; Walters *et al.*, 1987). (c) Dyskinesia may be elicited only by D2 agonists but prolonged stimulation of the D1 receptors may in some way alter the responsiveness of the D2 receptors leading to an abnormal response. If so it might be possible after chronic treatment with a selective D1 agonist to elicit dyskinesia with a selective D2 agonist alone and to observe changes in binding of striatal membranes to ^{3}H-spiperone, a D2 ligand.

We have therefore tested these hypotheses in a monkey model of parkinsonism in a series of acute and chronic pharmacological experiments involving treatment with selective D1 and D2 as well as mixed agonists. In some acute experiments, selective antagonists were also used.

METHODS AND RESULTS

In 36 female monkeys (*Macaca fascicularis*, weight approx. 3 kg) a parkinsonian syndrome was induced by injection of 1-methyl-4-phenyl-1,2,3,6-tetrahydropyridine (MPTP) (Aldrich Chemical) (Burns *et al.*, 1983) administered intravenously in doses of $0.3–1.5$ mg kg^{-1}.

The animals were used once they had developed a clear syndrome of akinesia. The response of the animals to the drugs was evaluated in terms of general behaviour, and side-effects such as abnormal movements. Two parameters were selected for quantification: (1) gross locomotor activity was evaluated with photocells placed on the sides of the cages and relayed to a microcomputer set to give a count every 15 min; (2) lingual dyskinesia were quantified by a trained observer using a time-event counter to record each abnormal tongue protrusion during a period of 60 min after injection of the drugs.

Chronic experiments

A total of 30 parkinsonian monkeys received one of the following treatments daily orally (by oesophagal tube) for periods ranging from 1 to 5 months.

Table 13.1   Effect of various drugs on locomotor activity and dyskinesia.

Drug	n	Relief of akinesia (maximum + + + +)	Dyskinesia (number of animals)
Sinemet	10	+ + + +	10/10
Bromocriptine	14	+ + + +	1/14
SKF 38393	3	0	0/3
Bromocriptine + SKF 38393	3	+ + + +	0/3

L-Dopa/carbidopa, $50/12.5 \, kg^{-1}$, bromocriptine $5 \, mg \, kg^{-1}$, SKF 38393, $5 \, mg \, kg^{-1}$ and a combination of bromocriptine and SKF 38393 (same doses). Table 13.1 summarizes our behavioural observations.

Both L-dopa and bromocriptine relieved the parkinsonian akinesia during the treatment period and the animals could move, eat and drink. All observers agreed however that the animals treated with bromocriptine appeared more normal. This was due to the fact the L-dopa induced at this dose marked stereotypies, agitation, shouts, biting, licking and abnormal movements especially of the face and tongue. In fact 10 animals treated chronically with L-dopa developed prominent dyskinesia of the face and tongue and three developed choreic movements of the limbs which were seen after each dose during the peak of the effect. Only one animal treated with bromocriptine alone developed mild facial dyskinesia. SKF 38393 had no apparent behavioural effect at the dose given and it did not appear to potentiate the response to bromocriptine in terms of locomotor activity or of dyskinesia.

Three days after the last dose of pharmacological agent, all animals were sacrificed with an overdose of pentobarbital, the brain was taken out and the caudate nucleus (anterior and posterior) and putamen (anterior and posterior) dissected out for determination of dopamine levels and ^{3}H-spiperone binding. Table 13.2 summarizes the results obtained in a group of 12 animals in one of the chronic experiments.

As can be seen, compared to control animals, denervation by MPTP which exceeded 90% in all groups as indicated by the dopamine levels, induced a 25% increase in the density of D2 receptors in the striatum which was however prevented or reversed by treatment with L-dopa, bromocriptine and surprisingly even by SKF 38393. Bromocriptine by itself induced a small increase in affinity.

Thus our chronic studies show that in monkeys as well as in humans, L-dopa is more likely than bromocriptine to induce dyskinesia and stereotype behaviour even though both drugs can correct parkinsonian deficits.

SKF 38393, although by itself had no behavioural effect and did not

Table 13.2

	Dopamine (μg g^{-1}) (% of control values)	^{3}H spiperone	
		K_d	B_{max}
Control	100 ± 15.2	100 ± 4.2	100 ± 2.6
MPTP	2.4 ± 0.6**	92.0 ± 2.9	125.8 ± 3.3**
MPTP + L-dopa	1.3 ± 0.7**	94.8 ± 4.7	$107.6 \pm 4.0^+$
MPTP + bromocriptine	7.9 ± 2.9**	81.9 ± 5.9*	$96.8 \pm 2.8^+$
MPTP + SKF 38390	1.1 ± 0.9**	93.3 ± 4.1	$102.4 \pm 3.6^+$

*$P < 0.05$; **$P < 0.01$ vs control; $^+P < 0.01$ vs MPTP.
Levels of dopamine in the striatum and ^{3}H-spiperone binding to striatal tissue after MPTP and chronic treatment with L-dopa/carbidopa, bromocriptine or SKF 38393.

potentiate the effect of bromocriptine, appeared to down-regulate the D2 receptor as well as the other two drugs. This suggests some interaction between the two receptors other than simple synergy or potentiation. The absence of behavioural effect by SKF 38393, as also observed by Nomoto *et al.* (1985) is surprising in view of observations in rats (Arnt and Hyttel, 1984) but may be due to species difference or to the fact that SKF 38393 does not elicit maximal stimulation of D1 receptors in monkeys.

Acute studies

A group of monkeys displaying a severe parkinsonian syndrome after MPTP were challenged with the following drugs: (1) SKF 38393, 5 mg kg^{-1} i.v.; (2) LY 171555 (Quinpirole), 0.5 mg kg^{-1} i.p.; (3) Quinpirole + SKF 38393 (same doses). All treatments were administered twice, once with the animals in their cage for recording of gross locomotor activity and once with the monkeys in a restraining chair for quantification of dyskinesia.

In addition, in some experiments the effect of the D1 antagonist SCH 23390, 0.6 mg kg^{-1} s.c. and of the D2 antagonist sulpiride 50 mg kg^{-1} i.p. was tested against Quinpirole and against the combination of Quinpirole and SKF 38393.

Figure 13.1 describes the effect of these various treatments on locomotion. SKF 38393 appeared to reduce locomotion, but Quinpirole had a strong activating effect which was somewhat reduced by the addition of SKF 38393. SCH 23390 antagonized the effect of Quinpirole.

In the case of lingual dyskinesia (Fig. 13.2), again SKF 38393 had little effect by itself as opposed to Quinpirole which induced a strong dyskinetic response. The addition of SKF 38393 to Quinpirole significantly increased the dyskinesia. SCH 23390 reduced slightly the effect of the combination of

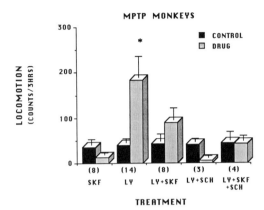

Fig. 13.1   Average counts of locomotor activity recorded in the cage by photocells during 3 h before (black bar) and 3 h after (hatched bar) injection of the experimental drugs in a group of MPTP monkeys. SKF = SKF 38393; LY = LY 171555 (Quinpirole); SCH = SCH 23390. * $P < 0.05$ compared to pre-drug count (Wilcoxon rank order test). The number of animals is given in parentheses above the name of the drug.

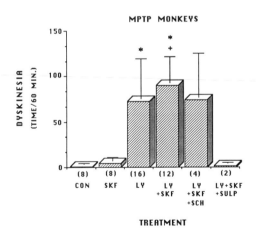

Fig. 13.2   Cumulative dyskinesia time recorded with a time–event counter during a period of 60 min after injection of a single or a combination of the following drugs: CON = vehicle; SKF = SKF 38393,   LY = LY 171555   (Quinpirole);   SCH = SCH 23390;   sulp. = sulpiride. * $P < 0.05$ compared to vehicle; + $P < 0.05$ compared to LY alone. Wilcoxon rank order test.

SKF 38393 and Quinpirole as if only the D1 part of the stimulation was blocked. Sulpiride on the other hand completely inhibited the effect of the combination.

CONCLUSIONS

From these acute experiments we conclude that D1 receptor stimulation by SKF 38393 has little behavioural effect by itself and that in these experiments, SKF 38393 appeared to decrease the locomotor effect of Quinpirole (LY 171555) while it potentiated the dyskinetic effect of the same drug. In the case of locomotion, it would appear that some cross blockade of the effect of a D2 agonist by a D1 antagonist occurs. In the case of dyskinesia however most of the effect was due to D2 stimulation and was antagonized only by the D2 antagonist.

Our results suggest that the response to D1 agonists and the interaction between D1 and D2 receptors may be different in primates and this will have to be taken into account before using new pharmacological agents in parkinsonian patients.

REFERENCES

Arnt, J. (1985) *Eur. J. Pharmacol* **113**, 79.
Arnt, J. and Hyttel, J. (1984) *Eur. J. Pharmacol.* **102**, 349.
Burns, R. S., Chiueh, C. C., Markey, S. P., Ebert, M. H., Jacobowitz, D. M. and Kopin, I. J. (1983) *Proc. Natl. Acad. Sci. USA* **80**, 4546.
Kebabian, J. W. and Calne, D. B. (1979) *Nature* **277**, 93.
Langston, J. W. and Ballard, P. (1984) *Can. J. Neurol. Sci.* **11**, 160–165.
Lees, A. J. and Stern, G. M. (1981) *J. Neurol. Neurosurg. Psychiat.* **43**, 1020.
Mashurano, M. and Waddington, J. L. (1986) *Neuropharmacology* **25**, 947.
Mones, R. J., Elizan, T. S. and Siegel, G. J. (1971) *J. Neurol. Neurosurg. Psychiat.* **34**, 668–673.
Nomoto, M., Jenner, P. and Marsden, C. D. (1985) *Neurosci. Lett.* **57**, 37–41.
Rascol, A., Guiraud, B., Montastruc, J. L., David, M. and Clavet, M. (1979) *J. Neurol. Neurosurg. Psychiat.* **42**, 143.
Rinne, V. K. (1987) *Neurology* **37**, 826–828.
Robertson, G. S. and Robertson, H. A. (1986) *Brain Res.* **384**, 387–396.
Walters, J. R., Bergstrom, D. A., Carlson, J. H., Chase, T. N. and Braun, A. R. (1987) *Science* **236**, 719–722.

14

Chronic Infusions of Dopamine Agonists in Animals

B. Costall, A. M. Domeney, S.-C. G. Hui, C. W. Ogle and R. J. Naylor

PREFACE

Attempts to improve the efficacy of dopamine agonists in the treatment of Parkinson's disease are investigating the value of continuous or delayed-release dosage administrations. It has been assumed that the clinical objective may be achieved by a continuous and lower drug level in the body tissues. We review the animal data reporting the consequences of the chronic administration of dopamine agonists and conclude that the changes in locomotor activity which occur during treatment are dependent on the dose of dopamine agonist, the nature of its daily administration, i.e. a single, pulsatile or continuous treatment, and the pattern of its administration, i.e. continuous or intermittent over the days of treatment. In addition, and dependent on the precise dose regimes employed, changes in motor activity may persist after the discontinuation of treatment. It is concluded that the chronic administration of dopamine agonists may cause changes in locomotor activity not observed in acute treatments but that there are insufficient data

DISORDERS OF MOVEMENT: CLINICAL, PHARMACOLOGICAL
AND PHYSIOLOGICAL ASPECTS   ISBN 0-12-569685-X

to enable a prediction of the consequences of continuous administration of dopamine agonists to man.

INTRODUCTION

The most effective therapy for the drug treatment of Parkinson's disease is L-dopa and other dopamine agonists. Yet a good initial response can frequently be followed by progressive deterioration with fluctuations in response, dyskinesias and loss of efficacy. The deterioration may be so great as to incapacitate the patient and attempts have been made to improve the treatment in two ways; by designing more potent dopamine agonists, or presenting the compound in a more "effective" manner (Jankovic, 1985).

The reasons for the deterioration in response to L-dopa and other dopamine agonists are not known but may involve changes in dopamine receptor sensitivity. If this is true, it is difficult to believe that the introduction of even more potent compounds will alleviate the problem, although the development of partial agonists may yield compounds worthy of clinical trial. The second approach, to present the compound in a more "effective" manner, is based on the hypothesis that fluctuations in drug levels after oral medication may contribute to the fluctuations in response and development of dyskinesias. It was considered that steady plasma drug levels may reduce such effects and indeed, the intravenous infusion of L-dopa was reported to abolish or reduce the on–off phenomenon in patients with chronic parkinsonism receiving long-term oral treatment with L-dopa (Shoulson *et al.*, 1975; Quin *et al.*, 1982).

Arising from the clinical findings the pharmacological challenge is clear: to provide from animal studies a rational basis for the use of persistent administration of dopamine agonists in the clinic. However, experiments performed over many years indicate the complexity of dopamine agonist action on motor behaviours. In the present chapter we review such studies and present data indicating that the nature of the presentation of a dopamine agonist may influence the effect on motor performance.

THE EFFECTS OF CHRONIC ADMINISTRATION OF DOPAMINE AGONISTS IN ANIMALS

Behavioural studies

The majority of studies have focused on the action of dopamine agonists to induce stereotyped behaviour and increase locomotor activity, effects related to drug interaction at dopamine receptors (see review by Kelly, 1977). An enhanced behavioural response has been observed in many studies following the chronic administration of dopamine agonists. Thus the administration of amphetamine to the cat (Ellinwood, 1971), to the rat (Segal and Mandell, 1974; Nelson and Ellison, 1978), to the mouse (Peachey et al., 1977) and guinea-pig (Klawans and Margolin, 1975) has been reported to enhance the stereotypic effects of the amphetamine. In addition, the chronic administration of amphetamine in the rat was also reported to enhance the locomotor stimulant effects (Segal and Mandell, 1974), similar results being recorded following the administration of amphetamine ($4 \, mg \, kg^{-1}$ i.p. once daily for 20 days) to mice (Bailey and Jackson, 1978). The latter authors also showed that apomorphine produced a significantly greater increase in locomotor activity in the amphetamine-treated mice and further, that a chronic treatment with apomorphine ($10 \, mg \, kg^{-1}$ i.p. day^{-1} for 20 days) enhanced the locomotor stimulant effect to subsequent challenge with apomorphine. L-Dopa ($200 \, mg \, kg^{-1}$ plus benserazide $50 \, mg \, kg^{-1}$, for 1, 4 or 10 days) pretreatment in mice also resulted in an enhanced locomotor response to apomorphine administered after the 10-day pretreatment (Bailey et al., 1979) and the administration of bromocriptine ($30 \, mg \, kg^{-1}$ daily for 30 days) similarly enhanced the stereotypic response to apomorphine and methylphenidate (Smith et al. 1979).

In the rat, repeated administration of methamphetamine ($6 \, mg \, kg^{-1} \, day^{-1}$ for 1, 3, 7 or 14 days) produced a long-term augmentation in the stereotyped behaviour induced by methamphetamine, apomorphine and nomifensine (Nishikawa et al., 1983). The repeated administration of lisuride (0.0125– $0.4 \, mg \, kg^{-1}$ i.p. daily for 28–31 days) caused a pronounced increase in motor activity and lowered the threshold dose of lisuride for causing stereotyped behaviour (Carruba et al., 1985). The chronic administration of amphetamine to cats (twice daily, in doses increasing from 5 to $15 \, mg \, kg^{-1}$ over a 10-day period) elicited a number of repetitive motor behaviours which were greatly potentiated by L-dopa and apomorphine (Trulson and Crisp, 1984). In the above studies the *enhanced* response to dopamine agonist challenge after the chronic administration of an agent such as amphetamine has frequently been interpreted to reflect an "innervation supersensitivity", "agonist-induced supersensitivity", "reverse tolerance", sensitization or "up regulation" of dopamine receptors.

However, there are also reports indicating that a chronic dopamine agonist treatment can *reduce* the response to subsequent challenge. In one of the earliest studies, Worms and Scatton (1977) found that a twice-daily treatment with $50 \, mg \, kg^{-1}$ i.p. apomorphine dipivaloyl ester to rats caused tolerance to the stereotypic effects of the aporphine. In mice, a 10-day treatment with L-dopa ($200 \, mg \, kg^{-1}$) plus benserazide ($50 \, mg \, kg^{-1}$) followed by 1 day of withdrawal reduced the locomotor stimulant effects of amphetamine, although enhancing that of L-dopa and apomorphine (Bailey *et al.*, 1979). Detailed observations of the stereotyped behaviour patterns induced by the repeated administration of amphetamine in rats revealed that although some components were enhanced, others were reduced. Thus whereas the repeated administration of amphetamine 2.5 mg kg^{-1} i.p. daily for 4 days produced a progressive increase in repetitive head and limb movements, long-term treatment with 5.5 mg kg^{-1} resulted in a reduction of the licking and biting behaviours induced by both amphetamine and apomorphine (Rebec and Segal, 1980). A 28-day treatment with a larger dose of amphetamine (10 mg kg^{-1} twice daily) in the rat markedly reduced the locomotor hyper-activity induced by this treatment and completely blocked the stereotyped licking and biting induced by apomorphine (Bendotti *et al.*, 1982). Similarly, a 7-day treatment with bromocriptine (15 mg kg^{-1} i.p. daily) caused a significant decrease in apomorphine induced stereotypy (Globus *et al.*, 1982). Treatments of rats with a very high dose of methylamphetamine (100 mg kg^{-1} day^{-1}) for 4 days also attenuated the ability of methylamphetamine and apomorphine to increase locomotor activity (Lucot *et al.*, 1980). In the cat, the administration of amphetamine (7.5 mg kg^{-1} i.p. twice daily for periods in excess of 3 days) decreased locomotor activity and stereotypical behaviour (Jacobs *et al.*, 1981).

It is clear from the results obtained in many studies that the repeated administration of a dopamine agonist can enhance or reduce the responsive-ness to subsequent dopamine agonist challenge. Yet in other studies investi-gators have found that such administrations fail to modify the response to subsequent challenge (Weston and Overstreet, 1976; Khanna and Loh, 1977; Flemenbaum, 1979). The differences in response may be due to the use of different species, drugs, doses and durations of treatment and it remains difficult from such experiments to predict the consequences of chronic dopamine agonist administration in animals, let alone in man.

Yet there remains a further most important variable which may influence the consequence of chronic dopamine agonist administration, namely the use of different treatment schedules, e.g. single daily treatments, multiple injections of infusion. Thus continuous but not intermittent exposure of rats to amphetamine (Nelson and Ellison, 1978) or L-dopa (Jenkins *et al.*, 1980) is reported to decrease the response to subsequent challenge. Rats exposed to i.p. injections of amphetamine showed an enhanced stereotypic response whereas animals exposed to continuous drug administration via a silicone pellet showed an attenuated response (Ellison and Morris, 1981). In a comparison of the behavioural effects induced by continuous administration of amphetamine from osmotically driven pumps, or by passive diffusion from silicone tubing implants (with a less constant release pattern), the pellet-implanted rats showed an earlier onset of locomotor enhancement and stereotyped responses (Eison *et al.*, 1983). An increased or decreased locomotor response in rats followed repeated administration of apomorphine and is reported to depend on dosage interval. Thus an enhanced response was obtained when a drug-free period was allowed between treatments whereas a reduced response was obtained when each dose was injected before metabolism of the previous dose was complete (Castro *et al.*, 1985). In a different behavioural model an enhanced acoustic startle response to amphetamine was facilitated when mice were exposed to long-term intermittent amphetamine treatment but attenuated when mice received a continuous exposure to amphetamine (Kokkinidis, 1984).

The above treatments refer to peripheral drug administrations, yet dosage schedules may also be important following intracerebral treatments. Thus the continuous infusion of dopamine into the nucleus accumbens of rat brain for 13 days causes a biphasic increase in locomotor activity with an absence of stereotyped behaviour. In contrast, 13 daily injections of dopamine caused a single phase of increased locomotor activity plus stereotyped biting (Costall *et al.*, 1983).

It becomes clear that the chronic administration of dopamine agonists to laboratory animals may modify the behavioural response to a subsequent challenge. Such changes may be important for dopamine agonists whose therapeutic response may diminish after prolonged therapy (Marsden and Parkes, 1977). Some reports indicate that chronic L-dopa treatment may induce down regulation of the dopamine receptors (Rinne *et al.*, 1980, 1981) which could contribute to a reducing therapeutic response. That a restoration of L-dopa efficacy in Parkinson's disease is achieved by a drug holiday period (Weiner *et al.*, 1980; Direngeld *et al.*, 1980) or by a drug treatment administered on alternate days (Goetz *et al.*, 1981; Koller, 1982) could be evidence to support such a hypothesis.

Biochemical studies

In animal studies attempts to obtain biochemical correlates of changing responsiveness to dopamine agonist challenge have measured changes in dopamine/metabolite levels and tyrosine hydroxylase activity (Scatton and Worms, 1978; Lucot *et al.*, 1980; Bardsley and Bachelard, 1981; Jacobs *et al.*, 1981; Papavasiliou *et al.*, 1981; Kuczenski *et al.*, 1983; Nishikawa *et al.*, 1983; Steranka, 1984; Trulson and Crisp, 1984) dopamine release (Robinson and Becker, 1982) and the number of dopamine receptors. Using the radioligand binding assays there are a number of reports that the chronic administration of amphetamine or L-dopa plus carbidopa may result in a decrease in ^{3}H-spiperone/^{3}H-ADTN binding in striatal and/or limbic rat brain areas (Howlett and Nahorski, 1979; Hitzemann *et al.*, 1980; Nielsen *et al.*, 1980; Reches *et al.*, 1982; Bendotti *et al.*, 1982; Ponzio *et al.*, 1984). Nevertheless, the repeated administration of amphetamine, methylphenidate and nomifensine failed to alter ^{3}H-haloperidol binding to rat striatal membranes (Algeri *et al.*, 1980), and the chronic administration of amphetamine in the Vervet monkey (Owen *et al.*, 1981) and the chronic administration of bromocriptine in the rat also failed to modify ^{3}H-spiperone binding (Globus *et al.*, 1982). Receptors in the rat striatum labelled with ^{3}H-haloperidol were also reported unchanged following amphetamine administration but showed a dramatic reduction following treatment with *N*-n-proplynorapomorphine (Riffee *et al.*, 1982).

It is difficult to draw conclusions as to the relevance of such findings to the behavioural development of enhanced responsiveness. Also, it is difficult to interpret the changes in dopamine receptor populations in the absence of data on the effect of chronic dopamine agonist treatment on other receptors. That chronic amphetamine treatment reduces the number of ^{3}H-dihydroalprenalol binding sites in the rat hippocampus (Bendotti *et al.*, 1982) emphasizes the importance of such studies. Nevertheless, it is noteworthy that in none of the above studies was a chronic treatment with a dopamine agonist reported to increase the number of dopamine receptors. An explanation for the enhanced behavioural response must be found elsewhere.

Electrophysiological studies

In electrophysiological experiments, it has been shown that intravenous injection of amphetamine will suppress the firing rate of dopamine neurones (see Heidenreich *et al.*, 1987). During long-term treatment with a dopamine agonist this response is attenuated (Jackson *et al.*, 1982; Kamata and Rebec 1983, 1984) with an increase in the release of dopamine (Robinson and Becker, 1982). Whilst such changes might be relevant to an increasing behavioural responsiveness to dopamine agonist challenge, it is then difficult to argue that such effects are relevant to an attenuation in behavioural response. Perhaps chronic amphetamine action to reduce the sensitivity of serotonergic dorsal raphé neurons might be relevant to such changes (Heidenreich *et al.*, 1987). In any event, it is obviously important to assess the actions of chronic dopamine agonist treatment on neurotransmitter systems other than dopamine (see Ritter *et al.*, 1984) to more adequately assess the importance of changes in the dopamine systems themselves.

Summary

It becomes apparent that the chronic administration of a dopamine agonist such as amphetamine can modify behaviour, brain biochemistry and neuronal activity to cause changes not observed to an acute challenge. Given the diversity of the behavioural response to increase or decrease locomotor activity, it is not possible to predict the response in animals, let alone clinical consequences of dopamine agonist challenge. Furthermore, the majority of the animal studies have used amphetamine, an indirectly acting dopamine agonist with little action in Parkinson's disease, and capable of influencing transmitter systems other than dopamine. Nevertheless, there are aspects of the animal studies that appear very important, particularly those preliminary experiments indicating that the dosage interval may dictate the nature of the behavioural change. In a series of experiments in the rodent we have further investigated the importance of the schedule of dopamine agonist administration and our findings are described below.

STUDIES IN THE RODENT

Experiments in the mouse

Reduction of locomotor activity in the mouse by an acute L-dopa–benserazide (50 mg kg^{-1} i.p.) treatment

L-Dopa was administered in a wide dose range (1.5–200 mg kg^{-1}) and failed to enhance the locomotor activity of the mice. Indeed, in doses greater than 6.25 mg kg^{-1} L-dopa reduced activity (Fig. 14.1).

Modification of locomotor activity in the mouse during and following L-dopa (6.25 mg kg^{-1} i.p.) plus benserazide (50 mg kg^{-1} i.p.) treatment

Using a 4 days on/4 days off, 4 days on/7 days off or 4 days on/14 days off treatment, mice received the L-dopa regimen or vehicle for a total of 32–56 days. The locomotor activity levels of mice receiving the 4 days on/off treatment remained at a constant level, with activity on the "off" days being slightly but not significantly higher throughout the 32-day period. Using the 4 days on/7days off schedule, whilst activity during the "off" days remained constant, activity during the periods of drug treatment decreased and this achieved significance during the third and fourth periods. Similar comments apply to the results obtained using the 4 days on/14 days off treatment (Fig. 14.2).

Modification of locomotor activity in the mouse during and following (−)N-n-propylnorapomorphine ((−)NPA) (0.4 μg kg^{-1} s.c. daily) treatment

The dose of (−)NPA was selected on the basis of initial studies showing that an acute administration failed to stimulate or reduce locomotor activity.

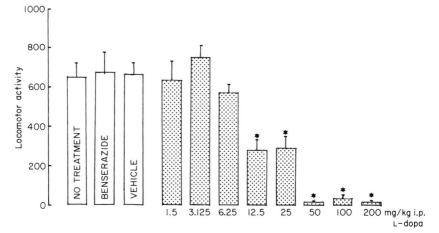

Fig. 14.1   Reduction in locomotor activity in the mouse by L-dopa plus benserazide (50 mg kg^{-1} i.p., 45 min pretreatment). Locomotor activity was measured in treadwheels and expressed as revolutions/20 min. Mice received no treatment, benserazide, vehicle or L-dopa (1.5–200 mg kg^{-1} i.p.) plus benserazide (stippled columns). Each value is the mean $\pm$ SEM of five determinations. A significant difference between the drug treatments and vehicle controls is indicated: * $P < 0.01$– $P < 0.001$ (one-way ANOVA followed by Dunnett's test).

Using a 4 days on/4 days off treatment, activity decreased during the 4-day period of drug administration on the second, third and fourth exposures. During the intervening 4 days "off" periods there was a clear trend to an increased activity but this failed to achieve significance. Similar results were obtained using the 4 days on/7 days off and 4 days on/14 days off regimens (Fig. 14.3).

Modification of locomotor activity in the mouse during and following the infusion of apomorphine (0.4 or 500 μg kg^{-1} daily) for 13 days

Alzet osmotic minipumps were primed with the apomorphine solution or vehicle and implanted subcutaneously in the scapula region. During the 13-day period of infusion the locomotor activity was assessed daily and animals receiving the drug and vehicle treatments were run concurrently. After discontinuation of the infusion locomotor activity was measured every 4 days and representative data are shown in Fig. 14.4. Mice receiving the infusion of apomorphine 0.4 μg kg^{-1} day^{-1} failed to demonstrate an increase in activity with a trend to reduce activity during infusion, and this became significant after discontinuing infusion. In contrast, the infusion of the high dose of 500 μg kg^{-1} day^{-1} apomorphine significantly increased locomotor activity during infusion and this persisted after discontinuation of the treatment (Fig. 14.4).

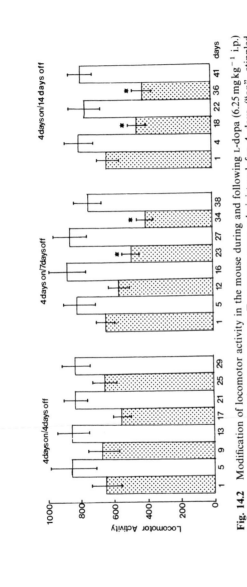

Fig. 14.2 Modification of locomotor activity in the mouse during and following L-dopa (6.25 mg kg^{-1} i.p.) plus benserazide (50 mg kg^{-1} i.p.) treatment. Treatments were administered for 4 days ("on", stippled histograms) followed by 4, 7 or 14 days "off" treatment (open histograms) before resuming the 4 days administration of L-dopa plus benserazide. Locomotor activity was measured in treadwheels and expressed as revolutions/20 min. Each value is the mean ± SEM of five determinations. A significant difference between drug treatments and the response on day 1 is indicated: * $P < 0.05 - P < 0.01$ (two-way ANOVA followed by Dunnett's test).

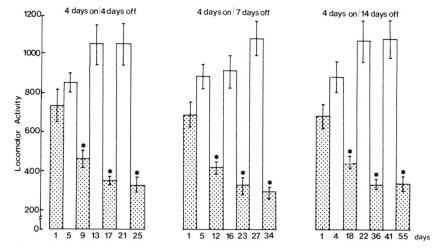

Fig. 14.3 Modification of locomotor activity in the mouse during and following $(-)N$-n-propylnorapomorphine (0.4 μg s.c. daily) treatment. Treatments administered for 4 days ("on", stippled histograms) followed by 4, 7 or 14 days "off" treatment (open histograms) before resuming the 4 days administration of $(-)N$-n-propylnorapomorphine. Locomotor activity was measured in treadwheels and expressed as revolutions/20 min. Each value is the mean ± SEM of five determinations. A significant difference between drug treatments and the response on day 1 is indicated: * $P < 0.01 - P < 0.001$ (two-way ANOVA followed by Dunnett's test).

The data show that the acute administration of L-dopa in the mouse reduces locomotor activity, unlike the locomotor stimulant effects observed in rodents with dopamine receptor "denervation supersensitivity" (Ungerstedt, 1971). The administration of a normally ineffective "acute dose" caused a decrease in locomotor activity on repeated treatment, dependent on the period allowed between drug administrations. The repeated administration of $(-)$NPA caused a similar profile of action, although even a short period of 4 days between $(-)$NPA treatment resulted in decreasing activity. It was also an interesting observation that there was a trend to an increased locomotor activity during the non-treatment days. Thus during and following L-dopa and $(-)$NPA treatment there are perturbations in the mechanisms controlling locomotor function which may involve pre- and post-synaptic changes in dopamine receptor function. Smaller doses of $(-)$NPA and shorter treatment times would be required to establish threshold doses and times required to trigger such changes.

The data obtained using the continuous infusion of a "low" and "high" dose of apomorphine indicate the importance of dosage. The low dose failed to enhance activity during infusion but caused a decreased activity after discontinuing infusion. However, a "high" dose enhanced activity during

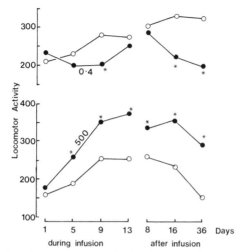

Fig. 14.4 Modification of locomotor activity in the mouse by subcutaneous infusion of apomorphine. Locomotor activity is expressed as wheel revolutions per 60 min assessed during the 13-day period of drug (closed symbols, doses indicated are μg of drug infused per 24-h period) or vehicle (open symbols) infusion and for 36 days post-infusion. $n = 5$. A significant difference between the drug treatments and vehicle controls is indicated: * $P < 0.05$–0.001 (two-way ANOVA followed by Dunnett's test).

infusion and animals maintained a raised level of activity after discontinuing the treatment. It is clear that the consequences of an infusion of apomorphine in the mouse can have long-term consequences, the response being dependent on the dose administered.

Experiments in the rat

Modification of spontaneous locomotor activity in the rat following a continuous or "pulsed" infusion of (−)NPA (0.05 mg kg^{-1} day^{-1})

For continuous infusion, Alzet osmotic minipumps containing (−)NPA or vehicle (N$_2$ bubbled 0.1% sodium metabisulphite solution) were implanted subcutaneously in the scapula region to deliver the solution at a continuous rate. For the pulsatile delivery, a coiled length of polythene tubing was loaded with 2 μl (equivalent to 2 h of infusion) pulses of (−)NPA solution (or vehicle) interspersed with 2 μl of nitrogen. The tubing was coiled tightly around and connected to the pump (which contained saline) to expel the contents of the tube. It was found in control experiments that the "pulses" of (−)NPA solution remained distinct for 4 days and the subsequent experiments were limited to this period. The dose of (−)NPA was chosen as a dose failing to significantly modify locomotor activity during the first day of treatment. The continuous infusion of (−)NPA over a 4-day period failed to modify locomotor activity, values were not significantly different from animals receiving the saline infusion (Fig. 14.5). However, animals receiving the same total daily dose of (−)NPA but administered as 2-h pulses showed increasing levels of activity on the second and third days of infusion. By the fourth day the level of activity was approximately 400% greater than observed on day 1. Such findings of a greater effectiveness of pulsatile

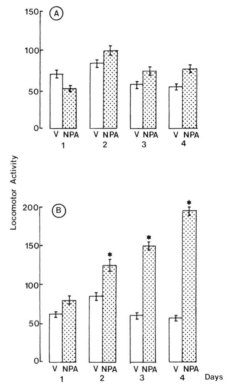

Fig. 14.5 Modification of locomotor activity in the rat during the subcutaneous infusion of (−)N-n-propylnorapomorphine ((−)NPA) for 4 days. Locomotor activity was measured in photocell cages and expressed as counts/60 min. Rats received (A) a continuous infusion of (−)NPA (0.05 mg kg^{-1} day^{-1}) and (B) a pulsatile infusion (2 h "on", 2 h "off") to deliver the same total daily dose of (−)NPA; rats receiving appropriate vehicle infusions (open histograms) were run concurrently with those receiving the drug treatments (stippled histograms). Each value is the mean ± SEM of five determinations. A significant difference between treatments and the values obtained on day 1 is indicated: * $P < 0.05$–$P < 0.001$ (two-way ANOVA followed by Dunnett's test).

injections to increase locomotor activity may reflect the greater amount of drug delivered over a given 2-h period. However, there remains the intriguing possibility that an additional and more important factor is the persistence or repetition of receptor stimulation and "recovery" following pulsatile treatments. To investigate this hypothesis, the treatment used in the present experiments of 2 h "on-and-off" could be modified to extended periods of regular or irregular length. Such lengthy experiments, with different dose regimes, might establish the optimal dose and time required for receptor "stimulation" and "recovery" to cause the maximal behavioural change. If such a hypothesis were to be confirmed, it would necessitate a revision of the classical pharmacological concepts of the relationship between a dose and response to include a variable for the time of receptor stimulation and "recovery".

CONCLUSIONS

It is apparent that the nature of the dopamine agonist regimen employed can determine the behavioural consequences during and following drug treatment. It is also clear that the present studies and those reported in the literature are of a preliminary nature. Further experiments are required to clarify the precise doses, the importance of a continuous, pulsatile or acute administration and the period for which they are given. Also, it is important to establish the patterns of administration that are required to modify motor response in predictable ways. In the absence of such data it is not possible to predict dosage schedules from animal studies to optimize the effects of dopamine agonist treatment in man. Further, there remains a major difference between the administration of dopamine agonists to man and animals: dopamine agonists are administered to animals with "normal" dopamine receptors whereas in man they are probably acting on denervated mechanisms of modified sensitivity. Nevertheless, the animal data indicate that a continuous infusion of a dopamine agonist to man might produce effects not observable to an acute administration, and this is clearly deserving of further study.

REFERENCES

Algeri, S., Brunello, N. and Vantini, G. (1980) *Pharmacol. Res. Comm.* **12**, 675–681.
Bailey, R. C. and Jackson, D. M. (1978) *Psychopharmacologia* **56**, 317–326.
Bailey, R. C., Jackson, D. M. and Bracs, P. U. (1979) *Psychopharmacology* **66**, 55–61.
Bardsley, M. E. and Bacheland, A. S. (1981) *Biochem. Pharmacol.* **30**, 1543–1549.

Bendotti, C., Borsini, F., Cotecchia, S., De Blasi, A., Mennini, T. and Samania, R. (1982) *Arch. Int. Pharmacodyn.* **206**, 36–49.

Carruba, M. O., Ricciardi, S., Chiesara, E., Spano, P. F. and Montegazza, P. (1985) *Neuropharmacology* **24**, 199–206.

Castro, R., Abreu, P., Calzadilla, C. H. and Rodriguez, M. (1985) *Psychopharmacology* **85**, 333–339.

Costall, B., Domeney, A. M. and Naylor, R. J. (1983) *Naunyn-Schmiedeberg's Arch. Pharmacol.* **324**, 27–33.

Direngeld, L. K., Feldman, R. G., Alexander, M. P. and Kelly-Hayses, M. (1980) *Neurology* **30**, 785–788.

Eison, M. S., Eison, A. S. and Iversen, S. D. (1983) *Neurosci. Lett.* **39**, 313–319.

Ellinwood, E. H. (1971) *Psychopharmacologia* **21**, 131–138.

Ellison, G. and Morris, W. (1981) *Eur. J. Pharmacol.* **74**, 207–214.

Flemenbaum, A. (1979) *Psychopharmacology* **62**, 175–179.

Globus, M., Bannet, J., Lerer, B. and Belmaker, R. H. (1982) *Psychopharmacology* **78**, 81–84.

Goetz, C. G., Tanner, C. M. and Nauseda, P. A. (1981) *Neurology* **31**, 1460–1462.

Heidenreich, B. A., Basse-Tomusk, A. E. and Rebec, G. V. (1987) *Neuropharmacology* **26**, 719–724.

Hitzemann, R., Wu, J., Hom, D. and Loh, H. (1980) *Psychopharmacology* **72**, 93–101.

Howlett, D. R. and Nahorski, S. R. (1979) *Brain Res.* **161**, 173–178.

Jackson, D. M., Walters, J. R. and Miller, L. P. (1982) *Brain Res.* **250**, 271–282.

Jacobs, B. L., Heym, J. and Trulson, M. E. (1981) *Eur. J. Pharmacol.* **69**, 353–356.

Jankovic, J. (1985) *Clin. Neuropharmacol.* **8**, 131–140.

Jenkins, O., Bailey, R., Crisp, E. and Jackson, D. M. (1980) *Psychopharmacology* **68**, 77–83.

Kamata, K. and Rebec, G. V. (1983) *Neuropharmacology* **22**, 1377–1382.

Kamata, K. and Rebec, G. V. (1984) *Life Sci.* **34**, 2419–2427.

Kelly, P. H. (1977) In *Handbook of Psychopharmacology*, Vol. 8 (edited by L. L. Iversen, S. D. Iversen and S. H. Snyder), pp. 295–331. Plenum Press, New York.

Khanna, C. D. and Loh, H. H. (1977) *Psychopharmacology* **54**, 295–302.

Klawans, H. W. and Margolin, D. I. (1975) *Arch. Gen. Psychiat.* **32**, 725–732.

Kokkinidis, L. (1984) *Pharmacol. Biochem. Behav.* **20**, 367–371.

Koller, W. C. (1982) *Neurology* **32**, 324–326.

Kuczenski, R., Leith, J. N. and Applegate, C. D. (1983) *Brain Res.* **258**, 333–337.

Lucot, J. B., Wagner, G. C., Schuster, C. R. and Seiden, L. S. (1980) *Pharmacol. Biochem. Behav.* **13**, 409–413.

Marsden, C. D. and Parkes, J. C. (1977) *Lancet* **i**, 345–349.

Nelson, L. R. and Ellison, G. (1978) *Neuropharmacology* **17**, 1081–1084.

Nielsen, E. B., Nielson, M., Ellison, G. and Braestrup, C. (1980) *Eur. J. Pharmacol.* **66**, 149–154.

Nishikawa, T., Mataga, N., Takashima, M. and Toru, M. (1983). *Eur. J. Pharmacol.* **88**, 195–203.

Owen, F., Baker, H. F., Ridley, R. M., Cross, A. J. and Crow, T. J. (1981) *Psychopharmacologia* **74**, 213–216.

Papavasiliou, P. S., Miller, S. T., Thal, L. J., Nerder, L. J., Houlihan, G., Rao, S. N. and Stevens, J. M. (1981) *Life Sci.* **28**, 2945–2952.

Peachey, J. E., Rogers, B. and Brien, J. F. (1977) *Psychopharmacology* **51**, 137–140.

Ponzio, F., Cimino, M., Achilli, G., Lipartiti, M., Perego, C., Vantimi, G. and Algeri,

S. (1984) *Life Sci.* **34**, 2107–2116.

Quinn, N., Marsden, C. D. and Parkes, J. D. (1982) *Lancet* **ii**, 412–415.

Rebec, G. V. and Segal, D. S. (1980) *Pharmacol. Biochem. Behav.* **13**, 793–797.

Reches, A., Wagner, H. R., Jiang, D.-H., Jackson, V. and Fahn, S. (1982) *Life Sci.* **31**, 37–44.

Riffee, W. H., Wilcox, R. E., Vaughn, D. M. and Smith, R. V. (1982) *Psychopharmacology* **77**, 146–149.

Rinne, U. K., Koskinen, V. and Lonnberg, P. (1980) In *Parkinson's Disease—Current Progress Problems and Management.* (edited by M. Klinger and G. Stamm), pp. 93–107. Elsevier, Amsterdam.

Rinne, U. K., Lonnberg, P. and Koskinen, V. (1981) *J. Neural. Transm.* **51**, 97–109.

Ritter, J. K., Schmidt, C. J., Gibb, J. W. and Hanson, G. R. (1984) *J. Pharmacol. Exp. Ther.* **229**, 487–492.

Robinson, T. E. and Becker, J. B. (1982) *Eur. J. Pharmacol.* **85**, 253–254.

Scatton, B. and Worms, P. (1978) *Naunyn-Schmiedeberg's Arch. Pharmacol.* **303**, 271–278.

Segal, D. S. and Mandell, A. J. (1974) *Pharmacol. Biochem. Behav.* **2**, 249–255.

Shoulson, I., Glaubiger, G. A. and Chase, T. N. (1975) *Neurology (Minneap.)* **25**, 1144–1148.

Smith, R. C., Shong, J. R., Hicks, P. B. and Samorajski, T. (1979) *Psychopharmacology* **60**, 241–246.

Steranka, L. R. (1984) *Naunyn-Schmiedebergs Arch. Pharmacol.* **325**, 198–204.

Trulson, M. E. and Crisp, T. (1984) *Eur. J. Pharmacol.* **99**, 313–324.

Ungerstedt, U. (1971) *Acta Physiol. Scand.* Suppl. 367, 49–68.

Weiner, W. J., Koller, W. C., Perlik, S., Nausieda, P. A. and Klawans, H. L. (1980) *Neurology* **29**, 1054–1057.

Weston, P. F. and Overstreet, D. H. (1976) *Pharmacol. Biochem. Behav.* **5**, 645–649.

Worms, P. and Scatton, B. (1977) *Eur. J. Pharmacol.* **45**, 395–396.

15

Delivery of (+)-4-Propyl-9-hydroxynaphthoxazine ((+)-PHNO) by a Novel Orally Administered Osmotic Tablet: Behavioural Effect in Animal Models of Parkinson's Disease

Celia Brazell, Gaylen M. Zentner, Colin R. Gardner,
Peter Jenner and Stephen M. Stahl

INTRODUCTION

The chemical family of naphthoxazines may be the most potent dopamine agonists *in vivo* that have yet been discovered, displaying a dopaminergic selectivity beyond that of the ergolines, pergolide and lisuride (Jones *et al.*, 1984; Martin *et al.*, 1984). The affinity shown by the (+)-enantiomer of 4-propyl-9-hydroxynaphthoxazine HCl [(+)-PHNO] (MK-458; Naxagolide HCl®) for the dopamine D2 receptor, its solubility and bioavailability following oral administration, suggest that (+)-PHNO may be a potential therapeutic agent in treating the dopaminergic deficit of Parkinson's disease. In animal models of dopamine depletion, (+)-PHNO is known to be active in the nigrostriatal lesioned rat (Martin *et al.*, 1984) and the 1-methyl-4-

DISORDERS OF MOVEMENT: CLINICAL, PHARMACOLOGICAL
AND PHYSIOLOGICAL ASPECTS   ISBN 0-12-569685-X

phenyl-1,2,3,6-tetrahydropyridine (MPTP) treated marmoset (Jenner *et al.*, 1986; Nomoto *et al.*, 1987). Limited clinical trials have confirmed the efficacy of a single dose of (+)-PHNO in acutely reversing parkinsonian rigidity, tremor and gait in both moderate and severely affected patients (Grandas-Perez *et al.*, 1986; Stoessl *et al.*, 1985).

However, both physician and patient have come to demand more of dopamine agonists than acute amelioration of motor deficits. The drug must provide continual symptomatic relief and avoid both peak-dose dyskinesias and the end of dose "wearing-off" effects seen in the long-term Parkinson's patient receiving L-dopa. In addition, as the disease progresses, patients experience debilitating motor fluctuations ("on–off" phenomenon) which appear unrelated to their medication schedule. In an endeavour to maintain effective therapy, intravenous infusions of L-dopa (Hardie *et al.*, 1984; Mouradian, *et al.*, 1987) and subcutaneous infusions of lisuride (Critchley *et al.*, 1986; Obeso *et al.*, 1986) and apomorphine (Stibe *et al.*, 1988) have been assessed. Slow-release oral preparations of L-dopa have also been formulated; Sinemet CR is a controlled-release type polymer, while Madopar HBS is a hydrodynamically balanced system which is retained in the upper gastro-intestinal (GI) tract and slowly releases drug from a matrix (Nutt *et al.*, 1986). The dramatic improvement in motor fluctuations associated with both "wearing-off" and "on–off" phenomena which have resulted, indicates that constant stimulation of central dopaminergic pathways is a prerequisite for the control of advanced Parkinson's disease (Fahn, 1982). Unfortunately, there remain numerous problems associated with sustained drug delivery that have yet to be overcome. Merely from a practical standpoint, the method of drug delivery should ideally be suitable for use in outpatients. Moreover, when considering direct-acting dopamine agonists, there is usually a relatively narrow window between doses giving the required therapeutic response and those producing the unwanted side-effects of dyskinesia and nausea. Thus, there is a definite need to advance the technology of drug delivery and attain an optimized state of drug homeostasis in a precise fashion.

Devices based on drug diffusion through a rate-limiting barrier, chemical or enzymic degradation of a drug carrier, and mechanical or osmotic pumping of drugs, all have the potential for controlled drug delivery and as such have been the subject of numerous reviews (Baker and Lonsdale, 1974; Robinson, 1980; Chien, 1983; Stahl, 1988; Stahl and Wets, 1988). Several devices for oral drug delivery have taken the form of simple tablets. Of the oral slow-release formulations, probably the best known is the Oros tablet which consists of a semipermeable membrane surrounding a core containing drug and osmotic agent. As the agent expands, the drug is expelled at a constant rate through a single laser-drilled hole. In this fashion levels of drug in the plasma are maintained. Such osmotic tablets have several advantages.

Firstly, the release rate of the drug is unaffected by changes in agitation or the acid/alkali balance of the GI tract. Secondly, the intestinal membranes receive the drug already in solution and so ready for absorption. Thus, the system offers precision of dosing and stability of drug usually only associated with the solid compound (see Stahl and Wets, 1988),

To apply the concept of controlled drug delivery to (+)-PHNO, we have studied the release of the compound from a novel orally administered osmotic pump (Zentner et al., 1985). The tablet-like pumps were designed to give zero-order release of (+)-PHNO via controlled porosity walls which coated the tablet. As with the Oros structure, release from the new tablet was independent of both pH and gut motility. However, unlike the earlier device, the drug was not released through a single orifice but from each unit of the wall. In this way concentration of the drug at points of direct contact between the pump surface and the absorbing mucosa was reduced, whilst the desired overall rate of release was maintained. The rate of (+)-PHNO release from the tablet was governed by the following features: (i) the level of leachable additive in the wall; (ii) the nature of the insoluble polymer component of the wall; (iii) the thickness and surface area of the wall; (iv) the total solubility and osmotic pressure of the tablet core; and (v) the drug load in the core.

The mechanism of drug release was principally osmotic pumping, with simple diffusion playing only a minor role. Delivery of (+)-PHNO from the tablet was ascertained initially in vitro and then in vivo in a rodent behavioural model of Parkinson's disease, the 6-hydroxydopamine (6-OHDA) unilaterally lesioned rat. To confirm the controlled-release of (+)-PHNO from the tablet in vivo, the turning behaviour induced in the rat was compared with that following direct infusion of (+)-PHNO through a gastric fistula. To further ascertain the clinical potential of delivery from these devices, the tablets were also tested in a primate model of Parkinson's disease, the MPTP-treated marmoset.

The final aspect of drug delivery which we considered was the development of tolerance in postsynaptic dopamine mechanisms which arise from repeated administration of dopamimetics (Guttman and Seeman, 1985). Recently, Martin-Iverson et al. (1987) have noted behavioural tolerance to (+)-PHNO in rats following 4 days of continuous subcutaneous delivery. To look for this phenomenon after oral drug administration, we gastrically infused (+)-PHNO to 6-OHDA lesioned rats for 3 days and then monitored their response to an additional dopaminergic challenge.

Our overall aim was therefore to assess the potential of using the novel osmotic tablet to provide sustained release of (+)-PHNO by monitoring the drug's effect in two different animal models of Parkinson's disease. Second, we sought to predict the possible therapeutic outcome of using such a drug regimen in parkinsonian patients with motor fluctuations.

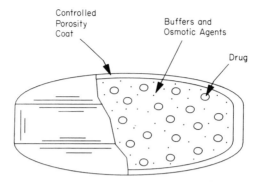

Fig. 15.1   Schematic diagram of the (+)-PHNO orally administered osmotic tablet (see text for details).

FORMULATION OF (+)-PHNO OSMOTIC TABLETS AND THE RELEASE OF (+)-PHNO *IN VITRO*

(+)-PHNO was formulated in the controlled porosity osmotic pump delivery system of Zentner *et al.* (1985) (Fig. 15.1). Core tablets were formed with 46.4 μg of (+)-PHNO HCl, crystalline mannitol and monosodium phosphate. Sucrose was used as a binder giving the core tablet a weight of 30.5 mg. These core structures were coated with cellulose acetate, sorbitol and diethylphthalate to a thickness of 75 μm. The final external diameter of the tablet was 3.2 mm. The *in vitro* release of (+)-PHNO from the tablet was evaluated in a paddle dissolution apparatus. Phosphate buffer, pH 7.4, 37°C, made isotonic to normal saline, served as the release media. At predetermined intervals, media samples were taken and chromatographically analysed for (+)-PHNO.

Confirmation of the controlled release of (+)-PHNO from the tablet was given by the drug release profile which we obtained (Fig. 15.2). The tablet released (+)-PHNO for approximately 5.5 h at a zero-order rate of 5.7 μg h^{-1}. A lag time of 1.4 h preceded zero-order delivery and approximately 40 μg of (+)-PHNO had been released after 11 h.

ANIMAL MODELS OF PARKINSON'S DISEASE AND THE RELEASE OF (+)-PHNO *IN VIVO*

The rat 6-hydroxydopamine model of Parkinson's disease

Unilateral dopamine depletion and rotational behaviour

The development by Ungerstedt (1968) of a chemical method to destroy the nigrostriatal dopamine system of the rat resulted in the 6-hydroxydopamine

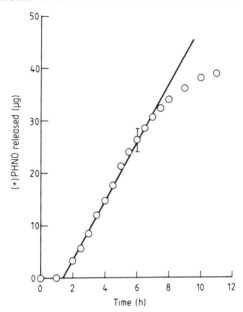

Fig. 15.2   Typical profile of (+)-PHNO release from a controlled porosity osmotic pump *in vitro*.

(6-OHDA) model becoming the mainstay of behavioural tests for drugs of potential use in Parkinson's disease. The method involves injection of 6-OHDA into the substantia nigra of the rat and results in a rapid degeneration of the nigrostriatal dopamine pathway. Unilateral 6-OHDA-induced degeneration results in postsynaptic supersensitivity of the dopamine receptors on the lesioned side of the brain, the outcome of which is an asymmetry in posture and movement. The direction of the turning behaviour which manifests itself indicates in which striatum dopamine function is prominent, the animal turning towards the least functional side. The result of administering a postsynaptic dopamine receptor agonist is that the animal rotates away from the lesioned side of the brain (contralateral rotation). The speed of rotation is an expression of the difference in dopamine function between the two striata. It is purely a relative measure and not indicative of the extent of dopamine receptor stimulation (Ungerstedt *et al.*, 1973).

Experimental methods

6-OHDA administration

To induce the 6-OHDA unilateral lesion, rats were injected in the right substantia nigra with 6-OHDA hydrobromide ($8\,\mu g$ in $4\,\mu l$) according to brain coordinates predetermined by Pellegrino and Cushman (1967). To ensure that the lesion was specific to dopamine neurons, animals were pretreated with the noradrenaline uptake inhibitor, desipramine.

Turning behaviour

Seven days after 6-OHDA, rats were tested for successful nigrostriatal lesions by administration of the dopamine receptor agonist apomorphine $(0.5 \, mg \, kg^{-1}, i.p.)$. Rats were then observed for turning behaviour, quantified by counting the number of full contralateral body turns every 10 min for a period of 1 min. Animals showing five or more rotations per minute were used in subsequent $(+)$-PHNO studies.

Implantation of gastric fistula

Successfully lesioned animals were implanted with flanged stainless steel cannula in the anterior wall of the forestomach. To allow unrestricted behaviour during drug infusion, animals were placed in a harness which allowed the cannula to be attached to a syringe pump via an overhead liquid swivel.

(+)-PHNO administration and antiparkinsonian effect

$(+)$-PHNO tablets were placed into the stomach entrance of the rat by means of a plastic gavage tube. Two pumps were administered simultaneously, the cumulative dose of $11.4 \, \mu g \, h^{-1}$ providing a measurable level of animal rotation without excessive activity. For comparable drug administration via the gastric fistula, $(+)$-PHNO HCl solution was delivered at $11.4 \, \mu g / 1.2 \, ml \, h^{-1}$.

$(+)$-PHNO administration by either method began at 3.30 a.m. Our *in vitro* results showed that $(+)$-PHNO would be released from the tablets in a zero-order fashion for approximately 6 h. To allow us to study the rats for a longer period, we maintained this pattern of drug release until 3.30 p.m. by administering two more tablets at 9.30 a.m. For behavioural comparison gastric infusion of $(+)$-PHNO was also terminated at 3.30 p.m., but both groups of animals continued to be recorded for turning behaviour until 8.30 p.m. During this time, the lights remained on and food and water were freely available.

Under these conditions oral administration of the $(+)$-PHNO tablets produced contralateral turning behaviour (Fig. 15.3). Mean onset of turning was $9.4 \pm 0.4 \, h$ after administration of the first two tablets and animals continued to turn throughout the recording period. During this time, the frequency of turning behaviour oscillated with a background activity of

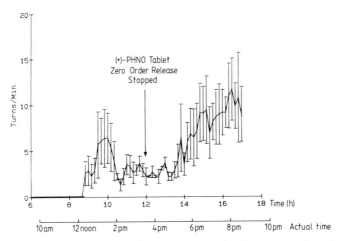

Fig. 15.3   Effect of oral administration of (+)-PHNO osmotic tablets on contralateral rotational behaviour in rats with unilateral 6-OHDA lesions. Two (+)-PHNO pumps ($2 \times 5.7\,\mu h\,h^{-1}$) were introduced into the stomach at 3.30 a.m. (time 0 h) and at 9.30 a.m. (time 6 h) and the number of contralateral rotations counted every 10 min for 60 s. Results are expressed as the number of turns min^{-1}, mean $\pm$ SEM of $n = 4$–9. Both the time after administration of the first (+)-PHNO tablets (h) and the actual time (a.m./p.m.) is shown.

around 2.5 turns min^{-1} being interrupted by periods of activity peaking at 1.20 p.m. and 8.00 p.m. (6.3 ± 2.7 and 11.5 ± 3.4 turns min^{-1}, respectively).

Similarly, gastric infusion of (+)-PHNO produced contralateral turning (Fig. 15.4). This time, the mean delay in onset was 7.47 ± 0.4 h after start of administration, 1.9 h prior to that seen with (+)-PHNO tablets. However, accounting for the 1.4 h lag time needed for the tablet to commence drug release, the two delay times are in good agreement. Once again, the frequency of animal rotation was seen to vary throughout the study period and a background activity of 2 turns min^{-1} was interrupted by activity peaks at 1.40 p.m. and 8.10 p.m. (5.6 ± 1.3 and 10.8 ± 2.3 turns min^{-1}, respectively). During the observation period, the frequency of rotation induced by (+)-PHNO gastric infusion was not significantly different from that induced by (+)-PHNO osmotic tablets, implying that the release of (+)-PHNO from the osmotic tablets was in a controlled fashion as predicted by our *in vitro* results.

However, during our study period, we were interested to note that the turning behaviour of the rats did not remain at a fixed frequency but appeared to follow a biphasic pattern. When the lesioned rats were given either placebo tablets or vehicle infusion alone, no significant contralateral rotational behaviour was seen. However, it was noted that between midday and 2.00 p.m. (± 1 h) and from approximately 6.30 p.m. onwards animal

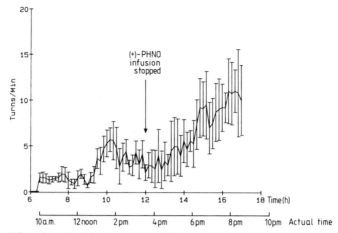

Fig. 15.4 Effect of gastric infusion of (+)-PHNO on the number of contralateral rotations in rats with unilateral 6-OHDA lesions. (+)-PHNO was infused at a rate of 11.4 μg h^{-1} between 3.30 a.m. (time 0) and 3.30 p.m. (time 12 h). The number of whole-body contralateral rotations was counted every 10 min for a duration of 60 s. Results are expressed as rotations min^{-1}, mean $\pm$ SEM of $n = 4$. Both the time after commencement of infusion (h) and the actual time (a.m./p.m.) is shown.

activity did increase. Unfortunately, because activity was not automatically monitored, it was impossible to score such movement. Despite this, our observations may indicate that the behavioural peaks of animals given (+)-PHNO may reflect additional excitation resulting from a normal circadian rhythm in rat locomotor activity. In this respect, the later activity peak occurred around 7.00 p.m. when both the animals' dark cycle and nocturnal activity would usually begin. Furthermore, the pattern of locomotor activity which we measured is in excellent agreement with that of O'Neill and Fillenz (1985) who used an *in vivo* technique to monitor endogenous dopamine release in the rat. Although the authors studied intact animals, the fact that unilaterally lesioned 6-OHDA rats respond to L-dopa by contralateral and not ipsilateral turning, indicates that such animals retain the capacity to synthesize and release dopamine in the lesioned striatum (Ungerstedt *et al.*, 1973), and therefore may also display a diurnal pattern of dopamine release. The fact that the increase in activity which we recorded was manifested as a greater number of contralateral rotations implies that stimulation of postsynaptic receptors on the lesioned striatum was still greater than those of the intact side of the brain. For this reason, a diurnal rhythm in the dopamine receptors themselves may also need to be considered (Naber *et al.*, 1980; Kafka *et al.*, 1983).

Effect of 3-day continuous (+)-PHNO infusion on turning behaviour

In the third behavioural test, we wanted to ascertain whether continual stimulation by (+)-PHNO would result in the development of tolerance. To look for this phenomenon, 6-OHDA lesioned animals were first challenged with an injection of (+)-PHNO (70 μg kg^{-1}, i.p., 9.00 a.m.) and their turning behaviour was scored (test day 0). Twenty-four hours later, animals were started on a gastric infusion of (+)-PHNO 7.5 μg h^{-1} which lasted for 3 days (test days 1–3). One and 5 days after cessation of infusion (test days 4 and 8), animals were again challenged with 70 μg (+)-PHNO i.p. (9.00 a.m.). Throughout the study, animals were maintained on a 12-h light cycle (lights on 7.00 a.m. until 7.00 p.m.).

During the 3-day infusion period, (+)-PHNO resulted in animal excitation indicative of dopaminergic stimulation, namely occasional rotational behaviour (1.5 ± 0.4 turns min^{-1}), excessive sniffing, licking and grooming. However, this behaviour was not always evident and the animals continued to feed, drink and rest. The result of this regimen on response to the (+)-PHNO challenge can be seen in Fig. 15.5a and b. The figures show that 24 h (test day 4) following the 3-day infusion, although the (+)-PHNO challenge still induced contralateral rotation, the duration of the response was both significantly shorter and of a lower frequency than that recorded on test day 0. Five days later (test day 8), the response to the challenge of (+)-PHNO was still blunted. Our results would therefore agree with those of Martin-Iverson *et al.* (1987) who determined tolerance to the motor stimulant effects of (+)-PHNO after 4 days of continuous subcutaneous administration. The notable feature of Martin-Iverson's study however, was that whilst the response to (+)-PHNO was monitored continuously, tolerance was only evident during the rats' light cycle. Furthermore, tolerance to (+)-PHNO could be reversed by stress-induced dopamine release or administration of a D1 receptor agonist, so suggesting the phenomenon to be mediated by changes in the activation of D1 receptors possibly by endogenous dopamine.

The 1-methyl-4-phenyl-1,2,3,6-tetrahydropyridine marmoset model of Parkinson's disease

It is now widely known that the inadvertant injection of 1-methyl-4-phenyl-1,2,3,6-tetrahydropyridine (MPTP) by young drug abusers produced irreversible symptoms of Parkinson's disease. The clinical features observed in those affected included tremor, rigidity and akinesia, ordinarily seen in the Parkinson's patient.

In the common marmoset, peripheral administration of MPTP causes selective bilateral loss of more than 80% of the dopamine-containing cell

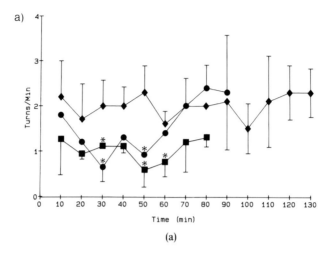

(a)

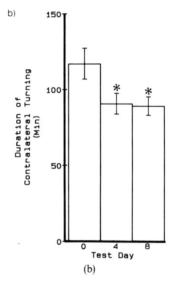

(b)

Fig. 15.5 The effect of 3-day (+)-PHNO gastric infusion on rotational behaviour. Rats with unilateral 6-OHDA lesions were infused for three consecutive days (test days 1–3) with (+)-PHNO (7.5 μg h^{-1}; 20 μl min^{-1}) via gastric fistula. One day prior to infusion (◆) and 1 (●) and 5 days (■) post-infusion, animals were injectd with (+)-PHNO (70 μg kg^{-1}, i.p.) and rotational behaviour scored. The number of contralateral whole body turns was counted every 10 min for 60 s until activity stopped. Results are expressed as turns min^{-1}, mean ± SEM of n = 4–6. Data were analysed by Mann–Whitney U-test; * $p < 0.05$ when (a) frequency of turns on test days 4 or 8 was compared with that shown on test day 0 and (b) when duration of turning behaviour was compared.

bodies in the pars compacta of the substantia nigra (Burns *et al.*, 1983; Jenner *et al.*, 1984; Langston *et al.*, 1984; Waters *et al.*, 1986). The outcome of such profound dopaminergic depletion is a persistent parkinsonian syndrome. Animal behaviour is characterized by akinesia or bradykinesia, rigidity of the limbs and trunk, stooped posture, postural tremor, loss of vocalization and blink reflex (Jenner *et al.*, 1984; Jenner and Marsden, 1987). When movement does occur, it is slow and deliberate and often incomplete and clumsy. In contrast to control animals, MPTP-treated marmosets show neither spontaneous social interaction nor the checking and evasive movements usually shown to the approaching investigator (Nomoto *et al.*, 1987). Moreover, MPTP-induced motor deficits respond to antiparkinsonian drugs and so can be used to discriminate drug actions of potential use in man. Thus, L-dopa/carbidopa combinations, bromocriptine and the selective D2 agonist LY 141865 have all been shown to alleviate the symptoms outlined (Jenner *et al.*, 1984; Nomoto *et al.*, 1985). Moreover, oral administration of (+)-PHNO to these animals results in a dose dependent reversal of akinesia and improvement in motor coordination (Nomoto *et al.*, 1987).

The relevance of studies in the MPTP marmoset to the oral administration of (+)-PHNO in man has already been demonstrated (Coleman *et al.*, 1987). Pharmacological investigation in the marmoset has proved predictive of (+)-PHNO's activity in Parkinson's disease in terms of dosage, duration of action and difference between oral and subcutaneous routes of administration. Therefore, following promising results in the rat, we sought to determine the response of MPTP marmosets to (+)-PHNO when delivered by the osmotic tablet.

Experimental methods

Administration and effect of MPTP in marmosets

In the present study, two adult common marmosets were treated with MPTP using a variable dosage schedule of between 1 and $4 \, mg \, kg^{-1}$ i.p. daily for 5–7 days. Such a regimen had to be employed because the individual response of animals differs for reasons that are still unknown. Each animal was treated in order to render it markedly akinetic as described. Nutrition and hydration were maintained by the subcutaneous injection of dextrose, amino acids, lipids, minerals, vitamins and sterile water. Following cessation of MPTP treatment, animals were allowed to recover for a period of 8 weeks.

(+)-PHNO administration and evaluation of antiparkinsonian effects

One (+)-PHNO osmotic tablet (release rate $5.7 \, \mu g \, h^{-1}$) was administered to the marmoset disguised in a banana mash. To prevent nausea and vomiting, but not influence the central effects of (+)-PHNO, animals were pretreated with the peripheral dopamine receptor antagonist, domperidone (Motilium) (1 mg kg^{-1} i.p.). The response to (+)-PHNO was gauged from the number of movements which the animal made across the four base segments of the cage, or in the vertical direction between the floor and perches. Movement was measured in 30-min periods for a maximum of 12 h.

Approximately 4 h following administration of the (+)-PHNO osmotic tablet to the MPTP-treated animals, a dramatic restoration in motor activity was evident. The quantitative effect of the drug was to increase the number of movements which the animal made in the test period (Fig. 15.6). In addition, the parkinsonian symptoms of bradykinesia and poor coordination of movement were alleviated. Furthermore, no dyskinetic phenomena were observed. Following the onset of activity, the number of animal movements continued to increase reaching a maximum after 6 h in one animal and 8 h in the second.

DISCUSSION

Our investigation was designed to examine the feasibility of delivering a behaviourally active concentration of (+)-PHNO from a novel orally administered osmotic tablet. The potential of using this method of adminis-

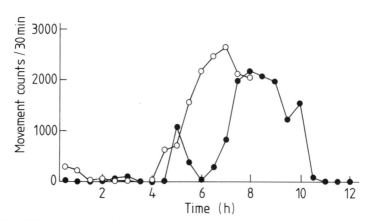

Fig. 15.6 Effect of oral administration of a (+)-PHNO tablet in MPTP-treated marmosets. One (+)-PHNO tablet (5.7 μg h^{-1}) was given to the marmoset between 8.00 and 11.00 a.m. and the number of movements counted over 30-min intervals for up to 12 h. The results from two animals are shown.

tration was demonstrated by stimulation of locomotor activity in the 6-OHDA lesioned rat and restoration of behavioural repertoire to MPTP-treated marmosets.

To date, sustained oral delivery of (+)-PHNO in man has been emulated by nasogastric infusion (Coleman *et al.*, 1987). In Parkinson's patients experiencing motor fluctuations ("on–off" phenomenon) in response to their normal L-dopa therapy, a 4–8 h (+)-PHNO infusion has been shown to increase the proportion of hours spent "on" despite the subjects receiving much smaller doses of L-dopa. However, this promising result may not be predictive of long-term therapeutic outcome if, as in our study, tolerance to the drug develops. Recent data by Martin-Iverson *et al.* (1987) suggest that in the case of (+)-PHNO, pulsatile delivery or the drug's combination with a D1 agonist may result in tolerance being delayed. Thus, future drug regimens will have to consider co-administration of D1 and D2 agonists, intermittent drug release and possibly manifestation of a normal diurnal variation in the dopaminergic system. The potential of using sophisticated orally administered devices to provide such controlled delivery remains a challenge to the technology of drug release.

REFERENCES

Baker, R. W. and Lonsdale, H. K. (1974) In *Controlled Release of Biologically Active Agents* (edited by A. C. Tanquary and R. E. Lacey), pp. 15–71. Plenum Press, New York.

Burns, R. S., Chiueh, C. C., Markey, S. P., Ebert, M. H., Jacobowitz, D. M. and Kopin, I. J. (1983) *Proc. Nat. Acad. Sci. USA* **80**, 4546–4550.

Chien, Y. W. (1983) *Drug Dev. Ind. Pharm.* **9**, 1291–1330.

Coleman, R. J., Grandas-Perez, F., Quinn, N., Jenner, P., Nomoto, M., Marsden, C. D. and Stahl, S. M. (1987) In *Parkinson's Disease: Clinical and Experimental Advances* (edited by F. C. Rose), pp. 179–187. John Libbey, London.

Critchley, P., Grandas-Perez, F., Quinn, N., Coleman, R., Parkes, D. and Marsden, C. D. (1986) *Lancet* **ii**, 349.

Fahn, S. (1982) In *Movement Disorders* (edited by C. D. Marsden and S. Fahn), pp. 123–145. Butterworth, London.

Grandas-Perez, F. J., Jenner, P. G., Nomoto, M., Stahl, S., Quinn, N. P., Parkes, J. D., Critchley, P. and Marsden, C. D. (1986) *Lancet* 906.

Guttman, M. and Seeman, P. (1985) *J. Neural Transm.* **64**, 93–103.

Hardie, R. J., Lees, A. J. and Stern, G. M. (1984) *Brain* **107**, 487–506.

Jenner, P. and Marsden, C. D. (1987) In *Current Problems in Neurology: 6, Parkinson's Disease: Clinical and Experimental Advances* (edited by F. C. Rose). John Libbey, London.

Jenner, P., Rupniak, N. M. J. and Rose, S. (1984) *Neurosci. Lett.* **550**, 85–90.

Jenner, P., Marsden, C. D., Nomoto, M. and Stahl, S. (1986) *Br. J. Pharmacol.* **89**, 626P.

Jones, J. H., Anderson, P. S., Baldwin, J. J., Clineschmidt, B. V., McClure, D. E.,

Lundell, G. F., Randall, W. C., Martin, G. E., Williams, M., Hirschfield, J. M., Smith, G. and Lumma, P. K. (1984) *J. Med. Chem.* **27**, 1607–1613.

Kafka, M. S., Wirz-Justice, A., Naber, D., Moore, R. Y. and Benedito, M. A. (1983) *Fed. Proc.* **42**, 2796–2801.

Langston, J. W., Forno, L. S., Rebert, C. S. and Irwin, I. (1984) *Brain Res.* **292**, 390–394.

Martin, G. E., Williams, M., Pettibone, D. J., Yarbrough, G. G., Clineschmidt, B. V. and Jones, J. H. (1984) *J. Pharm. Exp. Ther.* **230**, 569–576.

Martin-Iverson, M. T., Stahl, S. M. and Iversen, S. D. (1987) In *Parkinson's Disease: Clinical and Experimental Advances* (edited by F. C. Rose), pp. 169–177. John Libby, London.

Mouradian, M. M., Juncos, J. L., Fabbrini, G. and Chase, T. N. (1987) *Ann. Neurol.* **22**(4).

Naber, D., Wirz-Justice, A., Kafka, M. S. and Wehr, T. A. (1980) *Psychopharmacology* **68**, 1–5.

Nomoto, M., Jenner, P., and Marsden, C. D. (1985) *Neurosci. Lett.* **57**, 37–41.

Nomoto, M., Stahl, S., Jenner, P. and Marsden, C. D. (1987) *Movement Disorders* **2**, 37–45.

Nutt, J. G., Woodward, W. R. and Carter, J. H. (1986) *Neurology* **36**, 1206–1211.

Obeso, J. A., Luquin, M. R. and Martinez-Lage, J. M. (1986) *Lancet*, March 1.

O'Neill, R. D. and Fillenz, M. (1985) *Neuroscience* **16**, 49–55.

Pellegrino, L. J. and Cushman, A. J. (1967) *A Stereotaxic Atlas of the Rat Brain.* Appleton-Century Crofts, New York.

Robinson, J. R. (Editor) (1980) *Sustained and Controlled Release Drug Delivery Systems.* Marcel Dekker, New York.

Stahl, S. M. (1988). *J. Neural Trans.* (suppl.) **27**, 123–132.

Stahl, S. M. and Wets, K. M. (1988) *Clin. Neuropharmacol.* **11**, 1–17.

Stibe, C. M. H., Lees, A. J., Kempster, P. A. and Stern, G. M. (1988) *Lancet*, February 20.

Stoessl, A. J., Mak, E. and Calne, D. B. (1985) *Lancet* 1330–1331.

Ungerstedt, U. (1968) *Eur. J. Pharmacol* **5**, 107–110.

Ungerstedt, U., Avemo, A., Avemo, E., Ljungberg, T. and Ranje, C. (1973) In *Advances in Neurology*, Vol. 3, pp. 257–271. Raven Press, New York.

Waters, C. M., Hunt, S. P., Bond, A. B., Jenner, P. and Marsden, C. D. (1986) In *A Neurotoxin Producing a Parkinsonian Syndrome* (edited by S. P. Markey, J. N. Castagnoli, A. J. Trevor and I. J. Kopin), pp. 637–643. Academic Press, New York.

Zentner, G. M., Rock, G. S. and Himmelstein, K. J. (1985) *J. Controlled Rel.* **2**, 217–229.

16

Studies Towards Neural Transplantation in Parkinson's Disease

N. P. Quinn, S. B. Dunnett and W. Oertel

TRANSPLANTATION STUDIES IN RATS

The development of techniques for transplantation of neural and other catecholamine-secreting cells, and the characterization of the principles for achieving optimal graft viability, have been conducted almost exclusively in the laboratory rat.

Principles for viable neural transplantation

A number of principles have been determined for achieving successful transplantation of neural tissues.

Donor age

It has been found on pragmatic grounds that central nervous system (CNS) neurons only survive transplantation to the adult brain if taken during a particular time window in early development, within a few days of final mitotic cell division (Das *et al.*, 1980; Seiger, 1985). This corresponds to a stage when the phenotype is determined and neurite outgrowth is maximal,

DISORDERS OF MOVEMENT: CLINICAL, PHARMACOLOGICAL
AND PHYSIOLOGICAL ASPECTS   ISBN 0-12-569685-X

but extensive fibre connections are not yet formed. The presence of trophic factors within the embryonic environment and the ability of immature tissue to sustain more prolonged periods of anoxia almost certainly contribute to improved survival of embryonic donor tissue, but the relative importance of these additional factors has not been determined. Unlike CNS tissues, peripheral sources of neuronal and neuroendocrine cells can survive transplantation even when taken from mature donors. The age of the host appears to be relatively less important than the age of the donor tissue, and within adulthood relatively good survival of appropriate tissues can be achieved from youth through to old age (Hallas *et al.*, 1980; Gage *et al.*, 1983).

Transplantation site

Stenevi *et al.* (1976) have argued that a rich vascular supply is necessary at the site of transplantation to ensure the rapid incorporation of the grafted tissue into the host blood and cerebrospinal fluid circulation. In addition to naturally occurring rich vascular sites, such as the choroidal fissure and the lateral ventricle, techniques have been developed for artificial construction of vascularized beds which will sustain graft tissue (Stenevi *et al.*, 1976, 1980). Whereas this factor is important for "solid" grafts, grafts made by injection of dispersed cell suspensions or tissue fragments appear to have the capacity to survive at all sites in the host parenchyma without the provision of an enriched vascular environment (Björklund *et al.*, 1983; Schmidt *et al.*, 1981).

Intrinsic innervation

For many systems in the brain, graft viability is poor in the presence of an intact host innervation. Thus, for example, both survival and growth of cholinergic or sympathetic neurones implanted in the hippocampus is limited in the presence of an intact septohippocampal innervation (Björklund and Stenevi, 1981; Gage and Björklund, 1986). This restriction does not appear to apply to the nigrostriatal dopamine system, since good survival has been seen when nigral neurons are implanted into the intact neostriatum (Schmidt *et al.*, 1981), and the grafted cells give rise to a moderate ingrowth into the host neostriatum, albeit not as extensive as that seen following implantation in a completely deafferented striatum. This is important in considering clinical applications in Parkinson's disease patients where dopamine cell loss and striatal dopamine depletion, although extensive, is unlikely to be as complete as that obtained with model lesions in experimental animals.

Immunological factors

Immunological factors have turned out to be less important in the brain than might be expected from experience of transplantation in other organs of the body (Greene and Arnold, 1945; Mason *et al.*, 1986). The immunologically privileged status of the brain as a transplantation site is nevertheless not complete, and substantial rejection takes place when donor and host differ at both major and minor histocompatibility sites, unless prevented by immunosuppressant drugs such as cyclosporin (Brundin et al., 1985). Very little is known about the comparable status of immunological privilege in the human brain, and considerable care and attention to this factor will be necessary in any clinical application of transplantation of neural tissue from an unrelated donor.

Functional nigral grafts in rats

Following the identification of the principles for achieving viable neural transplantation, there is now a considerable literature assessing the functional efficacy of dopamine- or other catecholamine-rich tissues grafted to rats with dopamine-depleting lesions. In all the studies to be discussed, the denervation of striatal dopamine has been achieved with unilateral or bilateral injections of the toxin 6-hydroxydopamine (6-OHDA), which can produce in excess of 99% depletion of forebrain dopamine ipsilateral to the lesions, and results in well-characterized behavioural impairment in the rats.

Anatomical studies

The first studies of intrastriatal dopamine-rich nigral grafts in rats employed catecholamine fluorescence techniques to demonstrate good survival of the grafted cells and the development of extensive reinnervation of the denervated host caudate-putamen (Björklund and Stenevi, 1979; Björklund et al., 1980, 1983; Freed et al., 1980). Within the grafts, particularly of the "solid" type, the dopamine cells were frequently seen to aggregate in a layer with perpendicularly extending dendrites, reminiscent of a mini substantia nigra (Björklund et al., 1980). Fibre outgrowth can be up to 2–3 mm into the host neostriatum or other normal dopamine projection areas, but is extremely limited into areas that are not targets for the intrinsic innervation, such as the globus pallidus or neocortex (Björklund et al., 1983).

Subsequent studies have additionally employed tyrosine hydroxylase histochemistry to confirm the fibre outgrowth at the light microscopic level (Schultzberg et al., 1984; Strömberg et al., 1985b), and to identify terminals of graft origin under the electron microscope (Freund et al., 1985; Mahalik et al., 1985). In particular, the ultrastructural observations indicate that graft-derived axons make morphologically normal synaptic contacts at appropriate loci on the medium spiny neurones of the host striatum (Freund et al., 1985), providing the possibility of a functional interaction between the cells of the graft and the host neuronal circuitry.

Biochemical studies

The biochemical determination of an increased dopamine concentration in the host neostriatum proximal to the grafts reflects the fibre innervation observed anatomically (Freed *et al.*, 1980; Schmidt *et al.*, 1982, 1983). Additionally, compensatory changes in the rates of dopamine synthesis (determined by dopa accumulation following pharmacological inhibition of dopa decarboxylase), dopamine release (determined by measurement of dopac concentrations), and postsynaptic receptor supersensitivity (determined by spiroperidol binding assay) are all down regulated towards normal levels (Freed *et al.*, 1983b; Schmidt *et al.*, 1982, 1983), suggesting that the dopaminergic terminals seen under the microscope to reinnervate the host striatum are functionally active. Subsequent studies have employed *in vivo* monitoring of dopamine release by implantation of voltammetric electrodes (Hoffer *et al.*, 1987) and by microdialysis probes (Strecker *et al.*, 1987; Zetterström *et al.*, 1986) to demonstrate the return of spontaneous dopamine release and its pharmacological control from graft-derived dopamine terminals in the reinnervated neostriatum.

Electrophysiological studies

Wuerthele *et al.* (1981) first employed extracellular recording from cells in solid nigral grafts implanted into the lateral ventricle, and found that grafted neurons exhibited spontaneous electrophysiological activity with action potential waveforms and firing rates similar to those observed in cells of the intact substantia nigra pars compacta. The dopaminergic nature of the cells was confirmed by the demonstration that dopamine agonists inhibited, whilst dopaminergic antagonists excited, cellular activity. Subsequently, Arbuthnott *et al.* (1985) have observed that cells with both "dopaminergic" and "non-dopaminergic" waveforms and conduction velocities could be identified within nigral grafts. Both types of cells could be driven orthodromically from the host striatum, suggesting that the striatal reinnervation is not exclusively dopaminergic (as indeed is the case in the normal nigrostriatal projection). However, behavioural recovery in the animals was dependent on the presence of cells of the dopaminergic subpopulation projecting into the host striatum. Within the host brain, electrophysiological indices associated with dopamine denervation, such as the increased firing rate of striatal neurones and their insensitivity to local application of the indirect dopamine agonist phencyclidine, have been seen to be reversed to normal levels in a 2 mm zone proximal to nigral grafts but not more distally (Strömberg *et al.*, 1985b). Thus, grafted dopaminergic neurons do appear to exert a physiological influence on host neostriatal circuitry that may be functionally beneficial to the host animal.

Behavioural studies

The first studies of the behavioural effects of nigral grafts examined the pharmacologically induced motor turning asymmetry ("rotation") that was described by Ungerstedt and Arbuthnott (1970) in response to unilateral 6-OHDA lesions of the ascending nigrostriatal system. Numerous studies have demonstrated a substantial reduction of contralateral rotation induced by dopaminergic receptor agonists such as apomorphine, and a complete recovery of the ipsilateral rotation induced by dopaminergic stimulants such as amphetamine (Björklund and Stenevi, 1979; Perlow *et al.*, 1979; Freed *et al.*, 1980; Björklund *et al.*, 1980; Dunnett *et al.*, 1981a, 1983a; Freed, 1983). These observations suggest that the grafts give rise to a chronic spontaneous release of dopamine to reverse receptor supersensitivity and also that functional dopamine release can be pharmacologically stimulated from graft-derived nerve terminals in the host striatum.

Although these pharmacological tests still provide the clearest demon-stration of behavioural recovery, subsequent studies have also shown recovery in undrugged animals, in tests of the hypoactivity, contralateral sensory neglect, and some of the learning deficits induced by the lesions (Dunnett *et al.*, 1981b, 1983a, 1986, 1987, Fray *et al.*, 1983; Nadaud *et al.*, 1984). The neostriatum is a topographically heterogeneous nucleus, and graft placement has emerged as an important principle in functional recovery, with a distinct profile of recovery on some tests and not on others dependent on the particular location of the grafts and graft-derived reinnervation (Dunnett *et al.*, 1981a,b). In order to overcome such regional effects, multiple placements of suspension grafts have been found to provide more extensive reinnervation throughout the host neostriatum, and a corresponding additive profile of recovery on the different behavioural measures (Dunnett *et al.*, 1983a).

Limitations to behavioural recovery

In spite of the extensive or complete recovery that has been observed on some behavioural tests, there remain other measures on which we and others have been unable to detect any recovery whatsoever in lesioned rats receiving transplants. These include deficits in skilled manipulative abilities (Dunnett *et al.*, 1987), and in the aphagia and adipsia consequent upon bilateral lesions (Dunnett *et al.*, 1981c, 1983b). We have proposed that recovery is seen in those circumstances where reinnervation from grafts in an ectopic striatal site, and providing a tonic striatal activation, is sufficient to sustain the particular behaviour even though the grafted cells remain deafferented from the full pattern of connections that regulate intrinsic dopamine cells in the substantia nigra (Dunnett *et al.*, 1987). This will apply to those behavioural measures that reflect the activation of net striatal output, but not ones that are dependent on the nigrostriatal relay of specific sensory or motivational information. It has been suggested that if the dopaminergic grafts are implanted in neonatal animals, then sufficient reorganization is possible for the grafts to protect against the regulatory eating and drinking deficits that follow bilateral lesions of the nigrostriatal pathway in adulthood (Schwarz and Freed, 1987), but such neonatal prophylaxis is unlikely to have any general applicability in the development of a clinical transplantation programme for parkinsonism. More generally, in assessing the clinical prospects of a particular transplantation technique, care should be taken to address not only whether recovery can be demonstrated (in particular employing behavioural tests with optimal graft procedures), but also the extent of benefit to the host animal on the full syndrome of deficits induced by the model lesions.

Assessment of alternative donor tissues

Although embryonic substantia nigra can be an effective functional source of donor tissue in rats, the analogous use of embryonic human tissues for clinical transplantation both involves a controversial ongoing ethical debate and raises the still poorly understood question of the extent of immunological privilege in the brain with potentially serious consequences should a major graft rejection be mounted. A number of studies have therefore used the dopamine-depleted rat model of parkinsonism to assess the viability of alternative sources of graft tissues.

Adrenal graft

Freed *et al.* (1981) first reported that the catecholamine-secreting chromaffin cells of the adrenal medulla taken from young adult rats will survive transplantation to the lateral ventricle, and reduce apomorphine-induced rotation in host rats with unilateral 6-OHDA lesions. Further studies have indicated that whereas the normal adrenal medulla secretes primarily adrenaline, when isolated in the ventricle (or indeed in tissue culture) the adrenal graft tissue secretes predominantly dopamine and noradrenaline in approximately equal amounts (Freed *et al.*, 1983a).

The recovery in these animals is believed to be primarily due to diffuse release of dopamine and/or noradrenaline from the adrenal grafts, since (i) although the grafts were rich in catecholamine-fluorescent cells, they did not appear to give rise to any substantial fibre growth into the host brain, and (ii) a similar degree of recovery in apomorphine rotation is obtained by chronic infusion of dopamine into the denervated striatum using minipumps (Hargraves and Freed, 1987), whereas (iii) unlike nigral grafts, the adrenal grafts do not restore contralateral sensorimotor neglect (Freed, 1983) or amphetamine rotation (Brown and Dunnett, 1989). The chromaffin cell development of adrenal tissue can be diverted to a more neuronal phenotype in the absence of corticosteroids (which is achieved by the discrete dissection of the medulla in either tissue culture or prior to intracerebral implantation) in combination with the provision of nerve growth factor (NGF). Strömberg *et al.* (1985a) have found that infusion of NGF into the vicinity of intrastriatal grafts of adrenal medulla cells promotes not only graft survival and differentiation, but also the development of neurite outgrowth into the host neuropil. In these studies, the maintenance of this adrenal graft-derived reinnervation was seen to be dependent on the chronic provision of NGF, and extensive fibre retraction was seen as soon as the infusions were terminated. However, very recent studies by Pezzoli *et al.* (1988) in 6-OHDA-lesioned aged rats lend further support to NGF possibly playing a major role in promoting graft function. They found that continuous i/c/v infusion of NGF improves the ability of adrenal medullary implants to decrease apomorphine-induced rotations. However, NGF plus sciatic nerve or NGF plus fat were just as effective, and again more effective than an adrenal graft without NGF. NGF alone had no effect.

Thus, autografts (i.e. from the patient himself) or adrenal cells could provide catecholamine-rich tissue for clinical transplantation that bypasses both immunological and ethical concerns associated with other sources of tissue. However, the experimental studies in rats suggest that the general applicability of this approach may be limited by the fact that although adrenal tissues from young adult hosts appear to be viable, functional effectiveness is lost when taken from mature or older donor animals (Freed, 1983), and that neuronal regrowth rather than simply diffuse catecholamine secretion may be dependent on lifelong intracerebral infusion of NGF or other trophic factors to maintain the appropriate differentiation of the grafted cells.

Cultured cell lines

An alternative source of donor tissue may be provided by cultured cell lines, which have the advantage of constant availability of cells which can be well characterized and easily manipulated (Gash *et al.*, 1985). Several lines rich in catecholaminergic neurons are now available, the most extensively studied of which is the PC12 phaeochromocytoma cell line derived from a rat adrenal medullary tumour. Jaeger (1985) found that PC12 cells implanted in immature (4–12-day-old) rats survive and continue to proliferate in the immature brain, whereas initial observations in adult hosts suggested only temporary survival and functional influence (Hefti *et al.*, 1985). Subsequent studies suggest however that host age is not the critical factor; good and prolonged survival can be achieved in adult hosts (Freed *et al.*, 1986; Jaeger, 1987), and in many cases the grafted cells have been seen to give rise to some neurite extension into the host brain.

Implanted cells may continue mitotic cell division after transplantation, and Jaeger (1987) indicates 100-fold increases in graft volume so that the host animals had to be sacrificed within 2 months of implantation. Any clinical application of cell lines can only be possible if the tumourous growth of the grafted cells is inhibited. To this end, Jaeger (1987) reports preliminary success with X-ray irradiation of the cells prior to transplantation. An alternative approach has been to use chemical inhibition of mitosis in cultures of various human-derived cell lines prior to implantation in rat hippocampus (Gash *et al.*, 1985; Kordower *et al.*, 1987).

Cross-species grafts

A third source may be to obtain appropriate populations of nigral cells from the embryos of other donor species. Cross-species transplantation of nigral and other populations of embryonic cells has been studied with a view to characterizing the extent of immunological privilege in the brain. Even in the absence of immunosuppression, some limited survival of mouse nigral cells can be seen following implantation in the denervated rat neostriatum (Björklund *et al.*, 1982), and monkey adrenal chromaffin cells have also been seen to survive and show limited process formation in the rat brain for at least 48 h and possibly up to 28 days after implantation (Patel-Vaidya *et al.*, 1985). Both the survival and the functional efficacy is markedly enhanced when the recipients are treated with cyclosporin A (Brundin *et al.*, 1985). Nevertheless, the potential for immunogenic failure does not diminish once the protected graft becomes established, and termination of cyclosporin treatment even some months after surgery can result not only in rejection of the graft but also associated necrotic damage in the host striatum (Brundin *et al.*, 1988). The clinical use of xenogeneic graft tissues must take account of the fact that immunosuppressant drugs cause substantial side-effects, and prolonged treatment with such drugs would itself involve a substantial health risk.

Human donor cells

In spite of the alternative sources of cells that may be available for clinical transplantation, it may nevertheless prove to be the case that allogeneic grafts of the appropriate population of embryonic nigral cells remain the most suitable biological source of donor tissue for grafting in Parkinson's disease. The effectiveness of immunosuppression treatment in protecting xenogeneic grafts from rejection provides the possibility of determining the optimal procedures for obtaining and preparing human foetal neurons for transplantation, using the well-characterized rat as host. Two groups have shown that human foetal nigral cells survive transplantation well in the rat neostriatum following cyclosporin treatment, and are functionally effective in reversing lesion-induced rotation asymmetries using both the cell suspension (Brundin *et al.*, 1986) and solid graft (Strömberg *et al.*, 1986) techniques. Transplanted human nigral cells are larger, and fibre outgrowth is both coarser and more extensive than comparable grafts of rat nigral cells, and appears to develop over a much slower and more protracted time frame (Brundin *et al.*, 1988). Thus, transplanted cells appear to retain many developmental features of the donor, and are not phenotypically determined in full by the host environment. In both the Strömberg and the Brundin studies, the critical maximal donor age for viability of human nigral grafts was found to be approximately the eleventh week of gestation; above this age the grafts survived poorly and had little functional effect. In the absence of immunosuppression the grafts of embryonic nigra appear not to survive (Brundin *et al.*, 1988), although human foetal adrenal and sympathetic neurons taken through several weeks culturing *in vitro* do appear to be viable in rat brain without cyclosporin treatment (Kamo *et al.*, 1985, 1986).

TRANSPLANTATION STUDIES IN PRIMATES

The discovery that the neurotoxin MPTP induces bilateral parkinsonism in primates (Burns *et al.*, 1983; Langston *et al.*, 1983; Ballard *et al.*, 1985) has provided a closer model of human Parkinson's disease to use as a test-bed for transplantation studies. In contrast to the rodent (which, moreover, is not rendered parkinsonian by MPTP), non-human primates also possess a basal ganglia structure very similar to the human. Correlations between motor function and certain nuclei in the basal ganglia in normal and MPTP-treated non-human primates have been found (Alexander *et al.*, 1986; Miller and DeLong, 1988). For example, putaminal activity has been shown to be related to specific aspects of limb movements, including direction, amplitude (or velocity) and load. In addition, cells in the primate putamen may participate in preparation for movement (Alexander *et al.*, 1986). With appropriate neurosurgical techniques, dopamine-synthesizing cells can be focally grafted into the parenchyma of the putamen, into the body of the caudate nucleus, or into/onto the head of the caudate nucleus.

The MPTP primate model does, however, possess certain drawbacks. The first concerns the variable degree and rate of spontaneous partial functional recovery that occurs in many animals, and against which any apparent beneficial effect of transplantation must be set. The second concerns the much higher cost and limited availability of non-human primates, with consequent limitations on the number of animals studied. In addition, the MPTP primate model has become available only relatively recently, in comparison to the long-familiar 6-OHDA rodent model. The range of different experiments has thus been more limited in primates than in rats. Most of the primate experimental work considered below has been undertaken on animals rendered bilaterally parkinsonian by MPTP, but some has been done in the hemiparkinsonian model produced either by unilateral intracarotid injection of MPTP or by unilaterally placed 6-OHDA lesions.

Adult autografts (primate to primate)

Adrenal medulla

Rather surprisingly, in view of the large number of adrenal medullary autografts performed in humans, there are only two published studies of this procedure in non-human primates. Morihisa *et al.* (1984) studied four rhesus monkeys which had received a unilateral 6-OHDA lesion of substantia nigra 2 months prior to implant, and which were sacrificed 5–8 months after grafting of their own adrenal medulla into the denervated caudate nucleus. In the single animal receiving stereotaxic implantation, only two sites with surviving graft tissue, containing fewer than 10 catecholamine fluorescent cells per site, were found 5 months after transplant. In the other three animals which received their adrenal autografts under direct visualization, a total of nine sites of surviving graft were found; five contained fewer than 10 cells, and the other four sites (two each in two animals) contained 30, 160, 80 and 200 cells, respectively. However, there was no evidence of caudate reinnervation, and behavioural studies were not performed.

Porrino *et al.* (1988) transplanted adrenal medullary tissue into the dopamine-deficient caudate nucleus of unilaterally MPTP-treated rhesus monkeys. Drug-induced rotation was improved for 3 months, but by 6 months it had returned to preoperative levels.

Superior cervical ganglion

The superior cervical ganglion (SCG) represents an alternative source of catecholamine cells. It has been reported that its principal ganglion cells contain norepinephrine, whereas its small intensely fluorescent (SIF) cells contain dopamine, although this proposition remains controversial. Itakura *et al.* (1988) have recently performed autotransplantation of small pieces of SCG into caudate nucleus in five macaque monkeys pretreated with MPTP to render them parkinsonian. After sacrifice 2 weeks post-grafting, catecholamine cell bodies were said to have survived well, and histo-fluorescence studies showed many catecholamine fibres growing from the transplant into the dopamine-deficient caudate nucleus. No behavioural studies were performed, and the time course for fibre outgrowth seems remarkably rapid compared to that seen with foetal nigral implants.

Embryonic ventral mesencephalic cell allografts

Morihisa *et al.* (1984) also reported the results of stereotactically grafting six small pieces of foetal substantia nigra (from a 59-day and a 71-day foetus) into two unilaterally 6-OHDA lesioned rhesus monkeys. The remaining nigral tissue was injected into the ventricle. Both animals were sacrificed 3 months post-grafting. Neither animal demonstrated any evidence of surviving catecholamine-containing graft tissue, nor were grafts apparent in the ventricles.

Since 1986, the results of further grafts of allogeneic embryonic mesencephalic cells have been reported in several species of non-human primates including rhesus monkeys (Bakay *et al.*, 1988), African green monkeys (Redmond *et al.*, 1986; Sladek *et al.*, 1988) and common marmosets (Fine *et al.*, 1988) with bilateral MPTP-induced parkinsonism. Transplantation was mostly done unilaterally, into a cavity carved into the head of the caudate nucleus, where the grafted cells had access to the cerebrospinal fluid as well as to the parenchyma of the caudate nucleus. Bilateral improvement of rigidity, bradykinesia and tremor was observed post-transplantation at time intervals varying between a few days and 120 days in the rhesus monkey (Porrino *et al.*, 1988) and the African green monkey (Redmond *et al.*, 1986; Sladek *et al.*, 1988). Post-mortem measurements of dopamine and its metabolite homovanillic acid (HVA) showed a nearly normal ratio of HVA to dopamine in the vicinity of the graft, which contained tyrosine hydroxylase (TH) positive neurons. In contrast, areas distant from the graft contained a high HVA:dopamine ratio, indicating increased dopamine turnover (Sladek *et al.*, 1988). Thus, a unilateral graft appeared to improve a bilateral neurological deficit and to normalize dopamine turnover in the striatal area immediately surrounding the graft which was effectively devoid of endogenous innervation due to MPTP exposure. The bilateral clinical improvement observed after unilateral grafting may at least in part be due to a "pharmacological" action via the release of biologically active molecules such as dopamine into the CSF. Alternatively, the host parenchyma may in some way have been triggered to produce behavioural improvement either pharmacologically or through new patterns of neuronal activity, or both.

Bilateral grafts into the putaminal parenchyma of MPTP-treated marmosets have also been performed. Preliminary data suggest an improvement in spontaneous locomotor activity up to 6 months post-transplantation (Fine *et al.*, 1988).

In the unilateral MPTP-treated rhesus monkey, Porrino *et al.* (1988) found that grafting embryonic ventral mesencephalic cells onto/into the dopamine-deficient caudate nucleus diminished drug-induced rotations for at least 6 months. In a parallel group of recipients (v.s.), adrenal medullary tissue was effective for only 3 months. A similar pattern of recovery on monthly rotation tests has been observed by Annett *et al.* (1989) after transplantation of foetal nigral cells to the caudate and putamen of common marmosets in which unilateral lesions of the ascending dopamine fibres were made by stereotaxic injections of 6-OHDA.

Other primate experiments assessing the role of precavitation as an aid to graft survival have been performed in the macaque. Bakay *et al.* (1988) found a 240% average increase in the number of surviving TH-positive cells grafted into a tract left after removal of a blunt metal stylet which had been stereotactically placed into caudate–putamen 5–7 days prior to implantation.

Table 16.1   Different donor host combinations investigated

	Rat			Monkey			Human		
	Nigra	Adrenal	Lines	Nigra	Adrenal	Lines	Nigra	Adrenal	Lines
Source of donor tissue host species:									
Rat	1	2	3	—	4	—	5	6	7
Monkey	—	—	—	8	9	—	10	11	12
Human	—	—	—	—	—	—	13	14	—

(1) Björklund and Stenevi (1979); Björklund *et al.* (1980, 1983); Dunnett *et al.* (1981a–c, 1983a,b); Brundin *et al.* (1985, 1986); Perlow *et al.* (1979); Freed *et al.* (1980); Freed (1983).
(2) Freed *et al.* (1981, 1983a); Freed (1983); Strömberg *et al.* (1985a).
(3) Jaeger (1985); Hefti *et al.* (1985); Freed *et al.* (1986).
(4) Patel-Vaidya *et al.* (1985).
(5) Brundin *et al.* (1986); Strömberg *et al.* (1986).
(6) Kamo *et al.* (1985, 1986).
(7) Kordower *et al.* (1987); Gash *et al.* (1985).
(8) Bakay *et al.* (1985); Redmond *et al.* (1986); Sladek *et al.* (1986); Fine *et al.* (1988); Porrino *et al.* (1988).
(9) Morihisa *et al.* (1984); Porrino *et al.* (1988).
(10) Redmond *et al.* (1988).
(11) Pezzoli *et al.* (1987).
(12) Kordower *et al.* (1987); Gash *et al.* (1987).
(13) Madrazo *et al.* (1988); Hitchcock *et al.* (1988); Lindvall *et al.* (1988).
(14) Backlund *et al.* (1985); Lindvall *et al.* (1987a,b); Madrazo *et al.* (1987a,b); Burns *et al.* (1988); Lieberman *et al.* (1988); Watts *et al.* (1988); Goetz *et al.* (1988); Olanow *et al.* (1988); Tintner *et al.* (1988); Madrazo *et al.* (1988).

Embryonic adrenal medullary cell xenografts (human to primate)

Pezzoli *et al.* (1987) have reported their results from unilateral stereotactic grafting of human embryonic adrenal medullary cells from foetuses aborted at between 16 and 18 weeks gestation into the putamen of a single bilateral MPTP-treated macaque monkey. Improvement in drug-induced or stimulation-induced behavioural deficit was seen in this grafted monkey, but not in an unoperated control group, starting 7 days after the implant. The question of graft rejection was not addressed, and awaits further studies.

In summary, transplantation studies in non-human primates are at an early stage. In terms of biochemical and morphological indices of graft survival, they are in general agreement with the extensive experience in rodents, although the methods employed differ substantially from those used in rats. Data on behavioural improvement are limited, and further controlled studies are needed. Moreover, numerous questions remain unanswered (see human studies), and clearly further extensive basic research in non-human primates is needed.

TRANSPLANTATION STUDIES IN HUMANS

Adrenal autografts

The first human implants of catecholaminergic cells as an experimental treatment for Parkinson's disease date from 1982. It is worth noting that at that time the MPTP primate model had not yet been described so that the animal work upon which the human operation was based consisted entirely of experiments on unilaterally 6-OHDA lesioned rodents. These studies had already suggested that embryonic substantia nigra grafts should give better and more lasting benefits than either embryonic or adult adrenal medullary grafts. However, in man the use of adrenal autografts had the advantage of avoiding some important ethical and immunological problems.

The first two human patients (one each in 1982 and 1983) were given stereotactically placed implants of small fragments of the medulla of one of their own adrenal glands into the right caudate nucleus. Transient mild improvement was observed, but at 6 months the patients were judged to be no different from their pre-operative state (Backlund *et al.*, 1985). After these negative results, the same Swedish group, on the basis that dopamine loss in Parkinson's disease is greater in putamen than caudate (Kish *et al.*, 1988), and that the putamen is more important in terms of motor function (Alexander *et al.*, 1986), proceeded to two further operations, in 1985 and

1986, this time placing similar adrenal medullary autografts stereotactically into the right putamen. Again, despite transient minor improvement during the first 2 months, there was no evidence of any lasting benefit from the operation (Lindvall *et al.*, 1987a). Because of these disappointing results, and in view of the increasing weight of evidence from animal experimental studies in favour of embryonic grafts, the Swedes then launched the ethical and legal debate necessary as a prerequisite for proceeding to the first human embryonic substantia nigra implants.

Whilst these preparations were in hand, a paper from Mexico by Madrazo *et al.* (1987a) appeared in the *New England Journal of Medicine* in April 1987. This reported dramatic improvement in two parkinsonian patients who had had adrenal medullary autografts placed, by open operation with direct visualization, into a cavity made in the head of the right caudate nucleus, using a transfrontal approach. Clinically, these two patients were both highly unusual. The first, aged 35, had developed parkinsonism at age 32. He had been unable to tolerate L-dopa, and after only 3 years of disease was confined to a wheelchair and unable to perform even the most basic activities. The second, aged 39, had become severely disabled and incapable of writing, eating or performing other everyday activities on his own within 1 year of disease onset at age 33. He initially "responded satisfactorily" to L-dopa, "but soon intolerance and unresponsiveness to the drug developed". Clinical improvement was noted in both patients at 15 and 6 days, respectively after operation, and both were still dramatically, and bilaterally, improved 10 and 3 months later. The first patient was playing soccer with his son, and considering working.

Two months before this report appeared, the same group had in fact already undertaken 10 such operations, which were later reported in October 1987 to the 73rd Clinical Congress of the American College of Surgeons (Madrazo *et al.*, 1987b). Six of the recipients were judged to have dementia (three mild, three moderate), and two had died (from myocardial infarction and venous thrombosis). All 10 patients "showed objective improvement in the post-operative period", with follow-up ranging from 1.5 to 15 (median 6) months. Of the nine patients receiving L-dopa pre-operatively, four (750–2500, median 1375, mg day^{-1}) had been able to stop it completely. In the remaining five, the dose was reduced from 500–1750 (median 1750) mg day^{-1} to 250–375 (median 375) mg day^{-1}.

The authors discussed possible reasons for the results in their first two patients. They did not feel that the bilateral improvement could be due to the unilateral caudate lesion, but instead proposed that it resulted from the release of dopamine into the ventricles, from which it could then reach the contralateral caudate. They stressed the possible advantage of placing the graft in contact with the CSF, so allowing nutrients to promote graft survival, and "chemical substances secreted by the grafted cells to be transported to other structures in the central nervous system, thus acting as a biologic infusion pump".

Following the initial Mexican report, many more operations, most using an identical technique, were performed, principally in the United States and China, but also in other countries. By now, several hundred operations have probably been done, but very little information has appeared in the scientific literature. Much of the available data derives from experience in the United States. At the 40th Annual Meeting of the American Academy of Neurology in April 1988, the results of a total of 47 operations were reported from six centres: 12 from Nashville (Burns *et al.*, 1988), nine from NYU (Lieberman *et al.*, 1988), eight from Atlanta (Watts *et al.*, 1988), seven from Chicago (Goetz *et al.*, 1988), six from Tampa (Olanow *et al.*, 1988) and five from Dallas (Tintner *et al.*, 1988). Six patients suffered cerebral morbidity in terms of brain damage from haemorrhage, infarct or anoxia. Many had episodes of pneumonia, and most had psychiatric disturbance (Tanner *et al.*, 1988). One 69-year-old patient from Nashville had died 3 months post-operatively from a myocardial infarct. A representative from the Mexican group confirmed that four of 44 patients operated there had since died.

Autopsy findings in two cases were presented. Case 4 of Madrazo *et al.* (1987b) returned home to California. There, four months post-operatively, he had a generalized seizure. He was found to have extensor plantars and clonus, and a CT scan showed low density in the right frontal lobe. He then developed a urinary infection, arrested and died. Post-mortem showed necrotic substance in the graft site, which extended well into putamen. Only a few surviving adrenal cortical, but no medullary, cells were found (Peterson *et al.*, 1989). The brain of the Nashville case showed only a couple of areas of viable TH-negative cells at the graft site.

The clinical effect of the graft on the parkinsonian signs and symptoms in surviving patients was difficult to establish. In only a minority of patients had drug treatment been held constant, and here there was evidence after approximately 6 months follow-up that off-period severity and duration had modestly decreased. In many subjects, inadequate baseline monitoring and/or the fact that potent antiparkinsonian drug regimes had been altered, meant that it was impossible to assess the results of the experimental procedure. However, no group was able to replicate in even a single patient the dramatic results reported by Madrazo *et al.*

The overall picture to emerge was that the procedure could have a mortality rate as high as 10% and a serious morbidity rate of 10%, and that post-operative chest and/or psychiatric complications could be seen in the majority. Against this background, there is evidence of a modest improvement in response to L-dopa in a number of patients. However, if this benefit is confirmed, then the reason(s) for it occurring are far from clear, since the post-mortem evidence of adrenal medullary cell necrosis suggests that the original rationale for the operation is no longer tenable.

In many ways, the presentations threw up more questions and problems than they answered. First, in terms of patient selection, some subjects who almost certainly do not have idiopathic Parkinson's disease (and who are likely to have additional striatal pathology) as well as others who have never tolerated L-dopa and thus never given the best clinical indication of the idiopathic disease, have been operated. Some with dementia have been submitted to a dangerous experimental operation of unproven efficacy, and many patients not already known to transplant teams have been seen for the first time and then operated upon with apparent unseemly haste. This aspect is important not only for ensuring an adequate baseline, but also in terms of providing the continuing care of patients with a chronic neurological disability.

On the theoretical and technical side, controversy surrounds almost every aspect of the procedure. Steroids diminish the (already low) dopamine output of adrenal medullary cells. Should they be given pre- or post-operatively to cover possible adrenal insufficiency? Is it really possible to graft pure adrenal medullary tissue without including cortical, steroid-producing cells (current evidence suggests it is not)? What about the role of other tissue elements included in the graft, such as Schwann cells, endothelial cells, pericytes, fibroblasts and smooth muscle fibres? Should the right or left adrenal be removed, always assuming one has demonstrated that the patient does indeed have two functioning glands? Should the approach be abdominal, with ileus preventing early resumption of oral medication, or retroperitoneal? Is the adrenal gland itself affected because of the patient's Parkinson's disease? Carmichael *et al.* (1988) have found lowered catecholamine concentrations in adrenal medulla of three patients with "parkinsonism" relative to 15 controls without parkinsonism, and Riederer *et al.* (1978) have found low levels of tyrosine hydroxylase in patients with parkinsonism. Cervera *et al.* (1988), however, have found that adrenal gland levels of noradrenaline, adrenaline and tyrosine hydroxylase activity were only slightly but not significantly decreased in 12 patients with Parkinson's disease compared to nine age-matched controls. Furthermore, will the disease process further destroy the graft *in situ*?

Should the transfrontal approach, with inevitable damage to the frontal lobe, be used, or should the approach be transcallosal? What is the effect of open operation on the integrity of the blood–brain barrier? Positron emission tomography studies 6 weeks post-operatively suggest significant breakdown (Guttman *et al.*, 1988), with repair probably having occurred by 3 months (R. Frackowiak, personal communication). What would be the clinical consequences of such a breakdown? Would L-dopa preparations be functionally more potent (due to increased entry of L-dopa into brain) or less potent (due to decarboxylase inhibitors gaining unaccustomed access to brain)? Would stereotactic placement be preferable, producing less trauma to the brain and its barriers? Should the graft be placed in the caudate at all, or rather, as Kish *et al.* (1988) have suggested, the putamen? Would pre-lesioning improve graft viability, and to what extent is continuity with ventricular CSF of importance? Can the caudate lesion itself account for some of the clinical effects observed, as suggested by Meyers (1951) and Spiegel *et al.* (1965), but more recently largely discounted by Motti *et al.* (1988)? Could the reaction to the implant at the graft site be of importance? What about trophic factors such as nerve growth factor? Should grafts be placed unilaterally or bilaterally? How much tissue should be transplanted, and is it better to use tissue fragments or cell suspensions? Can an improved source of autologous donor catecholamine-producing cells be obtained by tissue culture techniques, or by using superior cervical ganglion?

The answers to most of these questions are unknown. If adrenal autografts are supplanted by embryonic substantia nigra grafts, some will become obsolete, only to be replaced by new and different questions considered below.

Embryonic nigral and adrenal grafts

In 1986 the Swedish Society of Medicine issued provisional ethical guidelines for human embryonic tissue grafting, paving the way for the first foetal substantia nigra implants in human patients. The work on grafting human embryonic substantia nigra into rats (heterologous transplantation) had suggested that the critical age of the embryo was 7–12 weeks, and that treatment with cyclosporin significantly enhanced graft survival. In addition, there was the question of how much nigral tissue was needed to replace the cells destroyed by the patient's Parkinson's disease. The normal human mesencephalon has approximately 225 000 dopaminergic cells on each side, of which some 60 000 innervate putamen. Since about 15 000 mesencephalic dopaminergic cells survive per embryo, the Swedes calculated that to unilaterally replenish the normal complement of nigral cells, tissue from four donor embryos should be needed. However, total replacement of missing cells may not be necessary. In untreated patients, 80% loss of striatal dopamine is necessary to produce the symptoms and signs of parkinsonism (Bernheimer *et al.*, 1975). Moreover, when exogenous L-dopa is supplied, the completeness of replacement of nigral cells containing aromatic acid decarboxylase may become less critical (Melamed, 1988).

Women seeking terminations in Lund University were informed about ongoing clinical trials using foetal material, and asked if they would consent to the aborted material being used in this way. Eighty per cent agreed, whilst 20% declined. All consenting women were tested for HIV, hepatitis, herpes and cytomegalovirus infection. The abortions, by a standard suction procedure, produced multiple fragments of embryonic material. Mesencephalic tissue was identified, rinsed several times in saline, and the substantia nigra dissected and subjected to trypsin digestion, resulting in a dissociated cell suspension suitable for injection.

The first two recipients, aged 48 and 55, both female, and both with a 14-year history of Parkinson's disease, had developed on–off fluctuations in 1980 (Lindvall *et al.*, 1988, 1989). Despite optimum therapy with L-dopa preparations, bromocriptine and anticholinergics, both were off for more than 50% of the waking day. Medication was held constant for 6 months before and after the implants, which took place in late 1987. During the whole period, the patients self-rated their on/off status every 30 min. In addition, the following test procedures were frequently repeated both pre- and post-operation: measurement of pronation/supination rates, fist clenching/unclenching, finger dexterity, foot lifting and walking. Single dose 100 mg oral L-dopa tests after overnight drug withdrawal with assessment every 15 min were performed on multiple occasions. Ipsilateral and contralateral neurophysiological measurements of pre-movement potential, simple arm movements, simple hand movements and combined arm and hand movements were also performed pre-operatively and 4–6 months later in both the on and off conditions. Finally, ^{18}F-fluorodopa PET scans were performed at the same time points.

Treatment with a combined regime of cyclosporin, azathiaprine and steroids was started at the time of operation and continued. Nigral cells were harvested from four 8–10-week embryos and stereotactically placed unilaterally in three sites in each patient: those from one embryo in putaminal site 1, those from a second in putaminal site 2, and those from two further embryos in the caudate nucleus.

At 4–6 months post-graft, none of the multiple indices that were measured showed any definite change. However, if one extrapolates from the human-to-rat experiments, where benefit only begins to appear 2 or 3 months post-implant, then it is possible that the time course for human-to-human grafts may be even more protracted, so that the 9- and 12-month clinical results may be more important.

Grafts of human foetal tissue have been reported from other centres. In September 1987 the substantia nigra and "adrenal medulla" from a spontaneous 13-week abortion were transplanted by Madrazo *et al.* (1988) by open operation into a cavity in the right caudate nucleus in a 50-year-old man and a 35-year-old woman, respectively. Both patients were maintained on oral cyclosporin ($2\,mg\,kg^{-1}\,day^{-1}$) and prednisone ($15\,mg\,day^{-1}$), but there was no mention in the report whether antiparkinsonian drug treatment was changed post-operatively. Both patients were said to show objective improvement 2 months after surgery, as evidenced by a lowering of scores on the United Parkinson Rating Scale (UPRS). However, in patients who were presumably experiencing response fluctuations on L-dopa, the relevance of these single scores is uncertain. A subsequent letter published in August 1988 (Madrazo and Drucker-Colin, 1988) reported that 3 months post-graft, the male nigral recipient's UPRS score had fallen from 59 pre-operatively (on $1000\,mg$ Sinemet day^{-1}) to 14 (on $375\,mg$ Sinemet day^{-1}), with freedom from on–off fluctuations and dyskinesias, imperceptible rigidity and no akinesia. The female adrenal recipient's condition was "not quite as good as it was earlier. She would now be considered a patient with a poor response". In the same issue, Dwork et al. (1988) questioned the feasibility of isolating adrenal medullary tissue from human foetuses of 12–13 weeks gestation.

In early 1988, foetal substantia nigra grafts were performed in Birmingham, England, in a 60-year-old woman with a 25-year history of Parkinson's disease and a 41-year-old man with a 6-year history (Hitchcock et al., 1988). Both had initially responded well to L-dopa, but subsequently developed response fluctuations. For each operation, mesencephalic tissue from a single aborted foetus of 14–16 weeks gestation was kept in a tissue medium with antibiotic, and disaggregated mechanically into a thick cell suspension shortly before injection. This suspension was implanted stereotactically into the head of the right caudate nucleus, without immunosuppression. The first patient "appeared to improve within hours of the operation". Preoperatively on L-dopa $700\,mg$ day^{-1} plus peripheral decarboxylase inhibitor (PDI) and bromocriptine $20\,mg\,day^{-1}$, she was on with severe involuntary movements for 8–12 h day^{-1}. Two months post-operatively, on L-dopa $150\,mg\,day^{-1}$ (plus PDI) "function had improved. There was little discernible fluctuation and very little dyskinesia". The second patient's treatment of L-dopa $1\,g\,day^{-1}$ (plus PDI and benzhexol) had been stopped for 2 weeks before surgery. Post-operatively he was said to show improvement in akinesia during the first 9 days, prior to restarting L-dopa $150\,mg\,day^{-1}$ (plus PDI and benzhexol). Clinical scores on the Hoehn and Yahr, Webster and Northwestern University Disability scales were difficult to interpret, since both patients had response fluctuations and their drug treatment was changed. No figures for on versus off time were given.

At the time of writing, a number of foetal nigral implants are believed to have also been performed in Cuba, East Germany and China, but details are lacking. As of December 1988, the Birmingham group had conducted 12 implant procedures and one patient had been operated in Denver, Colorado.

The difference in methodology and technique between the different groups complicates the assessment of their results. The Swedes placed cells from four 8–10-week embryos stereotactically into both caudate and putamen, gave immunosuppressants, and did not alter antiparkinsonian drug treatment. The Mexican group placed cells from one 13-week foetus by open operation into a cavity in the caudate and gave immunosuppressants. The English group placed cells from one 14–16-week foetus stereotactically into the head of caudate, without immunosuppression, and at the same time changed the patients' other drugs. The relative importance of these differences is at present uncertain. What is certain is that none of the three groups who have reported their results have produced conclusive evidence that any of the patients have actually been improved by the graft. Probably only the Swedish group have instituted an assessment protocol capable of demonstrating whether the graft actually works, and the results so far have been disappointing. After 6 months, the duration of follow-up in their patients may be too short for a definitive answer to have emerged. However, analysis of results after 1 year has also shown no clinically significant improvement, so that immuno-suppression has been discontinued in their two patients as from December 1988.

If the grafts have not helped, then what might be the reason(s)? Should one perhaps only use one embryo, is the optimum donor gestation period different when the host is also human, does the process of preparation damage the cells, and is immunosuppression really necessary? Does continuing L-dopa or agonist treatment kill or inactivate the donor cells, and will they, in the longer term, succumb to whatever causes the disease?

More animal experimental research is clearly needed, particularly in terms of grafting human embryonic substantia nigra into MPTP-treated primates. If operations in patients are to proceed, they must be planned in such a way that a real effect of the graft can be demonstrated, and not confounded by inadequate assessment, placebo effects, and juggling with the doses of powerful antiparkinsonian drugs. Ethical issues concerning the use of human embryonic material need to be aired and addressed by society and by parliament, and safeguards introduced to prevent the spectre of embryo farming. Cell culture techniques need to be exhaustively explored for, if ever the operation can be shown to be effective, demand will far outstrip supply, especially if tissue from four embryos (or even eight for bilateral grafting) is needed for a single patient.

The use of human embryonic tissue to restore function in degenerative neurological disease is, we believe, a noble endeavour with an impressive scientific basis. A cautious and thorough approach to the first human experiments is crucial. Idiopathic Parkinson's disease, with its restricted and well-defined histological lesions sparing the striatum, offers the best hope of success. If transplants do not work here, then the prospects for them helping people with other diseases (e.g. Alzheimer's, Huntington's) with more extensive pathology must be remote. Thus, those involved in brain tissue transplants for Parkinson's disease bear a heavy responsibility to establish optimum conditions and to demonstrate unequivocal clinical effects. These operations, though technically skilled, are relatively simple. To prove whether they work is altogether more problematic. Despite disappointing results so far, the scientific basis established in the laboratory gives ground for cautious optimism that in the coming years we may be able to harness these techniques to help patients with a variety of degenerative diseases of the brain.

REFERENCES

Alexander, G. E., DeLong, M. R. and Strick, P. L. (1986) *Ann Rev Neurosci.* **9**, 357–381.

Annett, L. E., Dunnett, S. B., Rogers, D. C., Ridley, R. M., Baker, H. F., Jenner, P. D. and Marsden, C. D. (1989) In *Neural Mechanisms in Disorders of Movement* (edited by A. R. Crossman and M. A. Sambrook). pp. 217–221. John Libbey, London.

Arbuthnott, G., Dunnett, S. B. and MacLeod, N. (1985) *Neurosci Lett.* **57**, 205–210.

Backlund, E.-O., Granberg, P.-O., Hamberger, B., Knutsson, E., Martensson, A., Sedvall, G., Seiger, A. and Olson, L. (1985) *J Neurosurg.* **62**, 169–173.

Bakay, R. A. E., Fiandaca, M. S., Barrow, D. L., Schiff, A. and Collins, D. C. (1985) *Appl Neurophysiol.* **48**, 358–361.

Bakay, R. A. E., Fiandaca, M. S., Sweeney, K. M., Colbassani, H. J. and Collins, D. C. (1988) *Prog. Brain Res* **78**, 463–472.

Ballard, P. A., Tetrud, J. W. and Langston, J. W. (1985) *Neurology* **35**, 949–956.

Bernheimer, H., Birkmayer, W., Hornykiewicz, O., Jellinger, K. and Sietelberger, F. (1975) *J. Neurol. Sci.* **30**, 415–455.

Björklund, A. and Stenevi, U. (1979) *Brain Res.* **177**, 555–560.

Björklund, A. and Stenevi, U. (1981) *Brain Res.* **229**, 403–428.

Björklund, A., Dunnett, S. B., Stenevi, U., Lewis, M. E. and Iversen, S. D. (1980) *Brain Res.* **199**, 307–333.

Björklund, A., Stenevi, U., Dunnett, S. B. and Gage, F. H. (1982) *Nature* **298**, 652–654.

Björklund, A., Stenevi, U., Schmidt, R. H., Dunnett, S. B. and Gage, F. H. (1983) *Acta Physiol. Scand.* suppl., **522**, 9–18.

Brown, V. and Dunnett, S. B. (1989) *Exp. Brain Res.* In press.

Brundin, P., Nilsson, O. G., Gage, F. H. and Björklund, A. (1985) *Exp. Brain Res.* **60**, 204–208.

Brundin, P., Nilsson, O. G., Strecker, R. E., Lindvall, O., Åstedt, B. and Björklund, A. (1986) *Exp. Brain Res.* **65**, 235–240.

Brundin, P., Strecker, R. E., Widner, H., Clarke, D. J., Nilsson, O. G., Åstedt, B., Lindvall, O. and Björklund, A. (1988) *Exp. Brain Res.* **17**, 192–208.

Brundin, P., Strecker, R. E., Clarke, D. J., Widner, H., Nilsson, O. G., Åstedt, B., Lindvall, O. and Björklund, A. (1989) *Prog Brain Res.* **78**, 441–448..

Burns, R. S., Chiueh, C. C., Markey, S. P., Ebert, M. H., Jacobowitz, D. M. and

Kopin, I. J. (1983) *Proc. Nat. Acad. Sci. USA* **80**, 4546–4550.

Burns, R. S., Allen, G. S. and Tulipan, N. B. (1988) *Neurology* **38** (suppl. 1), 143.

Carmichael, S. W., Wilson, R. J., Brimijoin, W. S., Melton, L. J., Okazaki, H., Yaksh, T. L., Ahlskog, J. E., Stoddard, S. L. and Tyce, G. M. (1988) *N. Engl. J. Med.* **318**, 254.

Cervera, P., Rascol, O., Ploska, A., Gaillard, G., Raisman, R., Duyckaerts, C., Hauw, J. J., Scherman, D., Montastruc, J. L., Javoy-Agid, F. and Agid, Y. (1988) *J. Neurol. Neurosurg. Psychiat.* **51**, 1104–1105.

Das, G. D., Hallas, B. S. and Das, K. G. (1980) *Am. J. Anat.* **158**, 135–145.

Dunnett, S. B., Björklund, A., Stenevi, U. and Iversen, S. D. (1981a) *Brain Res.* **215**, 147–161.

Dunnett, S. B., Björklund, A., Stenevi, U. and Iversen, S. D. (1981b) *Brain Res.* **229**, 209–217.

Dunnett, S. B., Björklund, A., Stenevi, U. and Iversen, S. D. (1981c) *Brain Res.* **229**, 457–470.

Dunnett, S. B., Björklund, A., Schmidt, R. H., Stenevi, U. and Iversen, S. D. (1983a) *Acta Physiol. Scand.* suppl., **552**, 29–37.

Dunnett, S. B., Björklund, A., Schmidt, R. H., Stenevi, U. and Iversen, S. D. (1983b) *Acta Physiol. Scand.* suppl., **522**, 39–47.

Dunnett, S. B., Whishaw, I. Q., Jones, G. H. and Isacson, O. (1986) *Neurosci. Lett.* **68**, 127–133.

Dunnett, S. B., Whishaw, I. Q., Rogers, D. C. and Jones, G. H. (1987) *Brain Res.* **415**, 63–78.

Dwork, A. J., Pezzoli, G., Silani, V., Fahn, S. and Hill, R. (1988) *N. Engl. J. Med.* **319**, 371.

Fine, A., Hunt, S., Oertel, W. H., Nomoto, M., Chong, P. N., Bond, A., Waters, C., Temlett, J. A., Annett, L., Dunnett, S., Jenner, P. and Marsden, C. D. (1989) *Prog. Brain Res.* **78**, 479–489.

Fray, P. J., Dunnett, S. B., Iversen, S. D., Bjorklund, A. and Stenevi, U. (1983) *Science* **219**, 416–419.

Freed, W. J. (1983) *Biol. Psychiat.* **18**, 1205–1267.

Freed, W. J., Perlow, M. J., Karoum, F., Seiger, A., Olson, L., Hoffer, B. J. and Wyatt, R. J. (1980) *Ann. Neurol.* **8**, 510–519.

Freed, W. J., Morihisa, J. M., Spoor, E., Hoffer, B. J., Olson, L., Seiger, A. and Wyatt, R. J. (1981) *Nature* **292**, 351–352.

Freed, W. J., Karoum, F., Spoor, H. E., Hoffer, B. J., Olson, L., Seiger, A. and Wyatt, R. J. (1983a) *Brain Res.* **269**, 184–189.

Freed, W. J., Ko, G. N., Niehoff, D. L., Kuhar, M. J., Hoffer, B. J., Olson, L., Cannon-Spoor, H. E., Morihisa, J. M. and Wyatt, R. J. (1983b) *Science* **222**, 937–939.

Freed, W. J., Patel-Vaidya, U. and Geller, H. M. (1986) *Exp. Brain Res.* **63**, 557–566.

Freund, T., Bolam, J. P., Björklund, A., Stenevi, U., Dunnett, S. B., Powell, J. F. and Smith, A. D. (1985) *J. Neurosci.* **5**, 603–616.

Gage, F. H. and Björklund, A. (1986) *Neuroscience* **17**, 89–98.

Gage, F. H., Björklund, A., Stenevi, U. and Dunnett, S. B. (1983) *Acta Physiol. Scand.* suppl., **552**, 67–75.

Gash, D. M., Collier, T. J. and Sladek, J. R. (1985) *Neurobiol. Aging* **6**, 131–150.

Gash, D. M., Notter, M. F. D., Okawara, S. H., Kraus, A. L. and Joynt, R. J. (1987) *Science* **233**, 1420–1422.

Goetz, C. G., Tanner, C. M., Penn, R. D., Shannon, K. M. and Klawans, H. L. (1988) *Neurology* **38** (suppl. 1) 142–143.

Greene, H. S. N. and Arnold, H. (1945) *J. Neurosurg.* **2**, 315–331.

Guttman, M., Peppard, R. F., Martin, W. R. W., Adam, M. J., Ruth, T., Calne, D. B., Walsh, E., Allen, G. and Burns, R. S. (1988) *Neurology* **38** (Suppl. 1), 144.

Hallas, B. S., Das, G. D. and Das, K. G. (1980) *Am. J. Anat.* **158**, 147–159.

Hargreaves, R. and Freed, W. J. (1987) *Life Sci.* **40**, 959–966.

Hefti, F., Hartikka, J. and Schlumpf, M. (1985) *Brain Res.* **348**, 283–288.

Hitchcock, E. R., Clough, C., Hughes, R. and Kenny, B. (1988) *Lancet* **i**, 1274.

Hoffer, B. J., Gerhardt, G. A., Rose, G. M., Stromberg, I. and Olson, L. (1987) *Ann. N.Y. Acad. Sci.* **495**, 510–527.

Itakura, T., Kamei, I., Nakai, K., Naka, Y., Nakakita, K., Imai, H. and Komai, N. (1988) *J. Neurosurg.* **68**, 955–959.

Jaeger, C. B. (1985) *Exp. Brain Res.* **59**, 615–624.

Jaeger, C. B. (1987) *Ann. NY Acad. Sci.* **495**, 334–349.

Kamo, H., Kim, S. U., McGeer, P. L. and Shin, D. H. (1985) *Neurosci. Lett.* **57**, 43–48.

Kamo, H., Kim, S. U., McGeer, P. L. and Shin, D. H. (1986) *Brain Res.* **397**, 372–376.

Kish, S. J., Shannak, K. and Hornykiewicz, O. (1988) *N. Engl. J. Med.* **318**, 876–880.

Kordower, J. H., Notter, M. F. D., Yeh, H. H. and Gash, D. M. (1987) *Ann. NY Acad. Sci.* **495**, 606–621.

Langston, J. W., Ballard, P., Tetrud, J. W. and Irwin, I. (1983) *Science* **219**, 979–980.

Lieberman, A. N., Ransohoff, J. and Koslow, M. (1988) *Neurology* **38** (Suppl. 1), 142.

Lindvall, O., Backlund, E.-O., Farde, L., Sedvall, G., Freedman, R., Hoffer, B., Nobin, A., Seiger, A. and Olson, L. (1987a) *Ann. Neurol.* **22**, 457–468.

Lindvall, O., Dunnett, S. B., Brundin, P. and Björklund, A. (1987b) In *Parkinson's Disease. Clinical and Experimental Advances* (edited by F. C. Rose), pp. 189–206. John Libbey, London.

Lindvall, O., Gustavii, B., Åstedt, B., Lindholm, T., Rehncrona, S., Brundin, P., Widner, H., Björklund, A., Leenders, K. L., Frackowiak, R., Rothwell, J. C., Marsden, C. D., Johnels, B., Steg, G., Freedman, R., Hoffer, B. J., Seiger, Å., Strömberg, I., Bygdeman, M. and Olson, L. (1988) *Lancet* **ii**, 1483–1484.

Lindvall, O., Rehncrona, S., Brundin, P., Gustavii, B., Åstedt, B., Widner, H., Lindholm, T., Björklund, A., Leenders, K. L., Rothwell, J. C., Frackowiak, R., Marsden, C. D., Johnels, B., Steg, G., Freedman, R., Hoffer, R. J., Seiger, Å., Bygdeman, M., Strömberg, I. and Olson, L. (1989) *Arch. Neurol.* (in press).

Madrazo, I. and Drucker-Colin, R. (1988) *N. Engl. J. Med.*, **319**, 371.

Madrazo, I., Drucker-Colin, R., Diaz, V., Martinez-Marta, J., Torres, C. and Becerril, J. J. (1987a) *N. Engl. J. Med.* **326**, 831–834.

Madrazo, I., Drucker-Colin, R., Leon, V. and Torres, C. (1987b) *Confin. Neurol.* **38**, 510–511.

Madrazo, I., Leon, V., Torres, C., del Aguilera, C., Varela, G., Alvarez, F., Fraga, A., Drucker-Colin, R., Ostrosky, F., Skurovich, M. and Franco, R. (1988) *N. Engl. J. Med.* **318**, 51.

Mahalik, T. J., Finger, T. G., Stromberg, I. and Olson, L. (1985) *J. Comp. Neurol.* **240**, 60–70.

Mason, D. M., Charlton, H. M., Jones, A. J., Lavy, C. B. D., Puklavec, M. and Simmonds, S. J. (1986) *Neuroscience* **19**, 685–694.

Melamed, E. (1988) *Clin. Neuropharmacol.* **11**, 77–82.

Meyers, R. (1951) *Acta Psychiat. Neurol. Scand.* suppl 67, 7–41.

Miller, W. C. and DeLong, M. R. (1988) *Ann. NY Acad Sci.* **515**, 287–302.

Morihisa, J. M., Nakamura, R. K., Freed, W. J., Mishkin, M. and Wyatt, R. J. (1984) *Exp Neurol.* **84**, 643–653.

Motti, E. D. F., Pezzoli, G., Silani, V. and Scarlato, G. (1988) *Lancet* **ii**, 346.

Nadaud, D., Herman, J. P., Simon, H. and LeMoal, M. (1984) *Brain Res.* **304**, 137–141.

Olanow, C. W., Cahill, D. and Cox, C. (1988) *Neurology* **38** (Suppl. 1), 142.

Patel-Vaidya, U., Wells, M. R. and Freed, W. J. (1985) *Cell Tiss. Res.* **240**, 281–285.

Perlow, M. J., Freed, W. J., Hoffer, B. J., Seiger, A., Olson, L. and Wyatt, R. J. (1979) *Science* **204**, 643–647.

Peterson, D. I., Price, M. L. and Small, C. S. (1989) *Neurology* **39**, 235–238.

Pezzoli, G., Fahn, S., Dwork, A., Truong, D. D., de Yebenes, J. G., Jackson-Lewis, V., Cadet, J. L. and Herbert, J. (1988). *Brain Res.* **459**, 398–403.

Pezzoli, G., Goodman, R., Ferrante, C., Silani, V., Yebenes, J., Truong, D., Jackson-Lewis, V. and Fahn, S. (1987) Schmitt Symposium, Rochester, NY, June 30–July 3 (Abstr.).

Porrino, L. J., Palombo, E., Bankiewicz, K. S., Viola, J., Jehle, J. and Kopin, I. J. (1988) *Soc. Neurosci. Abstr.* **14**, 9.

Redmond, D. E., Naftolin, F., Collier, T. J., Leranth, C., Robbins, R. J., Sladek, C. D., Roth, R. H. and Sladek, J. R. (1988) *Science* **242**, 768–771.

Redmond, D. E., Sladek, J. R., Roth, R. H., Collier, T. J., Elsworth, J. D., Deutch, A. Y. and Haber, S. (1986) *Lancet* **i**, 1125–1127.

Riederer, P., Rausch, W. D., Birkmayer, W., Jellinger, K. and Seemann, D. (1978) *J. Neural Transm.* (Suppl. 14), 121–131.

Schmidt, R. H., Bjorklund, A. and Stenevi, U. (1981) *Brain Res.* **218**, 347–356.

Schmidt, R. H., Ingvar, M., Lindvall, O., Stenevi, U. and Björklund, A. (1982) *J. Neurochem.* **38**, 737–748.

Schmidt, R. H., Björklund, A., Stenevi, U., Dunnett, S. B. and Gage, F. H. (1983) *Acta Physiol. Scand.* suppl., **522**, 19–28.

Schultzberg, M., Dunnett, S. B., Björklund, A., Stenevi, U., Hokfelt, T., Dockray, G. J. and Goldstein, M. (1984) *Neuroscience* **12**, 17–32.

Schwarz, S. S. and Freed, W. J. (1987) *Exp. Brain Res.* **65**, 449–454.

Seiger, A. (1985) In *Neural Grafting in the Mammalian CNS* (edited by A. Björklund and U. Stenevi), pp. 71–77. Elsevier, Amsterdam.

Sladek, J. R., Collier, T. J., Haber, S., Roth, R. H. and Redmond, D. E. (1986) *Brain Res. Bull.* **17**, 809–818.

Sladek, J. R., Redmond, D. E., Collier, T. J., Blount, J. P., Elsworth, J. D., Taylor, J. R. and Roth, R. H. (1988) *Prog. Brain Res.* **78**, 497–506.

Spiegel, E. A., Wycis, H. T., Szekely, E. G., Constantinovici, A., Egyed, J. J., Gildenberg, P., Lehman, R. and Werthan, M. (1965) *Confin. Neurol.* **26**, 336–341.

Stenevi, U., Björklund, A. and Svendgaard, N.-A. (1976) *Brain Res.* **114**, 1–20.

Stenevi, U., Björklund, A. and Dunnett, S. B. (1980) *Peptides* **1** (suppl. 1), 111–116.

Strecker, R. E., Sharp, T., Brundin, P., Zetterstrom, T., Ungerstedt, U. and Björklund, A. (1987) *Neuroscience* **22**, 169–178.

Strömberg, I., Herrera-Marschitz, M., Ungerstedt, U., Ebendal, T. and Olson, L. (1985a) *Exp. Brain Res.* **60**, 335–349.

Strömberg, I., Johnson, S., Hoffer, B. J. and Olson, L. (1985b) *Neuroscience* **14**, 981–990.

Strömberg, I., Bygdeman, M., Goldstein, M., Seiger, A. and Olson, L. (1986) *Neurosci. Lett.* **71**, 271–276.

Tanner, C. M., Goetz, C. G., Gilley, D. W., Shannon, K. M., Stebbins, G. T., Klawans, H. L., Wilson, R. S. and Penn, R. M. (1988) *Neurology* **38** (suppl. 1), 143–144.

Tintner, R., Clark, K., Hom, J., Peters, P., Fuchs, I., Speciale, S., Bebehani, K. and Kondraske, G. (1988) *Neurology* **38** (suppl. 1), 143.

Ungerstedt, U. and Arbuthnott, G. W. (1970) *Brain Res.* **24**, 485–493.

Watts, R. L., Bakay, R. A. E., Iuvone, P. M., Watts, N. and Graham, S. (1988) *Neurology* **38** (suppl. 1), 143.

Wuerthele, S. M., Freed, W. J., Olson, L., Morihisa, J., Spoor, L., Wyatt, R. J. and Hoffer, B. J. (1981) *Exp. Brain Res.* **44**, 1–10.

Zetterström, T., Brundin, P., Gage, F. H., Sharp, T., Isacson, O., Dunnett, S. B., Ungerstedt, U. and Björklund, A. (1986) *Brain Res.* **362**, 344–349.

PART II

Dystonia

17

The Pathophysiology of Dystonia

A. Berardelli

The term dystonia is used to describe a syndrome characterized by sustained muscle contractions, frequently causing twisting and repetitive movements, or abnormal postures. Dystonia can affect different parts of the body and is classified into focal, segmental, multifocal, hemidystonic and generalized forms. In most cases of dystonia, no brain lesions can be identified but in a small number of patients the cause has been traced to discrete focal lesions in the thalamus, striatum and globus pallidus (for references see Fahn *et al.*, 1987). Recently Marsden *et al.* (1985) studied a number of patients with symptomatic focal or hemidystonia in whom a computed tomography (CT) scan or anatomo-pathological studies clearly demonstrated a localized brain lesion. In these patients, the dystonia clinically resembled that seen in idiopathic torsion dystonia. The most common lesions were infarcts, arterio-venous malformations and haemorrhages. The CT brain scans localized the abnormalities to the caudate nucleus, lenticular nucleus (putamen) and thalamus. Jankovic and Pettigrew (1985) also described similar changes. In their series the hemidystonia was caused by cerebrovascular diseases, perinatal trauma, degenerative diseases and thalamotomy. In 73% of these patients the CT scans showed evidence of contralateral basal ganglia damage. The authors attributed the hemidystonia to a "disconnection" between the striatum and thalamus, with conservation of the corticospinal fibres. From both series of patients it appears that dystonia can be caused by an isolated

DISORDERS OF MOVEMENT: CLINICAL, PHARMACOLOGICAL
AND PHYSIOLOGICAL ASPECTS   ISBN 0-12-569685-X

lesion in the putamen, caudate, thalamus or globus pallidus. In idiopathic dystonia, these structures may also be affected by processes that are not demonstrated by CT scan. Even more recently, lesions in the putamen have been demonstrated in patients with dystonia by magnetic resonance imaging (Fross *et al.* 1987).

That the abnormalities of movement are not due to impaired cortico-motoneurone conduction has been demonstrated recently by Thompson *et al.* (1986), who studied the motor responses evoked by stimulation of the motor cortex using the technique of cortical stimulation in patients with dystonia. With this technique it is possible to activate the cortico-spinal tract (Rothwell *et al.*, 1987; Day *et al.*, 1987). The authors observed that arm muscle responses were evoked at a normal latency and concluded that the origin of movement disorder in dystonia lies not in abnormal conduction in the cortico-motoneurone pathways from the cerebral cortex to the spine but more probably in impaired conduction between the basal ganglia and motor cortex.

ELECTROMYOGRAPHIC RECORDINGS OF DYSTONIA

Dystonic movements produce continuous or semi-continuous involuntary electromyographic (EMG) activity, characterized by contraction of the agonist and antagonist muscles that increases during the execution of voluntary movements and disappears during sleep (Yanagisawa and Goto, 1971). Continuous EMG activity lasting several seconds or repeated, short, rhythmic activity may also be noted (Fig. 17.1). These EMG patterns are common to the different types of dystonia, including generalized dystonia and focal dystonias such as blepharospasm and writer's cramp (Yanagisawa and Goto, 1971; Rothwell *et al.*, 1983; Berardelli *et al.*, 1985a). In patients with writer's cramp Hughes and McLellan (1985) observed increased co-contraction activity and excessive activation of the triceps muscle.

Combinations of these abnormalities also occur, resulting in characteristic patterns. Another feature of dystonia is its association with other involuntary movements. In myoclonic dystonia, for example (Obeso *et al.*, 1983), long-lasting EMG activity is combined with shorter activity resembling that seen in myoclonus. Many dystonic patients also have a postural tremor similar in frequency and character to benign essential tremor. In addition, the performance of a voluntary movement or maintenance of a posture intensifies both the frequency and severity of the dystonia, which appears even in muscles not involved in that particular motor task ("overflow").

Polymyographic analysis has demonstrated that the characteristic finding in dystonia is co-contraction between agonist and antagonist muscles

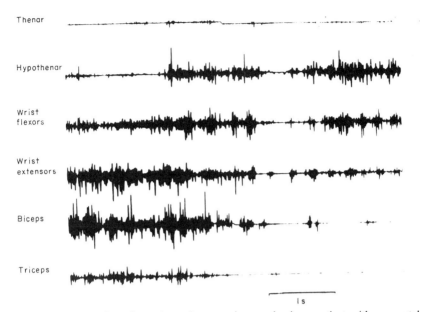

Fig. 17.1 Co-contraction of agonist and antagonist muscles in a patient with segmental dystonia during the execution of a movement. (From Rothwell *et al.*, 1983.)

(Yanagisawa and Goto, 1971; Rothwell *et al.*, 1983; Hughes and McLellan, 1985). This phenomenon suggests the presence in dystonia of an abnormality of the normal pattern of reciprocal inhibition between opposing muscle pairs.

RECIPROCAL INHIBITION IN DYSTONIA

The essence of reciprocal inhibition is that during contraction of synergist muscles the antagonist muscles are inhibited. The Ia inhibitory interneurons in the spinal cord mediate the reciprocal inhibition and receive two major inputs from the corticospinal tract and the axons of large diameter muscle fibres.

Reciprocal inhibition can be studied in man with a simple technique (Day *et al.*, 1984; Berardelli *et al.*, 1987). An H reflex is evoked in the forearm flexor muscles by stimulation of the median nerve at the elbow (control H reflex). By comparing the size of a series of control H reflexes with the size of H reflexes conditioned by radial nerve stimulation, the reciprocal inhibition between flexor and extensor muscles in the forearm can be calculated. In normal subjects the reciprocal inhibition between agonist and antagonist muscles consists of an initial short-lasting inhibitory phase, followed by a further longer-lasting inhibitory phase. The initial phase is disynaptic and postsynaptic onto spinal motoneurones. The second phase is probably produced by presynaptic inhibition of flexor Ia afferent terminals. Rothwell *et al.* (1983, 1988) report that in patients with dystonia the initial inhibitory phase is normal, whereas the later phase is absent or reduced (Fig. 17.2). Abnormalities of reciprocal inhibition in the forearm muscles have also been found in patients whose dystonia does not clinically involve the arm (Panizza *et al.*, in preparation).

Whether presynaptic inhibition from extensor muscle afferents onto flexor Ia terminals is reduced in dystonia or is actually normal but masked by a concurrent facilitation from other sources is not known. Whatever the true

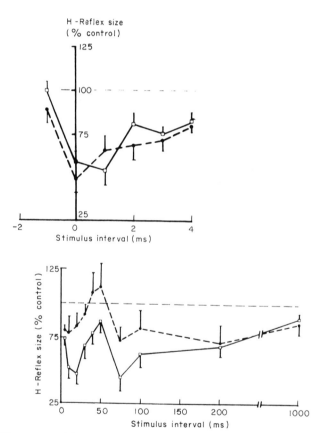

Fig. 17.2 Time course of first (top) and later (bottom) phases of reciprocal inhibition in eight normal (continuous line) and eight dystonic (dotted lines) subjects. The second phase of inhibition (from 10 to 50 ms) is much reduced in the patients. (From Rothwell *et al.*, 1988.)

explanation the existence of abnormalities in the control of spinal reflex circuits in these patients is beyond doubt.

STRETCH REFLEXES AND SHORTENING REACTION IN DYSTONIA

Tatton *et al.* (1984) and Rothwell *et al.* (1983) studied the stretch reflex in the upper limbs of patients with dystonia. A torque motor was used to elicit a stretch reflex while the subject voluntarily contracted the muscle being tested. In this experimental condition two components of a stretch reflex can be distinguished: the first is a short latency component evoked by way of a spinal monosynaptic circuit, the second is longer in latency and may travel via supraspinal circuits (Marsden *et al.*, 1983). Tatton *et al.* (1984) observed that in patients with dystonia musculorum deformans the short and long latency stretch reflexes in flexor carpi radialis have a normal amplitude modulation and a normal latency. On the other hand, the duration of the long latency stretch reflex is prolonged and there is a disturbance of normal temporal mechanisms that results in constant duration of the short and long latency reflexes, despite variability of imposed force step loads. Rothwell *et al.* (1983) noted, in addition, that in dystonic patients the stretch reflex not only occurs in the muscle being stretched but spreads to adjacent muscles.

Stretching in one muscle is usually accompanied by a shortening of the antagonist muscle. This phenomenon, known as Westphal's reaction, can be recorded in normal subjects in arm and tibialis anterior muscles (Berardelli and Hallett, 1984). This reaction is probably under the control of muscle and joint afferents coming from the shortened muscle. As first described by Yanagisawa and Goto (1971), dystonic patients often display such a paradoxical activation of the shortened muscle.

In summary abnormalities in the stretch reflex and in the shortening reaction may indicate dysfunction in the inhibitory mechanisms that limit the duration of the reflex responses.

FACIAL REFLEXES IN DYSTONIA

Blepharospasm and oromandibular dystonia are characterized by involuntary movements of the orbicularis oculi muscle and lower facial muscles. Marsden (1976) first pointed out that the clinical features of these movements resembled those observed in patients with torsion dystonia, so that blepharospasm is now considered a form of focal dystonia. Some of his patients later developed other forms of dystonia, such as torticollis and writer's cramp. Later, Jankovic and Ford (1983) provided further evidence that blepharospasm and oromandibular dystonia were of organic origin. They observed that many of these patients had a family history of dystonia, or had other associated movement disorders. Recently, Jankovic and Patel (1983) found evidence of a lesion involving the brain stem in six patients with blepharo-spasm. Berardelli *et al.* (1985a, 1988) observed that the EMG features of the involuntary facial movements in patients with blepharospasm and oromandibular dystonia resembled those described in patients with general-ized dystonia. They also found an abnormality of the orbicularis oculi reflexes.

The human orbicularis oculi reflex can be tested clinically or electro-physiologically by electrical or mechanical stimulation of the supraorbital nerve ("blink reflex") or by stimulation of the cornea ("corneal reflex"). The blink reflex is composed of an early homolateral response (R1) followed by a late bilateral response (R2). The corneal reflex shows only a late bilateral response which is similar but not equivalent to the R2 component of the blink reflex (Ongerboer de Visser, 1983; Berardelli *et al.*, 1985b). The R1 component is conducted through an oligosynaptic arc in the pons, from the supraorbital branch of the sensory trigeminal root to the ipsilateral facial nucleus. The corneal reflex and the R2 component of the blink reflex are conducted through polysynaptic pathways in the lateral reticular formation of the brain stem from the descending spinal fifth nerve nucleus to the ipsilateral and contralateral facial nucleus. Studies of the orbicularis oculi reflex in subjects with blepharospasm show that the latencies of the corneal reflex and of the R1 and R2 components of the blink reflex are normal, suggesting that in cranial dystonia the interneuronal pathways mediating these reflexes are intact. On the other hand, the amplitude and duration of the corneal reflex and of R1 and R2 are increased. In some patients, the R1 component is also present on the side opposite to stimulation, whereas in normal subjects it is usually present only on the stimulated side (Fig. 17.3). The excitability cycle of the R2 component of the blink reflex, tested with the paired shock technique, shows a facilitation of the response that is not related to contraction of the muscle being tested. These changes in the orbicularis oculi reflexes indicate an increased excitability of their reflex arcs. The most likely explanation is a facilitation of the interneuronal pathways of the corneal reflex and of the R2 of the blink reflex.

The orbicularis oculi reflexes are controlled by several regions of the brain. In patients with lesions of the sensorimotor cortex Berardelli *et al.* (1983) and Ongerboer de Visser (1983) have described depression of blink and corneal reflexes, caused by loss of facilitation inputs to the brain stem. Unlike voluntary blinking which is preceded by an electric potential of the "Bereitschaftspotential" type, no EEG activity precedes blepharospasm. The absence of a pre-movement potential suggests that blepharospasm is mediated by pathways other than those used in normal voluntary movements. Taken

Averaged rectified

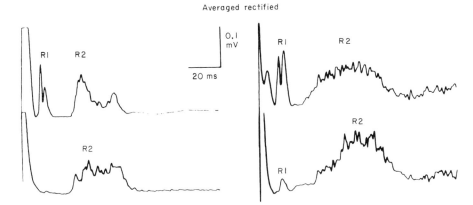

Fig. 17.3   Blink reflex in a normal subject (on the left) and in a patient with blepharospasm (on the right). Averages of 40 single full-wave rectified EMG records. The patient's response consists of a normal ipsilateral R1 (top) and, in this case, a small contralateral R1 (bottom), with prolonged bilateral R2 bursts. (From Berardelli *et al.*, 1985a.).

as a whole, these data suggest that the orbicularis oculi reflex circuit is itself intact but that a facilitation, probably originating in the basal ganglia, is exerted on the brain stem interneurons that mediate the orbicularis oculi reflex. It is known that basal ganglia influence the orbicularis oculi reflexes. A facilitation of the interneurons mediating the R2 component is present in Parkinson's disease (Kimura, 1973) which is related to the level of dopaminergic stimulation (Agostino *et al.*, 1987). In patients with Huntington's disease however, a depression of the R2 and corneal reflexes is evident (Agostino *et al.*, 1988). The finding of a facilitation of the R2 component of the blink reflex in patients with blepharospasm has been more recently confirmed by Tolosa *et al.* (1988). In addition they studied the blink reflex in patients with various forms of focal dystonia, observing that the R2 component was facilitated not only in patients with blepharospasm, but also with spasmodic torticollis and spasmodic dysphonia, yet was normal in patients with focal arm dystonia. The brain stem interneuron excitability is therefore increased in the dystonias mediated by the cranial nerves. We have recently confirmed the abnormal facilitation of the R2 component in patients with cranial dystonia and we have also seen an increased excitability of the brain stem interneurons mediating the late R2 component of the brainstem in patients with generalized dystonia (Fig. 17.4). Since the anatomical structures responsible for torticollis and spasmodic dysphonia are close to those of the orbicularis oculi reflex, an increased excitability could also affect the circuitry of the orbicularis oculi reflex. In generalized dystonia a diffuse facilitation may be present in segments not directly affected by the dystonia.

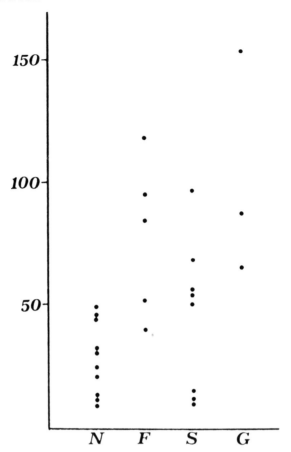

Fig. 17.4 Amplitude of the test R2 component when conditioned by a preceding shock at intervals of 500 ms in normal subjects and in patients with focal (F), segmental (S) and generalized (G) dystonia. (Unpublished observations.)

The enhanced excitability of brain stem interneurons in dystonia may be the substrate of the pathophysiology of cranial dystonia. The same may also be true of the abnormality of forearm reciprocal inhibition discovered in patients with focal arm dystonia. The picture that emerges is that of a substrate of disordered brain stem and spinal cord interneuronal function, which may be triggered by other factors into clinical manifestations.

ANALYSIS OF VOLUNTARY MOVEMENTS

Because EMG analysis of dystonic movements has shown that dystonia is exaggerated during the execution of rapid arm movements, we studied these movements in patients with dystonia. In a normal subject, a simple ballistic elbow movement is characterized by a typical pattern of di-/tri-phasic activity in the agonist and antagonist muscles. The initial agonist burst has the function of accelerating the limb through the distance; the antagonist burst, at least in part, stops the movement (Hallett *et al.*, 1975). The size and duration of these bursts change with the different types of movements (Berardelli *et al.*, 1984).

Studying the performance of rapid elbow flexion in patients with dystonia we have noted (Fig. 17.5): (1) reduced speed in movement performance with an increased duration of movement; (2) reasonably accurate execution of movements but more variable extent; (3) prolonged duration of the initial burst of activity in the agonist and antagonist muscles; (4) changes in control of the antagonist muscle varying from prolonged bursts to the presence of co-contraction; and (5) excess and inappropriate activity in other muscles of the upper limb. The movements, however, were normal in one respect; the velocity profile was bell-shaped, indicating that the acceleration and deceleration time were equal. The patients who were able to perform the task had only mild to moderate dystonia; abnormalities were more evident in patients with severe dystonia. The remaining patient with severe generalized dystonia was unable to make movements of different extent. Deterioration in simple movement therefore progresses with the severity of the clinical state of the patient.

Ghez *et al.* (1988) have recently studied the ability of patients with dystonia

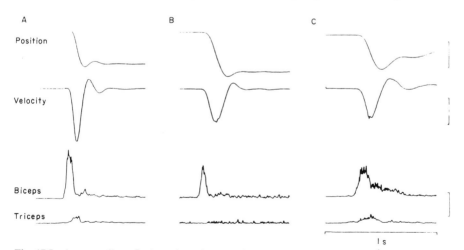

Fig. 17.5 Average elbow flexions through an angle of 10° in a normal subject (A) and two patients with mild (B) and moderate dystonia (C) affecting the arm. EMGs (bottom two traces) are rectified surface recordings from the agonist (biceps) and antagonist (triceps) muscles. (Unpublished observations.)

to contract antagonist muscles to produce different force trajectories. The authors observed that these patients show abnormalities of trajectory control and in recruitment and derecruitment of antagonist muscles. They have proposed that in dystonia there is loss of function in a segmental and suprasegmental loop which normally helps to terminate the motor commands.

In conclusion, several abnormalities are present in dystonia: (1) excessive co-contraction between agonist and antagonist muscles; (2) a defect of reciprocal inhibition between antagonist muscles; (3) a facilitation of stretch reflexes and shortening reaction; (4) an increased excitability of blink and corneal reflexes; and (5) a deficit in controlling voluntary movement.

These abnormalities indicate that in dystonia there is a facilitation or a reduced inhibition of reflexes and voluntary control that may be related to a reduced tonic inhibitory output from the basal ganglia (Denian et al., 1985; Hikosaka and Wurtz, 1985).

REFERENCES

Agostino, R., Berardelli, A., Cruccu, G., Stocchi, F. and Manfredi, M. (1987) *Movement Disorders* 2, 227–235.

Agostino, R., Berardelli, A., Cruccu, G., Pauletti, G., Stocchi, F. and Manfredi, M. (1988) *Movement Disorders*, 3, 281–289.

Berardelli, A. and Hallett, M. (1984) *Neurology* 34, 242–246.

Berardelli, A., Accornero, N., Cruccu, G., Fabiano, F., Guerrisi, V. and Manfredi, M. (1983) *J. Neurol. Neurosurg. Psychiatry* 46, 837–843.

Berardelli, A., Rothwell, J. C., Day, B. L., Kachi, T. and Marsden, C. D. (1984) *Brain Res.* 304, 183–187.

Berardelli, A., Rothwell, J. C., Day, B. L. and Marsden, C. D. (1985a) *Brain* 108, 593–608.

Berardelli, A., Cruccu, G., Manfredi, M., Rothwell, J., Day, B. and Marsden, C. D. (1985b) *Neurology* 35, 787–801.

Berardelli, A., Day, B. L., Marsden, C. D. and Rothwell, J. C. (1987) *J. Physiol.* 391, 71–83.

Berardelli, A., Rothwell, J. C., Day, B. L. and Marsden, C. D. (1988) *Adv. Neurol.* 50, 525–535.

Day, B. L., Marsden, C. D., Obeso, J. A. and Rothwell, J. C. (1984) *J. Physiol.* 519–534.

Day, B. L., Rothwell, J. C., Thompson, P. D., Dick, J. P. R., Cowan, J. M. A., Berardelli, A. and Marsden, C. D. (1987) *Brain* 110, 1191–1209.

Denian, J. M. and Chevalier, G. (1985) *Brain Res.* 334, 227–233.

Fahn, S., Marsden, C. D. and Calne, D. B. (1987) In *Movement Disorders* (edited by C. D. Marsden and S. Fahn), pp. 333–358. Butterworths, London.

Fross, R. D., Martin, W. R. W., Li, D., Stoessl, A. J., Adam, M. J., Ruth, T. J., Pate, B. D., Burton, K. and Calne, D. (1987) *Neurology* 37, 1125–1129.

Ghez, C., Gordon, J. and Hening, W. (1988) *Adv. Neurol.* 50, 141–155.

M., Shahani, B. T. and Young, R. R. (1975) *J. Neurol. Neurosurg. Psychiat.* 38, 1154–1162.

Hikosaka, O. and Wurtz, R. H. (1985) *J. Neurophysiol* **53**, 292–308.
Hughes, M. and McLellan, D. L. (1985) *J. Neurol. Neurosurg. Psychiat.* **48**, 782–787.
Jankovic, J. and Ford, J. (1983) *Ann. Neurol.* **13**, 402–411.
Jankovic, J. and Patel, S. C. (1983) *Neurology* **33**, 1237–1240.
Jankovic, J. and Pettigrew, L. C. (1985) *J. Neurol. Neurosurg. Psychiat.* **48**, 650–657.
Kimura, J. (1973) *Brain* **96**, 87–96.
Marsden, C. D. (1976) *J. Neurol. Neurosurg. Psychiat.* **39**, 1204–1209.
Marsden, C. D., Rothwell, J. C. and Day, B. L. (1983) *Adv. Neurol.* **39**, 509–539.
Marsden, C. D., Obeso, J. A., Zarranz, J. J. and Lang, A. E. (1985) *Brain* **108**, 461–483.
Obeso, J. A., Rothwell, J. C., Lang, A. E. and Marsden, C. D. (1983) *Neurology* **33**, 825–830.
Ongerboer de Visser, B. W. (1983) *Adv. Neurol.* **39**, 757–786.
Rothwell, J. C., Obeso, J. A., Day, B. L. and Marsden, C. D. (1983) *Adv. Neurol.* **39**, 851–863.
Rothwell, J. C., Thompson, P. D., Day, B. L., Cowan, J., Dick, J. P. R., Kachi, T. and Marsden, C. D. (1987) *Brain* **110**, 1173–1190.
Rothwell, J. C., Day, B. L., Obeso, J. A., Berardelli, A. and Marsden, C. D. (1988) *Adv. Neurol.* **50**, 133–140.
Tatton, W. G., Bedingham, W., Verrier, M. C. and Blair, R. D. G. (1984) *Can. J. Neurol. Sci.* **11**, 281–287.
Thompson, P. D., Dick, J. P. R., Day, B. L., Rothwell, J. C., Berardelli, A., Kachi, T. and Marsden, C. D. (1986) *Movement Disorders* **1**, 113–118.
Tolosa, E., Montserrat, L. and Bayes, A. (1988) *Movement Disorders* **3**, 61–69.
Yanagisawa, N. and Goto, A. (1971) *J. Neurol. Sci.* **13**, 39–65.

18

Pathophysiology of Cranial Dystonia: a Review

Eduardo S. Tolosa and Jaime B. Kulisevsky

The term cranial dystonia refers to a movement disorder characterized by the presence of dystonic spasms which involve predominantly the musculature of the head. The syndrome of cranial dystonia is usually idiopathic and is thought to represent an adult-onset form of torsion dystonia. It tends to remain focal or segmental in distribution (Tolosa *et al.*, 1988a). Blepharospasm, lingual dystonia, oromandibular dystonia, dystonic adductor dysphonia and dystonic dysphagia are examples of focal dystonia that may occur in a cranial distribution. When spasms occur in contiguous regions, a not uncommon situation, the term segmental cranial dystonia is at times used (Fahn *et al.*, 1987). The most common form of segmental cranial dystonia is Meige syndrome, which is characterized by the presence of spasms of the orbicularis oculi (blepharospasm) and of the lower facial or oromandibular muscles. It is not known whether a similar pathophysiology underlies the different types of cranial dystonia.

Reversible dystonic symptoms, mainly torticollis, have been produced by drugs or by experimental lesions involving various brain stem structures, but no animal model satisfactorily mimics the human condition. This circumstance has, in part, been responsible for slow advancement in our understanding of the pathophysiology of cranial dystonia.

DISORDERS OF MOVEMENT: CLINICAL, PHARMACOLOGICAL
AND PHYSIOLOGICAL ASPECTS   ISBN 0-12-569685-X

Etiological classification divides the causes of dystonia into two major categories: idiopathic, or primary, and symptomatic, or secondary. In the idiopathic group, the only clinical neurological abnormality is the presence of dystonic spasms. The role of genetic factors in this form of dystonia has not yet been clearly elucidated. Biochemical, neuroimaging and conventional electrophysiological studies are normal, although specialized physiological methods have revealed abnormalities in the excitability of brain stem interneurons. Clinical and pharmacological studies support the notion that dopamine and acetylcholine mechanisms play a role in the pathophysiology of cranial dystonia, although conclusive results are lacking.

Secondary causes have shed some light on the anatomical structures that may subserve cranial dystonia. Both basal ganglia and focal upper brain stem/diencephalic lesions can cause blepharospasm, and the coexistence of cranial dystonia and parkinsonism (idiopathic, post-encephalic and MPTP-induced), like the observed association of idiopathic blepharospasm with supranuclear dysfunction of eyelid activation ("apraxia" of lid opening), suggests an interesting role for the substantia nigra, both zona compacta and zona reticulata, in the generation of certain types of cranial dystonia.

In the text that follows we will review the available data, as outlined above, on the pathophysiology of cranial dystonia.

NEUROPATHOLOGY OF CRANIAL DYSTONIA

Few detailed post-mortem studies on cranial dystonia were available until recently (see Table 18.1). In 1981, García-Albea et al. reported that the brain of one typical case was essentially normal. Two years later, Altrocchi and Forno (1983) described the case of a 44-year-old man presenting with upper

Table 18.1   Pathological findings in primary cranial dystonia.

Authors/year	No. of cases	Findings
García Albea et al. (1981)	1	Non-specific abnormalities
Altrocchi and Forno (1983)	1	Patchy neuronal loss and gliosis in dorsal striatum and also mild in substantia nigra (SN)
Kulisevsky et al. (1988)	1	Moderate cell loss in SN, locus coeruleus, midbrain tectum and dentate nucleus. Frequent Lewy bodies in pigmented nuclei of brain stem. Intact striatum
Gibb et al. (1988)	4	No abnormalities detected in three cases. Small angioma (0.5 mm) in dorsal pons in one case
Jankovic et al. (1987)	1	Non-specific findings
Zweig et al. (1988)	1	Moderate to severe neuronal loss in SN, locus coeruleus raphé nuclei and pedunculopontine nucleus. Neurofibrillary tangles in substantia nigra

facial and oromandibular dystonia which later spread to the trunk and leg. This patient also developed rigidity of the arms and akathisia of all limbs. They found nerve cell loss and gliosis in the dorsal halves of the caudate and putamen, along with some involvement of the substantia nigra.

Within the last year, reports on seven additional cases have appeared in the literature. We examined the brain of a patient with a history of typical Meige syndrome of 4 years duration and observed morphological abnormalities in several brain stem nuclei (Kulisevsky et al., 1988). Lewy bodies were observed in about 1.7% of neurons per section of the zona compacta of the substantia nigra and in 14.4% of neurons per section of the locus coeruleus. Mild to moderate nerve cell loss was detected in both substantia nigra and locus coeruleus, and pigment granules (melanin) were found scattered in the neuropil of these structures. Mild to moderate nerve cell loss and gliosis were also present in the tectal region. In addition, the remaining neurons displayed rod-like intranuclear inclusions. Finally, moderate nerve cell loss and gliosis were present in the dentate nucleus of the cerebellum.

Jankovic et al. (1987) have also reported on the neuropathology of one case of Meige syndrome. No specific abnormalities were encountered in their case. Gibb et al. (1988) reported a histological study of the brains of four patients with cranial dystonia. No significant abnormalities were detected in three of the cases. In the fourth case, a patient afflicted with dystonic blepharospasm, an angioma was found in the dorsal pons, at the site of the central tegmental tract.

Recently Zweig et al. (1988) have reported the neuropathological study of the brains of four patients with "idiopathic" torsion dystonia. One of the patients had suffered typical Meige syndrome for 35 years. In his brain they found moderate to severe neuronal loss in several brain stem nuclei which included the substantia nigra pars compacta, the locus ceruleus, the raphé nuclei and the pedunculopontine nucleus. Infrequent neurofibrillary tangles were also noted in the substantia nigra. The authors draw attention to the fact that the pathological findings show some similarities to post-encephalitic parkinsonism.

Even though more detailed post-mortem examinations are needed to clarify the neuropathology of cranial dystonia, the reports reviewed here demonstrate histopathological abnormalities in the brain stem of several patients with cranial dystonia and support the notion that an alteration of the brain stem can cause a cranial dystonia syndrome. The distribution of the lesions in the cases of Altrocchi and Forno (1983), Kulisevsky et al. (1988) and Zweig et al. (1988) further suggest that involvement of the substantia nigra and the locus ceruleus may be of particular significance in some instances.

BIOCHEMICAL AND PHARMACOLOGICAL STUDIES

Little is known about the biochemistry of cranial dystonia. The recent report by Jankovic *et al.* (1987) on a post-mortem biochemical analysis of the brain of a 68-year-old woman with a 7-year history of cranial dystonia constitutes the only available information on that subject. In this case, the morphology of the brain revealed no abnormalities, but an increase in dopamine in the red nucleus and a marked increase in noradrenaline in the substantia nigra and red nucleus over control levels were found. It is noteworthy that similar noradrenergic changes were noted in the biochemical analysis of the brains of two patients with generalized torsion dystonia reported by Hornykiewicz *et al.* (1986). These authors, like Jankovic *et al.* (1987), comment that, in the absence of demonstrable cellular disease, the biochemical changes detected could be the result of persistent activation or compensatory overactivity of the midbrain noradrenergic systems in response to a primary neuronal loss in the locus coeruleus.

Several drugs known to modify basal ganglia neurotransmission modify the dystonic spasms in cranial dystonia. Tolosa and Lai (1979) studied the clinical response in Meige syndrome patients to apomorphine, physostigmine, haloperidol and L-dopa and concluded that a state of striatal dopamine preponderance and cholinergic hyperfunction underlies the dystonic spasms. Casey (1980) also proposed a predominant role for dopamine in the pathophysiology of Meige syndrome, possibly in the form of a dopamine receptor hypersensitivity. Since then, numerous trials with drugs known to modify central dopaminergic, cholinergic and GABAergic activity have been conducted in cranial dystonia. The results of these trials, recently reviewed elsewhere (Tolosa and Kulisevsky, 1988), seem to indicate that the disorder is pharmacologically heterogeneous, in that the response to drugs does not unequivocally indicate primary dysfunction of dopaminergic, cholinergic or GABAergic neurotransmission. However, based on the results of our own studies and those of others, and on our personal clinical experience, we believe that there is convincing evidence that drugs that lower central dopamine and acetylcholine activity, such as tetrabenazine or trihexiphenidyl, improve Meige syndrome, supporting the notion that both dopamine and acetylcholine mechanisms play an important role at least in a subgroup of cranial dystonia patients. Further biochemical studies of cerebrospinal fluid and autopsy material in cranial dystonia are needed to clarify these issues.

SECONDARY CRANIAL DYSTONIA

Dystonic spasms of the cranial musculature similar to those present in primary cranial dystonia occur on occasion in association with known neurological disorders or are produced by a variety of drugs. These secondary or symptomatic forms of Meige syndrome rarely represent a difficult diagnostic problem. The history of ingestion of the appropriate medications or the presence of various additional neurological signs allows for a correct diagnosis. An analysis of the different causes of symptomatic cranial dystonia, though, has contributed to our understanding of the primary forms.

The most common cause of secondary Meige syndrome is chronic neuroleptic administration producing a tardive dystonia syndrome (Burke *et al.*, 1982). Cases of tardive dystonia may be clinically identical to blepharospasm alone or in combination with oromandibular dystonia, and can only be differentiated by a history of neuroleptic use. Also relatively common is L-dopa-induced orofacial dystonia in Parkinson's disease (Weiner and Nausieda, 1982). More rarely, the syndrome can be induced by antihistaminic decongestants (Powers, 1982), anticholinergics (Martí *et al.*, 1986) and chronic amphetamine ingestion in patients without known neurological disease (Jankovic, 1981).

Dystonic spasms similar to those seen in primary Meige syndrome have been described as sequelae to focal hemispheric pathology (Fisher, 1963) and are encountered in the context of several neurodegenerative disorders. We have seen prominent cranial dystonia in head trauma, kernicterus, delayed-onset dystonia, progressive supranuclear palsy, Hallervorden–Spatz disease, olivopontocerebellar atrophy, Tourette syndrome, Huntington's disease, Wilson's disease and acquired hepatocerebral degeneration. It is also known to occur in ceroid lipofuscinosis, GM1 gangliosidosis, hexosaminidase A and B deficiency, juvenile dystonic lipidosis, glutamic acidemia, Joseph's disease and Leigh's disease, as well as in other secondary dystonias (Zeman and Whitlock, 1963).

Recently, Jankovic and Patel (1983) described secondary Meige syndrome in patients with upper brain stem strokes or demyelinating lesions of the brain stem from multiple sclerosis. Following their original observation, several case reports of patients with blepharospasm probably related to upper brain stem diencephalic lesions have been described (Keane and Young, 1985). The diagnosis in these additional reports has included focal ischaemic lesions, normal pressure hydrocephalus (Sandyk and Gillman, 1984; Jankovic, 1984), olivopontocerebellar atrophy (Sandyk and Gillman, 1984), bilateral thalamotomy (Jankovic, 1988), cerebral hypoxia and mass lesion. In these patients, the time between presumed cause and onset of blepharospasm has varied from 2 days to several years and the predominant clinical feature has consisted of blepharospasm (Jankovic, 1988).

Dystonic spasms similar to those of primary Meige syndrome thus occur in several neurological disorders that are known to affect the basal ganglia, such as Parkinson's disease, Wilson's disease, progressive supranuclear palsy and Huntington's disease, and also in patients receiving neuroleptics, antiemetics or dopaminergic drugs. This suggests that a dysfunction of the basal ganglia may underlie the production of the cranial muscle spasms and that a disturbance of normal functioning of the dopamine system is also present, at least in a subgroup of patients.

Cases of blepharospasm and oromandibular dystonia in patients with rostral brain stem diencephalic lesions suggest that a structural lesion just above the pontine facial nuclei can also cause orofacial dystonia (Jankovic and Patel, 1983). Because of the delay in some cases between the brain stem insult and the appearance of dystonia and the concomitant appearance of palatal myoclonus with the dystonic spasm in some cases of brain stem strokes, "denervation supersensitivity" of the facial nuclear complex has been proposed as the mechanism underlying this form of cranial dystonia (Jankovic and Patel, 1983; Keane and Young, 1985). Other mechanisms could be disinhibition or an abnormal excitatory drive to the facial nucleus and brain stem reflexes through interruption of descending pathways. As has been pointed out by Leenders et al. (1986), an alternative possibility is that such a lesion may interrupt ascending pathways from brain stem to basal ganglia so that blepharospasm could result from a functional disturbance rostral to the structural lesion.

BLINKING, VOLUNTARY EYELID CONTROL AND BLEPHAROSPASM

Several clinical and neurophysiological observations suggest that mechanisms underlying normal blinking play a role in the pathogenesis of cranial dystonia (Jankovic et al., 1982). Blepharospasm, the most frequent symptom in cranial dystonia, is frequently preceded by an increase in frequency and force of blinking (Jankovic and Ford, 1983), and Jankovic (1988) has further observed that relatives of patients with blepharospasm may suffer from excessive blinking. Studies on the electrically evoked blink response (Berardelli et al., 1985; Tolosa and Montserrat, 1985; Tolosa et al., 1988b) have found that the blink reflex is abnormal in cranial dystonia and that blink reflex habituation is reduced when compared to control subjects. This abnormality is particularly prominent in patients with blepharospasm, but has also been found in patients with oromandibular dystonia, spasmodic dysphonia and spasmodic torticollis who do not have blepharospasm (see Fig. 18.1), but not in patients with arm dystonia (Tolosa et al., 1988b).

The precise meaning of these abnormalities in spontaneous and reflex blinking in cranial dystonia is unclear. They suggest that mechanisms underlying normal blinking probably play a role in the genesis of cranial

BLINK   REFLEX   TO   PAIRED   STIMULI

200 msec   INTERVAL

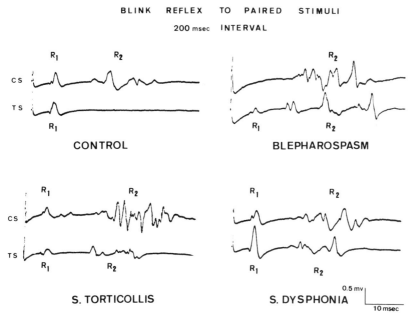

Fig. 18.1   Blink reflexes to pairs of stimuli given at an interstimuli interval of 200 ms in a normal control subject and in patients with focal dystonia. In the normal control subject, the conditioning stimulus (CS) completely inhibits the R2 of the test stimulus (TS). In the patients, on the other hand, a clear R2 follows the TS despite using the same short interstimuli interval. Stimuli were delivered at the supraorbital nerve and the reflex responses were recorded with surface electrodes from the ipsilateral orbicularis oculi (Tolosa *et al.*, 1988b).

dystonia syndromes and that in these disorders an enhanced excitability of those brain stem interneurons that mediate the late response of the blink reflex occurs. An enhanced excitability in brain stem interneurons may represent the substrate of the pathophysiology of cranial dystonia. As Berardelli *et al.* (1988) have pointed out, some trigger factor acting upon this abnormal substrate might be responsible for the clinical expression of the condition. One such trigger factor could be local eye disease: Grandas *et al.* (1988) have reported that in 12% of 264 patients with blepharospasm, the disease followed local eye injury or disease. Among 46 consecutive patients with blepharospasm referred for treatment with botulinum toxin we have found that local eye disease preceeded or accompanied the onset of dystonia in 21% of cases. The ocular disorders encountered included recurrent conjunctivitis, corneal ulcers and keratoconjunctivitis sicca (E. Tolosa and J. M. Martí, unpublished observations; see also Chapter 20).

An alteration in voluntary eyelid control, "apraxia" of lid opening, can also be present in patients with cranial dystonia. Patients with "apraxia" of lid opening cannot open their eyes because of difficulty in lid elevation, with no evidence of ongoing orbicularis oculi contraction and normal oculomotor or ocular sympathetic function (Goldstein and Cogan, 1965; Lepore and Duvoisin, 1985). This uncommon neurological condition occurs in progressive supranuclear palsy, where it can be associated with apraxia of lid closure (Lepore, 1988), and it has also been reported in Parkinson's disease (Brusa et al., 1986; Esteban and Giménez–Roldán, 1988), in patients with atypical parkinsonism (Goldstein and Cogan, 1965), in MPTP-induced parkinsonism in humans (Hotson et al., 1986), in neuroacanthocytosis (Bonaventura et al., 1986) and after stereotactic lesions of the upper brain stem/diencephalic region (Nashold and Gills, 1967). We have also described "apraxia" of lid opening in patients with Meige syndrome (Tolosa et al., 1987; Tolosa and Martí, 1988). "Apraxia" of lid opening is thought to result from an involuntary levator palpebrae inhibition of supranuclear origin (Lepore, 1988). Based on the known neuropathology of the conditions in which apraxia of lid opening occurs, it is thought that rostral brain stem disease provides a neuroanatomical basis for disruption of the extrapyramidal input to the facial nucleus and levator subnucleus that would result in such a neuro-ophthalmological anomaly. The association of levator palpebrae inhibition with dystonic blepharospasm suggests that a similar failure of extrapyramidal input occurs in cranial dystonia and again points to upper brain stem structures as being involved in the pathophysiology of cranial dystonia.

PARKINSONISM AND CRANIAL DYSTONIA

The relationship between dystonia and parkinsonism is a complex one to define. Clinically, both syndromes occur together often enough to suggest that the association is not mere chance. Before L-dopa was introduced in the treatment of Parkinson's disease, it had been emphasized that up to one-third of patients developed hand or foot dystonia as part of their illness (Marsden et al., 1982). Patients with post-encephalitic parkinsonism frequently suffered also from focal dystonia including torticollis and orofacial dystonia in conjunction with parkinsonism. Recently, also, a number of patients with otherwise typical Parkinson's disease have been reported to have developed blepharospasm or oromandibular dystonia or other focal dystonias preceding by a variable period of time the development of typical parkinsonism (Klawans and Paleologos, 1986; Katchen and Duvoisin, 1986; LeWitt et al., 1986; Poewe et al., 1988). Craniocervical dystonia can also accompany MPTP-induced parkinsonism in humans (Hotson et al., 1986).

The association of cranial dystonia with the various parkinsonian syndromes suggests that common pathophysiological mechanisms may underlie the two motor disorders and that an alteration of the nigrostriatal tract, and possibly also of the locus coeruleus, which are both lesioned in the parkinsonian syndromes described could play a role in cranial dystonia.

CONCLUSIONS

Neurophysiological studies have established an abnormality in brain stem interneuron excitability in patients with blepharospasm and in patients with other forms of cranial dystonia not associated with blepharospasm, such as oromandibular and laryngeal dystonia. This alteration could represent the substrate of the pathophysiology of cranial dystonia upon which, as Beradelli *et al.* (1988), have proposed a trigger mechanism would be required for the expression of the disease. In blepharospasm, local eye disease could be such a trigger factor, since it occurs in as many as 10% of patients prior to or at onset of the neurological symptoms, and local trauma is known to trigger other types of focal dystonia. Other trigger factors could be drugs, such as the neuroleptics, or immunological factors (Jankovic and Orman, 1984).

Biochemical and pharmacological studies in cranial dystonia have as yet added only inconclusive information about the mechanisms that underlie cranial dystonia. Biochemical analysis has been performed or reported in the brain of a single patient with cranial dystonia. It showed elevated levels of noradrenaline and dopamine in several subcortical nuclei. It is of interest that noradrenergic abnormalities have also been found in post-mortem studies of cases of childhood-onset torsion dystonia. The observation that the dystonic spasms are often sensitive to pharmacological manipulations of dopaminergic and cholinergic systems suggests that both dopamine and acetylcholine mechanisms play a role in cranial dystonia. Other lines of evidence which support the role for dopamine in cranial dystonia are: (a) normal blinking mechanisms are altered in cranial dystonia and dopamine mechanisms are important for normal blinking; (b) both chronic antipsychotic treatment and L-dopa administration can induce a clinical syndrome identical to Meige syndrome; and (c) cranial dystonia can precede or accompany Parkinson's disease, MPTP-induced parkinsonism in humans and post-encephalitic parkinsonism.

Cranial dystonia, like other forms of dystonia, remains an elusive disorder, of unknown pathophysiology and cause. Still, available evidence indicates that an alteration in brain stem interneuron excitability may be the main underlying pathophysiological abnormality. From the evidence reviewed, it is also clear that cranial dystonia is a heterogeneous disorder from a physiological, pharmacological and neuropathological standpoint.

REFERENCES

Altrocchi, P. H. and Forno, L. S. (1983) *Neurology* **33**, 802–805.
Berardelli, A., Rothwell, J. C., Day, B. L. and Marsden, C. D. (1985) *Brain* **108**, 593–609.
Berardelli, A., Rothwell, J. C., Day, B. L. and Marsden, C. D. (1988) *Adv. Neurol.* **50**, 525–535.
Bonaventura, I., Matías-Guiu, J., Cervera, L. and Puiggros, A. C. (1986) *Neurology* **36**, 1276.
Brusa, A., Mancardi, G., Meneghini, S., Piccardo, A. and Brusa, G. (1986) *Neurology* **36**, 134.
Burke, R. E., Fahn, S., Jankovic, J. *et al.* (1982) *Neurology* **32**, 1335–1346.
Casey, D. E. (1980) *Neurology* **30**, 690–695.
Esteban, A. and Giménez-Roldán, S. (1988) *J. Neurol. Sci.* **85**, 333–345.
Fahn, S., Marsden, C. D. and Calne, D. B. (1987) In *Movement Disorders 2* (edited by C. D. Marsden and S. Fahn), pp. 332–358. Butterworth, London.
Fisher, C. M. (1963) *Neurology* **13**, 77–78.
García-Albea, E., Franch, O., Muñoz, D. and Ricoy, J. M. (1981) *J. Neurol. Neurosurg. Psychiat.* **44**, 437–440.
Gibb, W. R. G., Lees, M. A. J. and Marsden, C. D. (1988) *Movement Disorders* **3**, 211–221.
Goldstein, J. E. and Cogan, D. F. (1965) *Arch. Ophthalmol.* **73**, 155–159.
Grandas, F., Elston, J., Quinn, N. and Marsden, C. D. (1988) *J. Neurol. Neurosurg. Psychiat.* **51**, 767–772.
Hornykiewicz, O., Kish, S. J., Becker, L. E., Farley, I. and Shannack, K. (1986) *N. Engl. J. Med.* **315**, 347–353.
Hotson, J. R., Langston, E. B. and Langston, J. W. (1986) *Ann. Neurol.* **20**, 456–463.
Jankovic, J. (1981) *Ann. Int. Med.* **94**, 788–793.
Jankovic, J. (1984) *Neurology,* **34**, 1523–1524.
Jankovic, J. (1988) *Adv. Neurol.* **49**, 103–116.
Jankovic, J. and Ford, J. (1983) *Ann. Neurol.* **13**, 402–411.
Jankovic, J. and Orman, J. (1984) *Ann. Ophthal.* **16**, 371–376.
Jankovic, J. and Patel, S. C. (1983) *Neurology* **33**, 1237–1240.
Jankovic, J., Havins, W. E. and Wilkins, R. B. (1982) *J. Am. Med. Assoc.* **248**, 3160–3164.
Jankovic, J., Svendson, C. N. and Bird, E. D. (1987) *N. Engl. J. Med.* **316**, 278–279.
Katchen, M. and Duvoisin, R. C. (1986) *Movement Disorders* **1**, 151–157.
Keane, J. R. and Young, J. A. (1985) *Arch. Neurol.* **42**, 1206–1208.
Klawans, H. L. and Paleologos, N. (1986) *Clin. Neuropharmacol.* **9**, 298–301.
Kulisevsky, J., Martí, M. J., Ferrer, I. and Tolosa, E. (1988) *Movement Disorders* **3**, 170–175.
Leenders, K. L., Frackowiak, R. S. J., Quinn, N. *et al.* (1986) *Movement Disorders* **1**, 51–58.
Lepore, F. E. (1988) *Adv. Neurol.* **49**, 85–90.
Lepore, F. E. and Duvoisin, R. C. (1985) *Neurology* **35**, 423–427.
LeWitt, A., Burns, R. S. and Newman, R. P. (1986) *Clin. Neuropharmacol.* **9**, 293–297.
Marsden, C. D., Parkes, J. D. and Quinn, N. (1982) In *Movement Disorders* (edited by C. D. Marsden and S. Fahn), pp. 96–122. Butterworth, London
Martí, M. J., Kulisevsky, J. and Tolosa, E. (1986) Presented at the IV International

Meeting of the Benign Essential Blepharospasm Research Foundation, Barcelona, Spain.

Nashold, B. S. and Gills, J. P. (1987) *Arch. Ophthalmol.* **77**, 609–618.

Poewe, W. H., Lees, A. J. and Stern, G. M. (1988) *Ann. Neurol.* **23**, 73–78.

Powers, J. M. (1982) *J. Am. Med. Assoc.* **247**, 3244–3245.

Sandyk, R. and Gillman, M. A. (1984) *Neurology* **34**, 1522–1523.

Tolosa, E. and Kulisevsky, J. (1988) *Adv. Neurol.* **49**, 433–431.

Tolosa, E. S. and Lai, C. (1979) *Neurology* **27**, 1126–1130.

Tolosa, E. and Martí, M. J. (1988) *Adv. Neurol.* **49**, 73–84.

Tolosa, E. and Montserrat, L. (1985) *Neurology* **35** (suppl. 1), 271.

Tolosa, E., Kulisevsky, J. and Martí, M. J. (1987) *Neurology* **37** (suppl. 1), 273.

Tolosa, E., Kulisevsky, J. and Fahn, S. (1988a) *Adv. Neurol.* **50**, 509–515.

Tolosa, E., Montserrat, L. and Bayés, A. (1988b) *Movement Disorders* **3**, 61–69.

Weiner, W. J. and Nausieda, P. A. (1982) *Arch. Neurol.* **39**, 451–452.

Zeman, W. and Whitlock, C. L. (1963) In *Handbook of Clinical Neurology, Vol. 6* (edited by P. J. Vinken and G. W. Bruyn), pp. 544–566. North-Holland, Amsterdam.

Zweig, R. M., Hedreen, J. C., Jankel, W. R., Casanova, M. F., Whitehouse, P. J. and Price, D. L. (1988) *Neurology* **38**, 702–706.

19

Positron Emission Tomographic Investigations of Dystonia

J. S. Perlmutter and K. L. Leenders

INTRODUCTION

Dystonia represents a spectrum of disorders characterized by sustained, involuntary muscle contractions that frequently twist parts of the body and produce abnormal postures (Marsden and Rothwell, 1987). One method of classifying different types of dystonia is by etiology. Those forms with an identifiable etiology are classified as secondary dystonias; the majority of cases, with no identifiable etiology, are known as the primary dystonias.

The pathophysiological basis of dystonia has remained elusive. Most investigators have attempted to correlate the site of an identifiable brain lesion with the development of symptoms. Numerous reports have described structural abnormalities in basal ganglia contralateral to the symptomatic side in patients with unilateral involvement (Messimy et al., 1977; Mauro and Fahn, 1980; Maki et al., 1980; Brett et al., 1981; Grimes et al., 1982; Russo, 1983; Demierre and Rondot, 1983; Burton et al., 1984; Narbonna et al., 1984; Pettigrew and Jankovic, 1985). Advances in computed tomography (CT) and magnetic resonance imaging (MRI) have permitted identification of putaminal lesions in patients with secondary dystonias (Fross et al., 1987; Rutledge et al., 1988). Furthermore, electrophysiological studies suggest that

DISORDERS OF MOVEMENT: CLINICAL, PHARMACOLOGICAL
AND PHYSIOLOGICAL ASPECTS   ISBN 0-12-569685-X

there might be abnormal output from basal ganglia structures. Patients with blepharospasm demonstrate hyperexcitability of lower brain stem interneurons involved in the blink reflex, and there is an abnormality of Ia reciprocal inhibition from extensors to flexors in the dystonic arm (Marsden and Rothwell, 1987). Although these functional abnormalities have been found, their role in the production of dystonia is unclear.

Positron emission tomogrpahy (PET) permits measurements of regional metabolic and pharmacological function of the brain in living humans and has the potential to provide unique insights into the pathophysiology of dystonia. Measurements of regional blood flow or metabolism, at least in the resting, normal state, are thought to reflect underlying neuronal activity (Mata et al., 1980; Fox and Raichle, 1985). It is important to note that ongoing controversy still remains concerning the exact relationships among flow, metabolism and neuronal function (Raichle et al., 1976; Fox and Raichle, 1986; Ginsberg et al., 1987; Lou et al., 1987), and this must be kept in mind when interpreting such PET studies. PET also permits evaluation of radioligand–receptor binding as well as of neurotransmitters themselves (Perlmutter, 1988). The methodology for these measurements is evolving and each study must be interpreted cautiously in light of a variety of technical considerations that will be reviewed briefly below.

POSITRON EMISSION TOMOGRAPHY

PET is a nuclear medicine imaging technique that measures the regional distribution of a previously administered radionuclide (Raichle, 1986). In effect, it produces an *in vivo* autoradiograph from a living subject. PET studies require appropriately labelled radiopharmaceuticals (Welch and Kilbourn, 1984). Positron-emitting isotopes have been incorporated into naturally occurring compounds (e.g. ^{15}O into $H_2^{15}O$), analogues of naturally occurring substances (e.g. ^{18}F into ^{18}F-fluorodeoxyglucose [FDG] and ^{18}F-6-fluorodopa [^{18}F-dopa]) or radioligands (e.g. ^{11}C into ^{11}C-raclopride or ^{11}C-N-methylspiperone [^{11}C-NMSP]). The development of a new radio-pharmaceutical for PET studies is a complicated process and can take several years.

After administration of a radiotracer, the amount of radioactivity in a brain region reflects several factors in addition to the physiological function of interest. These other factors include regional radiotracer delivery, intra-vascular content of radioactivity, accumulation of radiolabelled metabolites and radioactive decay. Each varies in importance depending upon the radiopharmaceutical administered, the tissue monitored and the timing of the PET scan after tracer administration. Mathematical models have been developed that consider these factors and convert PET data into relevant physiological measurements. Understanding the assumptions and implementation of models is necessary for the critical evaluation of many PET studies (Perlmutter et al., 1986).

Analysis of PET images also requires identification of anatomical structures that correspond to PET regions (Fox et al., 1985). Anatomical localization by visual inspection of a PET image can be inadequate because of the variable relationship between local physiology and "underlying" anatomy. Several approaches to this critical problem of anatomical localization have been proposed. Some methods use CT or MRI to obtain anatomical sections corresponding to the PET slices (Mazziotta et al., 1982). These methods require considerable care to identically reposition subjects in the different scanners. Head holders suitable for PET, CT and MRI have been developed for this purpose (Kearfott et al., 1984). Nevertheless, localization still requires subjective observer identification of regions of interest; a task that can be very difficult for cortical regions, in particular. Another method of localization is a stereotactic technique based upon a proportionate coordinate system similar to that used by neurosurgeons for many years (Fox et al., 1985). Its advantages include accuracy and lack of observer bias. PET regions of interest can be identified with standard stereotactic atlas coordinates, thus permitting comparisons of findings among different subjects, laboratories and imaging modalities. The method, however, does assume that the brain has a standard shape and is not suitable for subjects with distorted brain anatomy (Fox et al., 1985).

APPLICATION TO DYSTONIA

Initially, PET was used to identify regional abnormalities in patients with dystonia. In the first reported case, Perlmutter and Raichle (1984) described a 50-year-old man who suffered minor trauma to the right side of the head and neck. Within 20 min, he developed right-sided paroxysmal, intermittent dystonic posturing of the right face, forearm, hand and foot, with weaker contractions of the left foot, lasting several seconds and recurring every few minutes. Emotional stress exacerbated the frequency and severity of the posturing. Intentional movements decreased only the severity, and sleep abolished the movements. Conventional evaluation was normal including the remainder of the neurological examination, serum electrolytes, calcium, magnesium, arterial blood gases, cerebrospinal fluid, electroencephalography with nasopharyngeal leads, CT (initially and 4 weeks later) and cerebral angiography (left carotid and aortic arch injections). An abnormality of the contralateral basal ganglion was identified only with PET. Regional blood flow and volume were increased while oxygen extraction and utilization were decreased in the left basal ganglion (Fig. 19.1). The basal ganglia were located on the PET images by comparison with CT images made in the same plane. More precise localization techniques were not available at the time of the study. Nevertheless, this study did demonstrate the utility of PET to demonstrate a regional abnormality within the basal ganglia in a patient with a secondary form of dystonia (Perlmutter and Raichle, 1984).

Fross et al. (1987) suggested that a reduction in putaminal input to pallidum might be important in the development of dystonia. To support this notion, they described a 47-year-old woman with post-hemiplegic dystonia after cerebral infarction of the right putamen identified by CT and MRI. This lesion "spared all but a tiny segment of adjacent internal capsule and caudate nucleus". PET revealed diminished uptake of radioactivity in the right putamen after administration of ^{18}F-dopa, as one would expect from the CT findings. The authors noted that there was normal uptake in the caudate, which they said supported their contention that the "lesion was essentially confined to the right putamen", supporting their hypothesis. However, the much lower resolution of PET makes it difficult to know to what extent caudate was really spared.

Stoessl et al. (1986) measured glucose metabolism (CMRglu) in 16 patients with torticollis and compared the results to 11 normal subjects. The method for identification of regions of interest on the PET images was not stated. Nevertheless, regional CMRglu did not differ between patients and normals. Inter-regional correlational analysis in normal subjects demonstrated significant relationships between right- and left-sided caudate, lentiform nucleus and thalamus. Similarly, CMRglu in any one of these structures was significantly correlated with activity in the other two structures. In torticollis patients, the relationships across the midline were maintained, but the correlations between thalamic CMRglu and that in caudate and lentiform nucleus was no longer significant. They suggested that these findings indicate a disruption of the pallidothalamic projections in this form of dystonia.

Chase et al. (1988) used PET and FDG to study six patients with idiopathic torsion dystonia and reported left–right asymmetries of metabolism in the lenticular nucleus compared to nine normals. The degree of asymmetry correlated with the degree of clinical asymmetry of the dystonia. Similarly, Gilman et al. (1988) studied five patients with idiopathic torsion dystonia with marked asymmetry of clinical findings. Subcortical structures were identified on the PET images by positioning a region of interest over a local peak of radioactivity. Three patients had normal studies and two had asymmetries; one of caudate and the other of cerebellar metabolism. The authors noted that one of 14 normal subjects also had asymmetries of similar magnitude making the interpretation of the findings in the dystonic patients less clear.

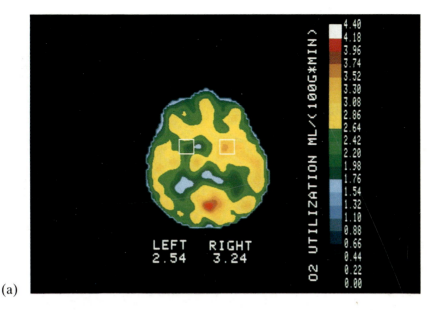

(a)

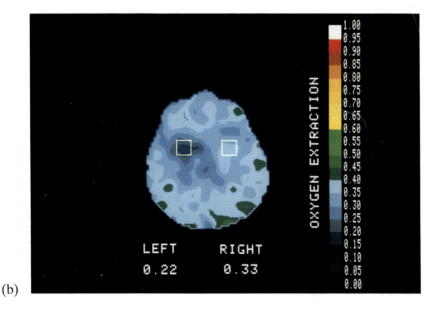

(b)

Fig. 19.1 Measurements of cerebral blood flow, blood volume, oxygen utilization and oxygen extraction in a patient with post-traumatic paroxysmal hemidystonia affecting the right side of the body. All four images are in the same anatomical plane and are oriented with the patient's left to reader's left and anterior up. The region of greatest abnormality is in the left putamen, identified using a stereotactic method of localization that is independent of the appearance of this image (Fox *et al.*, 1985). These data reveal a localized decrease in oxygen utilization (a) and extraction (b) with an increase in blood flow (c) and volume (d) (*see overleaf*). CT scan in the same plane revealed no abnormalities. This is a modified version of a similar figure from Perlmutter and Raichle (1984).

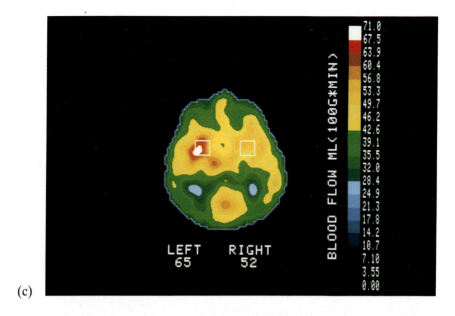

(c)

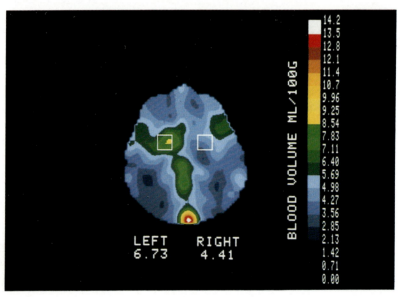

(d)

Fig. 19.1 – *continued*

Lang *et al.* (1988) studied nine patients with a variety of dystonic syndromes including two patients with symptomatic hemidystonia. These two patients had lesions identified on CT scans. Not surprisingly, radioactivity accumulation after either ^{18}F-dopa or FDG administration was diminished in the areas of the lesions. One patient with dopa-responsive dystonia had decreased radioactivity uptake in the left putamen after ^{18}F-dopa administration. Two patients with typical childhood-onset torsion dystonia had reduced accumulation of radioactivity in putamen, whereas two others, also with childhood-onset torsion dystonia, had normal scans. FDG scans were normal in these latter five patients. Two other patients, one with idiopathic right-sided hemidystonia and the other with cranial dystonia associated with right-sided writer's cramp had diminished accumulation in left putamen after ^{18}F-dopa but normal FDG scans. Further interpretation of these data is difficult since the method for identification of regions of interest was not stated, and results from normal control subjects were not described.

Martin *et al.* (1988), using PET and ^{18}F-dopa, studied three patients with dopa-responsive dystonia as well as another patient with dystonia who had not been treated with L-dopa. They analysed regional uptake of radioactivity using an adaptation of a kinetic model of unidirectional tracer uptake into tissue that included a correction for the presence of labelled metabolites of ^{18}F-dopa in blood. They found that the striatal uptake constant decreased with increasing age in 12 normal volunteers, and two of the dystonic patients had values below the normal range. The authors suggested that their findings indicated that generalized dystonia was a heterogenous disorder with presynaptic dopaminergic function abnormal in some but normal in others.

Leenders *et al.* (1988) used PET to study the integrity of the dopaminergic system in six patients with dystonia. The patients had a variety of clinical manifestations: one had unilateral parkinsonism and dystonia as well as blepharospasm associated with a focal lesion in the contralateral brain stem (see also Leenders *et al.*, 1986); two had unilateral parkinsonism and dystonia; one had unilateral, apparently idiopathic dystonia. This patient turned out to have multiple sclerosis and is published as case 2 in Coleman *et al.* (1988), and the remaining two had idiopathic torticollis. ^{18}F-dopa and ^{11}C-NMSP studies were performed on all subjects except the two patients with torticollis, who only had ^{18}F-dopa. Data analysis was limited to evaluation of ratios of radioactivity in striatum versus surrounding brain for ^{18}F-dopa and striatum versus cerebellum for ^{11}C-NMSP. The unilaterally affected patients had low striatal ratios on the side contralateral to symptoms. One of the patients with unilateral parkinsonism and dystonia had an increased striatal/cerebellar ratio after administration of ^{11}C-NMSP (Fig. 19.2). Interestingly this patient had to stop L-dopa after only 1 week of treatment because of intolerable dyskinesias on the symptomatic side only. The torticollis patients had diminished striatal ratios after ^{18}F-dopa. It should be noted that the latter two patients had 30 mg of diazepam during the course of the PET study to permit completion of the study with adequate control of head movement. The effect of this treatment on accumulation of striatal radioactivity is unknown. According to the authors, these preliminary results suggest that the dopaminergic system can be involved in some patients with dystonia.

Leenders *et al.* (in preparation) used PET and ^{11}C-NMSP to investigate dopaminergic receptors in seven patients with dystonia. The patients included four with torticollis, one with retrocollis, and two with writer's cramp. Regions of interest were identified on the PET images by visual inspection but confirmed by comparisons with CT images made in the same planes as the PET. Data were analysed by calculating radioactivity uptake normalized to the injected radioactivity per gram of body weight. Further analysis using a "multiple time graphical plot" assumes a unidirectional uptake of radioactivity into the tissue and yields an index of radioligand–receptor binding k_3 (which is equivalent to the product of the forward rate constant of radioligand and receptor and the maximal number of specific binding sites—Eckernas *et al.*, 1987). The investigation did not find a difference in striatal uptake or the calculated k_3 between patients and four normals, or

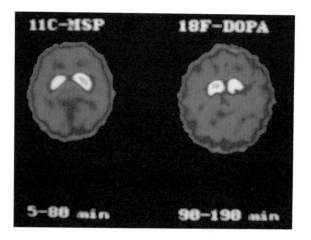

Fig. 19.2 ^{18}F-Dopa and ^{11}C-NMSP uptake is shown in a patient with hemiparkinsonism and hemidystonia on the left side of the body. Anatomical orientation is as for Fig. 19.1. The images represent the distribution of total radioactivity accumulated during the indicated times for that cross-section of brain. The image on the right shows diminished ^{18}F-dopa uptake (presynaptic dopaminergic activity) in the right (i.e. contralateral) striatum, but the image on the left demonstrates increased uptake of ^{11}C-NMSP (postsynaptic dopamine (D2) receptor binding) in the right striatum. (This figure appears in colour in the plate section facing page 360.)

between the dystonics and another group consisting of eight patients with Parkinson's disease and two with pituitary adenomas. There was a trend, although not statistically significant with the small number of patients, towards a higher k_3 in striatum contralateral to patients' most affected side (for torticollis patients the most affected side was defined as the direction of pulling of the chin). Further patients need to be studied to confirm these findings.

Perlmutter and Raichle (1988), using PET and ^{15}O-labelled water to measure regional blood flow, studied seven patients with asymmetric dystonia; one was the previously described patient with post-traumatic paroxysmal hemidystonia, one had familial paroxysmal dystonia, and the remaining five had idiopathic dystonia. Regions of interest were identified using a stereotactic localization technique that is independent of the appearance of the physio-logical PET images. Only two patients, one with post-traumatic hemidystonia and another with asymmetric idiopathic dystonia, had ratios of left–right putaminal blood flow beyond the normal range defined in 32 normal subjects. In both of these patients the putaminal blood flow was higher contralateral to the dystonic side. Additionally, two of the subjects had additional PET scans during vibrotactile stimulation of the fingerpads. This stimulation produced a robust and consistent increase in blood flow in contralateral sensorimotor cortex in normal subjects. One of the dystonics, however, had a diminished increase in blood flow after stimulation of the dystonc side. Follow-up preliminary data in a larger series of patients ($n = 11$) indicate consistently lower responses both contralateral and ipsilateral to the side of dystonia (Tempel and Perlmutter, 1988), suggesting that patients with dystonia might have an abnormality of sensorimotor integration.

CONCLUSIONS

Positron emission tomography has permitted identification of abnormal brain regions in a number of patients with dystonia as well as providing the means to demonstrate abnormal physiological responses in such patients. Further development of PET methodology promises to reveal additional insights into the pathophysiological basis of dystonia. New methods for assessment of neurotransmitter and receptor pharmacology will increase our understanding of these factors in the etiology and therapy of dystonia. However, great care must be exercised to avoid a number of methodological pitfalls. Patients must be selected carefully. Objective methods of anatomical localization for regions of interest in PET images must be used. Adequate validation of data analysis methods must be performed. The potential gains certainly justify the considerable efforts required to pursue these avenues of research.

ACKNOWLEDGEMENTS

This work has been supported by NIH grants HL13851, NS06833, AG03991, Teacher Investigator Development Award NS00929 (JSP), the Greater St Louis Chapter of the American Parkinson's Disease Association, and generous contributions from Mrs Donald Danforth and Mr and Mrs Jefferson Miller.

REFERENCES

Brett, E. M., Hoare, R. D., Sheehy, M. P. and Marsden, C. D. (1981) *J. Neurol. Neurosurg. Psychiat.* **44**, 460.

Burton, K., Farrell, K., Li, D. and Calne, D. B. (1984) *Neurology* **34**, 962–965.

Chase, T. N., Tamminga, C. A. and Burrows, H. (1988) *Adv. Neurol.* **50**, 237–241.

Coleman, R. J., Quinn, ·N. P. and Marsden, C. D. (1988) *Movement Disorders* **3**, 329–332.

Demierre, B. and Rondot, P. (1983) *J. Neurol. Neurosurg. Psychiat.* **46**, 404–409.

Eckernas, S. A., Aquilonius, S. M., Hartvig, P., Hagglund, J., Jundqvist, H., Nagren, K. and Langstrom, B. (1987) *Acta Neurol. Scand.* **75**, 168–178.

Fox, P. T. and Raichle, M. E. (1985) *Ann. Neurol.* **17**, 303–305.

Fox, P. T. and Raichle, M. E. (1986) *Proc. Natl. Acad. Sci. USA* **83**, 1140–1144.

Fox, P. T., Perlmutter, J. S. and Raichle, M. E. (1985) *J. Comput. Assist. Tomogr.* **9**, 141–153.

Fross, R. D., Martin, W. R. W., Stoessl, A. J., Adam, M. J., Ruth, T. J., Pate, B. D., Burton, K. and Calne, D. B. (1987) *Neurology* **37**, 1125–1129.

Gilman, S., Junck, L., Young, A. B., Hichwa, R. D., Markel, D. S., Koeppe, R. A. and Ehrenkaufer, R. L. E. (1988) *Adv. Neurol.* **50**, 231–236.

Ginsberg, M. D., Dietrich, W. D. and Busto, R. (1987) *Neurology* **37**, 11–19.

Grimes, J. D., Hassan, M. N., Quarrington, A. M. and D'Alton, J. (1982) *Neurology* **32**, 1033–1035.

Kearfott, K. J., Rottenberg, D. A. and Knowles, R. J. R. (1984) *J. Comput. Assist. Tomogr.* **8**, 1217–1220.

Lang, A. E., Garnett, E. S., Firnau, G., Nahmias, C. and Talalla, A. (1988) *Adv. Neurol.* **50**, 249–253.

Leenders, K. L., Frackowiak, R. J. S., Quinn, N., Brooks, D., Sumner, D. and Marsden, C. D. (1986) *Movement Disorders* **1**, 51–58.

Leenders, K. L., Quinn, N., Frackowiak, R. S. J. and Marsden, C. D. (1988) *Adv. Neurol.* **50**, 243–247.

Lou, H. C., Edvinsson, L. and MacKenzie, E. T. (1987) *Ann. Neurol.* **22**, 289–297.

Maki, Y., Akimoto, H. and Enomoto, T. (1980) *Childs Brain* **7**, 113–123.

Marsden, C. D. and Rothwell, J. C. (1987) *Can. J. Neurol. Sci.* **14**, 521–527.

Martin, W. R. W., Stoessl, A. J., Palmer, M., Adam, M. J., Ruth, T. J., Grierson, J.

R., Pate, B. D. and Calne, D. B. (1988) *Adv. Neurol.* **50**, 223–229.

Mata, M., Fink, D. J., Gainer, H., *et al.* (1980) *J. Neurochem.* **34**, 1132–1139.

Mauro, A. J. and Fahn, S. (1980) *Trans. Am. Neurol. Assoc.* **105**, 228–229.

Mazziotta, J. C., Phelps, M. E., Meadors, A. K., Ricci, A., Winter, J. and Bentson, J. R. (1982) *J. Comput. Assist. Tomogr.* **6**, 848–853.

Messimy, R., Diebler, C. and Metzger, J. (1977) *Rev. Neurol.* **133**, 199–206.

Narbonna, J., Obeso, J. A., Tunon, T., Martinez-Lage, J. M. and Marsden, C. D. (1984) *J. Neurol. Neurosurg. Psychiat.* **47**, 704–709.

Perlmutter, J. S. (1988) *Trends Neurosci.* **11**, 203–208.

Perlmutter, J. S. and Raichle, M. E. (1984) *Ann. Neurol.* **15**, 228–233.

Perlmutter, J. S. and Raichle, M. E. (1988) *Adv. Neurol.* **50**, 255–641.

Perlmutter, J. S., Larson, K. B., Raichle, M. E., Markham, J., Mintun, M. A., Kilbourn, M. R. and Welch, M. J. (1986) *J. Cereb. Blood Flow Metabol.* **6**, 154–169.

Pettigrew, L. C. and Jankovic, J. (1985) *J. Neurol. Neurosurg. Psychiat.* **48**, 650–657.

Raichle, M. E. (1986) *Trends Neurosci.* **9**, 525–529.

Raichle, M. E., Grubb, R. L., Gado, M. H. *et al.* (1976) *Arch. Neurol.* **33**, 523–526.

Russo, L. S. (1983) *Arch. Neurol.* **40**, 61–62.

Rutledge, J. N., Hilal, S. K., Silver, J., Defendini, R. and Fahn, S. (1988) *Adv. Neurol.* **50**, 265–275.

Stoessl, A. J., Martin, W. R. W., Clark, C., Adam, M. J., Ammann, W., Beckman, J. H., Bergstrom, M., Harrop, R., Rogers, J. G., Ruth, T. J., Sayre, C. I., Pate, B. D. and Calne, D. B. (1986) *Neurology* **36**, 653–657.

Tempel, L. W. and Perlmutter, J. S. (1988) *Neurology* **38**, S131.

Welch, M. J. and Kilbourn, M. R. (1984) In *Clinical Radionuclide Imaging* (edited by P. M. Freeman and L. M. Johnson), pp. 181–202. Grune and Stratton, London.

20

Blepharospasm: Clinical Aspects and Therapeutic Considerations

Francisco Grandas

Blepharospasm is an adult-onset focal dystonia characterized by spasms of contraction of the orbicularis oculi muscles. These cause eye closure and may render sufferers functionally blind.

Blepharospasm was first described by Henry Meige in 1910 as "spasme facial médian". He pointed out that frequently other facial muscles were involved and that there was an analogy between these spasms and torticollis. Marsden (1976a) drew attention to the relationship between blepharospasm and oro-mandibular dystonia and named this association Brueghel's syndrome, to give credit to Pieter Brueghel, The Elder, the artist who painted in the sixteenth century a man with apparent blepharospasm and dystonic jaw opening.

The cause of blepharospasm is unknown in the majority of patients. However, the clinical picture of blepharospasm can be seen in a variety of neurological conditions (Table 20.1).

The pathology of idiopathic blepharospasm remains unclear. Post-mortem examination of two cases has not been conclusive. One case did not reveal any abnormality in the central nervous system (García-Albea *et al.*, 1981); the other showed neuronal cell loss and gliosis in the dorsal halves of the caudate and putamen (Altrocchi and Forno, 1983; see also Chapter 18).

DISORDERS OF MOVEMENT: CLINICAL, PHARMACOLOGICAL
AND PHYSIOLOGICAL ASPECTS   ISBN 0-12-569685-X

Table 20.1   Etiology of blepharospasm.

I   Idiopathic
II   Secondary
   (A) Parkinsonism
      Parkinson's disease (Klawans and Enrich, 1970; Corin *et al.*, 1972)
      MPTP-induced (Hotson *et al.*, 1986)
      Post-encephalitic (Alpers and Patten, 1927)
      Progressive supranuclear palsy (Jackson *et al.*, 1983)
      Multiple system atrophy
      Late-onset Hallervorden–Spatz disease (Alberca *et al.*, 1987)
   (B) Structural lesions
      Rostral brain stem
         Infarction (Jankovic and Patel, 1983; Lang and Sharpe, 1984; Powers, 1985;
            Day *et al.*, 1986).
         Demyelination (Jankovic and Patel, 1983)
         Arterio-venous malformation (Leenders *et al.*, 1986)
      Basal ganglia (bilateral)
         Infarction (Keane and Young, 1985; Palakurthy and Iyer, 1987).
         Thalamotomy (Jankovic, 1986).
   (C) Drugs
      Neuroleptics (Weiner *et al.*, 1981; Burke *et al.*, 1982).
      L-Dopa (Weiner and Nausieda, 1982).
      Nasal decongestants (Powers, 1982)
   (D) Secondary generalized dystonia (e.g. Wilson's Disease) (Marsden, 1976b)

PREDISPOSING FACTORS AND CLINICAL FEATURES

Blepharospasm appears in women more than men with a female/male ratio
which varies from 1.8 to 3, according to the two largest reported series of
patients with this cranial dystonia (Jankovic and Orman, 1984; Grandas *et
al.*, 1988). The onset of symptoms occurs usually in the fifth or sixth decade.

A genetic predisposition for blepharospasm is suggested by the relatively
high incidence of family history of blepharospasm or other forms of dystonia
(Tolosa, 1981; Nutt and Hammerstad, 1981; Jankovic and Ford, 1982;
Grandas *et al.*, 1988). In addition, many patients have a positive family
history of other movement disorders such as Parkinson's disease or essential
tremor (Tolosa, 1981; Jankovic and Ford, 1982; Grandas *et al.*, 1988).

Psychiatric illness before or at onset of blepharospasm has frequently been
reported (Tolosa, 1981; Jankovic and Ford, 1982). However, some of these
patients were exposed to neuroleptic drugs before or at onset of their
blepharospasm, thus being examples of tardive dystonia.

Many patients experience local ocular disease preceding the onset of their
blepharospasm, such as blepharitis, conjunctivitis, Sjögren's syndrome or
other inflammatory or mechanical conditions. In addition, a great number of

patients complain of photophobia, dry eyes (other than Sjögren's syndrome), soreness in the eyes or ocular pain at the onset of their blepharospasm. However, only a small number of patients develop *new* eye problems during the course of their blepharospasm (Grandas *et al.*, 1988). Thus ocular "insults" may play a role in triggering the onset of this cranial dystonia, in the same way as trauma to other parts of the body may precede the onset of other forms of focal dystonia (Schott, 1985; Brin *et al.*, 1986).

In approximately 20% of patients the onset of blepharospasm is unilateral, becoming bilateral in the majority of cases after variable periods of time. Frequently excessive blinking precedes, often by years, the onset of spasms of orbicularis oculi. This suggests that a defect in the mechanisms of blinking is involved in blepharospasm (Jankovic *et al.*, 1982).

The intensity and frequency of spasms may be modified by several conditions. Bright lights, watching TV, reading, stress and driving are the commonest situations which worsen blepharospasm. On the other hand, blepharospasm often improves after sleep, relaxation, concentration, looking down or with movements involving the oro-facial area such as talking, yawning, singing or grimacing (Tolosa, 1981; Jankovic and Orman, 1984; Grandas *et al.*, 1988). In addition, many patients have special "tricks" like touching or pulling the eyelids slightly in order to open the eyes; these may be regarded as an equivalent of the "geste antagoniste", seen in other forms of dystonia.

Blepharospasm is very often associated with dystonia elsewhere, involving mainly the cranio-cervical region. Thus in our series of 264 cases of blepharospasm, only 58 (22%) experienced orbicularis oculi spasms as the only dystonic feature. In 206 (78%) patients blepharospasm was associated with dystonia in other areas. The lower facial and jaw muscles were the most commonly affected (71.2%). The neck (22.7%) and the laryngeal (17.4%) and respiratory muscles (14.8%) were the other areas most likely to be involved. Less than 10% of patients had dystonic features below the neck. In this series there tended to be an orderly temporal progression of dystonia in the cranio-cervical region (Grandas *et al.*, 1988).

A number of patients with blepharospasm show other movement disorders including parkinsonism and postural tremor of the arms, similar to that of benign essential tremor.

Some authors have stressed a relatively high incidence of autoimmune (Jankovic and Ford, 1982; Jankovic and Patten, 1987; Kurlan *et al.*, 1987) and thyroid diseases (Nutt *et al.*, 1984a) in patients with blepharospasm. However the frequency of thyroid disorders may not exceed that in other populations, the reported association being simply a consequence of the female preponderance of both conditions. Approximately one in ten patients may experience partial or complete spontaneous remissions. Unfortunately recurrences occur in most patients after variable periods of time (Grandas et al., 1988).

Blepharospasm is a disabling condition. Nearly two-thirds of patients are rendered functionally blind. Many have to give up work or cannot leave the house alone because they cannot see properly. Fear of crossing roads or bumping into objects often leads to individuals becoming recluses.

TREATMENT OF BLEPHAROSPASM

The treatment of idiopathic blepharospasm, like that of other forms of idiopathic dystonia, is exclusively symptomatic. Some drugs, several surgical approaches and, most recently, injections of botulinum toxin into the orbicularis oculi muscles may ameliorate this condition.

Drugs

Tolosa and Lai (1979) and Casey (1980) have suggested a striatal dopaminergic preponderance as the biochemical substrate of blepharospasm, based on the pharmacological response of a small number of patients. However, although improvement of blepharospasm in some subjects has been reported with dopaminergic antagonists (Jankovic, 1982) and cholinergic drugs (Miller, 1973; Casey, 1980; Karf and Sharpe, 1981), small series of patients or sporadic cases have also been described as improving with anticholinergic drug therapy (Lang et al., 1982; Tanner et al., 1982; Duvoisin, 1983; Fahn, 1983; Nutt et al., 1984b; Lang, 1986) dopaminergic agonists (Chakravorty, 1974; Micheli et al., 1982), clonazepam (Merikangas and Reynolds, 1979), baclofen alone (Fahn et al., 1983) or in combination with sodium valproate (Brennan et al., 1982), and even with cannabidiol (Snider and Consroe, 1984). From these studies it is difficult to draw general conclusions about the pharmacologic profile of blepharospasm since the majority were open studies involving small numbers of patients.

Jankovic and Ford (1982), Jankovic and Orman (1984) and Grandas et al. (1988), in series of 100, 250 and 264 patients with blepharospasm, respectively, did not find a consistent pharmacologic response. In our experience 20% of patients experienced some benefit with anticholinergics but improvement in occasional patients was also achieved with L-dopa, dopaminergic agonists (lisuride, bromocriptine), antidepressants, dopaminergic antagonists (tetrabenazine, haloperidol, pimozide, chlorpromazine), benzodiazepines and lithium.

Therefore the pharmacological manipulation of central dopaminergic and cholinergic systems does not seem to produce a consistent and sustained response (Marsden *et al.*, 1983).

Overall only about a fifth of patients with blepharospasm may achieve some benefit from drug treatment. However, if it is successful, drug therapy may be effective in restoring functional vision and avoid the need for more invasive treatment.

Surgical approaches

Because of the relatively poor response to drugs, different surgical approaches have been tried to relieve blepharospasm. All these techniques were intended to weaken the local effector muscles.

Injections of alcohol into the facial nerves was one of the earlier surgical attempts, being a common treatment of blepharospasm during the first half of this century (Gurdjian and Williams, 1928). Usually the benefit achieved by this therapeutic method (if it occurred) was short-lived.

Thermolytic lesioning of the branches of the facial nerves restricts the degree of possible facial paresis and causes less morbidity than other, more radical, surgical approaches. Patients can undergo facial thermolyosis under local anaesthesia and the procedure can be performed in stages (Battista, 1982). Unfortunately reinnervation, causing recurrence of blepharospasm, occurs very often.

Section and avulsion of the facial nerve is commonly performed in the region of the parotid gland. The upper branches of the nerve, which innervate the orbicularis oculi muscle, are resected for some length once they have been identified by means of an electrical stimulator, so that muscles of the lower face are spared (Gurdjian and Williams, 1928; McCabe and Boles, 1972; Coles, 1973; Weingarten and Putterman, 1976; Talbot *et al.*, 1982; McCord *et al.*, 1984a). Partial neurectomy of the branches of the facial nerves as they enter the orbicularis oculi muscles from the lateral side usually causes only slight and transient benefit (Talbot *et al.*, 1982).

In our own experience restoration of useful vision was achieved in 27 of 29 bilateral facial nerve avulsion operations. However 22 patients (75.9%) developed recurrence of spasms on average 1 year after surgery, due to reinnervation and recovery of at least partial muscle function. Unwanted side-effects were lower facial paralysis (11 cases), lagophthalmos (seven cases), persistent Bell's phenomenon (four cases), eyelid droop (three cases), ectropion (two cases), corneal exposure (one case) and accumulation of parotid secretion (one case). In addition, an inevitable consequence of denervation is an alteration in facial appearance. The Bell's phenomenon noted after surgery might be a dystonic epiphenomenon, more evident once spasms of eye closure have disappeared (Coles, 1977).

Orbicularis oculi myectomy involves the extirpation of all accessible orbicularis oculi and, in some circumstances, other eyebrow and eyelid muscles such as procerus and corrugator superciliaris (Fox, 1966; Castañares, 1973; Gillum and Anderson, 1981; McCord et al., 1984a,b). As it is virtually impossible to remove all orbicularis oculi, the potential benefit of this muscle stripping procedure depends on the degree of muscle resection achieved. Gillum and Anderson (1981) reported improvement in 10 patients treated with this technique with recurrence of blepharospasm in three cases. McCord et al. (1984a) found residual spasms combined with excessive blinking in nearly 25% of their patients treated by orbicularis oculi myectomy. In our series (Grandas et al., 1988) only two cases (25%) out of eight treated by the muscle-stripping procedure achieved some benefit, and recurrence of blepharospasm occurred in one of them. Side-effects were cosmetic problems due to changes in facial appearance, and local problems such as forehead numbness, exposure keratitis and corneal erosions.

Injections of botulinum toxin

Injection of type A botulinum toxin into the orbicularis oculi muscles has proved to be a major advance in the treatment of blepharospasm (Frueh et al., 1984; Scott et al., 1985; Elston and Russell, 1985; Mauriello, 1985; Tsoy et al., 1985; Dutton and Buckley, 1986; Jankovic and Orman, 1987). This procedure is simple, very effective and with little morbidity. Moreover, it can be performed in the out-patients clinic. Botulinum toxin injections may be considered as the treatment of choice for patients with disabling blepharospasm.

This topic is reviewed in depth in Chapter 25.

ADDENDUM

Since this Chapter was submitted results from a histological study in patients with cranial dystonia have been published (Gibb et al. (1988) *Movement Disorders* 3, 211–221). Post-mortem examination of a case of blepharospasm showed an angioma in the dorsal pons, but no abnormality was found in the basal ganglia and brainstem in two other cases.

REFERENCES

Alberca, R., Rafel, E., Chinchon, J., Vadillo, J. and Navarro, A. (1987) *J. Neurol. Neurosurg. Psychiat.* **50**, 1665–1668.

Alpers, B. J. and Patten, C. A. (1927) *Arch. Neurol. Psychiat.* **18**, 427–433.
Altrocchi, P. H. and Forno, L. S. (1983) *Neurology* **33**, 802–805.
Battista, A. F. (1982) In *Movement Disorders* (edited by C. D. Marsden and S. Fahn). pp. 319–321. Butterworths, London.
Brennan, M. J., Ruff, P. and Sandyk, R. (1982) *Br. Med. J.* **285**, 853.
Brin, M. F., Fahn, S., Bressman, S. B. and Burke, R. E. (1986) *Neurology* **36** (Suppl. 1), 119.
Burke, R. E., Fahn, S., Jankovic, J., Marsden, C. D., Lang, A. E., Gollomp, S. and Ilson, J. (1982) *Neurology* **32**, 1335–1346.
Casey, D. E. (1980) *Neurology* **30**, 690–695.
Castañares, S. (1973) *Plast Reconstr. Surg.* **51**, 248–253.
Chakravorty, N. K. (1974) *Postgrad. Med. J.* **50**, 521–523.
Coles, W. H. (1973) *South Med. J.* **66**, 1407–1411.
Coles, W. H. (1977) *Arch. Ophthalmol.* **95**, 1006–1009.
Corin, M. S., Elizan, T. S. and Bender, M. B. (1972) *J. Neurol. Sci.* **15**, 251–265.
Day, J. J., Lefroy, R. B. and Mastaglia, F. L. (1986) *J. Neurol. Neurosurg. Psychiat.* **49**, 1324–1325.
Dutton, J. J. and Buckley, E. G. (1986) *Arch. Neurol.* **43**, 380–382.
Duvoisin, R. C. (1983) *Clin. Neuropharmacol.* **6**, 63–66.
Elston, J. S. and Russell, R. W. (1985) *Br. Med. J.* **290**, 1857–1859.
Fahn, S. (1983) *Neurology* **33**, 1255–1261.
Fahn, S., Bresman, S., Burke, R., Hening, W., Ilson, J. and Walters, A. (1983) *Ann. Neurol.* **14**, 106.
Fox, S. (1966) *Arch. Opthalmol.* **76**, 318–321.
Frueh, B., Felt, D. P., Wojno, T. H. and Musch, D. (1984) *Arch. Ophthalmol.* **102**, 1464–1468.
García-Albea, E., Franch, O., Muñoz, D. and Ricoy, J. R. (1981) *J. Neurol. Neurosurg. Psychiat.* **44**, 437–440.
Gillum, W. and Anderson, R. L. (1981) *Arch. Opthalmol.* **99**, 1056–1062.
Grandas, F., Elston, J., Quinn, N. and Marsden, C. D. (1988) *J. Neurol. Neurosurg. Psychiat.* **51**, 767–772.
Gurdjian, E. S. and Williams, H. W. (1928) *J. Am. Med. Assoc.* **91**, 2053–2056.
Hotson, J. R., Langston, E. B. and Langston, J. W. (1986) *Ann. Neurol.* **20**, 456–463.
Jackson, J. A., Jankovic, J. and Ford, J. (1983) *Ann. Neurol.* **13**, 273–278.
Jankovic, J. (1982) *Ann. Neurol.* **11**, 41–47.
Jankovic, J. (1986) *Arch. Neurol.* **43**, 866–868.
Jankovic, J. and Ford, J. (1982) *Ann. Neurol.* **13**, 402–411.
Jankovic, J. and Orman, J. (1984) *Ann. Ophthalmol.* **16**, 371–376.
Jankovic, J. and Orman, J. (1987) *Neurology* **37**, 616–623.
Jankovic, J. and Patel, S. C. (1983) *Neurology* **33**, 1237–1240.
Jankovic, J. and Patten, B. M. (1987) *Movement Disorders* **2**, 159–163.
Jankovic, J., Havins, W. E. and Wilkins, R. B. (1982) *J. Am. Med. Assoc.* **248**, 3160–3164.
Karf, I. B. and Sharpe, J. A. (1981) *Lancet* **ii**, 957–958.
Keane, J. R. and Young, J. A. (1985) *Arch. Neurol.* **42**, 1206–1208.
Klawans, H. L. and Enrich, M. A. (1970) *Eur. Neurol.* **3**, 365–372.
Kurlan, R., Jankovic, J., Rubin, A., Patten, B., Griggs, R. and Shoulson, I. (1987) *Arch. Neurol.* **44**, 1057–1060.
Lang, A. E. (1986) *Can. J. Neurol. Sci.* **13**, 42–46.
Lang, A. E. and Sharpe, J. A. (1984) *Neurology* **34**, 1522.

Lang, A., Sheehy, M. and Marsden, C. D. (1982) *Can. J. Neurol. Sci.* **9**, 313–319.

Leenders, K. L., Frackowiak, R. S., Quinn, N., Brooks, D., Summer, D. and Marsden, C. D. (1986) *Movement Disorders* **1**, 51–58.

Marsden, C. D. (1976a) *J. Neurol. Neurosurg. Psychiat.* **39**, 1204–1209.

Marsden, C. D. (1976b) *Adv. Neurol.* **14**, 259–276.

Marsden, C. D., Lang, A. E. and Sheehy, M. P. (1983) *Neurology* **33**, 1100–1101.

Mauriello, J. A. (1985) *Neurology* **35**, 1499–1500.

McCabe, B. F. and Boles, R. (1972) *Ann. Otol.* **81**, 611–615.

McCord, C. D., Coles, W. H., Shore, J. W., Spector, R. and Putman, J. R. (1984a) *Arch. Ophthalmol.* **102**, 266–268.

McCord, C., Shore, J. and Putman, J. (1984b) *Arch. Ophthalmol.* **102**, 269–273.

Meige, H. (1910) *Rev. Neurol.* **21**, 437–443.

Merikangas, J. R. and Reynolds, C. H. (1979) *Ann. Neurol.* **5**, 401–402.

Micheli, F., Fernandez Pardal, M. M. and Leiguarda, R. C. (1982) *Neurology* **32**, 432–434.

Miller, E. (1973) *N. Engl. J. Med.* **289**, 697.

Nutt, J. G. and Hammerstad, J. P. (1981) *Ann. Neurol.* **9**, 189–191.

Nutt, J., Carter, J., De Garmo, P. and Hammerstad, J. (1984a) *Neurology* **34** (suppl. 1) 222.

Nutt, J. G., Hammerstad, J. P., De Garmo, P. and Carter, J. (1984b) *Neurology* **34**, 215–217.

Palakurthy, P. R. and Iyer, V. (1987) *Movement Disorders* **2**, 131–134.

Powers, J. M. (1982) *J. Am. Med. Assoc.* **247**, 3244–3245.

Powers, J. M. (1985) *Neurology* **35**, 282–284.

Schott, G. D. (1985) *J. Neurol. Neurosurg. Psychiat.* **48**, 698–701.

Scott, A. B., Kennedy, R. A. and Stubbs, H. A. (1985) *Arch. Ophthalmol.* **103**, 347–350.

Snider, S. R. and Consroe, P. (1984) *Neurology* **34** (Suppl. 1) 147.

Tablot, J. F., Gregor, Z. and Bird, A. C. (1982) In *Movement Disorders* (edited by C. D. Marsden and S. Fahn), pp. 322–329. Butterworths, London.

Tanner, C. M., Glantz, R. H. and Klawans, H. L. (1982) *Neurology* **32**, 783–785.

Tolosa, E. (1981) *Arch. Neurol.* **38**, 147–151.

Tolosa, E. S. and Lai, C. H. (1979) *Neurology* **29**, 1126–1130.

Tsoy, E. A., Buckley, E. G. and Dutton, J. J. (1985) *Am. J. Ophthalmol.* **99**, 176–179.

Weiner, W. J. and Nausieda, P. A. (1982) *Arch. Neurol.* **39**, 451–452.

Weiner, W. J., Nausieda, P. A. and Glantz, R. H. (1981) *Neurology* **31**, 1555–1556.

Weingarten, C. H. and Putterman, A. (1976) *Ear Nose Throat J.* **55**, 8–24.

21

The Treatment of Spasmodic Torticollis

Paul A. Cullis and Peter C. Walker

Spasmodic torticollis (ST) is the most common of the focal dystonias (Cullis, 1987). It is an involutary hyperkinesis manifesting itself by spasmodic or tonic contractions of the neck musculature, producing stereotyped abnormal deviations of the head, the chin being rotated to one side or the head bent forward or backward (Patterson and Little, 1943). Often there is an associated elevation of the shoulder, usually the one toward which the head deviates. Tremor occurs frequently in ST and it has been suggested that extranuchal dystonic symptoms are also common (Couch, 1976). The condition is not usually inherited, but familial cases do occur (Gilbert, 1977). Idiopathic ST must be differentiated from other conditions causing secondary torticollis, such as Wilson's disease, hyperthyroidism, tardive dyskinesia and syringomyelia, some of which are potentially treatable (Gilbert, 1972; Pagni et al., 1985; Landon and Cullis, 1987). The aetiology of ST is unknown, although abnormalities of cerebrospinal fluid coeruloplasmin have been reported (Kjellin and Stibler, 1975; Cullis et al., 1986). Good pathological studies have been rare, and the best one to date found neither macroscopic nor histological abnormalities (Tarlov, 1970). Certain neurophysiological abnormalities are now known to occur in patients with ST, but their significance is uncertain (Diamond et al., 1988). To some authors (Gildenberg, 1980) the sometimes bizarre nature of the involuntary movements in ST has

DISORDERS OF MOVEMENT: CLINICAL, PHARMACOLOGICAL
AND PHYSIOLOGICAL ASPECTS   ISBN 0-12-569685-X

suggested a psychogenic origin for the disorder, although most do not consider this to be the case (Marsden and Harrison, 1974; Matthews *et al.*, 1978). The low incidence of the disease, estimated at 11 per million per year, prevents generalists and most neurologists from becoming familiar with its treatment (Nutt *et al.*, 1988). Even those specializing in movement disorders still disagree to some extent on the pharmacotherapy of ST (Lal, 1979). For these reasons, it remains a special challenge to the primary care physician and even to most neurologists (Klassen, 1984).

In order to discover the oral pharmacotherapy favoured by most movement disorder specialists and to elucidate certain epidemiological factors such as the sex ratio, we sent questionnaires to 72 selected neurologists and neurosurgeons in North America and Great Britain who treat significant numbers of ST patients.

METHODS

We attempted to identify all neurologists and neurosurgeons in North America and Great Britain with a special interest in movement disorders. The process of developing a list began in 1984. Primarily by using current medical directories, the *Index Medicus* and all known bibliographies concerning torticollis and dystonia, an initial mailing list of 25 key physicians was created to be canvassed and to receive our first questionnaire. After developing an expanded address list of 72 physicians, a revised questionnaire was circulated. Each of the 53 neurologists on the list was asked the following questions: (1) how many patients with ST were seen over the 6-year period from 1979 to 1984; (2) how many of these had dystonia elsewhere; (3) how many were male and female; (4) what drugs were preferred; (5) what other treatments were recommended; (6) in how many cases was the condition thought to be psychogenic. Each of the 19 neurosurgeons on the list was asked the following questions: (1) how many patients with ST had surgery over the 6-year period from 1979 to 1984; (2) how many were male and female; (3) how many of these had dystonia elsewhere; (4) which surgical procedures were performed; (5) what was the rate of success; (6) how many patients had more than one operation; (7) how long was the follow-up; (8) what complications occurred. Once the replies were received, a more specific questionnaire on drugs was sent to the 10 neurologists who had seen the most ST patients over the survey period. Collectively they accounted for 2135 patients or 73% of the total. They were asked to rank from 1 to 5 the drugs or groups of drugs they found to be effective in decreasing order of effectiveness. The groups included: anticholinergics, benzodiazepines, muscle relaxants (baclofen), beta blockers (propranolol), tricyclic antidepressants, monoamine depleting agents (tetrabenezine and reserpine), anticonvulsants (carbamazepine and primidone), lithium, dopaminergic drugs (bromocriptine and L-dopa/carbidopa) and MAO inhibitors. If a reply was not received within a reasonable period of time, the physician was contacted by telephone to encourage a response. A weighted total was then developed for each group of drugs. Five points was given for each physicians's first choice; 4 for the second choice; 3 for the third choice; 2 for the fourth choice; and 1 for the fifth choice.

RESULTS

In all, 62 replies were received from 45 neurologists and 17 neurosurgeons. Ten physicians did not respond. Not every physician listed all five possible choices. Some listed more than five, but additional choices were discounted. Parenteral drugs were excluded, since the purpose of the study was to discover the oral pharmacotherapy favoured by most movement disorder specialists. Table 21.1 shows the responses received, in order of the number of ST patients seen during the study period. Physicians working together were grouped together for clarity. In total, 2923 patients were seen. Some of these, however, may have represented duplicates, since some patients may have been seen by two or more of the physicians involved. Sixty-four per cent had no dystonia elsewhere in the body. Fifty-eight per cent were female, giving a sex ratio of approximately 1.4:1 F:M. Only five cases (0.2%) were considered to be psychogenic.

Neurology

Table 21.2 shows the responses received from the 10 neurologists who had seen the most ST patients over the survey period. Five chose the anticholinergics as their first line treatment. Three chose the benzodiazepines. One chose the tricyclic antidepressants. One chose the beta blockers (propranolol). None chose the muscle relaxants (baclofen). The weighted totals are shown in Table 21.3. Anticholinergics were the favoured treatment. Benzodiazepines were a close second. Muscle relaxants (baclofen) and the tricyclic antidepressants were a distant third and fourth, respectively. All neurologists used drugs as their primary treatment. Except for botulinum toxin injections, no other treatments were recommended as being effective. There was wide agreement that drugs were effective in suppressing the dystonic movements in less than half, and probably only one-quarter–one-third, of patients. Drugs may be more effective than that in suppressing the painful dystonic spasms occurring in the trapezius muscles of many patients.

Table 21.1

Physician	Patients	Focal	Non-focal	Female	Male	Psychogenic
1	623	308	315	403	220	0
2	320	100	220	192	128	2
3	300	270	130	150	150	0
4	251	116	135	141	110	0
5	210	180	30	150	60	0
6 and 7	200	150	50	110	90	0
8	91	46	45	46	45	0
9	75	50	25	38	37	0
10	61	46	15	30	20	0
11	56	52	4	33	23	0
12	50	45	5	25	25	0
13	50	38	12	30	20	0
14	50	35	15	30	20	0
15 and 16	46	24	22	25	21	0
17 and 18	40	35	5	22	18	0
19	38	23	15	25	13	0
20	30	5	25	20	30	0
21	25	24	1	18	7	0
22	25	19	6	18	7	0
23	25	20	5	14	11	0
24	24	12	12	12	12	0
25	22	22	0	11	11	0
26	21	18	3	11	10	1
27	20	18	2	20	0	0
28	20	20	0	14	6	0
29	20	15	5	12	8	0
30	18	9	9	12	6	0
31	15	10	5	7	8	0
32 and 33	15	12	3	8	7	0
34	15	10	5	5	10	0
35	15	10	5	5	10	0
36	15	15	0	7	8	0
37	12	8	4	4	8	0
38	12	7	5	8	4	0
39	12	10	2	10	2	0
40	10	9	1	5	5	0
41	10	8	2	4	6	0
42	10	8	2	5	5	1
43	10	4	6	2	8	0
44	10	10	0	5	5	0
45	9	7	2	3	6	0
46	7	2	5	5	2	0
47	6	6	0	4	2	0
48	6	4	2	1	2	0
49	4	4	0	1	3	0
50	4	3	1	2	2	0
51	3	2	1	2	1	0
52	3	3	0	2	1	1
53	3	3	0	2	1	0
54	3	3	0	2	1	0
55	2	2	0	2	0	0
56	1	1	0	0	1	0
57	NA					
58	NA					
59	NA					
60	NA					
61	NA					
62	NA					
Totals	2923	1861	1062	1705	1218	5
		63.6%	36.3%	58.3%	41.6%	0.2%

NA, not available

Table 21.2

	Centre									
	1	2	3	4	5	6	7	8	9	10
Anticholinergics	4	1	1	1	1	1	2	2	2	4
Benzodiazepines	3	3	2	3	2	5	1	1	1	3
Baclofen		4	3	2		2			3	
Propranolol	1		4							
Antidepressants	2				3					1
Tetrabenazine		2								
Carbamazepine		5				4	3	4		
Primidone			5				4			
Reserpine				4	5					
Lithium					4					
Bromocriptine								3		
L-Dopa/carbidopa										
MAO Inhibitors										2

Table 21.3

Weighted totals	
Anticholinergics	41
Benzodiazepines	36
Baclofen	16
Antidepressants	12

Neurosurgery

Although the total was undoubtedly higher, we were able to identify only 17 neurosurgeons in North America and Britain who had performed invasive procedures for ST since 1978. In order of decreasing frequency, the procedures used were as follows.

(1) Anterior cervical rhizotomy (Fabinyi and Dutton, 1980; Maxwell, 1984). Approximately 100 procedures had been performed. The side-effects were heavily dependent on the exact details of the procedure. They included instability of the cervical spine, weakness of the diaphragm and reduced range of motion. Patient selection appeared to play a major role in determining success. A success rate of 50–90% was claimed.

(2) Selective cervical denervation (posterior ramicectomy) (Bertrand et al., 1982). Ninety-seven procedures had been performed (91 of which were by a single surgeon). The results appeared to be good with few side-effects. Eighty-eight per cent were rated "excellent or very good" results. The procedure, however, was technically very difficult to perform and required the cooperation of an electromyographer and considerable practice on cadavers.

(3) Spinal electrode implants (spinal cord stimulation). Sixty-two procedures had been performed (61 of which by a single surgeon). Good results were claimed: 51% marked improvement; 20% moderate improvement; and 12% mild improvement. However a more recent double-blind study has not been able to duplicate those results (Goetz et al., 1988). The side-effects are minimal.

(4) Stereotactic thalamotomy. Few were done. The exact numbers were not available. The procedure must be done bilaterally and is now considered too risky by most neurosurgeons, as the incidence of speech difficulty afterwards may be higher than 20%.

The other questions posed to the neurosurgeons remained largely unanswered.

CONCLUSIONS

Data were obtained on 2923 patients with ST in North America and Britain. Drugs were effective in suppressing the dystonic movements in less than half the cases. Anticholinergics were the favoured treatment. Benzodiazepines were a close second. Muscle relaxants (baclofen) and the tricyclic antidepressants were a distant third and fourth, respectively. The treatment varied substantially from centre to centre. The sex ratio was 1.4:1 F:M. The disease was rarely thought to be psychogenic. One-third of patients had dystonia elsewhere. Anterior cervical rhizotomy was the surgical procedure most commonly performed, but was not without considerable side-effects. Selective cervical denervation (posterior ramicectomy) was thought to be effective with few side-effects, but the procedure was technically difficult to perform and was not readily available.

POSTSCRIPT

Since this study was undertaken, the pharmacotherapy of ST has been changed somewhat by the more widespread, but still experimental use of botulinum toxin. The toxin is injected into the dystonic neck muscles, which become paralysed. The results have been encouraging and, on average, the effect appears to last about three months (Gelb et al., 1988; Kang et al., 1988). The long-term efficacy of the injections remains to be assessed, however. See also Chapter 25.

REFERENCES

Bertrand, C., Molina Negro, P. and Martinez, S. N. (1982) *Appl. Neurophysiol.* **45**, 326–330.

Couch, J. R. (1976) *Adv. Neurol.* **14**, 245–258.

Cullis, P. A. (1987) In *Current Therapy in Neurologic Disease, Vol. 2* (edited by R. T. Johnson), pp. 238–241. B. C. Decker, Toronto.

Cullis, P. A., Townsend, L., LeWitt, P., Pomara, N. and Reitz, D. (1986) *Movement Disorders* **1**, 179–186.

Diamond, S. G., Markham, C. H. and Baloh, R. W. (1988) *Arch. Neurol.* **45**, 164–169.

Fabinyi, G. and Dutton, J. (1980) *Aust. N. Z. J. Surg.* **50**, 155–157.

Gelb, D. J., Lowenstein, D. H. and Aminoff, M. J. (1988) *Neurology* **38** (Suppl. 1), 244.

Gilbert, G. J. (1972) *Arch. Neurol.* **27**, 503–506.

Gilbert, G. J. (1977) *Neurology* **27**, 11–13.

Gildenberg, P. L. (1980) *Appl. Neurophysiol.* **44**, 233–243.

Goetz, C., Penn, R. D. and Tanner, C. M. (1988) *Adv. Neurol.* **50**, 645–649.

Kang, U. J., Greene, P., Fahn, S., *et al.* (1988) *Neurology* **38** (Suppl. 1), 244.

Kjellin, K. G. and Stibler, H. (1975) *Eur. Neurol.* **13**, 461–475.

Klassen, A. (1984) *Postgrad. Med.* **75**, 124–125.

Lal, S. (1979) *Can. J. Neurol. Sci.* **6**, 427–435.

Landan, I. and Cullis, P. A. (1987) *Movement Disorders* **2**, 317–319.

Marsden, C. D. and Harrison, M. J. G. (1974) *Brain* **97**, 793–810.

Matthews, W. B., Beasley, P., Parry-Jones, W. and Garland, G. (1978) *J. Neurol. Neurosurg. Psychiat.* **41**, 485–492.

Maxwell, R. E. (1984) *Postgrad. Med.* **75**, 147–155.

Nutt, J., Muenter, M., Melton, L. J., Aronson, A. and Kurland, L. T. (1988) *Adv. Neurol.* **50**, 361–365.

Pagni, C. A., Naddeo, M. and Faccani, G. (1985) *J. Neurosurg.* **63**, 789–791.

Patterson, R. M. and Little, S. C. (1943) *J. Nerv. Ment. Dis.* **98**, 571–599.

Tarlov, E. (1970) *J. Neurol. Neurosurg. Psychiat.* **33**, 457–463.

22

Surgical Treatment of Spasmodic Torticollis

J. J. Maccabe

The cause of spasmodic torticollis is unknown. A focal manifestation of torsion dystonia usually presenting after 40 years of age, no satisfactory pathological or pharmacological explanation exists for the characteristic movements of head and neck. These usually begin insidiously, sometimes apparently after trauma, with complaints of stiffness or pulling or drawing sensations in the neck. Gradually jerking or more sustained turning movements of the head become established, the chin deviating to one side, perhaps with elevation of the shoulder. Though the sternomastoid on one side and the posterior cervical muscles on the other are usually the dominant groups this is by no means the only combination (Bertrand *et al.*, 1978). In this worker's extensive experience the combination of muscles involved varies almost *ad infinitum*. The severity of the movements also differs from patient to patient. Stress will usually aggravate the contractions, while in many patients placing the index finger on the chin will serve to reduce them (the *geste antagoniste*). In some patients severe sustained deviation of the head makes even eye to eye contact difficult. Ante-, and retro-collis are rare variants.

A detailed account of the natural history of the condition was first attempted by Meares (1971). In a prolonged study of 41 patients he identified three stages of the disorder. First a time of slow progression lasting some 5

DISORDERS OF MOVEMENT: CLINICAL, PHARMACOLOGICAL AND PHYSIOLOGICAL ASPECTS   ISBN 0-12-569685-X

years, followed by a stationary period of similar duration, with finally a phase where slight improvement might occur. Matthews *et al.* (1978) surveying 30 patients came to a similar conclusion. In a recent study of 25 patients conservatively treated and followed for some 9 years symptoms and signs remained unchanged in eight. Only one patient deteriorated whereas most of the remaining 16 improved significantly (Namba *et al.*, 1986). For the majority of patients it seems that after the initial deterioration symptoms persist largely unchanged over many years. Later some 10% or more will display evidence of other movement disorders — tremor, facial spasms, writer's cramp, or dystonia of the shoulder, arm or trunk.

Temporary complete remission is not unknown, especially in the early years of the illness, but relapse is also frequent. The highest frequency of sustained remission (23%) in a study of 26 patients was reported by Jayne *et al.* (1984). Relapse was recorded in two patients after intervals of 8 and 13 years.

The manifestations of the established condition vary widely from patient to patient. For patients with moderate symptoms a full life and even a normal working pattern are not impossible. For those severely affected the handicaps both economic and social may be severe (Matthews *et al.*, 1978). The adverse effects on a professional life have been described with feeling (O'Riordan, 1988). Pain in the shoulder and neck due to continuous muscular activity may also be a source of distress. Aggravated spondylotic change in the cervical spine may result in radiculo-myelopathy.

STEREOTAXIC SURGERY

Cooper (1965) was the first to report a series of patients with torticollis treated by a stereotaxic method. From observations on neck hyperkinesis made on patients with Parkinson's disease and dystonia treated by thalamotomy, he concluded that beneficial effects might also be observed in torticollis sufferers. The immediate results of surgery were often dramatic, but longer follow-up very soon revealed the difficulty of achieving permanent benefit. Bilateral thalamotomy, though more effective, left some 20% of patients with an unacceptable pseudobulbar deficit.

Following Cooper's lead, several targets for thalamotomy were proposed, each with its individual champion. Inconsistent results combined with serious side-effects however have served steadily to reduce the number of patients being referred for this form of surgery. The results of surgery in 22 patients have been reported in detail by Andrew *et al.* (1983). Few neurologists refer torticollis patients for stereotaxic treatment today. Many former advocates of this form of surgery now express different opinions.

PERIPHERAL SURGERY

Peripheral procedures designed to reduce or abolish the activity of the participating muscles, by myotomy or by denervation, have survived the test of time somewhat better. The operation to denervate the sternomastoid muscle is generally attributed to Campbell de Morgan (1866). Prompted by his neurological colleague Weir Mitchell, W. W. Keen (1891) described a procedure to denervate the equally important posterior muscles of the neck. The posterior divisions of the first three cervical nerves were divided on one side only to denervate selectively the more active posterior group. This was an extraspinal procedure, designed to complement denervation of the sternomastoid muscle.

Over the next several decades most procedures for torticollis followed the method pioneered by Keen, the most successful practitioners being Finney and Hughson. They were the first to section the nerves on both sides, a procedure they combined with accessory nerve section in the neck. Of 32 patients they reported 28 as relieved or improved, though some required more than one operation to achieve a satisfactory outcome (Finney and Hughson, 1925).

Impressed by these results, and believing that success owed everything to the bilateral nerve section, Walter Dandy advocated instead an intradural operation, sectioning the anterior motor roots C1–C3 and combining this with division of both accessory nerves in the neck (Dandy, 1930 *et seq*). Dandy justified this extensive denervation by pointing to the widespread muscle activity in torticollis. While admitting that the head might seem a little insecure afterwards, he believed this difficulty could be gradually overcome by post-operative re-training. Keen's operation was recognized as tedious and difficult. Dandy's intradural approach made the whole matter much simpler, and it became firmly established as the standard procedure, a position which it still holds today. The only modification was that by McKenzie (1955) who advocated selective section of the spinal accessory supply to the involved sternomastoid muscle as the initial procedure, to allow easier assessment of the activity of the posterior muscles.

The standard rhizotomy consists of sectioning the cervical anterior roots C1–C3 on both sides. An operating microscope is essential to see and preserve the fine vessels accompanying the rootlets, though a major vascular supply to the cord rarely accompanies an upper cervical root (Turnbull *et al.*, 1966). Identification of the first cervical root may be difficult. Failure to divide it will mean that the suboccipital muscles will continue to be active. There is an important contribution from the spinal accessory to the first root, present in some 50% of patients, which must also be divided (McKenzie, 1955). The same author advocated section of the fourth cervical root on the side showing greater posterior neck muscle activity. Bilateral section of C4 is also possible but may lead to greater neck instability and risk of respiratory complication from phrenic nerve denervation (Tasker, 1976). The spinal root of the accessory nerve may be divided at the level of the foramen magnum at the same operation. Bulbar tributaries of the vagus which travel with the accessory nerve may be at risk, however, and their inadvertent damage has been held responsible for post-operative dysphagia. Intraspinal section of the accessory nerve very often results in incomplete sternomastoid denervation (Hayward, 1985). The accessory nerve may be divided in the neck as Dandy preferred, though the infrequent involvement of the trapezius muscle suggests that McKenzie's refinement is better.

What can be said of this more or less standard operation? Many reports record beneficial effects from the widespread bilateral denervation. Hamby and Schiffer (1970) claimed that 79% of their patients considered themselves much improved, though residual head rotation was present in 53%. Arseni and Meretsis (1971) reported total recovery in 40%, while Tasker (1976) noted an immediate impressive amelioration of symptoms in 29 out of 37 patients followed in detail. Walsh (1976) had 79% excellent results in 33 patients available for later assessment. Scoville and Bettis (1979) claimed eminently satisfactory results with a majority of 23 patients. Fabinyi and Dutton (1980) had 18 good results out of 20, and Moossy and Nashold (1984) 10 out of 13. Speelman et al. (1987) found all dyskinetic movement abolished in seven out of eight patients.

Such claims need qualification. What is meant by an "excellent" or "satisfactory" result is not always sharply defined certainly in respect of involuntary movement — usually the principal reason for surgery. Walsh stated clearly that 26 patients had no involuntary movement. Nor are the results often set in relation to what must be variable degrees of pre-operative disability, except notably by Speelman who adapted a scoring scale from Marsden and Fahn. Criteria for recommending surgery are equally unclear — some operations being made for disability described as "real, severe or established", or for pain, while others require the patient to "earn the operation". Early surgery may be recommended to avoid the development of a fixed neck deformity.

The various combinations of cervical rhizotomy and accessory nerve section make comparison of results from different series difficult. Who assesses the results and when are also important considerations. Hamby and Schiffer found evaluation of their own results difficult. Some patients with what the authors regarded as an excellent objective result remained unhappy, while others more severely affected were satisfied. At long-term follow-up not all their patients were glad that they had submitted to surgery, an experience shared by other authors. The difficulty of summarizing the benefits against the pre-operative state can be appreciated from the individual patient details and comments provided by Scoville and Bettis (1979) and by Speelman *et al.* (1987).

That bilateral upper cervical denervation can lead to cessation or reduction of involuntary movement does seem certain but not all patients benefit. Some 75% experience improvement in the broadest sense. Five to 15% gain no benefit, their movements being unaffected. For a substantial number there are side-effects or complications to surgery. Dysphagia — noted in Dandy's first report, is not rare — Hamby and Schiffer reported over 30% with this complaint. Walsh mentions 10 out of 33, with a major problem in only one. Many authors do not comment on this problem. Various explanations have been offered for the occurrence of dysphagia — from inability to lift the head in swallowing to interference with the nerve supply to the infrahyoid muscles.

Weakness of the neck muscles is a regular observation. Most authors state the need for post-operative physiotherapy to re-educate remaining muscles. Some advocate a period wearing a collar for support, presumably in expectation of compensatory fibrosis and stiffness. Successful surgery often means limited neck mobility, with stiffness and pain. An unstable neck, occasionally requiring fusion, is reported by more than one author. Neck pain may be relieved by operation and also may result from it (Speelman *et al.*, 1987). The most serious complications have followed from what appear to be vascular causes, with effects at various levels from the pons to the upper cervical cord. While some have suffered temporary deficits others have sustained serious permanent disability, occasionally leading to death. The various mechanisms which may be responsible for this group of complications have been comprehensively reviewed by Adams (1984). Death has also occurred in the immediate post-operative period from other causes, listed as cerebral ischaemia, sinus thrombosis, pulmonary embolus, etc.

From the series available it is clear that bilateral upper cervical rhizotomy with accessory nerve division gives results in terms of relief of involuntary movement which are by no means uniform or certain. While movements can be reduced in the average patient their abolition is less likely except at the expense of an immobile, stiff and perhaps painful neck, with some swallowing difficulty. This was the basis of the criticism first levelled against the procedure by Meares, namely that not only did the operation often fail to abolish the distressing movements but that those submitted to surgery sometimes fared worse, sustaining additional disability not seen in the unoperated group.

Though the Dandy operation has probably provided the most relief so far, it is not unreasonable to question the use of a routine surgical approach for a condition whose clinical manifestations vary so widely. Many neck muscles are involved, as Dandy observed, but not equally so. Movements and posture vary. McKenzie was probably the first to suggest a more selective approach, when he recommended as a first step denervation of the more involved sternomastoid muscle, arguing that it was then easier to decide on the relative contribution of the posterior neck muscles. He also advised sparing the infrequently involved trapezius. Following a major complication Scoville also adopted a more cautious approach. Bilateral microsurgical lysis of the spinal accessory nerve roots was introduced in the belief that the cause of spasmodic torticollis was located peripherally in the craniocervical region and because of unhappy experience with cervical rhizotomy (Freckmann et al., 1981, 1986). The first to revert to a selective approach to denervation of the posterior neck muscles were Bertrand et al. (1978). Employing cervical posterior primary root division initially to complement unilateral stereotaxic surgery, within a few years this advocate of stereotaxis confirmed that selective denervation had become practically the only surgical treatment for spasmodic torticollis in his clinic (Bertrand et al., 1978, 1982).

The success of selective denervation depends on a thorough clinical and electromyographic assessment to identify accurately all the muscles involved in the abnormal movement. Those muscles not involved should be spared. In rotational torticollis the ipsilateral posterior cervical group and the contralateral sternomastoid muscle are usually responsible, whereas in laterocollis it is usually the posterior muscles ipsilateral to the inclination of the head. The sternomastoid muscle on either side may be involved. In retrocollis both posterior cervical groups may require denervation.

For unilateral denervation of the posterior cervical muscles the patient is placed either in the lateral or sitting up position. No paralysing agents are employed in anaesthesia, so that unipolar stimulation can be used to help locate neural elements. To display the suboccipital triangle and the posterior primary rami of C3–C6, the semispinalis is freed from its occipital attachment and reflected laterally (Fig. 22.1). Unilateral ramicectomy can go as far as C6 — providing more extensive relevant denervation especially of splenius and semispinalis than is possible by the intradural approach. The first cervical root is divided as it emerges over the posterior arch of the atlas, below the vertebral artery. The second cervical nerve is found immediately at its point of emergence between the atlas and the axis. The lower nerves are avulsed as they course round the lateral side of the facet joints. No laminectomy is required, so there is no risk to spinal cord function. Loss of the cutaneous distribution of the greater occipital nerve is the only side-effect of the operation. To denervate the sternomastoid muscle the accessory nerve is exposed throughout its course in the muscle, which is divided transversely. All branches supplying sternomastoid are identified by stimulation and divided, while the nerve to trapezius is preserved.

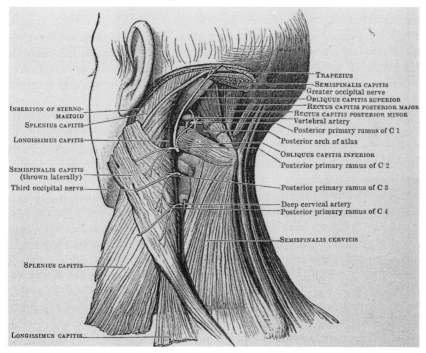

Fig. 22.1   Suboccipital triangle of the left side. (Reproduced from *Cunningham's Textbook of Anatomy*, 8th edition (eds. J. C. Brash and E. B. Jamieson). 1946. © Oxford University Press)

Achieving total denervation of the selected muscles is not as easy as a survey of anatomical texts might suggest. Several hours of painstaking dissection and stimulation are required to find all the supply to the posterior group, C1 being particularly important and difficult. Fixed neck deformity can create an additional problem of access. Post-operative electromyography may indicate surviving fibres and the need for re-exploration. Even denervation of the sternomastoid muscle can sometimes prove incomplete, due to reinnervation or to the presence of unsuspected aberrant nerve supply. Re-education of the remaining active muscles after surgery is always an essential part of treatment.

Bertrand *et al.* (1987) reported their results with 131 patients treated exclusively by selective denervation during the last 10 years. Abnormal movements were abolished completely in 48 patients and were reduced to insignificance in 67 — a good outcome for 115 (88%) while preserving posture and mobility. In those judged unsuccessful residual innervation or deliberate limitation of denervation were the explanations. Only two patients were unchanged. Six patients required re-operation. There were no complications and no serious sequelae. Reviewing the relative incidence of the various directions of abnormal movements these authors noted that flexion was only encountered in association with strong sternomastoid action in rotational problems. No patient was seen with pure anterocollis. They took this as evidence that the pre-vertebral muscles inserted close to the foramen magnum do not have an important role in spasmodic torticollis.

In a small series of 15 patients where a staged approach to selective denervation was adopted, the effectiveness of this technique in abolishing or significantly reducing the strength of the involuntary movements has similarly been demonstrated (J. J. Maccabe and C. D. Marsden, unpublished observations). Some patients gained what they regarded as acceptable relief from pain and movement after sternomastoid denervation alone. Denervation of posterior neck muscles, however, produced change convincing to both patient and observers, provided it was complete. Incomplete denervation, as judged by post-operative electromyography, resulted only in temporary or minimal benefit.

The prevailing conservative attitude to surgical treatment for spasmodic torticollis can be attributed not only to the inconsistent results of cervical rhizotomy but also to the well-known side-effects and complications of that procedure. An ideal operation should abolish abnormal movement or make its control easy consistently and without risk of imposing additional problems. There can be no justification for treating all patients by a routine method, when clinical and electromyographic evidence makes it clear that movement patterns vary widely. Nor is it reasonable to denervate muscles which do not take part in the involuntary movement. A selective approach to denervation offers the prospect of more certain benefit without major side-effects. The uninvolved muscles are left intact so that the neck remains stable. The important active muscle groups can be identified for each patient by careful clinical and electromyographic assessment. Denervation can be accomplished at one sitting though there are probably advantages to staging, with any lesser procedure coming first. The patient should understand that achieving complete denervation is not easy, and that reoperation may be required. Re-education of neck muscles must follow surgery.

If the management of patients with spasmodic torticollis is to improve, methods must be developed for measuring the degree of disability before treatment, and for objective evaluation afterwards. Speelman et al. (1987) adapted the scoring scale of Marsden and Fahn in an attempt to measure the seriousness of patient disability, before and after operation. Bertrand et al. (1978) employed a simple four-stage scale to assess outcome for involuntary movement. The value of such methods is dependent on minimal observer variability. An original development using computer analysis of pre- and post-operative video tape recordings of patients was described by Lye et al. (1987). The graphic presentation of oscillation of a reference marker on the patient's face enabled an assessment of the effect of surgery in reducing movement in the plane selected. The results of this analysis corresponded closely with the clinical impression of both patient and examiner. Techniques of this kind can supplement existing clinical and electromyographic methods.

REFERENCES

Adams, C. B. T. (1984) *J. Neurol. Neurosurg. Psychiat.* **47**, 990–994.
Andrew, J., Fowler, C. J. and Harrison, M. J. G. (1983) *Brain* **106**, 981–1000.
Arseni, C. and Maretsis, M. (1971) *Neurochirurgia* **14**, 177–180.
Bertrand, C., Molina-Negro, P. and Martinez, S. N. (1978) *Appl. Neurophysiol.* **41**, 122–133.
Bertrand, C., Molina-Negro, P. and Martinez, S. N. (1982) *Appl. Neurophysiol.* **45**, 326–330.
Bertrand, C., Moline-Negro, P., Bouvier, G. and Gorczyca, W. (1987) *Appl. Neurophysiol.* **50**, 319–323.
Cooper, I. S. (1965) *J. Neurol. Sci.* **2**, 520–533.
Dandy, W. E. (1930) *Arch. Surg.* **20**, 1021–1032.
de Morgan, C. (1866) *Br. For. Med.-Chir. Rev.* **38**, 218.
Fabinyi, G. and Dutton, J. (1980) *Aust. N. Z. J. Surg.* **50**, 155–157.
Finney, J. M. T. and Hughson, W. (1925) *Ann. Surg.* **81**, 255–269.
Freckmann, N., Hagenah, R., Herrmann, H.-D. and Muller, D. (1981) *Acta Neurochir. (Wien)* **59**, 167–175.
Freckmann, N., Hagenah, R., Herrmann, H. D. and Muller, D. (1986) *Acta Neurochir. (Wien)* **83**, 47–53.
Keen, W. W. (1891) *Ann. Surg.* **13**, 44–47.
Hamby, W. B. and Schiffer, S. (1970) *Clin. Neurosurg.* **17**, 28–37.
Hayward, R. (1986) *J. Neurol. Neurosurg. Psychiat.* **49**, 951–953.
Jayne, D., Lees, A. J. and Stern, G. M. (1984) *J. Neurol. Neurosurg. Psychiat.* **47**, 1236–1237.
Lye, R. H., Rootes, M. and Rogers, G. W. (1987) *Surg. Neurol.* **27**, 357–360.
Matthews, W. B., Beasley, P., Parry-Jones, W. and Garland, G. (1978) *J. Neurol. Neurosurg. Psychiat.* **41**, 485–492.
McKenzie, K. G. (1955) *Clin. Neurosurg.* **2**, 37–43.
Meares, R. (1971) *Lancet* **ii**, 149–151.
Moossy, J. J. and Nashold, B. S. (1984) *Neurosurgery* **15**, 269.
Namba, S., Wani, T., Masaoka, T., Nakamura, S. and Nishimoto, A. (1986) *No To Shinkei* **38**, 121–128.
O'Riordan, J. E. G. (1988) *Br. Med. J.* **296**, 1599.
Scoville, W. B. and Bettis, D. B. (1979) *Acta Neurochir.* **48**, 47–66.
Speelman, J. D., Van Manen, J., Jacz, K. and Van Beusekom, G. T. (1987) *Acta Neurochir.* Suppl. 39, 85–87.
Tasker, R. R. (1976) In *Current Controversies in Neurosurgery* (edited by T. P. Morley), pp. 443–447. W. B. Saunders, Philadelphia.
Turnbull, I. A., Brieg, A. and Hassler, O. (1966) *J. Neurosurg.* **24**, 951–965.
Walsh, L. S. (1976) *Proc. Inst. Neurol. Madras* **6**, 36–41.

23

Drug Treatment of Dystonia

Anthony E. Lang

INTRODUCTION AND GENERAL COMMENTS

Dystonia represents one of the most difficult management problems in movement disorders. In most cases, the abnormalities of movement cause significant disability to the patient. Multiple trials of diverse pharmacological agents are frequently necessary, and even then the response to treatment is often less than satisfactory. The literature abounds with reports of successful drug therapy in patients suffering from dystonic syndromes; however, several problems become evident when one attempts to interpret this literature.

First and foremost, it is important to remember that dystonia is not a single entity. This form of dyskinesia may occur in a wide variety of disorders (Calne and Lang, 1987). Although it is hoped that a better understanding of idiopathic or primary torsion dystonia will be obtained through the study of various forms of secondary dystonia, one cannot assume that the underlying biochemical disturbances and pharmacological responses are similar. The diagnosis of one form of secondary dystonia relies on a pharmacological marker, the response to L-dopa ("dopa-responsive dystonia"; see Chapter 24 by Nygaard in this volume). These patients also benefit from low doses of anticholinergic drugs and other dopamine agonists. The inclusion of such patients in reports of pharmacological trials in "idiopathic" dystonia complicates interpretation of the literature. Further confusion occurs when one considers the possibility that patients with disorders mimicking dystonia (Calne and Lang, 1987) might also be included in reported drug trials. This particularly applies to the uncommon but well-recognized problem of psychogenic dystonia. Finally, even if one can be assured of the diagnostic "purity" of the treated population as suffering from primary dystonia, further possible patient heterogeneity must be considered. Although strong arguments can be made for including various forms of generalized dystonia under a single diagnostic umbrella (Marsden, 1976), and there is little existing pharmacological evidence to suggest otherwise, it is still possible that pharmacological differences do exist between these subgroups. Pharmacological heterogeneity may even exist within individual forms of focal or segmental dystonia. For example, differing responses to drug therapy have encouraged the suggestion that there may be distinct pharmacological subtypes of cranial dystonia (Sechi et al., 1983). Finally, it may even be possible that individual movements in a single patient may have differing pharmacological bases. An example of this was seen in a patient of ours whose blepharospasm improved on a combination of tetrabenazine and α-methylparatyrosine, whilst his oromandibular dystonia worsened (Lang and Marsden, 1982).

Distinct from these factors of possible diagnostic and pharmacologic heterogeneity, several other problems exist in interpreting the literature on the treatment of dystonia. Most studies include only small numbers of patients. Considering the common tendency to report only therapeutic successes there is a strong potential for reviews which accumulate all previous case reports to be weighted in favour of drug response. Most of the studies with larger patient numbers are retrospective reviews of drug responses (Eldridge, 1970; Marsden, 1981; Lang et al., 1982; Greene et al., 1987) which suffer from all of the problems inherent to this approach. Most studies have been open-labelled, failing to take into account potential placebo response and the great natural variability possible in dystonia. The latter factor includes the well-recognized tendency for spontaneous remission in a certain proportion of these patients, most notably those with spasmodic torticollis. Another phenomenon observed in a small number of cases is what we have termed a "drug-induced remission" (Lang and Marsden, 1982). Here, the patient gradually improves in a dose-dependent fashion with the active drug (usually an anticholinergic or a dopamine antagonist). However, the benefit is maintained when the drug is withdrawn or a placebo is substituted. These potential confounding factors emphasize the need for placebo-controlled double-blind cross-over studies in larger numbers of well-characterized patients. One could even argue the need for the use of an "active" placebo in view of the difficulty maintaining the blind considering the high incidence of side-effects with drugs known to be effective in dystonia.

A response to a specific medication has often encouraged the formulation of hypotheses explaining the possible underlying biochemical disturbances in dystonia. A final source of confusion in the treatment of dystonia involves the possible mechanisms of action of various drugs. It is well known that sedation non-specifically improves dystonia, while anxiety and emotional upset often aggravate the movements. Several pharmacological conclusions have been drawn from acute studies in which patients were parenterally challenged with various drugs. However, the potential role of these non-specific effects often has not been taken into consideration. For example, we found that a certain proportion of patients benefitted from a challenge dose of an intravenous anticholinergic. However, the majority of those improving experienced concomitant sedation (Lang et al., 1983). Additional erroneous conclusions concerning the underlying biochemical disturbance in dystonia may occur if the effect of the drug in question is actually different from its primary mechanism of action. Aside from non-specific sedating and exciting effects, other examples of this might include the dopamine antagonist property of low doses of dopamine agonists and the possibility that extremely high doses of anticholinergics improve dystonia via effects on other transmitter systems distinct from those utilizing acetylcholine.

The reader must be cognizant of all of these potential sources of confusion in reviewing the literature of drug treatment in dystonia. Unfortunately, in recent years there have been more reviews of the treatment of dystonia published than carefully designed and implemented pharmacological trials. Because there are several recent comprehensive reviews of the field (Lang, 1985; Fahn and Marsden, 1987; Greene et al., 1987; Lang, 1987) I will limit the discussion in the following section on classes of drugs to selected controversial issues.

ANTICHOLINERGICS

The use of anticholinergics in dystonia dates back to a member of the first reported family with idiopathic generalized dystonia, Wolf Lewin, who improved with scopolomine (Regensburg, 1930; Jankowska, 1934). However, in low doses this treatment has rather disappointing results (Lang et al., 1982). A striking benefit obtained with the dosages usually used in Parkinson's disease should suggest the possibility of a diagnosis of dopa-responsive dystonia.

Based on the remarkable response of neuroleptic-induced acute dystonic reactions to parenteral anticholinergic therapy, Fahn introduced the concept of high-dosage anticholinergic treatment in dystonia (Fahn, 1979). Since then, a small number of studies have confirmed this response in a large number of patients using both open-labelled (Fahn, 1983; Marsden *et al.*, 1984; Lang, 1986; Greene *et al.*, 1987) and double-blind (Burke *et al.*, 1986) methodology. Approximately 50% of patients with childhood-onset dystonia and 40% of adult-onset patients obtain a moderate to marked improvement. Analysing a wide variety of factors which might have influenced the response to long-term open-labelled therapy in 227 patients with dystonia, Greene *et al.* (1987) found that only a duration of dystonia of less than 5 years significantly and positively affected the likelihood of improvement of anticholinergic medications.

Occasionally, dosages over 100 mg of trihexyphenidyl are required to obtain this response. Both peripheral and central adverse effects may limit the total dose and children frequently tolerate higher dosages than adults. The peripheral side-effects such as blurred vision and dry mouth may be controlled using pilocarpine eye drops or a concomitant peripherally acting anticholinesterase drug. However, central adverse effects such as drowsiness, confusion, altered sensorium, forgetfulness, behavioural changes, hallucinations and rarely even chorea (Lang, 1986) usually require a reduction in dosage. To date, no additional long-term side-effects have occurred with the high doses used, although this remains a significant concern. Also of concern is the potential for subtle effects on memory and cognition which may interfere with a child's learning capabilities and an adult's ability to function in the work place or at home. These cognitive effects, particularly with respect to memory, are currently under study.

In order to lessen adverse effects, it often takes considerable time and patience to reach the doses required for benefit. As mentioned above, Fahn's group has found little in the way of predictive factors aside from a shorter disease duration. One would hope that a response to a parenteral anticholinergic challenge might be useful in predicting the benefit from oral therapy, allowing those failing to improve to avoid the extensive time commitment necessary for this treatment. Indeed, Tanner and her colleagues have found that an intramuscular injection of scopolamine $0.01 \, \text{mg} \, \text{kg}^{-1}$ predicts the outcome of high-dosage chronic oral trihexyphenidyl in approximately 80% of patients (Tanner *et al.*, 1987). However, this still leaves some patients whose response cannot be predicted. On the other hand, as mentioned earlier, our studies found that most patients obtaining a benefit with intravenous anticholinergics did so at the expense of, and probably due to, sedation.

The mechanism of action of anticholinergics in dystonia remains a mystery. One of the most obvious obstacles here is our lack of understanding of the underlying pathological and/or biochemical disturbances in idiopathic dystonia. To date, there has been no evidence for a cholinergic disturbance in living patients in studies of platelet acetylcholinesterase function, CSF biochemical parameters (S. Fahn, personal communication) or erythrocyte membrane cholinesterase (Maltese *et al.*, 1985). In their detailed biochemical assessment of the brains of two dystonic patients, Hornykiewicz and his colleagues found no change in choline acetyltransferase in either cerebral cortex or basal ganglia (Hornykiewicz *et al.*, 1986).

A reliable animal model of dystonia is also lacking. Changes in central cholinergic systems have not been reported in the dystonic (dt) rat mutant. Large doses of scopolamine may reduce paw clasping and falls, however axial twisting worsens (Lorden, 1987). It has been suggested that mouthing movements in rodents may provide a useful testing ground for drugs which affect the cholinergic system (see Chapter 29 by Jenner in this volume). In the future, this model might be explored with respect to idiopathic dystonia.

DRUGS AFFECTING DOPAMINE SYSTEMS

There is more evidence, at both basic and clinical levels, to support an involvement of dopamine systems in the pathogenesis of dystonia than for any other neurotransmitter. Most compelling of all is the occurrence of dystonia as a complication of both dopamine agonist (in Parkinson's disease) and dopamine antagonist (acute and tardive dystonia) drugs and the occurrence of one form of juvenile-onset dystonia which is exquisitely sensitive to dopamine agonist therapy (see Chapter 24 by Nygaard in this volume). This information and the results of dopamine agonist and antagonist therapy in idiopathic dystonia have been reviewed previously (Lang, 1985, 1987). Only a few relevant points will be made regarding the use of each of these therapies here.

Dopamine agonists

All patients presenting with childhood-onset dystonia beginning in the legs, independent of whether or not they demonstrate diurnal fluctuations in the severity of dystonia, deserve an early trial of L-dopa therapy since 5–10% of these patients are dopa-responsive (Nygaard *et al.*, 1987). Unfortunately, in the larger population of patients with idiopathic torsion dystonia, favourable response to this therapy is uncommon. The response to direct dopamine receptor agonists such as bromocriptine seems equally disappointing. Because most reports of treatment with L-dopa and other dopamine agonists include a small number of dopa-responsive patients, it is impossible to obtain an accurate figure of the proportion of typical idiopathic torsion dystonia cases benefitting. There was initial enthusiasm in the use of lisuride for cranial dystonia (Micheli *et al.*, 1982). However, further experience has failed to support the earlier encouraging results (Nutt *et al.*, 1985; Quinn *et al.*, 1985).

Based upon the knowledge that L-dopa-induced dystonia in Parkinson's disease is usually an effect of low plasma levels in chronically treated patients, for example occurring in the morning before the first dose, in the off period, or as a beginning-of-dose or end-of-dose phenomenon rather than a peak-dose effect, it could be hypothesized that extremely high doses of L-dopa might be required for benefit in idiopathic torsion dystonia (Lang, 1985). However, since reviewing the topic in 1985, I have attempted to use L-dopa therapy in a group of patients with various forms of idiopathic dystonia in a schedule of gradually escalating doses (similar to the approach used for high-dose anticholinergic therapy) hoping to reach levels not previously used in this disorder. In all patients so treated the drug was eventually discontinued at doses of 1500–2500 mg (given with a decarboxylase inhibitor) because of various intolerable side-effects which occurred before any improvement in the underlying dystonia was seen.

Dopamine antagonists

A variety of drugs which antagonize the effects of dopamine have been utilized in the therapy of dystonia. These include both dopamine receptor blockers, particularly haloperidol, phenothiazines and pimozide, and dopamine storage depletors such as tetrabenazine and reserpine. There has been additional experience reported using α-methylparatyrosine (AMPT), an inhibitor of tyrosine hydroxylase, the rate-limiting enzyme in dopamine synthesis. The biochemical, animal and human support for this treatment has been reviewed elsewhere (Lang, 1987). The results of this treatment have been extremely variable with benefit reported in anywhere from 0 to 78% of patients with various forms of dystonia (see Table 2 in Lang, 1987). Overall, a beneficial response is seen in approximately a quarter to a third of patients, a response rate 2–3 times that seen with dopamine agonist therapy if dopa-responsive dystonia cases are excluded.

Trials comparing different dopamine antagonists are lacking. Most dopamine receptor blockers have additional effects on receptors for other bioamine neurotransmitters, especially noradrenaline. Of the drugs used in dystonia, pimozide is a notable exception with almost pure dopamine receptor blocking properties. Tetrabenazine and reserpine deplete serotonin and noradrenaline as well as dopamine. This non-selectivity is thought to account for many of the side-effects of these drugs, most notably depression. It is not known whether drugs with more restricted effects on the dopaminergic system are more or less effective than the less selective bioamine antagonists. Future drug trials will also have to compare selective blockers of specific dopamine receptor subcategories (i.e. D1 and D2) to determine the relative importance of these to dystonia.

In the hope of lessening the frequency of side-effects and taking advantage of different mechanisms of drug action, combinations of different dopamine antagonists have been used in treating dystonia. Examples of this have included the combination of a dopamine depletor (tetrabenazine) with a dopamine receptor blocker (pimozide) (Marsden *et al.*, 1984), and a dopamine depletor (tetrabenazine) with an inhibitor of tyrosine hydroxylase (AMPT) (Lang and Marsden, 1982). This methodology is particularly difficult to assess in a controlled fashion. Lang and Marsden (1982) treated patients in an open-labelled format with tetrabenazine and then added AMPT in a double-blind cross-over fashion, allowing adjustments in the dosage of tetrabenazine during this time. Although this combination had a definite additive effect in choreiform disorders (Huntington's disease and tardive dyskinesia) there was no obvious synergism in dystonia patients. Marsden *et al.* (1984) combined a fixed dose of tetrabenazine with variable doses of pimozide (and additional low doses of trihexyphenidyl to control the side-effects of this therapy). From this approach, it is not clear whether both drugs were required to obtain the recorded benefit or whether a higher dosage of either given individually might have been equally effective.

Other combinations of therapies require further study. To date, there has been no attempt to formally assess the usefulness of combined high-dosage anticholinergic therapy and dopamine antagonist treatment. Botulinum toxin therapy has become established in the treatment of movement disorders, particularly various forms of dystonia (see Chapter 25 by Stell and Elston in this volume). However, only a limited muscle area can be injected. In addition to potential toxicity, higher dosages may encourage antibody formation with subsequent limitation of efficacy (Brin *et al.*, 1987). In focal dystonia and even in patients with more widespread dystonia, it is possible that the effects of smaller doses of botulinum toxin might improve the response to additional drug therapy. The potential for synergism between these forms of local and systemic treatments requires further exploration.

OTHER DRUGS

A number of other pharmacological agents have been used sporadically in dystonia. Most have been totally ineffective or have improved only a very small proportion of the patients treated. Recently, Fahn and Marsden

(1987) have reviewed the literature concerning baclofen, benzodiazepines, carbamazepine, alcohol, clonidine, lithium and γ-vinyl GABA. Despite disappointing results, some patients do respond to one or more of these drugs in an unpredictable fashion and so trials of such agents are warranted in disabled patients who have not obtained benefit from previous therapy. The biochemical findings of Hornykiewicz *et al.* (1986) which indicated prominent changes in noradrenaline throughout the central nervous system, combined with one of the better, albeit limited, animal models of dystonia (the dystonic (dt) rat mutant (Lorden, 1987)) encourage further studies of drugs which interact with noradrenaline. Further trials of clonidine (Riker *et al.*, 1982; Greene *et al.*, 1987) are definitely indicated. However, Jacquet has found that clonidine increases the "dystonia" seen in rodents after injection of ACTH into the locus ceruleus (Jacquet, 1987).

Insights into the pathophysiology of dystonia are only recently becoming available. Studies of cranial (Berardelli *et al.*, 1985) and limb (Rothwell and Obeso, 1987) dystonia reveal evidence for either a lack of inhibition or excessive excitation in both reflex and voluntary movement. To date, pharmacological trials have been essentially empirical. As our physiological understanding improves, it may be possible to design drug trials of newer agents with specific effects directed towards these disturbed mechanisms. An example of this approach is a trial which we have recently undertaken using tizanidine, an anti-spasticity agent with effects on excitatory interneuronal function (Davies, 1982), in cranial dystonia.

CONCLUSION

Drug treatment of dystonia is extremely difficult and frustrating. High doses of anticholinergics and various dopamine antagonists remain the most effective therapy if one excludes patients with dopa-responsive dystonia and paroxysmal dystonia. As Fahn and Marsden (1987) have pointed out, many of the "negative" drug trials are limited by the use of low doses for short periods of time. This contrasts with the general experience using high doses of anticholinergics which are often required for 1–3 months before benefit is seen. Additional trials of combination therapy should be carefully under-taken and trials of newer agents should be guided by further progress in our understanding of the biochemical and physiological bases of dystonia.

REFERENCES

Berardelli, A., Rothwell, J. C., Day, B. L. and Marsden, C. D. (1985) *Brain* **108**, 593–608.

Brin, M. F., Fahn, S., Moskowitz, C. *et al.* (1987) *Movement Disorders* **2**, 237–254.

Burke, R. E., Fahn, S. and Marsden, C. D. (1986) *Neurology* **36**, 160–164.

Calne, D. B. and Lang, A. E. (1987) *Adv. Neurol.* **50**, 9–13.

Davies, J. (1982) *Br. J. Pharmacol.* **76**, 473–481.

Eldridge, R. (1970) *Neurology* **20**, 1–78.

Fahn, S. (1979) *Neurology* **29**, 605.

Fahn, S. (1983) *Neurology* **33**, 1255–1261.

Fahn, S. and Marsden, C. D. (1987) In *Movement Disorders 2* (edited by C. D. Marsden and S. Fahn), pp. 359–382. Butterworths, London.

Greene, P., Shale, H. and Fahn, S. (1987) *Adv. Neurol.* **50**, 547–556.

Hornykiewicz, O., Kish, S. J., Becker, L. E., Farley, I. and Shannak, K. (1986) *New Engl. J. Med.* **315**, 347–353.

Jacquet, Y. (1987) *Adv. Neurol.* **50**, 299–311.

Jankowska, H. (1934) *Neurolologia Pol.* **16–17**, 258–264.

Lang, A. E. (1985) *Clin. Neuropharmacol.* **8**, 38–57.

Lang, A. E. (1986) *Can. J. Neurol. Sci.* **13**, 42–46.

Lang, A. E. (1987) *Adv. Neurol.* **50**, 561–570.

Lang, A. E. and Marsden, C. D. (1982) *Clin. Neuropharmacol.* **5**, 375–387.

Lang, A. E., Sheehy, M. P. and Marsden, C. D. (1982) *Can. J. Neurol. Sci.* **9**, 313–319.

Lang, A. E., Sheehy, M. P. and Marsden, C. D. (1983) *Adv. Neurol.* **37**, 193–200.

Lorden, J. F. (1987) *Adv. Neurol.* **50**, 277–297.

Maltese, W. A., Bressman, S., Fahn, S. and DeVivo, D. C. (1985) *Arch. Neurol.* **42**, 154–155.

Marsden, C. D. (1976) *Adv. Neurol.* **14**, 259–276.

Marsden, C. D. (1981) In *Disorders of Movement* (edited by A. Barbeau), pp. 81–104. J. B. Lippincott, Philadelphia.

Marsden, C. D., Marion, M.-H. and Quinn, N. (1984) *J. Neurol. Neurosurg. Psychiat.* **47**, 1166–1173.

Micheli, F., Fernandez Pardal, M. M. and Leiguarda, R. C. (1982) *Neurology (NY)* **32**, 432–434.

Nutt, J. C., Hammerstad, J. P., Carter, J. H. and DeGarmo, P. L. (1985) *Neurology* **35**, 1242–1243.

Nygaard, T. G., Marsden, C. D. and Duvoisin, R. C. (1987) *Adv. Neurol.* **50**, 377–384.

Quinn, N. P., Lang, A. E., Sheehy, M. P. and Marsden, C. D. (1985) *Neurology* **35**, 766–769.

Regensburg, J. (1930) *Mschr. Psychiat. Neurol.* **75**, 323–345.

Riker, D. K., Hurtig, H., Kale, C. R., Copeland, P. and Roth, R. (1982) *Soc. Neurosci. Abstr.* **8**, 563.

Rothwell, J. C. and Obeso, J. A. (1987) In *Movement Disorders 2* (edited by C. D. Marsden and S. Fahn), pp. 313–331. Butterworths, London.

Sechi, G. P., Demontis, M. D. and Rosati, M. D. (1983) *Neurology* **35**, 1668–1669.

Tanner, C. M., Wilson, R. S., Goetz, C. S. and Shannon, K. M. (1987) *Adv. Neurol.* **50**, 557–560.

24

Dopa-responsive Dystonia: 20 Years into the L-Dopa Era

Torbjoern G. Nygaard

INTRODUCTION

Dopa-responsive dystonia (DRD) has recently been characterized as a distinct subset of dystonia (Nygaard et al., 1988). The clinical features of this disorder include: (1) onset with "dystonia", usually affecting gait, in childhood or adolescence; (2) the concurrent or subsequent development of "parkinsonian" signs; (3) a dramatic therapeutic response to L-dopa (and, perhaps, other similar medications).

Diurnal fluctuation of symptoms, worsening of dystonia with exercise and improvement following sleep are often reported. There is frequently a family history and autosomal dominant inheritance seems probable. The mode of presentation may suggest a variety of other diagnoses, including diplegic cerebral palsy, sporadic or familial spastic paraplegia, juvenile parkinsonism, and ataxic syndromes.

This review focuses on the clinical features of DRD based on data gathered from 123 cases (Castaigne et al., 1971; Segawa et al., 1971; Maekawa et al., 1972; Hongladarom, 1973; Rajput, 1973; Mel'nichuk and Sosnovskaia, 1973; Schenck and Kruschke, 1975; Winkelmann, 1975; Allen and Knopp, 1976; Segawa et al., 1976; Ouvrier, 1978; Montanini et al., 1979; Yokochi, 1979;

Bugiani and Gatti, 1980; Balottin et al., 1981; Garg, 1982; Gordon, 1982; Hakamada et al., 1982; Rolando and Cremonte, 1982; LeWitt et al., 1983; Martinius and Neuhauser, 1983; Richards et al., 1983; Chan-lui and Low, 1984a,b; Yokochi et al., 1984; Kumamoto et al., 1984; Bertelsmann and Smit, 1985; Deonna and Ferreira, 1985; Deonna, 1986; Nygaard and Duvoisin, 1986; Segawa et al., 1986; Shimoyamada et al., 1986; Torelli et al., 1986; Vogel, 1986; Costeff et al., 1987, Nomura et al., 1987; de Yebenes et al., 1988; Fink et al., 1988; Nygaard et al., 1988), 36 of whom have been examined by the author. Twelve cases personally examined by the author, but not yet described in the literature are included.

HISTORICAL BACKGROUND

The first apparent report of affected indivuals was by Beck (1947), who described a girl and her paternal uncle affected with "dystonia musculorum deformans". The girl, at age 8 and a half, began to "kick up her left heel" on walking. The other leg became similarly affected and within 6 months she could walk only with support. Rest tremor was seen in the hands. She stood and walked with hips and knees flexed, the legs adducted and internally rotated; the feet in equinovarus. Corner (1952) drew attention to diurnal fluctuation of symptoms in this patient and in her then affected brother. A dramatic response to trihexyphenidyl (benzhexol) was seen in this pair in doses of 8 and 7 mg day^{-1}, respectively. L-Dopa therapy was begun in the early 1970s with maintenance or improvement in prior symptomatic response.

After L-dopa was introduced for the treatment of Parkinson's disease in 1967 (Cotzias et al., 1967), it was tried in a variety of patients with dystonia with little positive effect (Barrett et al., 1970; Chase, 1970; Mandell, 1970), but there were a few exceptions (Coleman and Barnet, 1969; Chase, 1970; Mandell, 1970).

In 1971, Castaigne et al. (1971) reported two brothers with a "progressive extrapyramidal disorder", and Segawa et al. (1971), two cousins with "hereditary basal ganglia disease with marked diurnal fluctuation", who had a remarkable response to L-dopa therapy. Irving Cooper (1972), however warned that L-dopa rendered patients less responsive to thalamotomy, which was a major treatment modality for dystonia at the time. Eldridge et al. (1973) also urged caution in L-dopa use, but added that about 5% of patients with dystonia reported greatest therapeutic benefit from this treatment. Isolated reports of L-dopa-responsive dystonia continued to appear (Maekawa et al., 1972; Hongladarom, 1973; Mel'nichuk and Sosnovskaia, 1973; Rajput, 1973; Schenck and Kruschke, 1975; Winkelmann, 1975).

At the First International Symposium on Torsion Dystonia in 1975, Allen and Knopp (1976) reported a family afflicted with a disorder characterized by onset of dystonia in childhood with later development of parkinsonian features. These patients were responsive to L-dopa and anticholinergic medication in low doses. Although not noted in this report, the proband noted prominent diurnal variation of symptoms from the onset of the disorder and significant exacerbation of symptoms with menstruation. At the same symposium Segawa et al. (1976) presented further L-dopa-responsive patients. They stressed the progressive nature of the dystonia and the diurnal fluctuation of symptoms. They also described features of parkinsonism, including "cogwheel-like rigidity" and rest tremor in the older patients, "frozen gait", "pulsion", and "mask-like face". These authors concluded that "Some cases reported as juvenile parkinsonism may be the same disorder."

THE TYPICAL CASE

The onset is between 14 months and 12 years of age with a curious abnormality of gait. "Toe-walking", equinovarus or other abnormal postures of the leg are seen. Standing or walking often induces an accentuated lumbar lordosis or a crouched posture. The gait often has a long stride and wide base. Postural instability and a tendency to fall are common. Brisk reflexes in the legs, with unsustained clonus, and systonic extension of the big toe spontaneously or on plantar stimulation — "the striatal toe" (Duvoisin et al., 1972) which may be misinterpreted as a Babinski response — have often been observed.

The dystonia is progressive and more pronounced in the legs than the arms, but also affects the axial musculature with increased lumbar lordosis, scoliosis, and occasionally retrocollis or torticollis. In some of the earlier onset cases, generalization of the dystonia with severe disability did not occur and disability remained mild until more overt parkinsonism appeared in the fifth or sixth decade. No individual with onset in teenage progressed to severe generalized dystonia.

In addition to postural instability, other elements of parkinsonism such as cogwheel rigidity, hypomimia, bradykinesia, and, rarely, rest tremor may be seen. Rapid fatiguing of effort with repetitive or sustained motor tasks is a prominent feature. A variable increase in muscle tone, sometimes with "cogwheeling", and a marked effect of activation (contralateral motor activity during passive limb testing) is appreciated. Tremor appearing early in the disease is usually postural, resembling that of essential tremor, but typical parkinsonian rest tremor may occur later in the course.

There is no history of birth or developmental abnormality, no precipitating illness or drug exposure, no evidence of retinal, sensory, cerebellar, or intellectual disturbance, and no obvious etiology (i.e. abnormal copper metabolism).

INHERITANCE

In 37 families, representing 70 cases, more than one member was affected. Autosomal dominant inheritance is evident (male-to-male transmission) in 10 kindreds (Segawa *et al.*, 1976; Deonna, 1986; Nygaard and Duvoisin, 1986; Vogel, 1986; de Yebenes *et al.*, 1988; Nygaard *et al.*, 1988; S. Fahn, personal communication, 1987) and claimed in another (Bertelsmann and Smit, 1985). The pattern of inheritance in the remaining families is unclear, possibly due to the extremely variable degree of symptomatology in affected individuals. There are 53 cases with apparently "sporadic disease". Autosomal dominant inheritance with incomplete penetrance and variable expressivity appears likely.

A significant proportion of cases have Japanese or Anglo-Saxon heritage, but a wide variety of other ethnic backgrounds have been represented. In one affected family there was possibly a member of Jewish origin (de Yebenes *et al.*, 1988), but the trait could not be definitely linked to this lineage. Only two sporadic cases of Ashkenazi Jewish origin have been recognized (S. Fahn, unpublished data, 1986; Costeff *et al.*, 1987).

DIFFERENTIAL DIAGNOSES

The most common presentation was with insidious development of an inturning foot, often only on walking, with or without diurnal fluctuation. This generally progressed to a frankly dystonic gait, sometimes with unsteadiness and frequent falls. This clinical picture may be indistinguishable from that of childhood onset *idiopathic torsion dystonia* (ITD). However, impairment of balance and falls are not features of the early stages of ITD.

About 20% of this group presented with severe dystonia of the legs simulating paraplegia. Abnormalities found on examination included slight to moderate increases of muscle tone in the legs, unsustained ankle clonus and brisk knee jerks as well as variable extensor plantar responses. The gait appeared spastic with flexed knees, scissoring, and scuffing of the plantigrade feet. Initial impressions included *"diplegic cerebral palsy"* or *"sporadic (hereditary) spastic paraplegia"*. The absence of abnormal birth history, a normal development up to the time of first symptoms, and the progressive nature of disease should allow distinction from cerebral palsy. The presence of extrapyramidal features in the upper limbs would be unexpected in spastic diplegia.

The recognition of rigidity, bradykinesia, hypomimia, and tremor led to consideration of *juvenile parkinsonism* in 10% of these cases. The dystonic elements tended to predominate with early progression in these cases, but a significant degree of parkinsonism was apparent. A long-term response to therapy without complications may be necessary in some cases to allow this differentiation (see below).

A few patients have presented with a bizarre unsteady gait, not typically dystonic, with unexpected and unexplained falls and were diagnosed as an odd *"ataxic syndrome"*.

Thus, DRD needs to be considered in the differential diagnosis of a wide variety of clinical syndromes presenting with gait disorder in childhood or adolescence.

SUMMARY OF CLINICAL FEATURES OF 123 CASES

There was a 2.5:1 predominance of affected females. Although average age of onset was slightly earlier for females than males (6.0 vs 6.3 years) and occurred earlier in patients of either sex with diurnal fluctuation (5.8 vs 6.4 years for females; 6.2 vs 6.6 years for males), there was broad overlap among these groups. Only one case, presenting with torticollis at age 16 (A. H. Rajput, personal communication, 1986), had onset after age 12. A slightly greater proportion of females had diurnal fluctuation (75% vs 66% for males).

Among the 123 cases, the most frequent presenting features (Table 24.1) were "gait abnormality" or dystonia in the leg, typically equinovarus posture of a foot. Although presentation with one of these two features occurred in 117 cases (95%), it is significant to note that two patients presented with difficulty in an arm (Segawa *et al.*, 1976; De Yebenes *et al.*, 1988), two with torticollis (de Yebenes *et al.*, 1988; Nygaard *et al.*, 1988), and another with retrocollis and arm involvement. "Slowness dressing" (Segawa *et al.*, 1976) was the mode of presentation in the remaining patient. Only one patient had not had lower limb involvement by the time of treatment.

Some element of parkinsonism was present in 20% of cases at onset, and was appreciated in at least 67% by the time of treatment. Oculogyria (periodic involuntary upward gaze) was reported in four cases with advanced symptoms (Rajput, 1973; Deonna, 1986; Nygaard and Duvoisin, 1986) but did not recur following L-dopa therapy. Early treatment may have masked the appearance of these features in some patients.

DIURNAL FLUCTUATION AND SLEEP BENEFIT

Diurnal fluctuation and sleep benefit have been emphasized as distinguishing characteristics of this disorder (Segawa *et al.*, 1976). However, among the

Table 24.1 Clinical features in the 123 patients reviewed (sex: 88 female, 35 male; age of onset: 14 months–16 years (average 6.1 years)).

	At onset	At "full expression"
Gait disorder	9	122
Dystonia	101	123
Of leg	95	122
Of arm	5	88
Of neck*	5	31
Lumbar lordosis or scoliosis	3	52
Postural tremor	5	22
Parkinsonism	24	82
Rest tremor	0	14
Bradykinesia	7	48
Hypomimia	1	30
Postural instability	18	50
Cogwheel rigidity	?	49
Oculogyria	0	4
Diurnal fluctuation†	89	89

*Three patients with torticollis at onset. Fourteen patients with torticollis at "full expression", the remainder had retrocollis or anterocollis.

†One case had fluctuations only with her early symptoms, with subsequent disappearance of variation. Another case developed fluctuations after several years of illness.

123 cases included here, diurnal fluctuation was present in only 89 (72%). Fluctuations were reported slightly more frequently among females (75% vs 66%). The degree of fluctuation was quite variable, with some cases "normal" in the morning, and others merely "better" in the morning though still obviously affected. The degree of fluctuation diminished with the progression of the disorder in some cases, and fluctuations disappeared after a few years in at least one individual. Another case (Gordon, 1982) only experienced fluctuation after several years of illness. In four families (Yokochi, 1979; Deonna, 1986; de Yebenes et al., 1988) there were affected members with and others without diurnal fluctuation. Variability in the degree of fluctuation among affected members was noted in another family (Deonna, 1986).

Sleep benefit has been said to require REM sleep (Segawa et al., 1976) but others have noted the effectiveness of non-REM sleep (Kumamoto et al., 1984). Several personally examined cases have demonstrated significant benefit from a period of rest without sleep.

To assess the nosologic value of these phenomena, 51 patients with leg-onset ITD beginning before age 18 years and not responsive to ʟ-dopa were surveyed concerning the variability in severity of their dystonia (Table 24.2). This revealed 18 patients with "sleep benefit", 19 with exercise-induced exacerbation of their symptoms, and six who felt clearly worse in the afternoon. Thus, diurnal fluctuation, at least to some degree, sleep benefit, and worsening with exercise do not reliably discriminate patients with DRD from others with early onset dystonia and cannot be counted on to predict responsiveness to ʟ-dopa.

FREQUENCY OF DRD

Data from the London Dystonia Research Centre and Dystonia Clinical Research Center (New York) give some estimate of the frequency of DRD among those patients presenting with presumed ITD under age 18. In the past decade, 304 patients with non-dopa responsive ITD and 19 patients with DRD were seen. Recognizing that there is a referral bias in this sample, it is estimated that DRD represents 5–10% of those patients presenting with primary dystonia in childhood or adolescence.

Table 24.2 Diurnal variability and effect of exercise in 51 patients with ITD, onset in a leg.

	At onset	Currently
Symptom expression		
Entire day	38	45
Worse in a.m.	2	4
Worse in p.m.	4	7
Only in a.m.	0	0
Only in p.m.	2	2
Only after exercise	6	2
Variable, no pattern	5	2
Sleep and exercise effect		
Effect of exercise	22	25
Made better	3	7
Made worse*	19 (11)	18 (9)
Effect of sleep	21	21
Made better	18 (11)	18 (9)
Made worse	3	3

*Numbers in parentheses refer to patients with *both* exercise exacerbation and sleep benefit.

LABORATORY INVESTIGATIONS

Blood counts, routine blood chemistry, studies of copper metabolism, computerized axial tomography (CT) and magnetic resonance imaging (MRI) have been normal in all cases. Electromyography (EMG) has demonstrated typical co-contraction in agonist–antagonist muscles (Segawa et al., 1976; Richards et al., 1983). Sleep studies have demonstrated decreased movement during sleep and decreased time in REM sleep (Segawa et al., 1976) which reversed with L-dopa therapy.

In most patients studied, cerebrospinal fluid homovanillic acid (CSF–HVA) levels have been reduced (Table 24.3). Diurnal HVA levels were reported in seven patients (Ouvrier, 1978; Kumamoto et al., 1984; Shimoyam-ada et al., 1986; Fink et al., 1988) and showed mild rises in HVA, still below the normal range, or a fall in majority of patients. 5-Hydroxyindoleacetic acid (5-HIAA) levels have been elevated (Ouvrier, 1978), normal (Hakamada et al., 1982; Rolando and Cremonte, 1982; LeWitt et al., 1986; Shimoyamada et al., 1986; Nomura et al., 1987; Fink et al., 1988; Nygaard et al., 1988) and decreased (Allen and Knopp, 1976; Ouvrier, 1978; Fink et al., 1988; Nygaard et al., 1988). Decreased CSF tetrahydrobiopterin (BH_4, a tyrosine hydroxylase cofactor) and neopterin (a BH_4 precursor) levels were reported in six patients (LeWitt et al., 1983; Fink et al., 1988). Labelled fluoro-dopa positron emission tomography (PET) scanning was normal in two adult cases (Lang et al., 1988; Martin et al., 1988) but a decreased putaminal signal similar to the finding in early Parkinson's disease was found in a third case (Lang et al., 1988), a 12-year-old girl. The only neuropathologic study of this condition (Yokochi et al., 1984) reported to date demonstrated hypopigmented, round, "immature"-appearing cells in the substantia nigra. The nigral cells were felt to be "almost normal" in number, re-examination of the pathological material, in fact, revealed marked cell loss in ventrolateral position, similar to that observed in idiopathic Parkinson's disease (Gibb et al., 1989), but a few Lewy bodies were noted. Tyrosine hydroxylase (TH) activity was normal in the substantia nigra, but reduced in the striatum. The locus coeruleus presented similar changes of decreased pigment and TH activity, and "almost normal" number of cells.

RESPONSE TO TREATMENT

The dramatic response to L-dopa sets this group apart from ITD and the other disorders with which it may be confused. Patients with DRD have an immediate benefit from small doses of L-dopa. "Full benefit" has occurred within several days to a few months after initiation of therapy, with

Table 24.3 CSF homovanillic acid levels

Source	HVA in $ng\,ml^{-1}$	Normal values	Comments
Knopp and Allen, 1976			
Proband	19.7	35–66	12 days off Rx
Ouvrier, 1978			
Patient 1	0	9.3–32.3	Untreated
Mother of Pt. 1	27	9.3–32.3	Never treated
Patient 2	0 a.m.	9.3–32.3	Untreated
	0 p.m.		
Hakamada *et al.*, 1982			
8-year old female	11.2	38–46	Untreated
Kumamoto *et al.*, 1984			
Case 1	28.9 8 a.m.	38–46	Untreated
	73.1 8 p.m.	*	
Case 2	26.9 8 a.m.	38–46	14 days off Rx
	24.0 8 p.m.	*	
Case 3	13.5 8 a.m.	38–46	7 days off Rx
	25.0 8 p.m.	*	
Deonna, 1986			
Case 3	"Normal"	†	9 days off Rx
Case 19	"Normal"	†	Untreated
LeWitt *et al.*, 1986			
Patient 1	35.5	40.3–52.3	
Shimoyamada *et al.*, 1986			
9-year-old female	24.6 a.m.	68.6–111.8	
	36.5 p.m.	*	
Nomura *et al.*, 1987			
13-year-old female	57.9	?	Untreated
Fink *et al.*, 1988			
Case 1	25.9 a.m.	44.0–52.8	
	35.3 p.m.	*	
Case 2	56.7 a.m.	44.0–52.8	At least 7 days off Rx
	34.1 p.m.	*	
Case 3	49.1	73.9–115.5	
Case 4	38.2	35.0–117.8	
Nygaard *et al.*, 1988			
13-year-old female	19	40–120	Untreated
15-year-old female	20	40–120	Untreated
15-year-old male	14	23–47	Untreated

*Control p.m. levels not provided. Animal data suggests that p.m. levels may be about 10% greater than a.m. levels (Perlow *et al.*, 1977).

† Reported as "normal", values not provided.

patients returning to "normal" or "near normal". Hyperreflexia and apparent spasticity have resolved (Maekawa et al., 1972, Ouvrier, 1978; Bugiani and Gatti, 1980; Gordon, 1982; Costeff et al., 1987; Fink et al., 1988; Nygaard et al., 1988). A "disappearance of the Babinski response" (perhaps the response of the striatal toe to therapy) has been noted (Ouvrier, 1978; Montanini et al., 1979; Gordon, 1982; Rondot and Ziegler, 1983; Costeff et al., 1987; Nygaard et al., 1988). Minor abnormalities of gait may persist, but full functional capacity is generally achieved. Daily doses as small as 100 mg and not exceeding 3 g (average 500–1000 mg), have yielded maximum benefit. No dose of greater than 400 mg day^{-1} has been necessary when L-dopa has been given in combination with carbidopa (Sinemet). L-Dopa-induced dyskinesias were appreciated in at least 15% of cases at the initiation of therapy (Castaigne et al., 1971; Mel'nichuk and Sosnovskaia, 1973; Winkelmann, 1975; Allen and Knopp, 1976; Haidvogl and Stogmann, 1976; Segawa et al., 1976; Bugiani and Gatti, 1980; Gordon, 1982; Bertelsmann and Smit, 1985; Deonna, 1986; Nygaard and Duvoisin, 1986; Torelli et al., 1986; Costeff et al., 1987; Fink et al., 1988). These subsided with dose reduction and have not reappeared with long-term therapy.

The duration of response following discontinuation of chronic L-dopa therapy has been variable. Twelve hours to several days, and in one patient (T. Deonna, personal communication, 1986) several months, may pass before symptom reemergence. At least one patient remains symptom free on a single-dose of medication on alternate days.

Fourteen individuals in whom the illness had remained untreated for 20–45 years (Winkelmann, 1975; Allen and Knopp, 1976; Segawa et al., 1976; Yokochi et al., 1984; Kumamoto et al., 1984; Nygaard and Duvoisin, 1986; Fink et al., 1988; de Yebenes et al., 1988) still demonstrated an impressive response at initiation of therapy. In all 32 patients (Castaigne et al., 1971; Rajput, 1973; Schenck and Kruschke, 1975; Allen and Knopp, 1976; Segawa et al., 1976; Ouvrier, 1978; Deonna, 1986; Nygaard and Duvoisin, 1986; de Yebenes et al., 1988; Nygaard et al., 1988) for whom follow-up of over 10 years therapy is available, the benefit of L-dopa treatment has been sustained. A modest dose increase has been needed in a few patients on plain L-dopa. Fluctuations, "on–off" phenomena and "freezing episodes" have not appeared. This latter point is an important, but retrospective, differential from juvenile Parkinson's disease. The early appearance of L-dopa-induced dyskinesias in "juvenile parkinsonism" has been noted in several series (Gershanik and Leist, 1986; Narabayashi et al., 1986; Quinn et al., 1987). One case of "torsion dystonia" responding to L-dopa (Still and Herberg, 1976) developed "freezing" after 2.5 years of L-dopa therapy. These cases may have a clinical presentation that is virtually indistinguishable from DRD, including prominent diurnal fluctuation. Three sporadic male cases reported in an earlier series (Gordon, 1982; Nygaard *et al.*, 1988) have developed wearing-off or peak-dose dyskinesias during the first 5 years of L-dopa treatment and represent misclassified cases. Thus, the definition of DRD must be modified to include at least 5 years of stable response to L-dopa, if a family history of stable response is not available (Nygaard and Marsden, 1988).

Trihexyphenidyl (THP) and other anticholinergics were used in at least 22 of these patients with variable effect. A dramatic effect was seen in at least five cases (Corner, 1952; Segawa *et al.*, 1986; Fahn, personal communication, 1987), a moderate response in 12 others (Allen and Knopp, 1976; Nygaard *et al.*, 1988), and no response in the remaining five patients (Kumamoto *et al.*, 1984; Nygaard and Duvoisin, 1986; Nygaard *et al.*, 1988). Doses of 10 mg or less of THP were used in these latter cases and it is uncertain whether the higher doses currently used in typical ITD (Fahn, 1983) would have yielded greater benefit in those patients with poor response. Increasing doses of THP, as high as 100 mg day^{-1}, were needed because of decreasing efficacy in several patients as they progressed through their teenage years. Bradykinesia, rigidity, and postural instability have tended to become more obvious during this progression. L-Dopa was judged to be superior to THP in all cases except two, who already felt "normal" on THP, in which the response was equal.

Bromocriptine had an effect equivalent to that of L-dopa in two patients (Kumamoto *et al.*, 1984; Nomura *et al.*, 1987). Tetrahdyrobiopterin, given intravenously, was as effective as L-dopa in one patient (LeWitt *et al.*, 1986).

THE PATHOGENETIC BASIS — SPECULATION

The clinical data suggest a functional deficiency of central dopamine production. A deficiency in phenylalanine hydroxylase (Fig. 24.1) causes

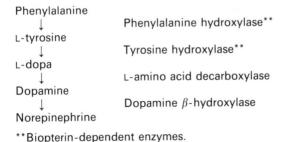

```
Phenylalanine
     ↓                Phenylalanine hydroxylase**
L-tyrosine
     ↓                Tyrosine hydroxylase**
L-dopa
     ↓                L-amino acid decarboxylase
Dopamine
     ↓                Dopamine β-hydroxylase
Norepinephrine
```

**Biopterin-dependent enzymes.

Fig. 24.1 Catecholamine biosynthetic pathway.

phenylketonuria and devastating clinical manifestations in the first years of life. Individuals with deficiencies of biopterin pathway enzymes. GTP cyclohydrolase I (Niederweiser *et al.*, 1984), dihydrobiopterin synthetase (Kaufman *et al.*, 1978; Tanaka *et al.*, 1987), dihydropteridine reductase (Kaufman *et al.*, 1975; Smith *et al.*, 1975, 1986; Butler *et al.*, 1978), and biopterin synthesis (Fink *et al.*, 1988), have generalized biopterin deficiency, phenylketonuria and clinical syndromes that include mental retardation, gross development delay, myoclonus and seizures. Two of these individuals had dystonic features and diurnal variation of their symptoms (Tanaka *et al.*, 1987; Fink *et al.*, 1988). No affected individual in any of the current families reviewed has had phenylketonuria or these other clinical manifestations. A central biopterin deficiency, however, has been suggested to play a role in pathogenesis of DRD (LeWitt *et al.*, 1986; Fink *et al.*, 1988). The pathologic study of Yokochi *et al.* (1984) demonstrated TH deficiency in the striatum and locus coeruleus with "almost normal" cell numbers. Could the dystonia seen in DRD be related to a consequent deficiency of norepinephrine in the locus coeruleus, as reported in ITD (Hornykiewicz *et al.*, 1988), with the parkinsonism due to the striatal dopamine deficiency? An intriguing thought is to look for genetic linkage with TH (on the short arm of chromosome 11) which has recently been implicated in another autosomal dominant condition, manic-depressive illness (Egeland *et al.*, 1987).

SUMMARY

Dopa-responsive dystonia (DRD) is a distinctive clinical entity and an unexpectedly common subgroup of torsion dystonia. A functional tyrosine hydroxylase deficiency seems probable though the pathogenesis of disease remains unknown. Clinically, diurnal fluctuation is often, but not always present and does not reliably distinguish the disorder from idiopathic torsion dystonia. DRD must be considered in the differential diagnosis of the child or adolescent presenting with a dystonic gait disorder, diplegic cerebral palsy, sporadic or familial spastic paraplegia, and ataxic syndromes. The response to L-Dopa is so dramatic and occurs so quickly that a diagnostic therapeutic trial should be considered in all patients presenting with these syndromes. The distinction from juvenile parkinsonism in sporadic cases may be difficult and may require a period of follow-up.

ADDENDUM

Since submission of this manuscript, linkage of DRD to the gene for byrosine hydroxylase on chromosome 11 has been excluded in several families (Fletcher *et al.*, 1989).

ACKNOWLEDGEMENTS

This work was supported in part by the Dystonia Medical Research Foundation.
The author is grateful to Drs M. Backonja (Madison, WI), T. Deonna (Lausanne, Switzerland), S. Fahn (New York, NY), D. Gardner-Medwin (Newcastle, UK), S. J. Horwitz (Cleveland, OH), M. R. Koenigsberger (Newark, NJ), A. E. Lang (Toronto, Ontario, Canada), N. F. Lawton (Southampton, UK), D. H. Mellor (Nottingham, UK), M. J. Noronha (Manchester, UK), J. G. Nutt (Portland, OR), R. A. Ouvrier (Sydney, Australia), S. G. Pavlakis (New York, NY), P. Procopis (Sydney), A. H. Rajput (Saskatoon, Saskatchewan, Canada), I. Rapin (New York, NY), E. Schenck (Freiburg, West Germany), W. Schutt (Bristol, UK), P. Siemens (Saskatoon, Saskatchewan, Canada), G. Wise, (NSW, Australia), G. Ziegler (Freiburg, West Germany) for providing further data or allowing examination of their patients.

REFERENCES

Allen, N. and Knopp, W. (1976). *Adv. Neurol.* **14**, 201–214.
Balottin, U., Lanzi, G. and Zambrino, C. A. (1981) *Riv. Neurobiol.* **27**, 584–590.
Barrett, R. E., Yahr, M. D. and Duvoisin, R. C. (1970) *Neurology* (Suppl. 2) **20**, 107–113.
Beck, D. K. (1947) *Proc. R. Soc. Med.* **40**, 551.
Bertelsmann, F. W. and Smit, L. M. E. (1985) *Clin. Neurol. Neurosurg.* **87**, 123–126.
Bugiani, O. and Gatti, R. (1980) *Ann. Neurol.* **7**, 93.
Butler, I. J., Koslow, S. H., Krumholz, A., Holtzman, N. A. and Kaufman, S. (1978) *Ann Neurol.* **3**, 224–230.
Castaigne, O. Rondot, P., Ribadeau-Dumas, J. J. and Said, G. (1971) *Rev. Neurol.* **124**, 162–166.
Chan-Lui, W. Y. and Low, L. C. K. (1984a) *Aust. Paediatr. J.* **20**, 143–146.
Chan-Lui, W. Y. and Low, L. C. K. (1984b) *Dev. Med. Child. Neurol.* **26**, 665–668.
Chase, T. N. (1970) *Neurology* (Suppl. 2) **20**, 122–130.
Coleman, M. and Barnet, A. (1969) *Trans. Am. Neurol. Assoc.* **94**, 91–95.
Cooper, I. S. (1972) *Lancet* **ii**, 1317–1318.
Corner, B. D. (1952) *Proc. Roy. Soc. Med.* **45**, 451–452.
Costeff, H., Gadoth, N., Mendelson, L., Harel, S. and Lavie, P. (1987) *Arch. Dis. Child* **62**, 801–804.
Cotzias, G. C., Van Woert, M. H. and Schiffer, L. M. (1967) *N. Engl. J. Med.* **276**, 374–379.
Deonna, T. (1986) *Neuropediatrics* **17**, 81–85.
Deonna, T. and Ferreira, A. (1985) *Dev. Med. Child Neurol.* **27**, 819–821.
Duvoisin, R. C., Yahr, M. D., Lieberman, J., Antunes, J. and Rhee, S. (1972) *Trans. Am. Neurol. Assoc.* **97**, 267.

Egeland, J. A., Gerhard, D. S., Pauls, D. L., Sussex, J. N., Kidd, K. K., Allen, C. R. et al. (1987) Nature 325, 783–787.

Eldridge, R., Kanter, W. and Koerber, T. (1973) Lancet ii, 1027–1028.

Fahn, S. (1983) Neurology 33, 1255–1261.

Fletcher, N. A., Holt, I. J., Harding, A. E., Nygaard, T. G., Mallet, J. and Marsden, C. D. (1989) J. Neurol. Neurosurg. Psychiatry 52, 112–114.

Fink, J. K., Barton, N., Cohen, W., Lovenberg, W., Burns, R. S. and Hallett, M. (1988) Neurology 38, 707–711.

Garg, B. P. (1982) Arch. Neurol. 39, 376–377.

Gershanik, O. S. and Leist, A. (1986) Adv. Neurol. 45, 213–216.

Gordon, N. (1982) Neuropediatrics 13, 152–154.

Haidvogl, M. and Stogmann, W. (1976). Jahrestag. Ost. Ges. Kinderheilkd. Millstatt. 24, 9.

Hakamada, S., Watanabe, K., Hara, K. and Miyazaki, S. (1982) No To Hattatsu 14, 44–48.

Hongladarom, T. (1973) Lancet i, 1114.

Hornykiewicz, O., Kish, S. J., Becker, L. E., Farley, I. and Shannak, K. (1988) Adv. Neurol. 50, 157–165.

Kaufman, S., Holtzman, N. A., Milstien, S., Butler, I. J. and Krumholtz, A. (1975) N. Engl. J. Med. 293, 785–790.

Kaufman, S., Berlow, S., Summer, G. K., Milstien, S., Schulman, J. D., Orloff, S., et al. (1978) N. Engl. J. Med. 299, 673–679.

Kumamoto, I., Nomoto, M., Yoshidome, M., Osame, M. and Igata, A. (1984) Rinsho Shinkeigaku 24, 697–702.

Lang, A. E., Garnett, E. S., Firnau, G., Nahmias, C. and Talalla, A. (1988) Adv. Neurol. 50, 249–253.

LeWitt, P. A., Newman, R. P., Miller, L. P., Lovenberg, W. and Eldridge, R. (1983) N. Engl. J. Med. 308, 157–158.

LeWitt, P. A., Miller, L. P., Levine, R. A., Lovenberg, W., Newman, R. P., Papavasiliou, A. et al. (1986) Neurology 36, 760–764.

Maekawa, K., Kitani, N. and Satake, Y. (1972) No To Hattatsu 4, 274–281.

Mandell, S. (1970) Neurology (Suppl. 2) 20, 103–106.

Martin, W. R. W., Stoessl, A. J., Palmer, M., Adam, M. J., Ruth, T. J., Grierson, J. R., et al. (1988) Adv. Neurol. 50, 223–229.

Martinius, J. and Neuhauser, G. (1983) Padiatr. Prax. 28, 45–49.

Mel'nichuk, P. V. and Sosnovskaia, L. S. (1973) Zh. Nevropatol. Psikhiatr. 73, 1495–1498.

Montanini, R., Basso, P. F. and Gasco, P. (1979) Minerva Med. 70, 1551–1553.

Narabayashi, H., Yokochi, M., Iizuki, R. and Nagatsu, T. (1986) In Handbook of Clinical Neurology (edited by P. J. Vinken, G. W. Bruyn and H. L. Klawans), Vol. 49, pp. 153–165. Elsevier Science Publ., Amsterdam.

Niederwieser, A., Blau, N., Wang, M., Joller, P., Atares, M. and Cardesa-Garcia, J. (1984) Eur. J. Pediat. 141, 208–214.

Nomura, K., Yamamoto, N., Takahashi, I., Furune, S., Aso, K., Negoro, T. et al. (1987) No To Hattatsu 19, 244–248.

Nygaard, T. G. and Duvoisin, R. C. (1986) Neurology 36, 1424–1428.

Nygaard, T. G. and Marsden, C. D. (1988) Neurology (Suppl. 1) 38, 130.

Nygaard, T. G., Marsden, C. D. and Duvoisin, R. C. (1988) Adv. Neurol. 50, 377–384.

Ouvrier, R. A. (1978) Ann. Neurol. 4, 412–417.

Perlow, M. J., Gordon, E. K., Ebert, M. E., Hoffman, H. J. and Chase, T. N. (1977) *J. Neurochem.* **28**, 1381–1383.

Quinn, N., Critchley, P. and Marsden, C. D. (1987) *Movement Disorders* **2**, 73–91.

Rajput, A. H. (1973) *Lancet* **i**, 432.

Richards, C. L., Bedard, P. J., Fortin, G. and Malouin, F. (1983) *Neurology* **33**, 1083–1087.

Rolando, S. and Cremonte, M. (1982) *Instituto Gaslini (Genova)* **14**, 176–179.

Rondot, P. and Ziegler, M. (1983) *J. Neural Trans.* (Suppl.) **19**, 273–281.

Schenck, E. and Kruschke, U. (1975) *Klin Wochenschr.* **53**, 779–780.

Segawa, M., Ohmi, K., Itoh, S., Aoyama, M. and Hayakawa, H. (1971) *Shinryo* **24**, 667–672.

Segawa, M., Hosaka, A., Miyagawa, F., Nomura, Y. and Imai, H. (1976) *Adv. Neurol.* **14**, 215–233.

Segawa, M., Nomura, Y. and Kase, M. (1986) *Adv. Neurol.* **45**, 227–234.

Shimoyamada, Y., Yoshikawa, A., Kashii, H., Kihira, S. and Koike, M. (1986) *No To Hattatsu* **18**, 505–509.

Smith, I., Clayton, B. E. and Wolff, O. H. (1975) *Lancet* **i**, 1108–1111.

Smith, I., Leeming, R. J., Cavanaugh, N. P. and Hyland, K. (1986) *Arch. Dis. Child* **61**, 130–137.

Still, C. N. and Herberg, K.-P. (1976) *South Med. J.* **69**, 564–566.

Tanaka, K., Yoneda, M., Nakajima, T., Miyatake, T. and Owada, M. (1987) *Neurology* **37**, 519–522.

Torelli, D., Lamontanara, G., Bracciolini, M. and Ciaravolo, G. A. (1986) *Acta Neurol. (Napoli)* **8**, 626–632.

Vogel, H. P. (1986) *Akt. Neurol.* **13**, 102–105.

Winkelmann, W. (1975) *J. Neurol.* **208**, 319–323.

Yokochi, M. (1979) *Adv. Neurol. Sci. (Tokyo)* **23**, 1048–1059.

Yokochi, M., Narabayashi, H., Iizuka, R. and Nagatsu, T. (1984) *Adv. Neurol.* **40**, 407–413.

de Yebenes, J. G., Moskowitz, C., Fahn, S. and Saint-Hilaire, M. H. (1988) *Adv. Neurol.* **50**, 101–111.

25

The Clinical Uses of Botulinum Toxin

Rick Stell and John S. Elston

INTRODUCTION

Clostridium botulinium is a ubiquitous organism present in 90% of random soil samples in the UK. In its inactive spore form it is resistant to drying and can survive up to 2 h of boiling at 100° C, but is killed at 120° C. The toxin it produces is however rapidly inactivated by heating at ordinary cooking temperatures.

Ingestion of the preformed toxin present in imperfectly sterilized canned or bottled food produces a paralytic illness called botulism. Seven antigenically different toxins, labelled A, B, C, D, E, F and G are produced though A, B and E account for most of the human cases. The disease severity depends upon the dosage of toxin ingested. Symptoms are heralded by dryness of the mouth, nausea and vomiting followed by involvement of the extraocular muscles and pupils, producing blurred vision and diplopia. There is subsequent involvement of other bulbar innervated muscles with dysphagia, dysarthria and respiratory insufficiency together with involvement of the limbs. Autonomic involvement is often prominent with the production of a paralytic ileus, cholestasis, urinary retention and severe postural hypotension. In severe, untreated cases the disease is fatal. With prolonged respiratory support muscle function slowly recovers, though mortality remains significant

DISORDERS OF MOVEMENT: CLINICAL, PHARMACOLOGICAL
AND PHYSIOLOGICAL ASPECTS   ISBN 0-12-569685-X

because of the complications of both artificial respiration and parenteral feeding.

Infantile botulism, first described in 1976 as a cause of the sudden infant syndrome (SIDS), is different from the adult disease in that the organism may actually colonize and multiply in the bowel whilst continuing to produce toxin which is then absorbed. The clinical picture is that of a floppy baby with poor feeding and head control. Ptosis and internal and external ophthalmoplegia may develop with loss of deep tendon reflexes.

PHYSIOLOGY

It was initially unclear where the site of action of the toxin lay, but in 1949 it was isolated and purified, and subsequent animal studies demonstrated that its mode of action was by neuromuscular blockade.

The toxin has a molecular weight of 150 000 kilodaltons and is synthesized as a single polypeptide which is broken by proteases into a two-chained structure connected by a disulphide bridge. The mode of action at the neuromuscular junction has now been established and related to the structure. A three-stage process of binding, internalization and intracellular activity appears to be involved. It is thought that the toxin binds, via its heavy chain, to a specific ganglioside or glycoprotein acceptor site on the terminal, non-myelinated axon (Dolly et al., 1984). Internalization is an active process, probably via micro-pinocytosis. The exact mechanism of action of the toxin once internalized is not known, but it appears to prevent calcium-dependent quantal release of acetylcholine at specific "active zones" on the pre-terminal membrane, thus blocking miniature end plate and end plate potentials (Drachman et al., 1976). Affected neuromuscular junctions are permanently inactivated; this produces a flaccid paralysis of the affected muscles which then undergo dennervation-like atrophy. Whilst neuromuscular blockade is permanent, clinical recovery occurs after weeks or months by collateral sprouting of motor nerve terminals and the formation of new neuromuscular junctions (Duchen, 1971).

Botulinum toxin is the most poisonous substance known, only minute amounts of toxin being required for a biological effect, and the lethal dose for mice being $15–50 \, \text{ng kg}^{-1}$. The toxic effect per molecule exceeds that of other clostridial toxins by a factor of 10. When the toxin is given systemically an antibody response rapidly develops but when injected in small amounts into the limb muscles of experimental animals no detectable immune response occurs.

OPHTHALMOLOGICAL APPLICATIONS

The concept of using this most potent naturally occurring toxin to treat patients with disorders of muscle function owes its origin to the work of

Scott *et al.* (1973). In an effort to develop a non-surgical treatment of strabismus, he evolved an elegant EMG technique for the injection of small doses of various agents into extraocular muscles. The ideal substance should have a predictable effect, last weeks to months without systemic toxicity or antibody response and be fully reversible with time. Botulinum toxin has come closest to this ideal — a single injection produces isolated extraocular muscle weakness without spread to other muscles and the effect slowly recovers over 8–10 weeks. The first clinical trial in the treatment of strabismus appeared in 1979 and since that time many thousands of cases of strabismus have been safely and effectively treated in this way. Publication of the results from various centres between 1981 and 1986 have confirmed its effectiveness and its safety but have also pointed out some of its limitations. Treatment of the lateral, medial and inferior rectus is possible but injection of the superior rectus always produces a ptosis. Spread of the toxin within the orbit is common in children, so that when horizontally acting muscles were treated, a vertical deviation and a partial ptosis would develop. The most important limitation to the routine use of botulinum toxin for strabismus, however, is that the effect is temporary. Though the visual axes can be realigned by injecting the appropriate external ocular muscle, the ocular deviation will recur unless the patient has the neurological substrate for binocular vision, i.e. sensory and motor fusion. Surgery remains the primary mode of treatment for most patients.

Further experience has shown that the ideal subjects for botulinum toxin are adults with a normal central nervous system who have an acquired peripheral strabismus such as a partly recovered lateral rectus palsy producing an esotropia with normal abducting saccadic velocity (Scott and Kraft, 1985). In lateral rectus palsy of variable cause, incomplete recovery of muscle function may occur producing a persistent esotropia in the primary position, yet with abduction beyond the midline. Such cases, which may be due to a contracture of the ipsilateral medial rectus, respond well to weakening of this muscle with botulinum toxin. The mild exotropia which follows injection usually corrects itself without the redevelopment of the esotropia. The same technique has been successfully used to treat partially recovered third nerve palsies. In more severe or unrecovered ocular deviations due to ocular motor dysfunction, botulinum toxin may be combined with surgical transposition of unaffected muscles to produce even better functional recovery than with conventional surgical techniques used alone.

Other conditions where pre-existing established binocularity has been disrupted may also be suitable for treatment with botulinum toxin. Diplopia associated with dysthyroid ophthalmopathy may be treated in this way during the evolution of the condition before definitive surgery is possible (Scott, 1984). Decompensating exophoria can be managed by injecting the lateral rectus, and the possibility of altering the relationship between the visual axes can be used as a prolonged pre-operative diplopia test. Unexpected binocular function may be revealed by this means and indicate the need for definite surgical treatment. Intolerable diplopia at all stages of recovery from the iatrogenic palsy argues against such treatment.

Thus, botulinum has an important role in the management of complex ocular motility problems, particularly those of neurogenic origin.

OTHER OPHTHALMOLOGICAL USES

Several other applications for botulinum in ophthalmology have been developed. Temporary relief of spastic ectropion may be achieved by weakening the pre-tarsal orbicularis of the lower lid. This may break a vicious cycle of ocular irritation leading to inflammation and lid spasm. Injection of botulinum toxin into the levator palpebrae superioris may be used to produce a temporary "neurogenic tarsorrhaphy" to treat conditions such as indolent corneal ulceration.

NEUROLOGICAL APPLICATIONS

Soon after its initial use it became clear that botulinum toxin was an ideal agent for the treatment of other disorders of motor control; indeed as far back as 1973 Scott suggested that blepharospasm could be treated in this way. Treatment trials in blepharospasm began in the early 1980s with excellent results (Scott and Kraft, 1985). Subsequently, other focal movement disorders have begun to be treated in this way including torticollis, spasmodic dysphonia, hemifacial spasm and most recently, writer's cramp, oromandibular and lingual dystonia.

BLEPHAROSPASM

Idiopathic blepharospasm is a focal form of dystonia involving the orbicularis oculi muscles producing involuntary eye closure which at its worst may produce functional blindness. It occurs more often in women than men in the sixth and seventh decades. Often it begins insidiously with irritation and discomfort in the eye in association with photophobia and an increased rate of blinking. Spasms of eye closure slowly develop and with time abnormal movements may spread from the orbicularis oculi muscles to involve other facial muscles as well as the jaw and neck.

Blepharospasm is typically variable, being aggravated by bright lights and tracking eye movements. As with all movement disorders it is exacerbated by stress though paradoxically it may improve when the patient is being examined. Once present it is usually permanent though occasional patients may have spontaneous remissions. A family history of abnormal blinking, blepharospasm, or other focal dystonic abnormalities may be obtained in up to 10% of patients. Though initially not thought to have an organic basis there is now compelling evidence that it is due to a neurochemical disorder of the basal ganglia, as is the case in other forms of dystonia.

The condition is refractory to medical treatment in the majority of patients. Anticholinergic treatment is the single most effective measure, but this has only a 10% chance of sustained improvement and then often at the cost of significant side-effects. Surgical treatments including local injections of alcohol, bilateral differential section of the facial nerve with or without avulsion, and myectomy have all been used with variable success (Anderson, 1985; Callahan, 1985; Battista, 1982) but as with all surgical procedures they are irreversible and may produce difficulty with eating, paralytic ectropion and an alteration in facial expression. In addition as a result of axonal regrowth and reinnervation any relief obtained may only be temporary.

Local injections of botulinum toxin into the orbicularis oculi mimic the effect of surgical denervation of the muscle, whilst confining it to the muscle in spasm. Fifty nanograms of botulinum toxin A-haemagglutenin complex* (containing 8 ng of pure neurotoxin) are diluted in 10 ml of sterile saline. The solution is then used to give a total of eight injections into the orbicularis oculi: 0.2 ml are injected laterally and 0.1 ml medially at the junction of the orbital and preseptal orbicularis, with the upper lid injections being angled away from the levator in the centre of the lid. A total of 6 ng of complex is thus injected. EMG control is not required as the muscles lie superficially and are effectively weakened by a subcutaneous injection. Since Scott et al.'s (1985) first trial several open trials on the effectiveness of botulinum toxin have been reported (Elston and Russell, 1985; Mauriello, 1985; Brin et al., 1987) and with two double-blind trials (Fahn et al., 1985; Jankovic and Orman, 1987), have consistently shown functional improvement in 70% of patients, the effect lasting on average 2.5 months. Following the injection detectable muscle weakness begins after 12–24 h with clinical improvement usually within 2–3 days. In a proportion of those patients responding, mid and lower facial spasms and even oromandibular dystonia may improve or completely resolve. In approximately 10% of patients weakening of the orbicularis produces little or no functional benefit to the patient. This may

---

*The dosages given in this chapter refer *only* to the UK supply from PHLS Porton Down. The nanograms and units of toxin, and the number of units per nanogram, all differ from those of the American supply.

result from persistence of the Bell's phenomenon and levator inhibition which are a normal part of forceful eye closure and which are not always reduced by the injections. Correct management of this subgroup of patients is not clear at the present time. The clinically useful effect of the toxin lasts for 8–10 weeks though some weakness is detectable for up to 6 months after a single injection. There is no loss of efficacy after repeated injections: indeed in some patients the period of benefit may increase. Reinjection is usually performed before muscle power significantly returns so that smaller doses can be used.

The mechanism by which injection of periocular muscles reduces the orofacial components, present in some patients, is not known. It is known that as with tetanus toxin, botulinum toxin is transported intra-axonaly (Black and Dolly, 1986) and possibly trans-synaptically and may thus modify brain stem interneuron function, though at the present time this is entirely speculative. However, there is no doubt that the major site of action of botulinum toxin is at the neuromuscular junction.

Side-effects, whilst relatively common, occurring in 50% of cases, are mild and reversible; they consist of partial ptosis due to involvement of levator palpebrae and slight diplopia usually due to weakening of one of the extraocular muscles, most often the superior rectus.

SPASMODIC TORTICOLLIS

Following the successful use of botulinum toxin in the treatment of blepharo-spasm several reports assessing its usefulness in spasmodic torticollis, another focal dystonia, were published (Tsui et al., 1986; Jankovic and Orman, 1987; Stell et al., 1988). Like blepharospasm, this condition has been refractory to medical therapy, and whilst surgical techniques including sectioning of anterior spinal roots and selective division of the cervical posterior primary rami have been effective in some patients, they are difficult and lengthy operations to perform, carry significant risk, and often do not lead to permanent benefit.

Patients receiving the toxin are usually studied with surface and needle electrodes in an effort to identify the two most active muscles. In the case of purely rotational torticollis these are the ipsilateral splenius capitis and contralateral sternomastoid. For laterocollis the most active muscles are usually the ipsilateral splenius and trapezius muscles, and for retrocollis the muscles selected are the two splenius capitis muscles. Fifty nanograms of the toxin complex are diluted in 10 ml of sterile saline and 2.5 ml of this solution is then injected into each of the two muscles selected, i.e. a total dose of 25 ng. Patients usually notice improvement by 5–7 days reaching a peak within 24 h of its onset. Improvement in posture is found in approximately 70% of patients and 90% improve with respect to pain. Benefit with regard to both posture and pain lasts from 9 to 15 weeks.

Side-effects which occur in up to 30% of cases are usually mild and well tolerated. The most frequent problem encountered is dysphagia, predominantly for solids, which usually occurs 4–7 days post-injection and lasts from days to several weeks. This is probably due to local diffusion from the injection site, in the sternomastoid muscle, into the subjacent pharyngeal muscles, as dysphagia has never been reported in patients in whom only the more posterior neck muscles are injected. Other side-effects include a slight deepening of the voice, local tenderness at the injection site, a transient feeling of weakness of the head upon the shoulders and a generalized feeling of asthenia, all of which subside within 2 weeks of onset. Despite repeated injections there has been only one report of the development of antibodies, in two patients, with subsequent loss of efficacy of the toxin (Brin et al., 1987). There are no reported cases of systemic toxicity despite EMG evidence of distant involvement of muscles (Sanders et al., 1986; Lange et al., 1987).

At the present time pure rotational torticollis and retrocollis are the most amenable to treatment; patients with antecollis and tremulous torticollis are more difficult to treat and often require injections into several muscles, and are thus more likely to develop side-effects.

HEMIFACIAL SPASM

This disorder which consists of involuntary rapid synchronous contraction of muscles innervated by the facial nerve on one side, has also been treated with botulinum toxin (Mauriello, 1985; Savino et al., 1985, Brin et al., 1987). Although it usually begins with contraction of the muscles around the eye, with time the twitches spread to involve the other ipsilateral facial muscles.

Hemifacial spasm is thought usually to result from microvascular compression of the facial nerve at its exit from the brain stem and there is surgical evidence to support this. Rarely it may be produced by irritative lesions in the posterior fossa such as arteriovenous malformations and cerebellopontine "tumours".

Drug treatment with carbamazepine though frequently used is rarely successful in abolishing the movements. Surgical procedures upon the facial nerve branches, or decompressing the nerve directly via a posterior fossa craniotomy, have been tried, and whilst good results have been achieved complications such as partial deafness or facial paresis may occur, and recurrences requiring reoperation are not infrequent. The recent introduction of microsurgical techniques for the posterior fossa operation are claimed to produce excellent results without the above side-effects. More recently unilateral orbicularis oculi muscle stripping has been advocated, and whilst this may reduce abnormal movements around the eye it will not influence the lower facial movement. Botulinum toxin is now frequently used to treat those patients who are unwilling or unfit to have a definitive surgical procedure. The toxin is injected locally into the muscles obviously contracting and produces major improvement in 90% of patients with few and infrequent side-effects consisting of partial ptosis, diplopia, or lower facial weakness.

OTHER NEUROLOGICAL USES

With the development of fine hollow EMG needles, other focal dystonias have been treated with encouraging results. These include spasmodic dysphonia due to spasm of the vocal cords (Blitzer et al., 1985), oromandibular dystonia (Jankovic and Orman, 1987), lingual dystonia (Brin et al., 1987) and writer's cramp (Cohen et al., 1987).

SUMMARY

Since its initial use as a therapeutic agent 10 years ago it has become evident that botulinum toxin, given in carefully controlled, minute doses is a safe and effective agent for the treatment of a number of disorders of motor control, previously refractory to all but surgical intervention, with the attendant operative risks and side-effects. Despite its well-recognized potency as a neurotoxin, small doses can be safely and easily injected locally into individual muscles, without systemic toxicity. Botulinum toxin is arguably the safest and most effective new treatment in the management of movement disorders since the use of L-dopa to treat Parkinson's disease over 20 years ago.

REFERENCES

Anderson, R. L. (1985) *Adv. Ophthal. Plastic Reconstruct. Surg.* **4**, 313–332.
Battista, A. F. (1982) In *Movement Disorders* (edited by C. D. Marsden and S. Fahn), pp. 319–321. Butterworths, London.
Black, J. D. and Dolly, J. O. (1986) *J. Cell Biol.* **103**, 521–534, 535–544.
Blitzer, A., Lovelace, R. E., Brin, M. F., Fahn, S. and Fink, M. E. (1985) *Ann. Otol. Rhino. Laryngol.* **94**, 591–594.
Brin, M. F., Fahn, S., Moskowitz, C., Friedman, A., Shale, H. M., Greene, P. E. *et al.* (1987) *Adv. Neurol.* **50**, 599–608.

Callahan, A. (1985) *Adv. Opthal. Plastic Reconstruct. Surg.* **4**, 379–384.

Cohen, L., Hallett, M., Geller, B., Dubinsky, R., Neer, J., Baker, M. and Hochberg, F. (1987) *Neurology* **37** (Suppl. 1) 123–124.

Dolly, J. O., Black, J., Williams, R. S. and Melling, J. (1984) *Nature* **307**, 457–460.

Drachman, D. B., Kao, I. and Price, D. L. (1976) *Science* **193**, 1256–1258.

Duchen, L. W. (1971) *J. Neurol. Sci.* **14**, 47–60, 61–64.

Elston, J. S. and Russell, R. W. R. (1985) *Br. Med. J.* **290**, 1857–1859.

Fahn, S., List, T., Moskowitz, C., Brin, M., Bressman, S., Burke, R. *et al.* (1985) *Neurology* **35**, Suppl. 1, 271–272.

Jankovic, J. and Orman, J. (1987) *Neurology* **37**, 616–623.

Lange, D. J., Brin, M. F., Warner, C. L., Fahn, S. and Lovelace, R. E. (1987) *Muscle Nerve* **10**, 552–555.

Mauriello, J. A. (1985) *Neurology* **35**, 1499–1500.

Sanders, D. B., Massey, E. W. and Buckley, E. C. (1986) *Neurology* **36**, 545–547.

Savino, P., Sergott, R., Bosley, T. and Schatz, N. (1985) *Arch. Opthalmol.* **103**, 1305–1306.

Scott, A. B. (1984) *Doc. Opthalmol.* **58**, 141–145.

Scott, A. B. and Kraft, S. P. (1985) *Opthalmology* **92**, 676–683.

Scott, A. B., Rosenbaum, A. and Collins, C. C. (1973) *Arch. Ophthalmol.* **12**, 924–927.

Scott, A. B., Kennedy, R. A. and Stubbs, M. A. (1985) *Arch. Ophthalmol.* **103**, 347–350.

Stell, R., Thompson, P. D. and Marsden, C. D. (1988) *J. Neurol. Neurosurg. Psychiat.* **51**, 920–923.

Tsui, J. K., Eisen, A., Stoessl, A., Calne, S. and Calne, D. B. (1986) *Lancet* August 2, 245–246.

26

Dystonia: Where next?

Stanley Fahn

INTRODUCTION

The term dystonia refers both to a specific type of involuntary movements and to the clinical syndromes in which these movements occur. It is helpful for understanding the spectrum of the pattern of abnormal movements comprising dystonia to be aware of the seminal papers of Oppenheim (1911) and Flatau and Sterling (1911) describing the clinical entity of the syndrome known today as idiopathic generalized torsion dystonia (dystonia musculorum deformans), and the more current paper of Marsden (1976) linking the focal dystonias with generalized dystonia.

The syndrome described by Oppenheim and by Flatau and Sterling shows a variety of abnormal movements, with a wide range in speed, amplitude, rhythmicity, torsion, forcefulness, distribution in the body, and relationship to rest or voluntary activity. The diversity of these movements is reflected in the terms that have been used to refer to them, for example tetanoid chorea, tic-like, myoclonia, tonic spasms, and myorhythmia (Fahn et al., 1987). To this day, dystonic movements are the most likely abnormal involuntary movements to be misdiagnosed as some other type of movement disorder, usually chorea. Moreover, these movements are also more often considered to be psychogenic than other types of movement disorders. The

DISORDERS OF MOVEMENT: CLINICAL, PHARMACOLOGICAL AND PHYSIOLOGICAL ASPECTS   ISBN 0-12-569685-X

increase of the movements with voluntary motor activity was one of the first characteristic patterns described for dystonia (Destarac, 1901). Yet, this feature has probably been responsible for misleading clinicians into the misdiagnosis of a psychogenic disorder. The most distinctive feature of the involuntary movements is the sustained nature of the contractions. The other characteristic patterns of dystonic movements are the twisting and the repetitive pattern of the movements, hence the term coined by Flatau and Sterling (1911) of "progressive torsion spasm". These distinctive features have led to the modern definition of dystonia as "a syndrome of sustained muscle contractions, frequently causing twisting and repetitive movements, or abnormal postures" (Fahn et al., 1987).

Dystonic states have been classified in three ways: by age at onset, by body distribution of the abnormal movements, and by etiology (Fahn et al., 1987) (Table 26.1). Classification by age at onset is useful because this is the most important single factor related to prognosis of idiopathic dystonia. Marsden et al. (1976) analysed 72 cases of idiopathic dystonia and found that, in general, the younger the age at onset, the more likely that the dystonia would become severe and would also spread to involve multiple parts of the body. This pattern was also seen by Cooper et al. (1976) and more recently confirmed in the patient population reported by Fahn (1986).

Since dystonia usually begins by affecting a single part of the body (focal dystonia), and since dystonia can either remain focal or spread to involve other body parts, it is useful to classify dystonia according to its distribution of involvement of the body (Table 26.1). Marsden (1976) linked those dystonias beginning in adulthood and remaining limited to one part of the body (i.e. the focal dystonias), such as blepharospasm, torticollis, oromandibular dystonia, spastic dysphonia, and writer's cramp, to the well-

Table 26.1   Classification of dystonia.

I. By age at onset	
A. Childhood onset:	0–12 years
B. Adolescent onset:	13–20 years
C. Adult onset:	> 20 years
II. By distribution	
A. Focal	
B. Segmental	
C. Multifocal	
D. Generalized	
E. Hemidystonia	
III. By etiology	
A. Idiopathic	
Sporadic	
Familial	
B. Symptomatic	

recognized childhood-onset generalized dystonias. Arguments in favour of the concept that generalized, segmental and focal dystonias are related include the following.

(1) Idiopathic generalized dystonia almost always begins as a focal dystonia before it spreads to involve other parts of the body. It is possible that in some individuals the spread may not take place or may be limited and then plateau; the dystonia may remain as a focal dystonia or at most as a segmental dystonia.

(2) In families wth dystonia, various members may have generalized, segmental or focal dystonia, suggesting that they are related, giving rise to the concept that the less involved individuals are formes frustes of generalized dystonia (Zeman et al., 1960).

(3) Sensory tricks often ameliorate dystonic movements and postures and this manoeuver can be effective in different parts of the body (Fahn, 1984, 1988). Such a common phenomenology suggests a common pathophysiology.

(4) The drugs commonly used to reduce the severity of generalized dystonia are virtually identical to those useful in controlling the focal dystonias (Fahn and Marsden, 1987). These consist of anticholinergics, benzodiazepines, baclofen, carbamazepine, and antidopaminergics.

The third factor by which dystonia is classified is etiology. The causes of dystonia are divided into two major categories idiopathic (or primary) and symptomatic (or secondary). The idiopathic group consists of familial and non-familial (sporadic) types. The symptomatic group is subdivided into those conditions associated with various hereditary neurologic disorders, those due to environmental causes, dystonia associated with parkinsonism, and psychogenic dystonia (Fahn and Williams, 1988).

With the advance in understanding the clinical expression and classification of dystonia, there have also been advances in therapeutics (Greene, et al., 1988). However the pharmacotherapeutics of dystonia remain non-specific and symptomatic, i.e. not addressed to the cause of dystonia. In fact, even if a patient responds to such non-specific therapy, the benefit may not last, and progression may occur at any point.

There is much to be done in advancing the understanding and treatment of dystonia. Those areas that are currently being actively investigated are genetics, imaging, biochemistry, anatomy and physiology, animal models, and therapy. These areas, and probably others, can be considered as where the field of study on dystonia is heading.

GENETICS

As mentioned above, idiopathic torsion dystonia consists of familial and non-familial (sporadic) types. Although most patients with torsion dystonia give a negative family history for this disorder, Zeman and Dyken (1967) have emphasized the importance of personal examination of family members to be absolutely certain about the presence or absence of dystonia. These investigators showed that a pattern of autosomal dominant transmission exists in many families (Zeman and Dyken, 1968). Because of consanguinity, an autosomal recessive pattern may be present in gypsies (Gimenez-Roldan *et al.*, 1988). Although Eldridge (1970) argued for an autosomal recessive pattern of inheritance of dystonia in the Ashkenazi Jewish population, this concept has been challenged. Korczyn and his colleagues (Korczyn *et al.*, 1980; Zilber *et al.*, 1984), studying the hereditary pattern in Israel, found evidence supporting the concept that this population also inherits dystonia in an autosomal dominant pattern. Bressman *et al.* (1988) in the United States have supported this concept. Major arguments by Eldridge favouring autosomal recessive inheritance in the Ashkenazi population are: (1) the lack of any large pedigrees in contrast to those seen in the non-Jewish population; (2) the phenomenon of pseudodominance could explain dystonia in both parent and child (pseudodominance occurs in close population groups in which intermarriage allows a carrier to marry an affected individual, thus producing affected offspring); and (3) a different clinical picture between the Ashkenazi and non-Jewish populations. The third point has been challenged (Burke *et al.*, 1986). The other points need rebuttal by conclusive studies on larger numbers of the Ashkenazi population; these are being pursued, and a definitive answer will be available in the near future.

Besides mode of inheritance, several studies are proceeding in their search for the gene(s) causing dystonia, using molecular genetic techniques. Breakefield and her colleagues have been actively pursuing such investigations, particularly in a large North American and a large Swedish family (Breakefield *et al.*, 1986; Kramer *et al.*, 1987, 1988). Standard linkage analysis with blood markers has also been utilized (Bressman *et al.*, 1984; Falk *et al.*, 1988) to supplement DNA techniques. We can anticipate that ultimately, the locus for the abnormal gene(s) causing dystonia will be found. This will lead to linkage analysis, identification of the abnormal gene product, and resolution of the question of relationships among the clinical variants of dystonia, including the different ethnic groups and the different clinical varieties of dystonia, such as the dopa-responsive dystonias (Nygaard *et al.*, 1988).

Insight into the workings of dystonia may also come about from studies searching for the abnormal gene in other ethnic groups with dystonia. Lee *et al.* (1976) described an X-linked recessive form of dystonia in Filipinos on the Island of Panay. A recent study on a large family from this island found that there is a mixture of parkinsonism and dystonia in this population (Fahn and Moskowitz, 1988). Interestingly, if dystonia is the mode at onset, it is eventually superseded by the appearance of parkinsonism, with elimination of dystonia. It would seem that with progressive loss of striatal dopamine, dystonia is no longer present. It is not clear if dystonia requires the presence of striatal dopamine or if parkinsonian signs are dominant over signs of dystonia.

IMAGING

A field that is receiving considerable attention is neuroimaging. Magnetic resonance imaging (MRI) of patients with idiopathic torsion dystonia has not revealed any abnormalities (Rutledge et al., 1988). In contrast, many of the symptomatic dystonias show CT and MRI abnormalities, particularly in the putamen (Burton et al., 1984), but also in areas of the brain connecting with the basal ganglia (Obeso and Gimenez-Roldan, 1988). Unfortunately, some symptomatic dystonias also have normal CT and MRI scans, including many patients with perinatal injury. Thus, a normal MRI is not evidence for idiopathic dystonia. But an abnormal MRI is evidence against a diagnosis of idiopathic dystonia. At present, most MRI studies monitor hydrogen atoms and thereby measure water content. When the magnetic power of future scanners increases, other atoms will be able to be detected, including phosphorus. It is likely that some day high energy phosphates will be measurable in brain, and this may divulge abnormalities in idiopathic torsion dystonia.

Positron emission tomography (PET) is another imaging technique that can evaluate regional brain chemistry in the living subject. To date, with measurements of glucose metabolism and dopamine storage sites, no consistent abnormalities have been detected in idiopathic dystonia (Martin et al., 1988; Gilman et al., 1988; Chase et al., 1988; Leenders et al., 1988; Lang et al., 1988; Perlmutter and Raichle, 1988). However, the future of PET scanning seems limitless, and one can predict that as new positron-emitting ligands become available, they will be utilized in the study of dystonia patients. One can anticipate that some day a ligand will be found that detects an abnormality in idiopathic torsion dystonia.

BIOCHEMISTRY

A major advance in understanding dystonia will come when its biochemical pathology is completely understood. The publication by Hornykiewicz et al. (1986) was the first report in this area. They described results obtained in two patients with idiopathic dystonia. This was followed by a report by Jankovic et al. (1987) in another patient, showing some similarities in results. Abnormalities in norepinephrine and serotonin were found in some regions of the brain. Similar findings in the symptomatic dystonia occurring in neuroacanthocytosis have also been reported (de Yebenes et al., 1988a,b). One can anticipate further biochemical studies in dystonia. These will either confirm or refute previous findings. By increasing the number of available reports, an accurate and reliable picture of the biochemistry of dystonia will unfold.

ANATOMY AND PHYSIOLOGY

Concepts on the anatomical pathology of dystonia have traditionally placed the suspected pathology in the basal ganglia since this is the region involved in almost all cases of symptomatic dystonia (Burton et al., 1984). Lesions producing dystonia have been discovered to involve not only the basal ganglia, but also connections to and from these nuclei, such as the thalamus and the cortex (Marsden et al., 1985). However, the basal ganglia are not always the site of physiologic pathology in idiopathic dystonia. The rostral brain stem has been found to be pathologically damaged in some cases of secondary blepharospasm (Jankovic and Patel, 1983). Blepharospasm can result from lesions from stroke, multiple sclerosis and encephalitis in this region. Furthermore, this region is associated with the abnormal blink reflexes present in idiopathic cranial and cervical dystonias (Berardelli et al., 1985; Tolosa et al., 1988). These abnormal reflexes imply excess physiologic excitatory drive to the midbrain region. Clearly, the mechanism by which such excitatory drive occurs in the cranial dystonias needs clarification by physiological studies.

ANIMAL MODELS

Animal models of dystonia have not yet been productive in elucidating the mechanism of human dystonia. It has been questioned whether the genetically dystonic mouse (Duchen, 1976) has abnormal sustained limb postures as a result of dystonia or as a result of sensory deprivation of the limbs due to the peripheral nerve lesions found in this rodent. The genetically dystonic rat (Lorden et al., 1988) more closely resembles human dystonia. Future studies on this species may some day lead to biochemical, physiological and anatomical findings that may provide insight into human dystonia. The animal model of dystonia in rats following injection of ACTH and its analogues (Jacquet, 1988) into the locus coeruleus has suggested that dystonia may be related to an abnormality of brain norepinephrine. Whether this animal model will prove to be helpful in uncovering new information about human dystonia will depend on what future studies will unfold. Another animal model that could be used to shed light on dystonia is the acute dystonic reaction seen after neuroleptic administration. Stewart *et al.* (1988) have evaluated rats after neuroleptics in an attempt to derive biochemical understanding of acute dystonic reactions. Other studies on animals exposed to agents or environmental insults such as hypoxia, chronic neuroleptics or toxins could be used in future studies and these could lead to more clues on dystonia.

THERAPY

Despite progress in developing newer symptomatic therapies for torsion dystonia (Fahn and Marsden, 1987; Greene *et al.*, 1988), including elucidating the phenomenon of dopa-responsive dystonia (Nygaard *et al.*, 1988) and the use of botulinum toxin for focal dystonias (Brin *et al.*, 1987), many patients do not respond satisfactorily. There is a desperate need to develop new therapies. Stereotaxic surgery with more sophisticated computer guidance might prove more successful because this approach may lead to fewer complications than were seen with older techniques. Continued experimental therapeutic investigations will continue, and this will undoubtedly lead to new useful therapies. Hopefully, these kinds of studies will also lead to more specific therapy, rather than just symptomatic medications.

REFERENCES

Berardelli, A., Rothwell, J. C., Day, B. L. and Marsden, C. D. (1985) *Brain* **108**, 593–609.
Breakefield, X. O., Bressman, S. B., Kramer, P. L., Ozelius, L., Moskowitz, C., Tanzi, R., Brin, M. F., Hobbs, W., Kaufman, D., Tobin, A., Kidd, K. K., Fahn, S. and Gusella, J. F. (1986) *J. Neurogenet.* **3**, 159–175.
Bressman, S. B., Fahn, S., Falk, C., Allen, F. H. Jr and Suciu-Foca, N. (1984) *Neurology* **34**, 1490–1493.
Bressman, S. B., de Leon, D., Brin, M. F., Risch, N., Shale, H., Burke, R. E., Greene, P. E. and Fahn, S. (1988) *Adv. Neurol.* **50**, 45–56.
Brin, M. F., Fahn, S., Moskowitz, C., Friedman, A., Shale, H. M., Greene, P. E., Blitzer, A., List, T., Lange, D., Lovelace, R. E. and McMahon, D. (1987) *Movement Disorders* **2**, 237–254.
Burke, R. E., Brin, M. F., Fahn, S., Bressman, S. B. and Moskowitz, C. (1986) *Movement Disorders* **1**, 163–178.

Burton, K., Farrell, K., Li, D. and Calne, D. B. (1984) *Neurology* **34**, 962–965.

Chase, T. N., Tamminga, C. A. and Burrows, H. (1988) *Adv. Neurol.* **50**, 237–241.

Cooper, I. S., Cullinan, T. and Riklan, M. (1976) *Adv. Neurol.* **14**, 157–169.

Destarac, (1901). *Rev. Neurol.* **9**, 591–597.

Duchen, L. W. (1976) *Adv. Neurol.* **14**, 353–365.

Eldridge, R. (1970) *Neurology* **20**, 1–78.

Fahn, S. (1984) *Neurol. Clin. North Am.* **2**, 541–554.

Fahn, S. (1986) *Clin. Neuropharmacol.* **9**, Suppl 2, S37–S48.

Fahn, S. (1988) *Adv. Neurol.* **49**, 125–133.

Fahn, S. and Marsden, C. D. (1987) In *Movement Disorders 2* (edited by C. D. Marsden and S. Fahn), pp. 359–382. Butterworths, London.

Fahn, S. and Moskowitz, C. (1988) *Ann. Neurol.* **24**, 179.

Fahn, S. and Williams, D. T. (1988) *Adv. Neurol.* **50**, 431–455.

Fahn, S., Marsden, C. D. and Calne, D. B. (1987) In *Movement Disorders 2* (edited by C. D. Marsden and S. Fahn) pp. 332–358. Butterworths, London.

Falk, C. T., Bressman, S. B., Allen, F. H. Jr, Moskowitz, C., Fahn, S. and Brin, M. (1988) *Adv. Neurol.* **50**, 67–72.

Flatau, E. and Sterling, W. (1911) *Z. Ges. Neurol. Psychiat.* **7**, 586–612.

Gilman, S., Junck, L., Young, A. B., Hichwa, R. D., Markel, D. S. *et al.* (1988) *Adv. Neurol.* **50**, 231–236.

Gimenez-Roldan, S., Delgado, G., Marin, M., Villanueva, J. A. and Mateo, D. (1988) *Adv. Neurol.* **50**, 73–81.

Greene, P., Shale, H. and Fahn, S. (1988) *Movement Disorders* **3**, 46–60.

Hornykiewicz, O., Kish, S. J., Becker, L. E., Farley, I. and Shannak, K. (1986) *N. Engl. J. Med.* **315**, 347–353.

Jacquet, Y. (1988) *Adv. Neurol.* **50**, 299–311.

Jankovic, J. and Patel, S. C. (1983) *Neurology* **33**, 1237–1240.

Jankovic, J., Svendsen, C. N. and Bird, E. D. (1987) *N. Engl. J. Med.* **316**, 278–279.

Korczyn, A. D., Kahana, E., Zilber, N., Streifler, M., Carasso, R. and Alter, M. (1980) *Ann. Neurol.* **8**, 387–391.

Kramer, P. L., Ozelius, L., Gusella, J. F., Fahn, S., Kidd, K. K. and Breakefield, X. O. (1987) *Genet. Epidemiol.* **4**, 377–386.

Kramer, P. L., Ozelius, L. O., Brin, M. F., Bressman, S. B., Moskowitz, C. B., Fahn, S., Kidd, K. K., Gusella, J. and Breakefield, X. O. (1988) *Adv. Neurol.* **50**, 57–66.

Lang, A. E., Garnett, E. S., Firnau, G., Nahmias, C. and Talalla, A. (1988) *Adv. Neurol.* **50**, 249–253.

Lee, L. V., Pascasio, F. M., Fuentes, F. D. and Viterbo, G. H. (1976) *Adv. Neurol.* **14**, 137–151.

Leenders, K. L., Quinn, N., Frackowiak, R. S. J. and Marsden, C. D. (1988) *Adv. Neurol.* **50**, 243–247.

Lorden, J. F., Oltmans, G. A., Stratton, S. and Mays, L. E. (1988) *Adv. Neurol.* **50**, 277–297.

Marsden, C. D. (1976). *Adv. Neurol.* **14**, 259–276.

Marsden, C. D., Harrison, M. J. G. and Bundey, S. (1976) *Adv. Neurol.* **14**, 177–187.

Marsden, C. D., Obeso, J. A., Zarranz, J. J. and Lang, A. E. (1985) *Brain* **108**, 463–483.

Martin, W. R. W., Stoessl, A. J., Palmer, M., Adam, M. J., Ruth, T. J. *et al.* (1988) *Adv. Neurol.* **50**, 223–229.

Nygaard, T. G., Marsden, C. D. and Duvoisin, R. C. (1988) *Adv. Neurol.* **50**, 377–384.

Obeso, J. A. and Gimenez-Roldan, S. (1988) *Adv. Neurol.* **50**, 112–122.

Oppenheim, H. (1911) *Neurol. Centrabl.* **30**, 1090–1107.

Perlmutter, J. S. and Raichle, M. E. (1988) *Adv. Neurol.* **50**, 255–264.

Rutledge, J. N., Hilal, S. K., Silver, A. J., Defendini, R. and Fahn, S. (1988) *Adv. Neurol.* **50**, 265–275.

Stewart, B. R., Rupniak, M. N. J., Jenner, P. and Marsden, C. D. (1988) *Adv. Neurol.* **50**, 343–359.

Tolosa, E., Montserrat, L. and Bayes, A. (1988) *Movement Disorders* **3**, 61–69.

de Yebenes, J. G., Vazquez, A., Martinez, A., Mena, M. A., del Rio, R. M. *et al.* (1988a) *Adv. Neurol.* **50**, 167–175.

de Yebenes, J. G., Brin, M., Mena, M. A., de Felipe, C., Del Rio, R. M., Bazan, E., Martinez, A., Fahn, S., Del Rio, J., Vazquez, A. and Varela de Seijas, E. (1988b) *Movement Disorders* **3**, 300–312.

Zeman, W. and Dyken, P. (1967) *Psychiat. Neurol. Neurochir.* **70**, 77–121.

Zeman, W. and Dyken, P. (1968) *Handbook of Clinical Neurology*, Vol. 6, pp. 517–543. North-Holland Publ. Co., Amsterdam.

Zeman, W., Kaelbling, R. and Pasamanick, B. (1960) *Neurology* **10**, 1068–1075.

Zilber, N., Korczyn, A. D., Kahana, E., Fried, K. and Alter, M. (1984) *J. Med. Genet.* **21**, 13–20.

PART III

Neuroleptic-induced Movement Disorders

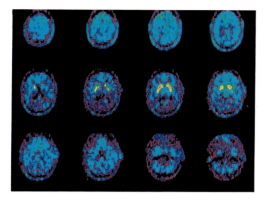

Fig. 7.1 L-[^{18}F]-6-Fluorodopa uptake in a healthy volunteer's brain. The images are contiguous transaxial cross-sections of the radioactivity from the top of the brain (left upper image) to the level cutting through the cerebellum (right lower image). The top of the image is frontal and the head is seen from above. The activity shown is the cumulative activity from 30 to 90 min after administration of the tracer.

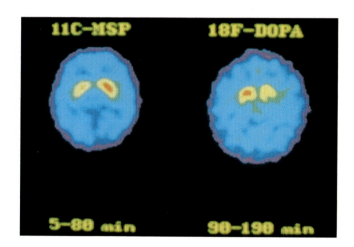

Fig. 19.2 ^{18}F-Dopa and ^{11}C-NMSP uptake is shown in a patient with hemiparkinsonism and hemidystonia on the left side of the body. Anatomical orientation is as for Fig. 19.1. The images represent the distribution of total radioactivity accumulated during the indicated times for that cross-section of brain. The image on the right shows diminished ^{18}F-dopa uptake (presynaptic dopaminergic activity) in the right (i.e. contralateral) striatum, but the image on the left demonstrates increased uptake of ^{11}C-NMSP (postsynaptic dopamine (D2) receptor binding) in the right striatum.

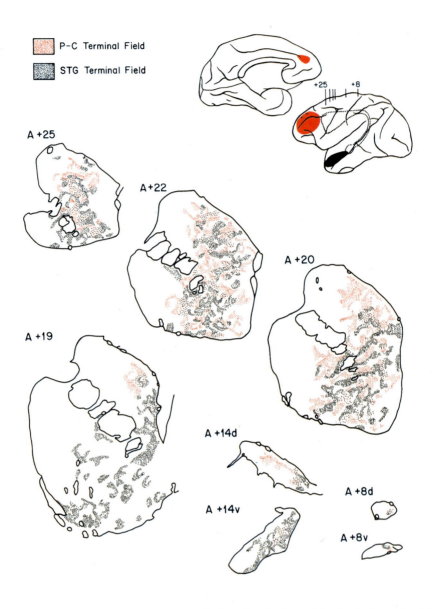

Fig. 2.2 Diagram illustrating the interdigitation of projections originating in the prefrontal cortex (red) and superior temporal cortex (black). HRP was injected into frontal cortex and tritiated amino acids were injected into temporal cortex, and adjacent sections reacted histochemically for localization of HRP and processed as autoradiograms, respectively. (Reproduced, with permission, from Selemon and Goldman-Rakic, 1985.)

27

Neuroleptic-induced Movement Disorders

Daniel Tarsy

INTRODUCTION

Extrapyramidal reactions to neuroleptic, antipsychotic drugs (APD) have been a subject of great interest ever since they were described shortly after the introduction of APDs in the 1950s. The panoply of extrapyramidal syndromes appearing at that time had not been seen in such large numbers due to a single cause since the pandemic of encephalitis lethargica and post-encephalitic parkinsonian syndromes of the early part of this century. Identification of dopamine (DA) blockade as the probable mechanism of action of APD occurred at the same time as L-dopa was successfully introduced for the treatment of Parkinson's disease and contributed greatly to understanding the pathophysiology and pharmacology of parkinsonism. Moreover, the observation that APD and L-dopa were both capable of producing choreoathetotic dyskinesias gave promise that the pharmacology of hyperkinetic movement disorders might also be better understood.

Early classifications of APD-induced extrapyramidal reactions suggested a relatively simple breakdown into early and late manifestations which were thought to be easily identifiable and for which effective pharmacologic treatment could be provided. Thus, acute dyskinesias, parkinsonism, and akathisia were considered early side-effects attributable to acute DA receptor blockade and responsive to anticholinergic treatment while tardive dyskinesia (TD) was considered a late manifestation possibly related to some form of DA overactivity and best treated by DA antagonists of one sort or another.

As time has passed it has become increasingly clear that APD-induced syndromes are not as phenomenologically distinct as had been thought while, except for parkinsonism, well-established pathophysiological mechanisms and rational pharmacologic treatments have not been forthcoming. Nonetheless, interest in APD-induced extrapyramidal reactions has served to provide a great stimulus for investigations into the physiology and pharmacology of the basal ganglia and will undoubtedly continue to motivate further productive research in this area.

ACUTE DYSKINESIA (ACUTE DYSTONIA)

Clinical features

Acute dyskinesias occur immediately after initiating treatment with APD and consist of intermittent or sustained muscular spasms and abnormal postures of the eyes, face, neck and throat. Oculogyric crises, blepharospasm, trismus, oromandibular dystonia, grimacing, protrusion or twisting of the tongue, distortions of the lips, or glossopharyngeal contractions occur and in severe cases dysarthria, dysphagia, jaw dislocation, and respiratory stridor with cyanosis may result. The neck is often involved producing spasmodic torticollis or retrocollis. The trunk is more severely affected in children who may show various forms of truncal dystonia including opisthotonos, scoliosis, lordosis, and tortipelvis. Torsional movements of the extremities with dystonic hyperpronation and adduction postures may also occur. Contrary to popular belief, some acute dyskinesias may be indistinguishable in appearance from the dyskinesias usually associated with TD. Although in the older literature it is probable that some acute dyskinesias were unrecognized manifestations of TD, tongue protrusions, lip-smacking, blinking, athetosis of the fingers and toes, shoulder-shrugging, and a variety of myoclonic muscle contractions of the face, neck and extremities were described among acute dyskinesias (Sarwer-Foner, 1960: Delay and Deniker, 1968). More subtle forms of orofacial dystonia such as cramping sensations of the jaw and tongue with subjective difficulty in chewing or speaking may occur alone or precede more obvious manifestations.

The acute dyskinesias may be painful and are usually frightening. At a time when they were less familiar they were occasionally confused with seizures, tetanus, rabies, encephalitis, meningitis, or subarachnoid haemorrhage. Because dystonic signs sometimes remit and exacerbate or appear to respond to reassurance and suggestion, they have also been mistaken for hysterical reactions (Angus and Simpson, 1970).

Acute dyskinesias are the earliest APD syndrome to appear and may begin within hours of a single dose of a neuroleptic agent. There may be a delay of 12–36 h between drug administration and the appearance of dyskinesia. They may either remit or continue to fluctuate spontaneously over several hours or days and will usually reappear if other APDs of equal potency are introduced. Long-acting parenteral phenothiazines such as fluphenazine decanoate may produce acute dyskinesias in a cyclical pattern within 72 h of each administration and in this setting are often confused with TD. Persistent dyskinesia has not been reported following acute and brief exposure to ordinary doses of antipsychotic drugs. There also appear to be no well-documented cases of acute dyskinesia following the use of reserpine. Following routine clinical use of neuroleptic agents, the overall incidence of acute dyskinesia is approximately 10% although there is considerable variation depending on the potency and dose of APD which is used (Swett, 1975).

Mechanisms

The pathophysiology of acute dyskinesia has remained uncertain. Although acute dyskinesias are regarded as idiosyncratic their incidence appears to be especially common after treatment with agents of high antidopamine and low anticholinergic potency, and to be somewhat dose dependent (Swett, 1975). In one study patients with acute dyskinesia had higher erythrocyte concentrations of phenothiazines than patients without dyskinesia (Garver et al., 1976). Although it is reasonable to assume that interference with synaptic DA mechanisms is responsible, several clinical observations have raised doubts about this mechanism. The fact that reserpine, despite its ability to produce acute presynaptic depletion of DA, does not produce acute dyskinesia suggests that acute inhibition of DA transmission is insufficient for their production. Although tetrabenazine, also a presynaptic depletor of catecholamines, may produce acute dyskinesia (Burke et al., 1985) this may be due to its properties as a blocker of DA receptors (Login et al., 1982). The clinical similarity between APD-induced acute dyskinesias, dystonias produced by L-dopa in patients with Parkinson's disease, and the various dystonias associated with tardive dyskinesia has suggested the alternative possibility that activation rather than blockade of DA mechanisms may be responsible (Marsden et al., 1975; Marsden and Jenner, 1980).

Acute dystonic reactions have been described in baboons and monkeys

(Meldrum *et al.*, 1977; Liebman and Neale, 1980; Heintz and Casey, 1987). Similar to humans, baboons show individual susceptibility as only three of 25 animals developed acute dystonias following haloperidol treatment (Meldrum *et al.*, 1977). However, all three also developed dystonias following treatment with a number of other APDs, including chlorpromazine and pimozide, but not thioridazine, thereby paralleling common clinical experience. In this study, catecholamine depletion produced by pretreatment with reserpine or reserpine plus α-methyptyrosine greatly reduced or abolished haloperidol-induced dystonia consistent with a dependency of acute dystonias on presynaptic catecholaminergic mechanisms. However, in Cebus monkeys a monoamine oxidase inhibitor reduced haloperidol-induced acute dystonia (Heintz and Casey, 1987) consistent with the more traditional view that acute dystonia is due to DA inhibition rather than facilitation.

The synthesis and release of DA in the nigrostriatal system increases acutely in response to blockade of DA receptors. This compensatory increase in DA synthesis and metabolic turnover is a complex function mediated by long–loop striatonigral feedback connections, local intrastriatal feedback connections, and effects on DA autoreceptors (Marsden and Jenner, 1980). It has been suggested that activation of nigrostriatal DA turnover produced by DA receptor blockade may account for acute dyskinesia (Marsden *et al.*, 1975; Meldrum *et al.*, 1977; Marsden and Jenner, 1980). Similar to acute dyskinesia, the activation of DA turnover following DA receptor blockade is a relatively transient phenomenon which declines with repeated drug exposure. This may correspond to the clinical observation that patients who experience acute dyskinesia often develop "tolerance" to this effect following continued APD exposure even in the absence of anticholinergic treatment (Black *et al.*, 1985). One study of prophylactic anticholinergic drugs and incidence of acute dyskinesia has shown that incidence of acute dyskinetic reactions diminishes greatly after 1 week of APD exposure (Winslow *et al.*, 1986). In man acute dyskinesias may occur after a delay of as much as 24–36 h after a single dose although they appear more acutely in non-human primates (Heintz and Casey, 1987). In one human study acute dyskinesias occurred 23–56 h after a single dose of butaperazine, at a time when plasma and red cell butaperazine concentrations were falling (Garver *et al.*, 1976). If acute dyskinesias are to be attributed to an acceleration of DA turnover their appearance in the presence of a DA receptor antagonist is difficult to explain. However, in animal studies striatonigral DA turnover peaks earlier than the appearance of acute dyskinesia in humans (Matthysse, 1973; Marsden and Jenner, 1980) so that delayed onset of dyskinesia may be due to a combination of falling striatal APD concentrations coupled with a continued increase in DA availability. Since a single dose of APD can produce supersensitivity of DA receptors evident 24–48 h after acute drug administration (Kolbe *et al.*, 1981), enhanced DA release acting on incompletely blocked and increasingly sensitive DA receptors may be responsible for acute dyskinesias (Marsden and Jenner, 1980; Kolbe *et al.*, 1981).

The fact that anticholinergic drugs abort acute dyskinesia is traditionally interpreted as evidence for acute DA deficiency with restoration of dopamine–acetylcholine balance by cholinergic blockade. That apomorphine, a direct DA receptor agonist, also reverses acute dyskinesia (Gessa *et al.*, 1972) provides additional evidence that acute dyskinesia results from acute DA deficiency. However, in view of the formulation discussed above, the capacity of anticholinergic drugs and apomorphine to inhibit DA turnover may be more relevant mechanisms. Anticholinergic drugs block the increase in DA turnover produced by APD while apomorphine reduces nigrostriatal DA turnover (Nyback *et al.*, 1970) and suppresses nigrostriatal firing rates by activation of presynaptic autoregulatory DA receptors (Skirboll *et al.*, 1979).

Acute dyskinesias are far more common in children and young adults than older individuals (Bateman *et al.*, 1986). In addition, paralleling idiopathic torsion dystonia, children frequently show more severe trunk and extremity dystonia while adults usually show more restricted involvement of the neck, face, tongue, or arms (Ayd, 1961). If acute dyskinesia is due to activation of DA mechanisms then children and young adults may respond to APD with a brisker activation of nigrostriatal DA turnover due to the larger numbers of nigrostriatal neurons and higher concentrations of striatal DA and tyrosine hydroxylase found in their brains (McGeer and McGeer, 1977).

Treatment

Acute dyskinesia is easily recognized and readily treated. However, it is important to be aware of minor but uncomfortable dystonic manifestations which often escape diagnosis. The frequent delay of several hours between administration of a single dose of APD and the appearance of dyskinesia is a particular cause of misdiagnosis. Intravenous or intramuscular benzotropine is highly effective for severe acute dyskinesias while oral anticholinergics may suffice for milder forms. If the APD is discontinued repeated administration of benzotropine may remain necessary for a period of 24–48 h due to the briefer action of anticholinergic drugs. Diphenydramine is similarly effective, presumably because of its anticholinergic properties, and may be given parenterally or by mouth. Intravenous diazepam has also been recommended (Korczyn and Goldberg, 1972; Gagrat *et al.*, 1978) but it should be used with caution because of the potential risk of respiratory depression.

The value of prophylactic anticholinergic drugs for prevention of extrapyramidal reactions is uncertain and controversial. Some studies have shown no reduction in the incidence of extrapyramidal syndromes although the severity of symptoms appears to be reduced (DiMascio and Demirgian, 1970). However, the incidence of acute dyskinesia is reduced by prophylactic treatment (Keepers et al., 1983; Winslow et al., 1986) and young patients in whom acute dyskinesia is common and anticholinergic toxicity is rare, may benefit from prophylactic anticholinergic treatment (Keepers et al., 1983; Winslow et al., 1986).

PARKINSONISM

Clinical features

All of the cardinal neurologic signs of Parkinson's disease including akinesia, rigidity, postural abnormalities, and tremor may be produced by APD. Bradykinesia or akinesia is the most common and sometimes the only manifestation of drug-induced parkinsonism, and accounts for the facial masking, absent arm swing, slow initiation of motor activity, soft and monotonous speech, and flexed posture that are so commonplace among APD-treated psychiatric patients. In some cases mutism and dysphagia are particularly prominent manifestations of akinesia (Behrman, 1972; Solomon, 1977). Reduced facial expression and arm swing are useful early signs of akinesia and may be accompanied by subjective symptoms such as muscle fatigue or weakness. Drug-induced parkinsonism usually produces symmetrical akinesia and rigidity from the onset of symptoms, while early idiopathic Parkinson's disease is usually asymmetrical in distribution. It is obviously important to distinguish drug-induced akinesia from the psychomotor retardation commonly associated with underlying psychiatric illnesses such as depression or schizophrenia (Rifkin et al., 1975). APD may produce behavioural inhibition ranging in severity from subtle akinesia to catatonic stupor (Gelenberg and Mandel, 1977). If these signs are mistaken for psychiatric decompensation, more rather than less APD is likely to be given, and the problem made worse.

Rigidity of the extremities, neck or trunk, may appear days to weeks after the onset of akinesia. The presence or absence of cogwheel rigidity is not especially useful and simply reflects the presence or absence of an underlying tremor (Lance et al., 1963). Characteristic parkinsonian postural and gait abnormalities including flexed posture, impairment of righting responses, and propulsive or retropulsive gait often occur as well.

The characteristic, pill-rolling rest tremor of Parkinson's disease is uncommon in APD-induced parkinsonism. Instead, the tremor of drug-induced parkinsonism is usually a moderate to high frequency action tremor similar to essential tremor or parkinsonian action tremor. The "rabbit syndrome" refers to a focal, perioral tremor which may also occur in Parkinson's disease, is reversible where APD are discontinued, and responds to antiparkinson medication (Villeneuve, 1972). In some patients atypical action tremors occur which appear different from the classical tremors associated with other extrapyramidal disorders.

Signs of parkinsonism may begin within the first several days of APD treatment and gradually increase in incidence so that most cases appear within the first 3 months of treatment (Ayd, 1961). Following piperazine phenothiazines or other potent APDs, the progression of signs is often telescoped so that acute akinesia with mutism may appear within 48 h followed within days by rigidity and a resultant "akinetic hypertonic" syndrome (Delay and Deniker, 1968). Severe parkinsonian reactions may also occur following sudden discontinuation of anticholinergic drugs even if APDs have been discontinued at the same time. This may be due to cholinergic rebound together with the continued DA blocking effects of slowly excreted APD in the absence of protection by the more rapidly excreted anticholinergic drug (Simpson and Kunz-Bartholini, 1968). After discontinuation of APD most patients are free of extrapyramidal signs within a few weeks, but in some elderly individuals signs may persist for as much as several months to a year (Klawans et al., 1973; Marsden and Jenner, 1980; Stephen and Williamson, 1984). Prolonged drug-induced parkinsonism is usually distinguishable from Parkinson's disease by improvement over time. Anecdotal reports of permanent parkinsonism after drug discontinuation are usually limited to elderly individuals in whom underlying Parkinson's disease has been made clinically manifest by APD and which usually continues to progress despite discontinuation of APD treatment (Wilson et al., 1987).

It is often stated that tolerance develops for the parkinsonian effects of antipsychotic drugs but few prospective studies have adequately dealt with this phenomenon. Some observations that withdrawal of anticholinergic drugs after several months of administration leads to the appearance of a relatively small incidence of parkinsonism may be artifact due to the substantial number of patients who are placed on anticholinergic drugs prophylactically or to treat acute dyskinesia rather than parkinsonism (Gelenberg, 1982). Other anticholinergic withdrawal studies have shown a greater than 50% incidence of parkinsonism even after months to years of treatment (Korczyn and Goldberg, 1976; Manos et al., 1981).

Parkinsonism occurs following the use of reserpine, phenothiazines, thioxanthenes, and the butyrophenones. Following phenothiazines or butyrophenones, the incidence has varied between 5 and 60%, although incidence in routine psychiatric practice is about 10–15% (Marsden *et al.*, 1975). The incidence of drug-induced parkinsonism is determined by the potency of the APD which is used and the sensitivity of clinical examination. In spite of nearly universal susceptibility, individual predisposition does play an important role. In one study, the total dose of trifluoperazine required to produce a similar degree of parkinsonism ranged from 20 to 480 mg (Simpson and Kunz-Bartholini, 1968). Dramatic examples of severe parkinsonism appearing in some individuals within several days of treatment with small doses of APD are not uncommon.

There are few clues to the basis of individual susceptibility to parkinsonian effects of APD. In large studies, the age distribution of APD-induced parkinsonism closely parallels that of Parkinson's disease, with a sharp rise in incidence after age 40 (Ayd, 1961). However this has not been a universal finding (Moleman *et al.*, 1986) and occasional cases are also reported in children and neonates. Drug-induced parkinsonism is more frequent in patients with enlarged ventricles (Luchins *et al.*, 1983). The fact that levels of DA, tyrosine hydroxylase, and nigral cell counts all decline with advancing age (McGeer and McGeer, 1977; Smith and Baldessarini, 1980) may account for increased susceptibility to the effects of DA antagonists in older individuals. In two studies (Goetz, 1983; Wilson *et al.*, 1987) elderly patients developed idiopathic parkinson's disease within 1–4 years of having experienced reversible drug-induced parkinsonism. This suggests that APD can precipitate parkinsonism in elderly individuals with pre-clinical Parkinson's disease. Chase *et al.* (1970) reported that in patients with APD-induced parkinsonism, cerebrospinal fluid (CSF) levels of homovanillic acid (HVA) were lower than in those without parkinsonism, suggesting that the ability to mount a compensatory increase in DA turnover following DA receptor blockade was less efficient in affected patients. Van Praag and Korf (1976) measured responses of HVA levels in CSF to probenecid as a measure of DA turnover before and after treatment with APD in a group of patients with acute psychosis. Patients with more severe parkinsonism had a greater increase in HVA response than patients without parkinsonism suggesting more effective DA receptor blockade in those patients. Although this appears to contradict the conclusions of Chase *et al.* (1970), Van Praag and Korf (1976) also found that HVA responses to probenecid prior to APD treatment were significantly lower among patients who later developed parkinsonism suggesting that incidence of APD-induced parkinsonism may depend on pretreatment DA turnover more than post-treatment HVA responses to APD treatment.

Mechanisms

Many anatomical and physiological studies have demonstrated a dopaminergic nigrostriatal pathway, whose importance in relation to Parkinson's syndrome has been reviewed extensively (Marsden et al., 1975; Marsden and Jenner, 1980). Reserpine depletes brain DA, norepinephrine, and serotonin by interfering with presynaptic vesicular storage mechanisms, thereby permitting increased degradation by monoamine oxidase. Reserpine and tetrabenazine, a synthetic reserpine analogue with similar effects, can produce parkinsonism. In animals, reserpine induces a state of catalepsy characterized by profound akinesia and rigidity that is dramatically reversed by L-dopa. The phenothiazines, thioxanthenes, and butyrophenones produce a state of functional deficiency of DA by blockade of postsynaptic D1 and D2 DA receptors. Blockade of D2 receptors appears to be involved in the production of extrapyramidal syndromes (Black et al., 1985). In laboratory animals, these drugs induce catalepsy and antagonize the behavioural effects of the dopamine agonists, L-dopa, apomorphine, and amphetamine. Although a significant role for norepinephrine blockade has not been entirely excluded by these observations, the weight of evidence indicates that APD-induced parkinsonism is primarily due to interference with extrapyramidal dopaminergic mechanisms (Marsden and Jenner, 1980).

The effects of drugs used to treat APD-induced parkinsonism are consistent with this mechanism. L-Dopa reverses the effects of reserpine in animals and humans, presumably by replacement of depleted catecholamines. L-Dopa is rarely used in the management of drug-induced parkinsonism in psychotic patients because of potential for exacerbating psychosis but appears to be only partially effective (Bruno and Bruno, 1966), perhaps because of persistent blockade of postsynaptic DA receptors by continued use of APD. Amantadine has been used to treat APD-induced parkinsonism with some success, particularly in patients who fail to respond to anticholinergic drugs, and does not appear to aggravate psychiatric symptoms (Gelenberg, 1978; Borison, 1983).

Because of the reciprocal relationship between DA and acetylcholine in basal ganglia (Tarsy, 1979), drug-induced parkinsonism, similar to Parkinson's disease, should be characterized by cholinergic sensitivity. Physostigmine does in fact exacerbate phenothiazine-induced parkinsonism (Ambani et al., 1973; Gerlach et al., 1974) and anticholinergic drugs are usually therapeutic, although much less so than in the treatment of acute dyskinesias. APDs with potent anticholinergic properties, such as thioridazine or clozapine, produce a lower incidence of acute extrapyramidal reactions than APDs with little or no anticholinergic effect such as piperazine phenothiazines, thioxanthenes, or haloperidol. However, because co-administration of an anticholinergic drug with a phenothiazine of low anticholinergic potency still produces more akinesia in laboratory animals than thioridazine or clozapine, other factors, such as relative potency as DA receptor antagonists or regional selectivity in basal ganglia, are probably also important.

Treatment

Skilful management of drug-induced parkinsonism begins with awareness of minor clinical manifestations of the syndrome. Subtle forms of akinesia are often mistaken for negative symptoms of schizophrenia, depression, or social withdrawal (Rifkin *et al.*, 1975). If the dose of APD is inappropriately increased or a more potent APD is introduced, symptoms will be exacerbated rather than improved. Prophylactic anticholinergic drugs are more effective in preventing acute dyskinesia in young individuals than parkinsonism in older individuals (DiMascio and Demirgian, 1970; Keepers *et al.*, 1983). Since the incidence of anticholinergic toxicity in the form of confusion, hallucinations, memory disturbance, and urinary retention is greater in older individuals their routine use for prophylactic purposes is not recommended in older patients.

In patients known to be sensitive to parkinsonian effects, a relatively low potency APD such as thioridazine or molindone should be considered. If high-potency APDs are required in such a patient, short-term prophylactic administration of an antiparkinsonian drug is reasonable. If parkinsonism appears the dose of APD should be reduced rather than adding an antiparkinsonian drug in order to avoid introducing peripheral or central anticholinergic toxicity. Once patients have been maintained on anti-parkinsonian drugs for 3–6 months, the drugs should be tapered and discontinued as many patients will no longer require their use (Manos *et al.*, 1981).

In current practice, anticholinergic drugs are the most commonly used agents for treatment of APD-induced parkinsonism. L-Dopa is of unproved value in this situation and introduces the potential hazard of exacerbating psychiatric symptoms. Amantadine has had variable effects in treatment of APD-induced parkinsonism (Gelenberg, 1978; Borison, 1983; McEvoy *et al.*, 1987) but offers the advantage of avoiding anticholinergic side-effects.

NEUROLEPTIC MALIGNANT SYNDROME

Neuroleptic malignant syndrome (NMS) was first described in the French literature of the early 1960s where it was referred to as "akinetic hypertonic" syndrome (Delay and Deniker, 1968), but until recently had received only scant attention in the English language literature (Meltzer, 1973) and was unfamiliar to most clinicians. As a result, in many cases there has been delay in diagnosis with extensive and unnecessary medical evaluation for systemic disease. The clinical and laboratory features of NMS should allow for prompt diagnosis. Signs usually appear within several days or weeks after beginning therapy but sometimes appear after an increase in dose or change in APD in a patient treated for a more prolonged period of time. High-potency APDs such as the piperazine phenothiazines, particularly when given in long-acting parenteral form, haloperidol, and the thioxanthenes have been incriminated, However, rare cases have also been reported following less potent phenothiazines such as thioridazine. Incidence has been estimated at 0.5–1.5% (Delay and Deniker, 1968; Caroff, 1980; Pope et al., 1986). Men are affected more frequently than females, and the syndrome has occurred in psychiatrically normal individuals as well as in patients with Parkinson's disease (Henderson and Wooten, 1981) and Huntington's chorea (Burke et al., 1981). Clinical features evolve rapidly over the first 24–72 h to include acute extrapyramidal signs, major disturbances in autonomic function, and stupor. Extrapyramidal signs include widespread muscular rigidity, diffuse tremor, and fixed dystonic postures of the trunk, face and extremities. The neck, trunk, arms, and legs become hyperextended, and the hands and feet become fixed in carpopedal spasm. Persistent orthopaedic deformities of the extremities may result. Oropharyngeal hypertonicity and akinesia results in trismus, sialorrhea, dysarthria, dysphagia, and mutism. Mental status is typically difficult or impossible to assess and may be characterized by varying degrees of stupor or coma. Many patients remain aware and alert but are unable to communicate because of mutism. Autonomic disturbances include hyperthermia, diaphoresis, tachycardia, hypertension, hypotension, and urinary incontinence. Severe cases may be complicated by dehydration progressing to acute renal failure with myoglobinuria due to muscle rhabdomyolysis. Dyspnoea, tachypnoea, and frank respiratory failure are common late manifestations that may be due to pneumonia or pulmonary emboli. The most characteristic laboratory abnormalities in NMS are leukocytosis and marked elevation of muscle enzymes. The duration of untreated NMS following discontinuation of APD varies from one to several weeks and usually requires intensive care unit support for much of that time. Mortality is estimated at 20–25% (Caroff, 1980) and is usually due to respiratory failure, pulmonary embolism, pneumonia, cardiovascular collapse, or acute renal failure. There is some suggestion that mortality is higher among patients developing NMS from long-acting parenteral fluphenazine than in patients receiving only oral APDs (Caroff, 1980).

Differential diagnosis should include encephalitis or meningitis, intercurrent systemic infection associated with parkinsonian rigidity (Levinson and Simpson, 1986), malignant hyperthermia, heat stroke, and anticholinergic toxicity. Lethal catatonia (Stauder, 1934; Fricchione, 1985) is a condition which was described well before the introduction of APD and was characterized by high fever, stupor, catatonic posturing, autonomic lability, exhaustion, and sudden death. Lethal catatonia bears a remarkable similarity to NMS (Caroff, 1980) but usually occurred on a background of prolonged psychotic agitation and was unaccompanied by the muscle rigidity and dystonia which characterizes NMS (Mann *et al.*, 1986). Lethal catatonia is currently a very rare condition and probably was due to untreated extreme agitated psychotic states complicated by dehydration, exhaustion, and hypothalamic failure.

The pathophysiology of NMS is unknown but is probably attributable to central more than peripheral mechanisms. An intriguing observation is that many patients with NMS have been exposed to similar APDs both before and afterward without developing NMS. It has been suggested that concomitant factors, such as psychiatric agitation, physical exhaustion, or dehydration, may play a role in predisposition to NMS at a particular time (Itoh *et al.*, 1977). The syndrome often seems to appear in patients who are resistant to APD treatment and have required rapid escalation of dosage but these factors are not consistent (Smego and Durack, 1982; Pearlman, 1986). NMS is frequently compared with malignant hyperthermia (MH) because of the common features of hyperpyrexia and muscular rigidity. MH is a hereditary muscle disorder in which muscle-relaxing agents such as succinylcholine and inhalant anaesthetics produce muscle contraction in susceptible individuals by triggering influx of calcium into muscle cytoplasm (Wilner and Nagakawa, 1983). Dantrolene effectively reduces MH and recently has also been found to reverse the hyperpyrexia and muscle necrosis of NMS (Granato *et al.*, 1983). Although NMS does not appear to be associated with the same defect in muscle metabolism that occurs in MH, the possibility that they are related in pathophysiology is still considered (Wilner and Nagakawa, 1983; Caroff *et al.*, 1983). Note, however, that patients with NMS tolerate succinylcholine and anaesthetics without risk (Lotstra *et al.*, 1983) and, conversely, that patients with MH have been given APDs as preanaesthetic agents without precipitating MH.

It has been suggested that DA blockade is responsible for most of the features of NMS. This is supported by the appearance of an identical syndrome in patients with Parkinson's disease following acute withdrawal of L-dopa (Toru *et al.*, 1981) and in Huntington's disease following the use of DA-depleting drugs such as α-methylpyrosine and reserpine (Burke *et al.*, 1981). Clinically, the extrapyramidal signs give the appearance of severe parkinsonism occurring in combination with dystonia. Although extreme muscular rigidity may contribute to the hyperpyrexia of NMS, disturbances in temperature regulation and other autonomic manifestations are probably more closely related to DA blockade in hypothalamus or hypothalamic connections (Cox *et al.*, 1978; Pearlman, 1986). Anterior hypothalamic neuronal changes have recently been described at post-mortem examination in one patient with NMS (Horn *et al.*, 1988). The ability of bromocriptine to reverse the extrapyramidal and autonomic signs in NMS and the precipitation of NMS in occasional patients with Parkinson's disease are both consistent with the formulation that central DA receptor blockade plays a major role in its pathophysiology.

Prevention of NMS may be impossible due to its unpredictability but avoidance of rapid introduction of high dose APD, cautious drug treatment of refractory patients, avoiding treatment of medically ill patients, and avoiding acute withdrawal of anticholinergic drugs have been recommended (Sternberg, 1986). Treatment of NMS consists of prompt discontinuation of APDs, intensive support of cardiovascular, respiratory, and renal function, and treatment of intercurrent infection. Anticholinergic drugs, diphenhydramine, and diazepam are of no value. Although controlled treatment studies are lacking it appears from a series of individual case reports that dantrolene improves rigidity, hyperthermia, and rhabdomyolysis and bromocriptine improves extrapyramidal and autonomic signs (Coons *et al.*, 1982; Mueller *et al.*, 1983; Granato *et al.*, 1983). Although many survivors of NMS have subsequently received APDs without reappearance of the syndrome, recurrence of NMS has been reported in at least 25% of patients rechallenged with APD (Pearlman, 1986; Caroff *et al.*, 1987) and in 50% of patients rechallenged with the same APD or APD of equivalent potency (Shalev and Munitz, 1986). Therefore, diagnostic re-evaluation, alternative therapies, and if necessary gradual reintroduction of a different low potency APD are indicated (Caroff *et al.*, 1987).

AKATHISIA

Clinical features

The term akathisia means "not sitting" and was first introduced by Haskovec (1901) to describe motor restlessness in neurotic individuals. Subsequently, other authors (Bing, 1939; Wilson, 1940) described akathisia in encephalitis lethargica, post-encephalitic parkinsonism, and Parkinson's disease. With the introduction of APDs akathisia was observed in large numbers of patients for the first time since the epidemic of encephalitis lethargica. Although early descriptions described motor restlessness and movements of the feet (Ayd, 1961), later accounts emphasized subjective distress and a secondary need to move in order to relieve uncomfortable sensations (Van Putten, 1975). Recently this dichotomy has been expressed in the form of a dilemma as to whether akathisia is a movement disorder, a mental disorder, or both (Stahl, 1985).

The diagnosis of acute akathisia is based on a combination of subjective distress and motor restlessness following administration of an antipsychotic drug and should be suspected in any patient with restlessness or psychiatric agitation appearing shortly following APD treatment, a change to a more potent APD, or an increase in dose. Cyclically recurrent motor restlessness in patients on parenteral fluphenazine esters should also suggest the diagnosis.

Patients usually report subjective anxiety, inner tension, being driven to move, pulling or drawing sensations in their legs, and inability to tolerate inactivity in a seated or lying position. Subjective complaints include intense feelings of internal discomfort, tension, turmoil, irritability, impatience or unease. Strong affective components of fright, terror, anger, rage and anxiety may be included (Van Putten, 1975) but it is not clear if these are primary or secondary features. Such subjective sensations may be associated with a desire to escape a closed environment and in some cases may result in suicidal or violent behaviour. The compelling need to move usually serves to distinguish akathisia from pretreatment anxiety, dysphoria, or psychosis. Patients with psychotic agitation, may also exhibit excessive movements but usually in the form of semipurposeful activities such as hand-wringing and other repetitive behaviours. Psychotic patients with akathisia may have difficulty describing their symptoms but will usually admit to internal restlessness when that possibility is suggested to them (Van Putten *et al.*, 1984). Objective motor manifestations of akathisia include a variety of restless motor patterns. By definition, there should be inability to remain seated but this is commonly associated with repetitive shifting of weight, pacing or marching in place, and rocking of the trunk. Continuous shuffling or tapping movements of the feet while sitting may also be present but these are less specific and may represent choreoathetotic dyskinesia. Although in some cases the distinction will be arbitrary, an attempt should be made to decide whether the movements are variants of normal behaviour, psychotic behaviours, restless movements, or choreoathetotic dyskinesias (Stahl, 1985).

There are few causes for akathisia other than administration of neuroleptic drugs. Historically, akathisia was attributed to psychoneurosis or neurasthenia (Hascovec, 1901) but in current clinical practice this is an obscure and obsolete cause. Since encephalitis lethargica and its varied manifestations no longer occur this is not a practical consideration. Akathisia has also been described as a manifestation of Parkinson's disease. Although akathisia was a more common manifestation of post-encephalitic parkinsonism than idiopathic paralysis agitans (Bing, 1939; Wilson, 1940) a recent study in which patients with Parkinson's disease were carefully questioned about symptoms of akathisia indicates a relatively high incidence of this symptom although usually much less prominent than following neuroleptic treatment (Lang and Johnson, 1987). Although in most cases the diagnosis of idiopathic Parkinson's disease will be apparent, the frequent combination of parkinsonism and akathisia in patients on neuroleptic drugs may create a problem in differential diagnosis. Restless legs syndrome has strong clinical similarities to akathisia but can usually be distinguished by subjective discomfort confined to the legs, the absence of strong internal feelings of anxiety or restlessness, the prominence of symptoms when reclining, and the absence of the other motor manifestations of akathisia described above.

Akathisia must also be distinguished from choreoathetotic dyskinesias which it may resemble. Mild forms of chorea may be limited to fidgety, restless movements of the extremities and apparent inability to keep still. Sydenham's chorea is especially characterized by this type of motor restlessness. Akathisia should be distinguishable from akathisia-like movements seen in tardive dyskinesia ("tardive akathisia") by the presence of subjective internal discomfort and the relief produced by movement and changes of body position (Munetz and Cornes, 1983). Differentiation of akathisia from akathisia-like manifestations of tardive dyskinesia may be difficult however and at times becomes quite arbitary (Munetz and Cornes, 1982, 1983; Barnes and Braude, 1985; Stahl, 1985). Munetz and Cornes (1983) have proposed several clinical criteria to help make this differentiation. Firstly, akathisia should be subjectively distressing. Although tardive dyskinesia may cause distress this is secondary to the presence of involuntary movements rather than the cause. Care should be taken that the patient's appearance is not the sole basis for inferring subjective distress since severely dyskinetic patients often appear to be in distress without necessarily experiencing this subjectively. The movements of akathisia appear to be voluntary although very compelling attempts to relieve distressing internal stimuli while those of tardive dyskinesia are involuntary movements which may cause distress.

Mechanisms and treatment

The pathophysiology of akathisia is obscure. Because of its occurrence in Parkinson's disease and its occasional response to anticholinergic drugs it has been assumed to be an extrapyramidal disturbance although there is actually no compelling evidence for this. The absence of suitable animal models for akathisia has been a limiting factor in the understanding of this disorder. Marsden and Jenner (1980) have suggested that akathisia may be a result of postsynaptic DA receptor blockade in cerebral DA containing regions of the brain other than the corpus striatum. Thus, while blockade of DA receptors in the striatum and mesolimbic regions produces akinesia or catalepsy, locomotor hyperactivity occurs following blockade of mesocortical DA systems. Iversen (1971) showed that bilateral lesions of frontal cortex in rats cause enhancement of amphetamine-induced behavioural hyperactivity in contrast to bilateral lesions of substantia nigra which impair the locomotor response to amphetamine. Carter and Pycock (1978, 1980) showed that bilateral electrolytic or 6-hydroxydopamine lesions in medial prefrontal cortex of rats enhanced the stereotypic effects of amphetamine. By contrast, intracortical injection of DA into the same region of frontal cortex produced a dose-dependent state of catalepsy. These results were interpreted to indicate that catecholamines play an inhibitory role in frontal cortex but the failure of intracortical fluphenazine injection to enhance amphetamine-induced stereotypic or locomotor activity (Carter and Pycock, 1978) fails to support the concept that pharmacologic mesocortical DA receptor blockade enhances behavioural hyperactivity. LeMoal et al. (1969) described a behavioural syndrome in rats produced by high frequency lesions of the ventral tegmental area characterized by locomotor hyperactivity, hyperreactivity, and other disturbances of complex behaviours. The ventral tegmental area contains DA cells of origin for mesolimbic and mesocortical neurons which innervate the frontal and cingulate cerebral cortex. Tassin et al. (1978) also found that locomotor hyperactivity produced by bilateral ventral tegmental area lesions was dependent on ascending DA projections.

Non-human primate models for locomotor hyperactivity produced by neuroleptic drugs are scarce. Gore and Hadley (1959) observed a phase of excitation or restlessness in rhesus monkeys who were given piperazine phenothiazines. An initial sedative phase characterized by akinesia was followed by a phase of restlessness with bursts of motor activity including rolling on their backs, crawling on the bellies, leaping, attempting to escape from the observation table, assuming odd or bizarre positions, and retropulsion. This phase of restlessness was brief and followed by catalepsy. Although this phenomenon may be related to akathisia subsequent studies of the acute effects of phenothiazines in non-human primates describe acute dyskinesias and dystonias without motor restlessness (Meldrum et al., 1977; Liebman and Meale, 1980; McKinney et al., 1980).

Recent studies showing a therapeutic effect of α and β adrenergic blockers in akathisia (Lipinski et al., 1984; Zubenko et al., 1984; Adler et al., 1986, 1987) raise the possibility that akathisia is due to excessive noradrenergic activity. Norepinephrine nuclei located in the midbrain, some of which project to the spinal cord, may be important in some forms of locomotor activity (Bartels et al., 1981). In one human study urinary excretion of norepinephrine metabolites was reduced in patients with akathisia compared to controls suggesting the possibility of reduced norepinephrine turnover due to blockade and supersensitivity of postsynaptic norepinephrine receptors (Bartels *et al.*, 1981). Finally, the reported therapeutic effect of opioid drugs in akathisia (Walters *et al.*, 1986) should stimulate investigations of opioid mechanisms in the pathophysiology of akathisia.

Treatment of akathisia is notoriously difficult. For many years anticholinergic drugs were purported to be effective (Van Putten *et al.*, 1974) but in clinical practice have been disappointing. In part, this has probably been due to failure to distinguish acute and more reversible forms of akathisia from tardive and more resistant forms of the disorder. Recently, a series of studies has indicated that β blockers (Lipinski *et al.*, 1984; Adler *et al.*, 1986), clonidine (Zubenko *et al.*, 1984; Adler *et al.*, 1987), and opioids (Walters *et al.*, 1986) may be effective in the treatment of akathisia.

TARDIVE DYSKINESIA

Definition

Tardive dyskinesia (TD) is a choreoathetotic and dystonic disorder which appears relatively late in the course of treatment with APD. Although similar dyskinesias have been reported in small numbers following the use of other drugs including tricyclic antidepressants, antihistamines, benzodiazepines, and anticonvulsants, they usually occur acutely, are typically seen in elderly patients, and are transient phenomena which disappear with discontinuation of medication. TD associated with metoclopramide is comparable to APD-induced TD and is related to the DA receptor-blocking properties of this drug. By contrast with APD-induced TD, metoclopramide TD usually appears earlier in treatment, often occurs simultaneously with parkinsonism, and is rarely permanent when the drug is discontinued (Kataria *et al.*, 1978; Board, 1986). The term "tardive" was introduced in the early 1960s (Faurbye *et al.*, 1964) to emphasize the distinction from acute dyskinesia which by that time was a well-recognized phenomenon. Early reports emphasized the orofacial distribution and referred to it as a "buccolinguomasticatory syndrome" (Sigwald *et al.*, 1959). The term "persistent dyskinesia" was introduced to emphasize the frequently irreversible course of TD (Crane, 1973). However, currently it is recognized that many late appearing dyskinesias are not permanent (Tarsy, 1983). Subtypes of TD such as transient TD, withdrawal emergent TD, and persistent TD have been used to suggest clinical subtypes categorized according to their time course (Schooler and Kane, 1982). Although other criteria for subtyping have been proposed on the basis of pathophysiology, pharmacologic characteristics, topographic distribution, or type of involuntary movements it is not currently clear that these are sufficient to permit useful and valid subclassifications (Schooler and Kane, 1982).

Clinical features

TD is characterized by a variable mixture of orofacial dyskinesia, athetosis, dystonia, chorea, tics and facial grimacing. Rhythmic tremor is not part of the syndrome. Orofacial and lingual dyskinesias are the most characteristic and well-recognized features of TD. Such movements are typically insidious in onset and at first may be detectable as subtle restless movements of the tongue. Tic-like movements of facial muscles and increased blink frequency may also be early manifestations. Later, more obvious protruding, twisting, and curling movements of the tongue, pouting, puckering, sucking or smacking movements of the lips, retraction of the corners of the mouth, bulging of the cheeks, chewing jaw movements, blepharospasm, and facial grimaces may occur in various combinations.

In older patients, orofacial dyskinesia is typically the most conspicuous feature of the syndrome. However, involuntary movements of the extremities and trunk are usually present as well. Restless, choreiform and athetotic limb movements may include twisting, spreading, and "piano-playing" finger movements; tapping motions of the feet; and dorsiflexion postures of the great toe. Extremity movements are often more severe in younger individuals and may include dystonic and ballistic postures and movements. Dystonias of the neck and trunk include torticollis, retrocollis, truncal dystonia, rocking and swaying, shoulder shrugging, and rotatory or thrusting pelvic movements. Respiratory dyskinesias such as periodic tachypnoea, irregular respiratory rhythms, and grunting may occur and often lead to extensive pulmonary investigations. Severe and disabling axial dystonias referred to as tardive dystonia usually occur in individuals under 50 years of age (Tarsy and Bralower, 1977; Gerlach, 1979; Burke et al., 1982). A syndrome resembling TD has also been reported in children, primarily when APDs are withdrawn, known as withdrawal emergent symptoms (Polizos et al., 1973; Gualtieri et al., 1984). Like other choreoathetotic syndromes, involuntary movements of TD usually worsen with emotional stress, diminish with drowsiness or sedation, and disappear in sleep. Although patients are often unaware of the more subtle manifestations of TD, failure to complain of severe dyskinesia is usually limited to chronically institutionalized or severely psychotic patients. Some orofacial dyskinesias may interfere with speech, eating, or respiration while the more generalized truncal dystonias interfere with gait and mobility and are extremely distressing. Although attempts have been made to divide manifestations of TD into topographical subtypes (Kidger *et al.*, 1980) which may display distinctive pharmacologic responsiveness (Casey and Denney, 1977, Moore and Bowers, 1980; MacKay 1982) it is not currently established that these represent distinctive forms of TD.

Clinical course

TD may appear after as little as 3–6 months of APD exposure although in most of the early literature dyskinesia was observed after 2 or more years of treatment. Onset is insidious and usually occurs while the patient is still receiving APD. Frequently, however, onset is first noted following a reduction in dose or discontinuation of treatment carried out for psychiatric reasons or because of a suspicion of early TD. Under these circumstances the dyskinesia may be remarkably severe from the outset. This "unmasking" effect of APD withdrawal is due to the hypokinetic and parkinsonian effects of APDs which often result in a delay of recognition of TD.

Sudden withdrawal of APD may be followed by a self-limited dyskinesia that lasts for several days or weeks before subsiding. This is known as "withdrawal dyskinesia" (Jacobson *et al.*, 1974; Gardos *et al.*, 1978; Levine *et al.*, 1980) and is probably similar to withdrawal emergent symptoms described in children (Polizos *et al.*, 1973). The relationship of withdrawal dyskinesia to more persistent forms of TD is uncertain. However, since even patients with persistent TD may show slow improvement and disappearance of dyskinesia over several months to years following discontinuation of APD (Klawans *et al.*, 1984; Casey *et al.*, 1986; Kane *et al.*, 1986) it seems prudent to consider withdrawal dyskinesia a precursor of more persistent forms of TD (Baldessarini *et al.*, 1980).

Early studies of the natural course of TD following drug withdrawal were carried out among institutionalized patients with prolonged exposure to APD and long duration of TD. Remission rates in such studies varied between 5 and 40% (Baldessarini *et al.*, 1980). Currently, TD is identified earlier and in younger out-patient populations. In such patients a remission rate of 50–90% has been observed, usually occurring within several months, but sometimes requiring as long as 1–2 years after drug withdrawal (Baldessarini *et al.*, 1980; Kane *et al.*, 1986). Although it remains uncertain whether transient withdrawal dyskinesias constitute a reversible phase in the evolution toward persistent TD, these reports indicate the need for careful surveillance, early diagnosis and, when appropriate, prompt discontinuation of APD treatment.

Since patients with chronic psychosis may decompensate psychiatrically when an APD is discontinued, it is often impractical to terminate treatment in patients with TD. The prognosis of TD in patients who continue to receive APD is not known. However, in the vast majority of cases, TD either remains fairly static or becomes suppressed by the hypokinetic effects of continued APD treatment (Baldessarini et al., 1980). Whether the chance for remission is compromised by continued APD treatment is currently unknown.

Differential diagnosis

Although the diagnosis of TD is usually straightforward, other possibilities need to be considered in patients receiving an APD who also manifest signs of dyskinesia. In Wilson's and Huntington's diseases, for example, extrapyramidal signs may appear some time after psychiatric symptoms have already been treated with APD.

Initially TD must be distinguished from other APD-induced extrapyramidal syndromes. In patients still receiving APD, this task may be complicated by the coexistence of more than one drug-induced syndrome. Tremor is the only involuntary movement disorder unrelated to TD. This is a reversible sign which is usually associated with bradykinesia or rigidity. Rabbit syndrome sometimes occurs late in treatment unaccompanied by other parkinsonian signs but is a form of parkinsonian tremor which responds to anticholinergic agents and is entirely reversible. Akathisia is usually a reversible early phenomenon but may also appear late in treatment in the form of tardive akathisia and persist after discontinuation of APD. Acute dyskinesias are readily identified by their dramatic character and onset immediately following APD treatment. However, they sometimes are relatively mild and focal in distribution making a distinction from TD on the basis of appearance alone impossible. Since acute dyskinesia sometimes appears late in treatment when an APD is changed or dose increased, or may recur during treatment with long-acting fluphenazine esters, confusion with TD is possible.

TD may be difficult to distinguish from stereotyped movements and psychotic mannerisms traditionally associated with chronic schizophrenia (Marsden et al., 1975). Although it has been suggested that the prevalence, severity, and distribution of abnormal involuntary movements is not greater among APD-treated chronic schizophrenics than untreated patients (Owens et al., 1982), this is a minority view based on a relatively small sample of untreated patients. The mean prevalence of dyskinesia in 56 studies involving 35 000 APD-treated patients was 20% compared with a mean prevalence of 5% among 19 untreated patient samples totalling 11 000 patients (Kane and Smith, 1982). As a general rule, stereotyped movements of schizophrenia are usually less rhythmic, more variable and complex, and rarely choreoathetotic or dystonic. Spontaneous mouthing movements of the elderly, usually associated with dementia and edentulism, may also be difficult to distinguish from TD. Recent surveys indicate a substantial prevalence of spontaneous orofacial dyskinesias in untreated institutionalized geriatric patients (Kane and Smith, 1982). The common association with dementia, the usual restriction to the mouth and face, and lack of drug exposure should help differentiate these from TD.

Meige syndrome is an idiopathic focal dystonia which usually begins in middle age and is manifested by blepharospasm and oromandibular dystonia indistinguishable in appearance from some forms of orofacial TD (Tolosa, 1981). Differentiation from TD requires documenting the lack of APD exposure prior to its appearance. When TD is manifested by truncal dystonia in a young individual, it may be difficult to distinguish from idiopathic torsion dystonia (Burke et al., 1982). In addition to the drug history, helpful differentiating features are the slowly progressive course in most idiopathic dystonias contrasted with the rapid onset and static or slowly resolving course seen in tardive dyskinesia.

Huntington's disease is usually identifiable by the family history, progressive course, marked gait abnormality, and associated dementia. Unlike Huntington's disease, TD is rarely a pure chorea, but more commonly includes repetitive and rhythmic athetotic and dystonic movements or postures. In some patients with TD, facial tics and grimacing are particularly prominent and may resemble those of Tourette syndrome. Although irregular respiratory patterns and grunting can occur in TD the vocalizations of Tourette syndrome rarely occur. The childhood onset, lack of antecedent APD exposure, and a fluctuating course also helps to identify Tourette syndrome. Recent reports of symptoms strongly resembling Tourette syndrome appearing after prolonged APD treatment (Klawans et al., 1978) and orolingual dystonia appearing in a patient with Tourette syndrome who was treated with APD (Mizrahi et al., 1980) introduce additional complexities for differential diagnosis of the two conditions. Other differential diagnostic possibilities include choreoathetosis associated with chronic liver disease, hyperthyroidism, hypoparathyroidism, calcification of basal ganglia, drug intoxication with L-dopa, amphetamines, anticholinergics, antidepressants, and anticonvulsants, rheumatic or lupus chorea, and rare instances of dyskinesia associated with brain tumor or arteriovenous malformation in basal ganglia.

A helpful clue in the diagnosis of TD is that, except for the time of onset and the period during or shortly after APD withdrawal, TD does not appear to be a progressive disorder (Baldessarini et al., 1980) and is unaccompanied by tremor, rigidity, akinesia (unless the patient is still receiving APD), cerebellar signs or pyramidal weakness.

Risk factors and prevalence rates

Although it seems intuitively reasonable that prolonged treatment with large doses of APD would predispose to TD, it has not been possible to establish this in retrospective studies (Kane and Smith, 1982). It is likely, however, that a dose–response relationship does exist at low cumulative doses which is not readily identifiable in retrospective studies of patients treated for many years.

Nearly all studies evaluating age and TD have indicated that patients above age 50 are at a higher risk and may also have a poorer prognosis for eventual remission (Smith and Baldessarini, 1980). The ageing brain may be more susceptible to the appearance of oral dyskinesias under a variety of adverse circumstances including prolonged exposure to APD. Orofacial dyskinesias and minor choreoathetotic syndromes occur spontaneously and commonly among elderly individuals. Most American studies of spontaneous dyskinesia have indicated prevalences of less than 10% (Kane and Smith, 1982; Kane et al., 1982; Klawans and Barr, 1982) while several European studies have arrived at rates of 30% or more (Delwaide and Desseilles, 1977; Blowers, 1981). Studies confined to non-institutionalized, normal elderly individuals indicate a prevalence of 4–8% with an increase with advancing age (Kane and Smith, 1982; Kane et al., 1982). Evidently there is some risk of spontaneous choreoathetosis among elderly persons manifest primarily by orofacial dyskinesia which increases with advancing age. Exposure to APD may further increase the risk of persistent dyskinesia in this population of patients. Female sex, especially in the elderly, also appears to be a risk factor although less consistent than age (Kane and Smith, 1982; Kane et al., 1982).

Whether or not acute extrapyramidal side-effects are predictive for subsequent TD is not clearly established. Studies which suggest that APD-induced parkinsonism predisposes to subsequent appearance of TD (Chouinard et al., 1986) are difficult to interpret due to variability in APD exposure. In their prospective study Kane et al. (1986) report that TD incidence was higher among patients with than without early extrapyramidal symptoms but the type of symptoms was not defined. Incidence of TD has also been reported to be higher in patients with affective disorder than other psychoses (Rosenbaum et al., 1977; Kane et al., 1986).

There is no unequivocal evidence that any APDs in routine clinical use are more or less likely than others to produce TD. Attempts to relate class of APD to TD are limited by the fact that most patients available for study have been exposed to multiple APDs as well as by the frequent unreliability of drug histories. According to one review (Kane and Smith, 1982) five of 15 studies analysing relationships between risk of TD and type of APD reported significant associations. Most of these implicated depot forms of fluphenazine suggesting that compliance rather than drug type may have been responsible. Although low potency neuroleptics such as thioridazine or molindone produce a lower incidence of acute extrapyramidal reactions, there is no evidence that they are also associated with a lower incidence of TD. The suggestion that thioridazine causes fewer extrapyramidal effects due to selectivity for limbic rather than striatal DA receptors (Borison *et al.*, 1981) is not supported by all studies (Reynolds *et al.*, 1982). In fact, at least one clinical study has indicated a greater exposure to low potency APD among patients with TD compared to those without TD (Gardos *et al.*, 1977). The search continues for new APDs with less risk for all extrapyramidal reactions. There is evidence that clozapine, a dibenzazepine compound, and some of its congeners which remain investigational in the US, may be unique among APD in producing few, if any, acute or persistent extrapyramidal effects.

It has been difficult to establish prevalence rates of TD since criteria and standards for diagnosis vary widely. Although some epidemiologic studies have suggested a prevalence rate approaching 50%, these probably reflect increased sensitivity to minor degrees of adventitious movements. Based on retrospective studies it has been estimated that clinically appreciable cases occur in 10–20% of patients exposed to APD for more than a year, with the rate being higher in the elderly (Baldessarini *et al.*, 1980). An ongoing prospective study of incidence of TD in a cohort of patients, many of whom were placed on APD for the first time, has shown a cumulative incidence rate of 18.5% after 4 years and 40% after 8 years of treatment. The incidence of TD persistent for at least 6 months was 11% at 4 years and 22% at 8 years (Kane *et al.*, 1986).

Mechanisms

The prolonged and frequently irreversible course of TD suggests that permanent structural alterations of the brain are responsible for this disorder. However, neuropathologic studies in laboratory animals treated with chronic APD have not demonstrated consistent, specific, or localized pathologic changes (Baldessarini and Tarsy, 1979). Studies using quantitative, objective methods of assessment of the effects of haloperidol given to rats daily for several months suggest that complex changes may occur in synaptic profiles at the electron microscopic level (Benes *et al.*, 1983). In man, post-mortem changes following chronic APD treatment in patients without chronic extrapyramidal syndromes have usually consisted of scattered areas of neuronal degeneration and gliosis in the basal ganglia and other brain regions without consistent anatomic specificity (Tarsy and Baldessarini,

1984). Since neuropathological findings in spontaneous dystonias such as spasmodic torticollis and torsion dystonia are lacking, while one patient with spontaneous orofacial dyskinesias was found to have gliosis and neuronal changes limited to the caudate nucleus (Altrocchi and Forno, 1983), it is likely that neuropathological changes in TD will be relatively subtle if demonstrable at all.

It seems clear that an important and relatively selective action of APDs is to block the action of DA as a neurotransmitter in the basal ganglia and several other regions of the central nervous system. Parkinsonism and increased prolactin secretion reflect these actions in the basal ganglia and the anterior pituitary, respectively. Numerous clinical observations support the proposal that TD may be associated with relative dopaminergic over-activity. Thus, TD tends to worsen on withdrawal of APD; closely resembles the dyskinesias induced by L-dopa and amphetamines, usually worsens following administration of such DA agonists, and is often suppressed by treatment with drugs that deplete or block the action of DA. A state of relative excess of DA function in TD could come about through several mechanisms: presynaptic dyscontrol of DA synthesis and release; increased quantity or effectiveness of postsynaptic DA receptors due to denervation supersensitivity; or decreased availability of other modulating systems including those using GABA, acetylcholine, serotonin, substance P, endor-phins, or other peptides.

With respect to presynatic mechanisms, the increased turnover of DA and its metabolites induced in animals by APDs is short lasting (Asper et al., 1973) and unlikely to account for persistent behavioural effects following withdrawal of APD. In human studies, levels of homovanillic acid in cerebrospinal fluid do not remain elevated in schizophrenic patients exposed to prolonged APD treatment (Post and Goodwin, 1975) or in those with signs of TD (Bowers et al., 1978).

Evidence for the existence of DA receptor supersensitivity in animals following repeated treatment with an APD or other DA antagonists has been reviewed in detail elsewhere (Tarsy and Baldessarini, 1984). There is a gradual evolution of tolerance to many behavioural and biochemical effects after repeated administration of APD (Asper et al., 1973). Such effects are not found with neuroendocrine responses to APD due to differences in the regulation of pituitary and forebrain DA systems (Friend et al., 1978). In addition many animal species show increased behavioural sensitivity to acute treatment with a DA agonist upon withdrawal from repeated treatment with APD (Tarsy and Baldessarini, 1974; Muller and Seeman, 1978). None the less, these examples of supersensitivity to DA agonists after APD treatment are gener-ally short lived even in animals treated for as long as 1 year (Clow et al., 1980), raising questions about their pertinence to long-lasting cases of TD.

There thus remain serious problems and shortcomings of the hypothesis that TD is due to disuse supersensitivity of the DA receptor. The duration of supersensitive responses in most laboratory animals is relatively brief, usually involving a return to baseline status within a few weeks after discontinuation of APD (Tarsy and Baldessarini, 1974; Muller and Seeman, 1978). While TD can persist for many months or years, transient withdrawal dyskinesia may more closely resemble the experimental situation. On the other hand, newer primate models are more promising in this regard, as long-lasting supersensitivity to DA agonists as well as spontaneous dyskinesias have been produced, particularly in Cebus monkeys (Gunne and Barany, 1976). Since most biochemical and behavioural manifestations of presumed DA receptor supersensitivity involve relatively small increments (Tarsy and Baldessarini, 1984) it is not clear whether these are sufficient to be relevant for the pathophysiology of TD. Brain tissue from patients with untreated Parkinson's disease may have increased numbers of binding sites for neuroleptic agents in basal ganglia (Lee et al., 1978) supporting the concept that DA depletion may lead to proliferation of DA receptors in man. Post-mortem brain tissues of chronically psychotic patients may also contain an increase in the total number of neuroleptic binding sites in basal ganglia (Owen et al., 1978; Lee et al., 1980). Since this increase in putative DA receptor sites has been found only in patients in which APD treatment has been continued until the time of death (MacKay et al., 1982), the change in neuroleptic binding site density is presumably due to APD treatment rather than the underlying psychosis (Snyder, 1982). However, since there has been no difference in binding between tissues of psychotic patients with and without movement disorders the relationship between increases of DA receptors and pathophysiology of TD remains uncertain (Crow et al., 1982).

The possibility that changes in other neuronal systems may account for increased sensitivity to DA or make independent contributions to TD has also been considered (Baldessarini and Tarsy, 1979; Fibiger and Lloyd, 1984; Thaker et al., 1987). For example, TD could represent toxic or destructive effects on striatal interneurons which exert a physiologic feedback influence on nigrostriatal DA neurons or serve as part of an output pathway for nigrostriatal projections. Such interneurons may utilize GABA, acetylcholine, or peptides such as substance P as their neurotransmitter. Of these possibilities, the GABA hypothesis has been given the most consideration. Chronic treatment of Cebus monkeys with APD over a period of 3–6 years produced persistent dyskinesia in six of 12 animals (Gunne et al., 1984). Dyskinetic monkeys had reduced glutamic acid decarboxylase and GABA levels in substantia nigra, subthalamic nucleus, and globus pallidus. In addition, cerebrospinal fluid GABA levels were found to be significantly reduced in schizophrenic patients with TD compared with non-dyskinetic schizophrenic controls (Thaker et al., 1987).

Prevention and treatment

Although over the years there have been a large number of reports concerning drug treatment for TD, few have proven to be consistently useful in clinical practice. No treatment to date uniformly benefits all signs of dyskinesia and most treatments cause only slight to moderate benefit in fewer than half of patients treated (Jeste and Wyatt, 1982a). Emphasis, therefore, must rest on prevention, early detection, and management of early and potentially reversible cases.

The use of APD for more than 6 months requires careful evaluation of indications and risks and should be restricted to clinical situations in which other therapies are inadequate (Baldessarini et al., 1980). Long-term APD treatment in psychoneurosis, anxiety states, personality disorders, affective disorders, and chronic pain syndromes has little scientific basis and should be discouraged. The value of maintenance APD treatment for more than 6 months in psychiatric conditions is scientifically supported only for schizophrenia. Even in schizophrenia, however, there should be efforts to maintain patients on lowest effective doses of APD with periodic re-evaluation of the need for continued treatment.

Particular care is indicated in the treatment of patients above age 50. Although at this time a clear relationship between drug dose and risk for TD has not been proven, there is growing evidence that many chronically psychotic patients require doses of APD much lower than are commonly used (Baldessarini et al., 1980). There is no compelling evidence that APDs associated with a lower incidence of acute extrapyramidal side-effects are less likely to produce TD than high potency drugs. Recent reports that depot esters of fluphenazine are associated with a higher prevalence of TD are of uncertain significance and may relate to compliance rather than specific pharmacologic characteristics (Kane and Smith, 1982).

Since early extrapyramidal symptoms have been shown to correlate with subsequent appearance of TD in a prospective study (Kane et al., 1986) parkinsonism and akathisia should not be allowed to persist for an extended period. If possible, it is preferable to manage early extrapyramidal effects by lowering the dose of APD or changing to a less potent drug rather than introducing anticholinergic agents. Anticholinergic drugs do not prevent TD and, at least in some studies, appear to be associated with higher incidence of TD (Kane and Smith, 1982). In addition, an important practical reason for avoiding parkinsonism is that it masks the signs of underlying dyskinesia.

Since there is increasing evidence that early withdrawal of APD affords a better prognosis for recovery (Kane et al., 1986), patients should be carefully examined for dyskinesia at regular intervals with particular attention to the tongue, facial muscles, blink rate, fingers and toes. Use of standard dyskinesia rating scales (Baldessarini *et al.*, 1980; Jeste and Wyatt, 1982a) are useful to heighten awareness of the signs of TD, to promote careful clinical observation and documentation, and for follow-up comparisons. The value of "drug holidays" in reducing the risk of TD is uncertain (Kane and Smith, 1982). However, they have major potential value in allowing for unmasking of underlying dyskinesia.

When a patient develops dyskinesia while receiving an APD, ideal management is discontinuation of the medication. The manifestations of dyskinesia should be documented and the patient evaluated to exclude alternatives to TD. APD should then be withheld indefinitely in the hope that the dyskinesia will disappear, either slowly or rapidly. In some cases, the dyskinesia becomes even more severe when the drug is stopped. If it initially appears upon drug discontinuation it can be considered a withdrawal dyskinesia. If the dyskinesia fades within 4–8 weeks it is regarded as a transient dyskinesia. If it continues unchanged for 3 or more months off APD, it is considered a persistent form of TD, although even persistent cases may eventually remit over months to years of follow-up without APD (Klawans *et al.*, 1984). Since transient forms of TD may be precursors of persistent TD, patients should not be re-exposed to APD unless absolutely necessary. Psychiatric re-evaluation to determine whether alternative psychiatric diagnoses or treatments are available is indicated.

Before treatment of TD is attempted, the need for suppressing the dyskinesia should be carefully assessed. Although cosmetically undesirable, the dyskinesia may not be sufficiently disturbing to require specific therapy. If the dyskinesia produces mild symptoms then low doses of a benzodiazepine such as clonazepam or phenobarbital may reduce both the dyskinesia or associated anxiety (Bobruff *et al.*, 1981). A large number of other drugs have been used to treat TD (Jeste and Wyatt, 1982b; Alphs and Davis, 1982) including putative cholinergic agents such as choline and lecithin, GABA agonists such as benzodiazepines, γ-acetylenic GABA. γ-vinyl GABA, muscimol, and tetrahydroisoxazolopyridinol, drugs used because of presumed GABA agonist properties such as baclofen and sodium valproate, noradrenergic blocking drugs such as propranolol and clonidine, and a variety of miscellaneous agents including narcotics and calcium channel blockers. However efficacy of most of these drugs has only been studied in short-term studies of small numbers of patients, they have often had relatively slight and inconsistent effects, and have sometimes appeared to be useful only in conjunction with continued APD treatment. It is commonly stated that anticholinergic drugs exacerbate TD. Although this is often the case, they commonly have no effect when given orally, and may sometimes be effective in severe dystonic forms of TD (Wolf and Koller, 1984). Novel experimental approaches such as treatment with low doses of a DA agonist which presumably reduce DA transmission by activating presynaptic DA autoreceptors (Carroll *et al.*, 1977; Tarsy *et al.*, 1979) and attempts to desensitize supersensitive postsynaptic DA receptors with gradually elevated doses of L-dopa (Friedhoff, 1977; Casey *et al.*, 1982; Shoulson, 1983) have not led to practical therapies.

In patients with persistent, distressing and disabling dyskinesia, it may be necessary to resort to cautious use of DA antagonist drugs for effective treatment. Before resorting to APD a trial with a presynaptic DA depleting agent such as reserpine should be considered. Since reserpine does not bind to postsynaptic DA receptor sites and has rarely been incriminated in TD, it is theoretically preferable to treatment with postsynaptic DA blocker. Although the same has been said for tetrabenazine there is recent evidence that this drug also acts as a postsynaptic DA receptor blocker (Reches *et al.*, 1982).

It has long been recognized that when reintroduced to patients with TD, APDs suppress dyskinesia, either with or without induction of parkinsonism. Such treatment carries with it the theoretical risk of reducing the likelihood of eventual remission of the dyskinesia or further aggravating its severity. At present, however, there is no evidence that once TD appears it will continue to increase in severity with continued APD exposure (Baldessarini *et al.*, 1980; Casey *et al.*, 1986). Since patients with severe dyskinesia or psychosis unresponsive to other treatments typically require reintroduction of APD, the apparently non-progressive character of TD is of some consolation to the patient and family and may be of potential medical–legal significance. In patients for whom reintroduction of APD is necessary for psychiatric indications rather than suppression of dyskinesia the use of a low-potency APD is recommended, although not proven to be safer.

SUMMARY

The extrapyramidal movement disorders produced by APD continue to be a significant clinical problem. Acute dyskinesias, parkinsonism, neuroleptic malignant syndrome, akathisia and tardive dyskinesia have been reviewed with emphasis on their clinical features, pathophysiologic mechanisms, management and treatment. Although all are a direct result of treatment with APD, other epidemiologic factors may be important in accounting for their variable incidence in various patient populations. Most concepts of their pathophysiology have emphasized effects on DA neurotransmission, but other mechanisms have only recently been explored and deserve further investigation.

REFERENCES

Adler, L., Angrist, B., Peselow, E. et al. (1986) Br. J. Psychiat. 149, 42–45.
Adler, L. A., Angrist, B., Peselow, E. et al) (1987) Am. J. Psychiatr. 144, 235–236.
Alphs, L. and Davis, J. M. (1982) J. Clin. Psychopharmacol. 2, 380–385.
Altrocchi, P. H. and Forno, L. S. (1983) Neurology 33, 802–805.
Ambani, L. H., Van Woert, M. H. and Bowers, M. B., Jr. (1973) Arch. Neurol. 29, 444–446.
Angus, J. W. S. and Simpson, G. W. (1970) Acta Psychiatr. Scand. Suppl. 212, 52–58.
Asper, H., Baggiolini, M., Burki, H. R. et al. (1973) Eur. J. Pharmacol. 22, 287–294.
Ayd, R. F., Jr. (1961) J. Am. Med. Assoc. 175, 102–108.
Baldessarini, R. J. and Tarsy, D. (1979) Int. Rev. Neurobiol. 21, 1–45.
Baldessarini, R. J., Cole, J. O., Davis, J. M. et al. (1980) Am. J. Psychiat. 137, 1163–1172.
Barnes, T. R. E. and Braude, W. M. (1985) Arch. Gen. Psychiat. 42, 874–878.
Bartels, M., Gaertner, H. J. and Golfinopoulos, G. (1981) J. Neural. Transm. 52, 33–39.
Bateman, D. N., Rawlins, M. D. and Simpson, J. M. (1986) Quart. J. Med. 59, 549–556.
Behrman, S. (1972) Br. J. Psychiat. 121, 599–604.
Benes, F. M., Paskevich, P. A. and Domesick, V. B. (1983) Science 221, 969–971.
Bing, R. (1939) Textbook of Nervous Diseases (translated by W. Haymaker), p. 169. C. V. Mosby, St. Louis.
Black, J. L., Richelson, E. and Richardson, J. W. (1985) Mayo Clin. Proc. 60, 777–789.
Blowers, A. J. (1981) Neuropharmacology 20, 1339–1340.
Board, A. W. (1986) N. Engl. J. Med. 315, 518–519.
Bobruff, A., Gardos, G., Tarsy, D. et al. (1981) Am. J. Psychiat. 138, 189–193.
Borison, R. L. (1983) Clin. Neuropharmacol. 6, Suppl. 1, S57–S63.
Borison, R. L., Fields, J. Z. and Diamond, B. I. (1981) Neuropharmacology 20, 1321–1322.
Bowers, M. B., Jr., Moore, D. and Tarsy, D. (1978) Psychopharmacology 61, 137–141.
Bruno, A. and Bruno, S. C. (1966) Acta Psychiat. Scand. 42, 264.
Burke, R. E., Fahn, S., Mayeux, R. et al. (1981) Neurology 31, 1022–1026.
Burke, R. E., Fahn, S., Jankovic, J. et al. (1982) Neurology 32, 1335–1346.
Burke, R. E., Reches, A., Traub, M. M. et al. (1985) Ann. Neurol. 17, 200–202.
Caroff, S. N. (1980) J. Clin. Psychiat. 41, 70–83.
Caroff, S., Rosenberg, H. and Gerber, J. C. (1983) J. Clin. Psychopharmacol. 3, 120–121.
Caroff, S. N., Mann, S. C. and Lazarus, A. (1987) Arch. Gen. Psychiat. 44, 838–839.
Carroll, B. J., Curtis, G. C. and Kokmen, A. D. (1977) Am. J. Psychiat. 134, 785–789.
Carter, C. J. and Pycock, C. J. (1978) Br. J. Pharmacol. 42, 402P.

Carter, C. J. and Pycock, C. J. (1980) *Brain Res.* **192**, 163–176.

Casey, D. and Denney, D. (1977) *Psychopharmacology* **54**, 1–8.

Casey, D. E., Gerlach, J. and Bjorndal, N. (1982) *Psychopharmacology* **78**, 89–92.

Casey, D. E., Povisen, U. J., Meidahl, B. *et al.* (1986) *Psychopharmacol. Bull.* **22**, 250–253.

Chase, T. N., Schur, J. A. and Gordon, E. K. (1970) *Neuropharmacology* **9**, 265–275.

Chouinard, G., Annable, L., Mercier, P. *et al.* (1986) *Psychopharmacol. Bull.* **22**, 259–263.

Clow, A., Theodorou, A., Jenner, P. *et al.* (1980) *Eur. J. Pharmacol.* **63**, 135–144.

Coons, D. J., Hillman, F. J. and Marshall, R. W. (1982) *Am. J. Psychiat.* **139**, 944–945.

Cox, B., Kerwin, R. and Lee, T. F. (1978) *J. Physiol.* **282**, 471–483.

Crane, G. E. (1973) *Br. J. Psychiat.* **122**, 395–405.

Crow, T. J., Cross, A. J., Johnstone, E. C. *et al.* (1982) *J. Clin. Psychopharmacol.* **2**, 336–340.

Delay, J. and Deniker, P. (1968) In *Diseases of the Basal Ganglia* (edited by P. J. Vinken and G. W. Bruyn), pp. 248–266. North Holland, Amsterdam.

Delwaide, P. J. and Desseilles, M. (1977) *Acta Neurol. Scand.* **56**, 256–262.

DiMascio, A. and Demirgian, E. (1970) *Psychosomatics* **11**, 596–601.

Fann, W. E. and Lake, G. R. (1976) *Am. J. Psychiat.* **8**, 940–943.

Faurbye, A., Rasch, P. J., Bender Petersen, P. *et al.* (1964) *Acta Psych. Scand.* **40**, 10–27.

Fibiger, H. C. and Lloyd, K. G. (1984) *Trends Neurosci.* **7**, 462–464.

Fricchione, G. L. (1985) *Biol. Psychiat.* **20**, 304–313.

Friedhoff, A. J. (1977) *Comp. Psychiat.* **18**, 309–317.

Friend, W. C., Brown, G. M., Jawahir, G. *et al.* (1978) *Am. J. Psychiat.* **135**, 839–841.

Gagrat, D., Hamilton, J. and Belmaker, R. H. (1978) *Am. J. Psychiat.* **135**, 1232–1233.

Gardos, G., Cole, J. O. and Labrie, R. (1977) *Prog. Neuropsychopharmacol.* **1**, 147–154.

Gardos, G., Cole, J. O. and Tarsy, D. (1978) *Am. J. Psychiat.* **135**, 1321–1324.

Garver, D. L., Davis, J. M. and Dekirmenjian, H. (1976) *Arch. Gen. Psychiat.* **33**, 862–866.

Gelenberg, A. J. (1978) *Curr. Ther. Res.* **23**, 375–380.

Gelenberg, A. (1982) *Biol. Ther. Psychiat.* **5**, 10–11.

Gelenberg, A. J. and Mandel, M. R. (1977) *Arch. Gen. Psychiat.* **34**, 947–950.

Gerlach, J. (1979) *Dan. Med. Bull.* **26**, 209–245.

Gerlach, J., Reisby, N. and Randrup, A. (1974) *Psychopharmacologia* **34**, 21–35.

Gessa, R., Tagliamonte, A. and Gessa, G. K. (1972) *Lancet* **ii**, 981–982.

Goetz, C. G. (1983) *Arch. Neurol.* **40**, 325–326.

Gore, E. M. and Hadley, F. V. (1959) *Fed. Proc.* **18**, 397.

Granato, J. E., Stern, B. J., Ringel, A. *et al.* (1983) *Ann. Neurol.* **14**, 89–90.

Gualtieri, C. T., Quade, D., Hicks, R. E. *et al.* (1984) *Am. J. Psychiat.* **141**, 20–23.

Gunne, L. M. and Barany, S. (1976) *Psychopharmacology* **50**, 237–240.

Gunne, L. M., Haggstrom, J. E. and Sjoquist, B. (1984) *Nature* **309**, 347–349.

Hascovec, L. (1901) *Rev. Neurol.* **9**, 1102–1109.

Heintz, R. and Casey, D. E. (1987) *Psychopharmacology* **93**, 207–213.

Henderson, V. W. and Wooten, G. F. (1981) *Neurology* **31**, 132–137.

Horn, E., Lach, B., Lapierre, Y. *et al.* (1988) *Am. J. Psychiat.* **145**, 617–620.

Itoh, H., Ohtsuka, N., Ogita, K. *et al.* (1977) *Folia Psychiat. Neurol. Jpn.* **31**, 565–576.

Iversen, S. (1971) *Brain Res.* **31**, 295–311.

Jacobson, G., Baldessarini, R. J. and Manschreck, T. (1974) *Am. J. Psychiat.* **131**, 910–913.

Jeste, D. V. and Wyatt, R. J. (1982a) *Understanding and Treating Tardive Dyskinesia.* Guilford Press, New York.

Jeste, D. V. and Wyatt, R. J. (1982b) *Arch. Gen. Psychiat.* **39**, 803–816.

Kane, J. M. and Smith, J. M. (1982) *Arch. Gen. Psychiat.* **39**, 473–481.

Kane, J. M., Weinhold, P., Kinon, B. *et al.* (1982) *Psychopharmacology* **77**, 105–108.

Kane, J. M., Woerner, M., Borenstein, M. *et al.* (1986) *Psychopharmacol. Bull.* **22**, 254–258.

Kataria, M., Traub, M. and Marsden, C. D. (1978) *Lancet* **ii**, 1254–1255.

Keepers, G. A., Clappison, V. J. and Casey, D. E. (1983) *Arch. Gen. Psychiat.* **40**, 1113–1117.

Kidger, T., Barnes, T. R. E., Traver, T. *et al.* (1980) *Psychol. Med.* **10**, 513–520.

Klawans, H. L. and Barr, A. (1982) *Neurology* **32**, 558–559.

Klawans, H. L., Bergen, D. and Bruyn, G. W. (1973) *Confin. Neurol.* **35**, 368–377.

Klawans, H. L., Falk, D. K., Nausieda, P. A. *et al.* (1978) *Neurology* **28**, 1064–1068.

Klawans, H. L., Tanner, C. M. and Barr, A. (1984) *Clin. Neuropharmacol.* **7**, 153–159.

Kolbe, H., Clow, A., Jenner, P. *et al.* (1981) *Neurology* **31**, 434–439.

Korczyn, A. D. and Goldberg, G. J. (1972) *Br. J. Psychiat.* **121**, 75–77.

Korczyn, A. D. and Goldberg, G. J. (1976) *J. Neurol. Neurosurg. Psychiat.* **39**, 866–869.

Lance, J. W., Schwab, R. S. and Peterson, E. A. (1963) *Brain* **86**, 95–108.

Lang, A. E. and Johnson, K. (1987) *Neurology* **37**, 479–481.

Lee, T., Seeman, P., Rajput, A. *et al.* (1978) *Nature* **237**, 59–61.

Lee, T., Seeman, P., Tourtelotte, W. W. *et al.* (1980) *Nature* **274**, 897–900.

LeMoal, M., Cardo, B. and Stinus, L. (1969) *Physiol. Behav.* **4**, 567–574.

Levine, J., Schooler, N. R., Severe, J. *et al.* (1980) In *Long-term Effects of Neuroleptics* (*Adv. Biochem. Psychopharmacol., Vol. 24*) (edited by F. Cattebeni, G. Racagni and P. F. Spano), pp. 483–493. Raven Press, New York.

Levinson, D. F. and Simpson, G. W. (1986) *Arch. Gen. Psychiat.* **43**, 839–848.

Liebman, J. and Meale, R. (1980) *Psychopharmacology* **68**, 25–29.

Lipinski, J. F., Zubenko, G., Cohen, B. M. *et al.* (1984) *Am. J. Psychiat.* **141**, 412–415.

Login, I. S., Cronin, M. J. and MacLeod, R. M. (1982) *Ann. Neurol.* **12**, 257–262.

Lotstra, F., Linkowski, P. and Mendlewicz, J. (1983) *Biol. Psychiat.* **18**, 243–247.

Luchins, D. J., Jackman, H. and Meltzer, H. (1983) *Psychiat. Res.* **17**, 7–14.

Mann, S. C., Caroff, S. N., Bleier, H. R. *et al.* (1986) *Am. J. Psychiat.* **143**, 1374–1381.

MacKay, A. V. P. (1982) In *Movement Disorders* (edited by C. D. Marsden and S. Fahn), pp. 249–262. Butterworth, London.

MacKay, A. V. P., Iversen, L. L., Rossor, M. N. *et al.* (1982) *Arch. Gen. Psychiat.* **39**, 991–997.

Manos, N., Gziouzepas, J. and Logothetis, J. (1981) *Am. J. Psychiat.* **138**, 184–188.

Marsden, C. D. and Jenner, P. (1980) *Psychol. Med.* **10**, 55–72.

Marsden, C. D., Tarsy, D. and Baldessarini, R. J. (1975). In *Psychiatric Aspects of Neurologic Disease* (edited by D. F. Benson and D. Blumer), pp. 219–266. Grune and Stratton, New York.

Matthysse, S. (1973) *Fed. Proc.* **32**, 200–205.

McEvoy, J. P., McCue, M. and Freter, S. (1987) *Clin. Ther.* **9**, 429–433.

McGeer, P. L. and McGeer, E. G. (1977) *Arch. Neurol.* **34**, 33–35.

McKinney, W. T., Moran, E. C., Kraemer, G. W. *et al.* (1980) *Psychopharmacology* **72**, 35–39.

Meldrum, B. S., Anlezark, G. M. and Marsden, C. D. (1977) *Brain* **100**, 313–336.

Meltzer, H. Y. (1973) *Psychopharmacologia* **29**, 337–346.

Mizrahi, E. M., Holtzman, D. and Tharp, B. (1980) *Arch. Neurol.* **37**, 780.

Moleman, P., Janzen, G., Von Bargen, B. A. *et al.* (1986) *Am. J. Psychiat.* **143**, 232–234.

Moore, D. and Bowers, M. B., Jr (1980) *Am. J. Psychiat.* **137**, 1202–1205.

Mueller, P., Vester, J. W. and Fermaglich, J. (1983) *J. Am. Med. Assoc.* **249**, 386–388.

Muller, P. and Seeman, P. (1978) *Psychopharmacology* **60**, 1–11.

Munetz, M. R. and Cornes, C. L. (1982) *Comp. Psychiat.* **23**, 345–352.

Munetz, M. R. and Cornes, C. L. (1983) *J. Clin. Psychopharmacol.* **3**, 343–349.

Nyback, H., Schubert, J. and Sedvall, G. (1970) *J. Pharm. Pharmacol.* **10**, 622–624.

Owen, F., Cross, A. J., Crow, T. J. *et al.* (1978) *Lancet* ii, 223–225.

Owens, D. G. C., Johnstone, E. C. and Frith, C. D. (1982) *Arch. Gen. Psychiat.* **39**, 452–461.

Pearlman, C. A. (1986) *J. Clin. Psychopharmacol.* **6**, 257–273.

Polizos, P., Engelhardt, D. M., Hoffman, S. P., *et al.* (1973) *J. Autism Child Schizo.* **3**, 247–253.

Pope, H. G., Keck, P. E., Jr and McElroy, S. C. (1986) *Am. J. Psychiat.* **143**, 1227–1233.

Post, R. M. and Goodwin, F. K. (1975) *Science* **190**, 488–489.

Reches, A., Burke, R. E., Kuhn, C. *et al.* (1982) *Ann. Neurol.* **12**, 94.

Reynolds, G. P., Cowey, L., Rossor, M. N. *et al.* (1982) *Lancet* ii, 499–500.

Rifkin, A., Quitkin, F. and Klein, D. F. (1975) *Arch. Gen. Psychiat.* **32**, 672–674.

Rosenbaum, K. M., Niven, R. G., Hanson, N. P. *et al.* (1977) *Dis. Nerv. Syst.* **38**, 423–427.

Sarwer-Foner, G. J. (1960) *Can. Med. Assoc. J.* **83**, 312–318.

Schooler, N. R. and Kane, J. M. (1982) *Arch. Gen. Psychiat.* **39**, 486–487.

Shalev, A. and Munitz, H. (1986) *Acta Psychiat. Scand.* **73**, 337–347.

Shoulson, I. (1983) In *Experimental Therapeutics of Movement Disorders* (*Adv. Neurol. Vol. 37*) (edited by S. Fahn, D. B. Calne and I. Shoulson), pp. 259–266. Raven Press, New York.

Sigwald, J., Bouttier, D., Raymondeaud, C. *et al.* (1959) *Rev. Neurol.* **100**, 751–755.

Simpson, G. M. and Kunz-Bartholini, E. (1968) *Dis. Nerv. Syst.* **29**, 269–274.

Skirboll, L. R., Grace, A. A. and Bunney, B. S. (1979) *Science* **206**, 80–82.

Smego, R. A. and Durack, D. T. (1982) *Arch. Int. Med.* **142**, 1183–1185.

Smith, J. M. and Baldessarini, R. J. (1980) *Arch. Gen. Psychiat.* **37**, 1368–1373.

Snyder, S. H. (1982) *Lancet* ii, 970–974.

Solomon, K. (1977) *Am. J. Psychiat.* **134**, 308–311.

Stahl, S. (1985) *Arch. Gen. Psychiat.* **42**, 915–917.

Stauder, K. H. (1934) *Arch. Psychiat. Nervenkr.* **102**, 614–634.

Sternberg, D. E. (1986) *Am. J. Psychiat.* **143**, 1273–1275.

Stephen, P. J. and Williamson, J. (1984) *Lancet* ii, 1082–1083.

Swett, Jr, C. (1975) *Am. J. Psychiat.* **132**, 532–534.

Tarsy, D. (1979) In *Acetylcholine and Neuropsychiatric Disease* (edited by K. L. Davis and P. A. Berger), pp. 395–424. Plenum Press, New York.

Tarsy, D. (1983) *Clin. Neuropharmacol.* **6**, 91–100.

Tarsy, D. and Baldessarini, R. J. (1974) *Neuropharmacology* **13**, 927–940.

Tarsy, D. and Baldessarini, R. J. (1984) *Ann. Rev. Med.* **35**, 605–623.

Tarsy, D. and Bralower, N. (1977) *Am. J. Psychiat.* **134**, 1032–1034.

Tarsy, D., Gardos, G. and Cole, J. O. (1979) *Neurology* **29**, 606.

Tassin, J. P., Stinus, L., Simon, H. *et al.* (1978) *Brain Res.* **141**, 267–281.

Thaker, G. K., Tamminga, C. A., Alphs, L. D. *et al.* (1987) *Arch. Gen. Psychiat.* **44**, 522–529.

Tolosa, E. S. (1981) *Arch. Neurol.* **38**, 147–151.

Toru, N., Matsuda, O., Makiguch, K. *et al.* (1981) *J. Nerv. Ment. Dis.* **169**, 324–327.

Van Praag, H. M. and Korf, J. (1976) *Am. J. Psychiat.* **133**, 1171–1176.

Van Putten, T. (1975) *Comp. Psychiat.* **16**, 43–47.

Van Putten, T., Mutalpassi, C. R. and Malkin, M. D. (1974) *Arch. Gen. Psychiat.* **30**, 102–105.

Van Putten, T., May, P. R. A. and Marder, S. R. (1984) *Arch. Gen. Psychiat.* **41**, 1036–1039.

Villeneuve, A. (1972) *Can. Psychiat. Assoc. J.* **17**, SS69–SS72.

Walters, A., Henning, W., Chokroverty, S. *et al.* (1986) *Movement Disorders* **1**, 119–127.

Willner, H. and Nagakawa, M. (1983) *Sem. Neurol.* **3**, 275–282.

Wilson, J. A., Primrose, W. R. and Smith, R. G. (1987) *Lancet* **i**, 443–444.

Wilson, S. A. K. (1940) *Neurology* (edited by G. N. Bruce), pp. 118, 793. Williams and Wilkins, Baltimore.

Winslow, R. S., Stillner, V., Coons, D. J. *et al.* (1986) *Am. J. Psychiat.* **143**, 706–710.

Wolf, M. E. and Koller, W. C. (1984) *Neurology* **34** (Suppl. 1), 129.

Zubenko, G. S., Cohen, B. M., Lipinski, J. F. *et al.* (1984) *Psychiat. Res.* **13**, 253–259.

28

The Consequences of Chronic Exposure to Neuroleptics in the Rat

Angela Clow

INTRODUCTION

Neuroleptic drugs possess antipsychotic activity which is generally attributed to their capacity to block cerebral dopamine receptors (Carlsson and Lindqvist, 1963). Indeed, on acute administration the ability of neuroleptics to act as cerebral dopamine receptor antagonists in man and animals is well documented. However, two major problems plague this basic theory. First, chronic neuroleptic therapy in man is known to lead to tardive dyskinesias in a significant proportion of patients. This syndrome has long been associated with cerebral dopamine receptor overactivity and can worsen on drug withdrawal (Marsden et al., 1975). It is difficult to reconcile the occurrence of tardive dyskinesia with persistent cerebral dopamine receptor blockade. Second, animal evidence shows that repeated neuroleptic administration leads to tolerance to its acute dopamine receptor blocking effects (Asper et al., 1973) and withdrawal of the drug leads to dopamine receptor supersensitivity (Tarsy and Baldessarini, 1974). So clinical evidence and animal studies suggest a loss of cerebral dopamine receptor blockade during chronic neuroleptic treatment.

To shed light on this problem, the effects of long-term (up to 18 months) neuroleptic drug administration and subsequent withdrawal on cerebral dopamine systems and the function of related neuronal pathways in the rat have been investigated. The studies have looked at typical and atypical neuroleptics with the ability to induce a greater or lesser incidence of tardive dyskinesia. In this way it was hoped to differentiate the actions of the drugs responsible for antipsychotic activity from those precipitating tardive dyskinesia.

CHRONIC ADMINISTRATION OF PHENOTHIAZINE DRUGS

In an initial study the effect of the two commonly used phenothiazine drugs trifluoperazine and thioridazine were examined (Clow et al., 1979). These compounds were selected because of the potent antidopaminergic properties of trifluoperazine compared to the low dopamine antagonist action but high anticholinergic activity of thioridazine (Miller and Hiley, 1974). Hence the comparison of these two drugs was of interest as the severity of extrapyramidal side-effects may depend on the inherent anticholinergic potency of individual neuroleptics (Marsden et al., 1975).

Administration of trifluoperazine or thioridazine by incorporation into drinking water for up to 12 months initially caused substantial dopamine receptor blockade as evidenced by reduced spontaneous locomotor activity, marked catalepsy and inhibition of apomorphine-induced stereotyped behaviour. However, after 3 months of continuous drug administration most signs of dopamine receptor blockade had disappeared, to be replaced after 6–12 months' drug administration by an enhanced behavioural response to apomorphine. In particular, the high intensity components of apomorphine-induced stereotypy, namely gnawing and biting, were exaggerated after long-term trifluoperazine or thioridazine administration (Fig. 28.1). Such high-intensity stereotyped responses are believed to be due to stimulation of striatal dopamine receptors (Costall et al., 1977).

Biochemical analysis showed a similar sequence of events (Clow et al., 1980a). Striatal dopamine turnover was increased by the acute administration of trifluoperazine and thioridazine but this effect disappeared on continued administration of the drugs for a month or more. Thus, striatal HVA and DOPAC concentrations were increased in the first 2 weeks of drug administration, but thereafter generally were no different from those in control animals. Striatal dopamine concentrations were not altered consistently during the administration of neuroleptic drugs for 12 months.

Long-term administration of trifluoperazine or thioridazine caused alterations in the dopamine specific binding of ^{3}H-spiperone to striatal preparations (Fig. 28.1). The number of striatal binding sites for ^{3}H-spiperone

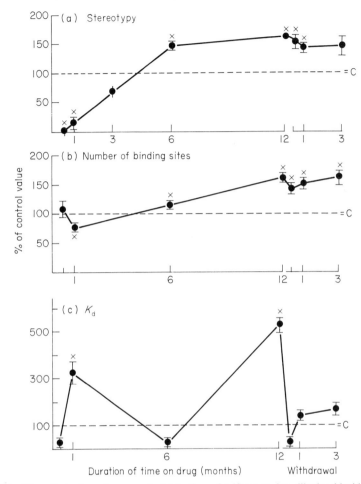

Fig. 28.1 The effect of continuous administration of trifluoperazine dihydrochloride (2.5–3.5 mg kg^{-1} day^{-1} p.o.) to rats and its subsequent withdrawal for 3 months on: (a) apomorphine (0.5 mg kg^{-1} s.c.)-induced stereotyped behaviour; (b) specific striatal ^{3}H-spiperone binding sites (B_{max}); and (c) the dissociation constant (K_{d}) for specific striatal ^{3}H-spiperone binding.

increased progressively by comparison with controls, over the year of drug administration. This increase in B_{max} was paralleled by similar changes in the sensitivity of striatal adenylate cyclase to dopamine stimulation. Thus, initial inhibition of dopamine stimulation of striatal adenylate cyclase after 1 month's drug administration by comparison with control animals, was progressively reversed after 3–12 months' drug intake such that dopamine stimulated striatal adenylate cyclase more in drug-treated than control animals.

So it appeared that during chronic administration of trifluoperazine or thioridazine to rats for 1 year, initial dopamine receptor blockade was replaced by striatal dopamine receptor supersensitivity despite continued drug administration. But are these changes specific to phenothiazines or do other classes of neuroleptics produce similar effects?

CHRONIC ADMINISTRATION OF THIOXANTHENES AND BUTYROPHENONES

The *cis* and *trans* isomers of flupenthixol (a thioxanthene) stereoselectively inhibit brain dopamine receptors. *cis*-Flupenthixol is the more pharmacologically active compound possessing antipsychotic activity while the *trans* isomer is devoid of such effects (Johnstone *et al.*, 1978). So administration of both isomers was a useful means of determining whether the active isomer alone was able to induce changes in dopamine receptor sensitivity.

Haloperidol was selected as a representative of the butyrophenone class of neuroleptics. Flupenthixol or haloperidol were administered continually to rats for 18 months in their drinking water and indices of brain dopamine function were assessed compared to control animals (Murugaiah *et al.*, 1983, 1984; Rupniak *et al.*, 1985b).

Administration of *cis*-flupenthixol or haloperidol initially caused a slight depression of apomorphine-induced stereotyped behaviour but after 6 months or more of continuous drug intake the stereotyped response was markedly enhanced.

Biochemical correlates of cerebral dopamine function also showed progressive alterations during continuous administration of *cis*-flupenthixol or haloperidol. The initial increase in striatal and mesolimbic dopamine turnover had abated by 6 months. Again by 12 months of haloperidol and 18 months of *cis*-flupenthixol administration the number (B_{max}) of specific binding sites for ^{3}H-spiperone in the striatum had increased. Chronic administration of *cis*-flupenthixol or haloperidol also led to enhanced dopamine stimulation of striatal adenylate cyclase.

So, other classes of neuroleptics also induce dopamine receptor supersensitivity during continuous drug administration. Interestingly, *trans*-flupenthixol was without effect indicating the specificity of the drug effect concerned.

CHRONIC ADMINISTRATION OF ATYPICAL NEUROLEPTIC DRUGS

The substituted benzamide drug sulpiride differs from classical neuroleptic agents in causing few extrapyramidal side-effects (Alberts, 1985). Sulpiride also differs from classical neuroleptic agents in its acute pharmacological profile possessing generally weak activity in certain dopamine-mediated behaviours (Costall and Naylor, 1975; Elliott *et al.*, 1977). Sulpiride is also unable to antagonize dopamine stimulation of adenylate cyclase activity via D1 receptors (Trabucchi *et al.*, 1975) and only weakly displaces specific ^{3}H-fluphenthixol or ^{3}H-piflutixol binding to D1 sites. So, are the long-term effects of sulipride administration on cerebral dopamine function the same as those produced by typical neuroleptic drugs?

In contrast to all previously investigated neuroleptics the continuous oral administration of sulpiride for up to 1 year did not induce any changes in apomorphine-induced stereotypy or striatal ^{3}H-spiperone binding indicating that under these conditions sulpiride was without effect on D2 receptor function (Rupniak *et al.*, 1984; Table 28.1). Totally unexpectedly, treatment for 12 months with sulpiride altered striatal dopamine-stimulated adenylate cyclase activity, although not specific ^{3}H-piflutixol binding to D1 sites. These findings are difficult to explain through a direct action of sulpiride on D1 receptors. The most likely explanation for these findings appears to lie in the ability of sulpiride to potentiate dopamine-induced cyclic AMP formation in striatal prisms, an effect presumably mediated via inhibition of D2 receptors (Stoof and Kebabian, 1981).

Clozapine, like sulpiride, is claimed not to induce tardive dyskinesias in man (Simpson and Varga, 1974). Clozapine is weakly active on cerebral dopamine D1 and D2 receptors alike, and possesses a comparable or

Table 28.1   The effect of continuous chronic administration for 12 months of haloperidol, clozapine or sulpiride on dopamine function in the striatum.

Drug treatment (dose mg kg^{-1} day^{-1})	Apomorphine-induced (0.5 mg kg^{-1} s.c.) stereotypy score	Binding of ^{3}H-spiperone B_{max} (pmol g^{-1})	K_d (nM)
Controls (distilled water alone)	3.09 ± 0.21	13.8 ± 1.0	0.13 ± 0.03
Haloperidol (1.6–1.7)	3.91 ± 0.29*	26.4 ± 2.2*	0.20 ± 0.04
Sulpiride (102–109)	3.50 ± 0.22	17.2 ± 1.4	0.13 ± 0.02
Clozapine (24–27)	3.16 ± 0.16	15.3 ± 0.9	0.15 ± 0.02

Values are presented as the mean (± 1 SEM) of determinations made on three separate occasions using different tissue pools each from 3 to 7 animals.

Specific binding was defined by the incorporation of 10^{-5} M (−)-sulpiride.

*$P < 0.05$ compared with age-matched control animals; two-tailed Student's *t*-test for parametric data, Mann–Whitney U-test for non-parametric stereotypy scores.

greater affinity for non-dopamine receptors such as acetylcholine, serotonin, histamine and noradrenaline.

Like sulpiride, chronic administration of clozapine did not alter any component of stereotyped behaviour induced by apomorphine or striatal D2 receptors identified by ^{3}H-spiperone (Rupniak et al., 1985a; Table 28.1). Chronic clozapine administration did not alter dopamine stimulation of striatal adenylate cyclase but it did alter D1 receptors as indicated by an increase in sites identified by ^{3}H-piflutixol after 9 and 12 months. So again an atypical neuroleptic drug affects D1 rather than D2 receptor function.

MESOLIMBIC DOPAMINE FUNCTION DURING CHRONIC NEUROLEPTIC DRUG ADMINISTRATION

Most of the biochemical changes reported so far with all groups of neuroleptics have been centred on the striatum. A series of studies was undertaken to determine whether chronic neuroleptic administration also caused functional dopamine receptor supersensitivity in the mesolimbic areas. In the rat the mesolimbic dopamine system is primarily concerned with the control of locomotor activity. Acute administration of neuroleptic drugs antagonizes the effects of dopamine agonists applied focally in the nucleus accumbens (Costall and Naylor, 1976). However, chronic administration of haloperidol, sulpiride or clozapine for 12 months did not result in any enhancement of the locomotor hyperactivity caused by direct injection of dopamine into the nucleus accumbens (Rupniak et al., 1985b). So, behavioural supersensitivity does not develop but nor was there evidence for continuing dopamine receptor blockade. Measurement of specific ^{3}H-spiperone binding in mesolimbic tissue confirmed the lack of changes in dopamine receptor sensitivity at this time. However, following chronic administration of both trifluoperazine and haloperidol initial increases in B_{max} from ^{3}H-spiperone binding in the mesolimbic region were seen after 1 month of treatment but these gradually disappeared as drug administration continued (Clow et al., 1980c; Rupniak et al., 1985b).

EFFECTS OF CHRONIC TREATMENT WITH NEUROLEPTICS ON ACETYLCHOLINE LEVELS IN THE STRIATUM

An important question which is raised by these long-term studies with neuroleptics is whether the changes observed are of functional significance. The probing of animals using apomorphine to induce stereotyped behaviour, or the in vitro examination of receptor binding sites for D2 ligands, may reveal an underlying supersensitivity of dopamine receptors, but this may not be of relevance to the whole animal. Dopamine receptors might remain blocked by the large concentrations of the neuroleptic compounds present in the tissues of these animals. In other words, although there is an underlying supersensitivity of dopamine receptors in the striatum during long-term therapy with neuroleptics, this may not be manifest in behavioural terms. To examine this problem the effects of chronic treatment with neuroleptics on the function of acetylcholine in the striatum were looked at since it is believed that many postsynaptic dopamine receptors lie on the cell bodies of cholinergic interneurons in the striatum (Sethy and van Woert, 1974).

After treatment for 12 months with haloperidol or clozapine, but not with sulpiride, the levels of acetylcholine in the striatum were increased (Rupniak et al., 1986; Table 28.2). The activity of choline acetyltransferase in the striatum was not altered by any treatment with drug. The B_{max} for specific binding of ^{3}H-quinuclidinyl benzilate (QNB) to muscarinic receptors in the striatum was not altered by treatment with haloperidol, sulpiride or clozapine.

These findings suggest that alterations in the content of acetylcholine in the striatum caused by chronic treatment with some, but not all, neuroleptics reflect changes in cholinergic neuronal activity rather than in synthesis or destruction of neurotransmitter. The lack of effect of sulpiride reflects its failure to change the parameters of D2 receptors in the striatum on chronic administration, in contrast to the effects of haloperidol on both dopamine receptors and on levels of acetylcholine in the striatum. The failure of clozapine to alter the function of D2 receptors, but to increase the content of acetylcholine in the striatum, is not unexpected; this drug possesses more inherent anticholinergic activity than dopamine receptor antagonist properties.

Table 28.2 The effect of continuous chronic administration of haloperidol, sulpiride or clozapine for 12 months on parameters of cholinergic function in the striatum.

Drug treatment	Acetylcholine content (nmol g^{-1})	CAT (nmol acetylcholine h^{-1} mg^{-1} protein)	B_{max} ^{3}H-QNB (pmol mg^{-1} protein)
Control	19.9 ± 2.1	96.9 ± 6.4	1.45 ± 0.81
Haloperidol	35.7 ± 3.3*	93.5 ± 4.2	1.31 ± 0.33
Sulpiride	26.5 ± 2.6	95.7 ± 1.6	1.43 ± 0.45
Clozapine	29.5 ± 3.4*	93.8 ± 3.4	1.53 ± 0.28

The results are the mean (±1 SEM) of values obtained from striatal tissue from 6 to 8 individual animals. Data are subjected to analysis of variance. Where F ratios were associated with $P < 0.05$, groups were compared by two-tailed Student's t-test.
*$P < 0.05$ compared to control animals; Student's t-test.
CAT = choline acetyltransferase.

EFFECTS OF LONG-TERM TREATMENT WITH NEUROLEPTICS ON OTHER NEURONAL SYSTEMS

The concept that the chronic administration of neuroleptic drugs leads to the development of a functional supersensitivity of dopamine receptors fits well with the idea of tardive dyskinesia being the result of dopaminergic overactivity, as suggested by the clinical pharmacology of the disorder (Marsden et al., 1975). However, there are a number of important discrepancies between the evidence available to support this hypothesis and the clinical facts (Fibiger and Lloyd, 1984; Waddington, 1985). These discrepancies lead to the idea that supersensitivity of dopamine receptors alone may not be the only factor responsible for the induction of tardive dyskinesia.

The ability of chronic treatment with neuroleptics to alter the content of acetylcholine in brain illustrates the principle that neuroleptics do not alter the function of dopamine receptors alone. There are two reasons for change in other neuronal systems: many neuroleptic drugs do not act specifically on dopamine receptors but will also interact with α-adrenergic, 5-hydroxytryptamine (5HT) and histamine sites. This type of effect may be responsible for the down-regulation of cortical 5HT2 sites observed after chronic treatment with neuroleptics. Also, dopamine systems in the basal ganglia are intimately connected with acetylcholine, γ-aminobutyric acid (GABA) and peptide neurons such that compensatory changes in the response to altered function of dopamine receptors are to be expected. So changes in other neuronal systems induced by treatment with neuroleptics may contribute to the induction of tardive dyskinesia.

A number of different concepts have been put forward to explain tardive dyskinesia, including the idea that such patients may have structural abnormalities in the brain, as shown by enlarged ventricular size (Crow et al., 1984). Recently, biochemical evidence has been provided to indicate other alterations in brain function. Gunne and his colleagues reported that, in monkeys treated for prolonged periods with neuroleptic drugs, all animals showed acute dystonic reactions to treatment with neuroleptics, whereas only a proportion exhibited tardive dyskinesia (Gunne et al., 1984). At postmortem, a decrease in the activity of the enzyme glutamic acid decarboxylase (GAD) was found in the substantia nigra, but only in those animals exhibiting tardive dyskinesia. One interpretation of this change is that the treatment with the neuroleptic was toxic to neurons of the strionigral GABA-containing pathway. Indeed, previously Fog et al. (1980) reported a decrease in the number of neuronal cell bodies in the striatum of rats treated for prolonged periods with neuroleptic drugs.

Recently, changes in the nature of the GABA receptor complex in the striatum and substantia nigra of rats receiving chronic treatment with haloperidol, sulpiride or clozapine was examined (Rupniak et al., 1987). Administration of haloperidol, but not of sulpiride or clozapine, for 12 months increased the activity of GAD in the striatum nigra but not in substantia (Table 28.3). So, neuroleptic treatment alone would not appear to be sufficient to induce the decrease in the activity of GAD in the nigra, which appears to be associated with the production of tardive dyskinesia in primates. The number of specific binding sites for ^{3}H-flunitrazepam in membrane preparations of striatum was not altered by administration of any of these drugs for 12 months. But surprisingly, the B_{max} for the binding of ^{3}H-flunitrazepam to cerebellar membranes was decreased by administration of all drug treatments for 12 months. The reason for this change is not clear, especially since the cerebellum is not an area associated with dopaminergic innervation. However, previously, Lloyd and his colleagues found treatment of animals with clozapine or haloperidol for 6 months decreased the B_{max} for specific binding of ^{3}H-GABA in cerebellar tissue (Lloyd et al., 1977). Although the cause of this change is unknown it may be significant that atypical and typical neuroleptic compounds appear able to induce an identical alteration.

The basal ganglia contains large concentrations of a variety of neuropeptides so it is perhaps not surprising that altered dopamine function can lead to changes in some of these substances. For example, treatment of rats for 18 months with trifluoperazine increased the content of met- and leu-enkephalin and neurotensin in the striatum and nucleus accumbens. However, in the substantia nigra the content of met-enkephalin was decreased, leu-enkephalin unchanged and that of neurotensin increased (De Ceballos et al., 1986). The changes in the level of enkephalin are in agreement with those observed after short periods of treatment with neuroleptics. The data suggest

Table 28.3 The effect of continuous chronic administration of haloperidol, sulpiride or clozapine for 12 months on parameters of GABA function in brain.

Drug treatment	Activity of GAD (nmol CO_2 h^{-1} mg^{-1} protein)		^{3}H-flunitrazepam B_{max} (pmol mg^{-1} protein)	
	Substantia nigra	Striatum	Striatum	Cerebellum
Control	1065 ± 75	300 ± 23	0.83 ± 0.06	1.84 ± 0.12
Haloperidol	1092 ± 83	357 ± 11*	0.67 ± 0.03	1.36 ± 0.05*
Sulpiride	1018 ± 40	328 ± 12	0.77 ± 0.07	1.36 ± 0.07*
Clozapine	883 ± 47	314 ± 12	0.71 ± 0.04	1.29 ± 0.18*

For GAD activity, values are the mean (±1 SEM) values obtained in tissue from 3 to 6 animals per treatment group. For ^{3}H-flunitrazepam binding, values are the mean (±1 SEM) values obtained from three separate tissue pools per treatment group. Data were subjected to analysis of variance. Where F ratios were associated with $P < 0.05$, groups were compared using a two-tailed Student's t-test.

*$P < 0.05$ compared to control animals; Student's t-test.

that these changes in the content of enkephalin therefore persist for long periods. The results also suggest that treatment with neuroleptics can differentially alter levels of peptides in different regions of the brain. This may at least partially reflect the altered involvement of different enkephalin precursors. Again, the significance of changes in the levels of enkephalins and neurotensin after chronic treatment with neuroleptics is unclear but emphasizes the complexity of the changes which can occur as a result of the action of drugs on dopamine systems in brain.

NEUROLEPTIC-INDUCED PERIORAL MOVEMENTS

If changes in striatal dopamine function are of relevance to the emergence of tardive dyskinesias, it might be expected that rats undergoing chronic neuroleptic treatment should display spontaneous dyskinesias. Rats treated chronically with a wide range of neuroleptic drugs do display an increase in perioral behaviours (Clow *et al.*, 1979; Howells and Iversen, 1979). However, there are several reasons for considering that these movements do not represent an animal model of tardive dyskinesia. The mouth movements are not an abnormal phenomenon but their incidence is increased by neuroleptic drugs. Indeed an increase in mouth movements occurred even after a single dose of neuroleptic drug, and compounds such as sulpiride induced mouth movements much like typical neuroleptic drugs (Rupniak *et al.*, 1983). Moreover, the appearance of neuroleptic-induced chewing did not correlate with changes in striatal dopamine or acetylcholine function suggesting that they are not associated with striatal dopamine receptor supersensitivity (Rupniak *et al.*, 1985a). The increased incidence of mouth movements disappeared rapidly following drug withdrawal and the movements were suppressed by anticholinergic drugs. So if anything these movements more closely resemble acute dystonia than tardive dyskinesia.

WITHDRAWAL FROM CHRONIC NEUROLEPTIC DRUG ADMINISTRATION

By the end of 12 months continuous high-dose phenothiazine intake rats showed an exaggerated stereotyped response to apomorphine, increased stimulation of striatal adenylate cyclase by dopamine and increased striatal dopamine receptor density for ^{3}H-spiperone binding, but no change in the concentration of dopamine HVA or DOPAC. No dramatic new behavioural or biochemical changes occurred in the 6 months after withdrawal (Clow *et al.*, 1980b). Quite surprisingly, there was no exaggeration of the dopamine receptor supersensitivity observed at the end of chronic treatment. Instead there was a gradual reversal of all elevated parameters towards normal. The exaggerated response to apomorphine returned to normal by 3 months after withdrawal. The increased dissociation constant for ^{3}H-spiperone had reverted to control values after just 2 weeks but the increased B_{max} was not normal until 6 months after drug withdrawal. The only persistent change which lasted throughout the 6 months' withdrawal period was the increased stimulation of striatal dopamine-sensitive adenylate cyclase.

These results differ markedly from those of withdrawal after a short period of drug administration. This suggests that different types of supersensitivity of cerebral dopamine systems occur after acute short-term and long-term neuroleptic administration.

DRUG HOLIDAYS

The use of intermittent neuroleptic treatment in the form of drug holidays has been tried both clinically and in animal experiments in an effort to prevent the development of dopamine receptor supersensitivity.

In a study using discontinuous versus continuous administration of *cis*-flupenthixol or trifluoperazine no difference was observed in the development of enhanced apomorphine-induced stereotyped behaviour or increased striatal ^{3}H-spiperone receptor binding sites (Murugaiah et al., 1985). This failure of drug holidays to prevent dopamine receptor changes appears to be in relatively good agreement with the clinical literature which suggests that drug holidays are of little benefit in limiting tardive dyskinesias.

SUMMARY

Tardive dyskinesias may occur during and after chronic neuroleptic treatment and their pathophysiology suggests they may be a result of overstimulation of cerebral dopamine receptors (Marsden et al., 1975). The only common mechanism of action for neuroleptic compounds which induce this movement disorder is the ability to block brain dopamine receptors. So neuroleptic drugs, which on acute intake block dopamine receptors, appear to cause overstimulation of central dopamine systems when given over long periods. We have shown that continuous administration of typical neuroleptic drugs (which commonly cause tardive dyskinesia) but not atypical neuroleptic agents (which do not induce this disorder) to rats for periods of up to 18 months causes striatal dopamine receptor supersensitivity. The relevance of these changes to the onset of tardive dyskinesia in man is not clear. It does seem that changes in dopamine receptor sensitivity alone are not sufficient to induce tardive dyskinesias. It may be that the development of dopamine receptor supersensitivity is just the first step in a chain of neuronal events which may be more relevant to the final production of tardive dyskinesia in primates. It has been shown that the chronic administration of typical neuroleptics also causes alterations in striatal acetylcholine or GABA parameters consistent with the development of dopamine receptor super-sensitivity. An explanation for why brain dopamine receptors are not consistently blocked by the large concentrations of neuroleptic drugs received is not clear. Extra receptors may be produced in response to neuroleptic treatment which have different characteristics from the existing population. These may be of relevance in tardive dyskinesia.

REFERENCES

Alberts, J. L. (1985) *Sem. Hosp.* **61**, 1351–1357.

Asper, H., Baggiolini, M., Burki, H. R., Launer, H., Ruch, W. and Stille, G. (1973) *Eur. J. Pharmacol.* **22**, 287–294.

Carlsson, A. and Lindquist, M. (1963) *Acta. Pharmacol. Toxicol.* **20**, 140–144.

Clow, A., Jenner, P. and Marsden, C. D. (1979) *Eur. J. Pharmacol.* **57**, 365–375.

Clow, A., Theodorou, A., Jenner, P. and Marsden, C. D. (1980a) *Eur. J. Pharmacol.* **63**, 135–144.

Clow, A., Theodorou, A., Jenner, P. and Marsden, C. D. (1980b) *Eur. J. Pharmacol.* **63**, 145–157.

Clow, A., Theodorou, A., Jenner, P. and Marsden, C. D. (1980c) *Psychopharmacology* **69**, 227–233.

Costall, B. and Naylor, R. J. (1975) *Psychpharmacology* **43**, 69–74.

Costall, B. and Naylor, R. J. (1976) *Eur. J. Pharmacol.* **40**, 9–19.

Costall, B., Marsden, C. D., Naylor, R. J. and Pycock, C. J. (1977) *Brain Res.* **123**, 89–111.

Crow, T. J., Bloom, S. R., Cross, A. J., Ferrier, I. N., Johnstone, F., Owens, D. G. C. and Roberts, G. W. (1984) In *Catecholamines, Neuropharmacology and Central Nervous System — Therapeutic Aspects*, pp. 61–66. Alan R. Liss, New York.

De Ceballos, M. L., Boyce, S., Jenner, P. and Marsden, C. D. (1986) *Eur. J. Pharmacol.* **130**, 305–309.

Elliott, P. N. C., Jenner, P., Huizing, G., Marsden, C. D. and Miller, R. (1977) *Neuropharmacology* **16**, 333–342.

Fibiger, H. C. and Lloyd, K. G. (1984) *Trends Neurosci.* **7**, 462–464.

Fog, R., Pakkenberg, H., Nielsen, E. B., Munkvad, I., Lyon, M. and Randrup, A. (1980) In *Phenothiazines and Structurally Related Drugs: Basic and Clinical Studies* (edited by E. Usdin, E. Eckert and I. Forrest), pp. 357–359, Elsevier North Holland, Amsterdam.

Gunne, L.-M., Haggstrom, J. E., and Sjoquist, B. (1984) *Nature* **309**, 347–349.

Howells, R. B. and Iversen, S. D. (1979) *Neurosci. Lett. Suppl.* **3**, 210.

Johnstone, E. D., Crow, T. J., Frith, C. D., Curney, M. W. P. and Price, J. S. (1978) *Lancet* April 22, 848–851.

Lloyd, K. G., Shibuya, M., Davidson, L. and Hornykiewicz, O. (1977) In *Advances in Biochemical Psychopharmacology*, Vol. 16 (edited by E. Costa and G. L. Gessa), pp. 409–415. Raven Press, New York.

Marsden, C. D., Tarsy, D. and Baldessarini, R. J. (1975) In *Psychiatric Aspects of Neurologic Disease* (edited by C. F. Benson and D. Blumer), pp. 219–266. Grune & Stratton Inc., New York.

Miller, R. J. and Hiley, C. R. (1974) *Nature* **248**, 596–597.

Murugaiah, K., Theodorou, A., Jenner, P. and Marsden, C. D. (1983) *Neuroscience* **10**, 811–819.

Murugaiah, K., Fleminger, S., Hall, M. D., Theodorou, A., Jenner, P. and Marsden, C. D. (1984) *Neuropharmacology* **23**, 599–609.

Murugaiah, K., Theodorou, A., Clow, A., Jenner, P. and Marsden, C. D. (1985) *Psychopharmacology* **86**, 228–232.

Rupniak, N. M. J., Jenner, P. and Marsden, C. D. (1983) *Psychopharmacology* **79**, 226–230.

Rupniak, N. M. J., Mann, S., Hall, M. D., Fleminger, S., Kilpatrick, G., Jenner, P. and Marsden, C. D. (1984) *Psychopharmacology* **84**, 503–511.

Rupniak, N. M. J., Hall, M. D., Mann, S., Fleminger, S., Kilpatrick, G., Jenner, P. and Marsden, C. D. (1985a) *Biochem. Pharmacol.* **34**, 2755–2763.

Rupniak, N. M. J., Hall, M. D., Kelly, E., Fleminger, S., Kilpatrick, G., Jenner, P. and Marsden, C. D. (1985b) *J. Neural Transm.* **62**, 249–266.

Rupniak, N. M. J., Briggs, R. S., Petersen, M. M., Mann, S., Reavill, C., Jenner, P. and Marsden, C. D. (1986) *Clin. Neuropharmacol.* **9**, 282–292.

Rupniak, N. M. J., Prestwich, S. A., Horton, R. W., Jenner, P. and Marsden, C. D. (1987) *J. Neural Transm.* **68**, 113–125.

Sethy, V. H. and van Woert, M. H. (1974) *Nature* **251**, 529–530.

Simpson, G. M. and Varga, E. (1974) *Curr. Ther. Res.* **16**, 679–686.

Stoof, J. C. and Kebabian, J. W. (1981) *Nature* **294**, 366–368.

Tarsy, D. and Baldessarini, R. J. (1974) *Neuropharmacology* **13**, 927–940.

Trabucchi, M., Longoni, R., Fresia, P. and Spano, P. F. (1975) *Life Sci.* **17**, 1551–1556.

Waddington, J. L. (1985) *Trends Neurosci.* **8**, 200.

29

Experimental Models of Acute Dystonia

P. Jenner

INTRODUCTION

In contrast to Parkinson's disease, dystonia is a condition for which there is no recognized pathology in brain, only sparse data on biochemical change and little in the way of effective treatment. Dystonia can take many forms but idiopathic torsion dystonia or in its familial form, dystonia musculorum deformans, is the most important (see Fahn *et al.*, 1987). Dystonia is characterized by sustained muscle contraction, and by twisting and repetitive movements or abnormal postures involving the head, neck, trunk and limbs which contort the body.

No consistent pathology has been reported in the brains of patients dying with idiopathic dystonia. Often no pathological abnormalities are apparent but some cases appear to be associated with lesions of the putamen or thalamus (Marsden *et al.*, 1985). However, damage to these brain regions does not necessarily result in the onset of dystonia. Only recently have any biochemical changes in brain been reported in dystonia. Hornykiewicz *et al.* (1986) found no significant histological change in the basal ganglia, cerebral cortex, higher brain stem nuclei, locus coeruleus or raphé nuclei of two patients dying with generalized childhood-onset dystonia musculorum deformans. Similarly, choline acetyltransferase activity and GABA and glutamic acid levels were normal in the cerebral cortex and basal ganglia. However, there were marked changes in noradrenaline levels in many regions of the brain including the hypothalamus, dorsal raphé nucleus, red nucleus, superior colliculus and thalamus. In addition, 5HT levels were altered in the globus pallidus, subthalamic nucleus, locus coeruleus and raphé nuclei and there were some changes in brain dopamine content. The authors concluded that some of these monoamine changes and particularly those in noradrenaline may represent a basic neurochemical abnormality in dystonia. However, confirmation of these findings is required by a detailed analysis of further brains from dystonic patients.

The therapeutic treatment of the dystonias is poor and patient response to drugs unpredictable (see Fahn and Marsden, 1987). A wide variety of drugs have been used including neuroleptics, dopamine agonists, anticholinergics, cholinomimetics and drugs affecting brain GABA function. Of these approaches only high-dose anticholinergic treatment was found to be consistently effective particularly in children with dystonia (Fahn, 1983; Marsden et al., 1984; Burke et al., 1986).

The paucity of information on the cause and treatment of dystonia has resulted in attempts to establish an experimental model of this disorder. Some of the models examined are described below and in particular attention is focused on purposeless chewing behaviour in rats as a pharmacological test bed for assessing potential drug treatments for dystonia (updating Stewart et al., 1988b).

SOME EXPERIMENTAL MODELS OF DYSTONIA

Many animal models of dystonia have been described but few mimic the disease as it occurs in man. A widely used model of idiopathic torsion dystonia is the mutant dystonic (dt) rat which exhibits sustained twisting movements involving the axial musculature (Lordon et al., 1984). These animals also show a paddling gait, hyperflexion of the trunk and self-clasping of the fore- and hind-limbs. Frequent falls to the side are accompanied by rigid extension of the limbs. In this position the limbs resist passive movement and the motor abnormalities disappear when the animals are at rest. The behaviour observed closely resembles ataxia induced by damage to the superior cerebellar peduncle. Indeed, while no morphological lesion is apparent, present evidence strongly favours a defect in the cerebellum of the dt rat involving Purkinje cells and an abnormality in GAD activity in the deep cerebellar nuclei (Oltmans et al., 1984a; Lorden et al., 1988). The behavioural syndrome observed in the dt rat is worsened by cholinomimetics such as physostigmine, while some components are improved by administration of the anticholinergic scopolamine (Lorden *et al.*, 1988). This suggests an involvement of cholinergic systems in the motor abnormalities exhibited by the dt rat and a drug response resembling that occurring in man. The administration of diazepam or clonidine also causes some behavioural improvement (Oltmans *et al.*, 1984b; Lutes *et al.*, 1985).

A variety of other models of dystonia have been proposed which involve either focal lesions or focal injections of drugs and transmitter substances into brain. For example, the unilateral injection of ACTH 1-24 into the region of the noradrenaline cell bodies of the locus coeruleus induces postural asymmetry and impaired locomotion that is likened to dystonia (Jacquet and Abrams, 1982; Jacquet 1984; Abrams and Jacquet, 1984). Similar bilateral injections of ACTH 1-24 induce retrocollis. The effect is site and drug specific and limited to only a few ACTH fragments (Jacquet, 1988; Adams and Foote, 1988). Similarly, intracerebroventricular (i.c.v.) injection of chlorpromazine methiodide induces swivelling or barrel rotation which is claimed to be dystonia (Rotrosen *et al.*, 1980). However, the effects of chlorpromazine methiodide are thought to be due to a muscarinic receptor antagonist action since the behavioural syndrome was enhanced by atropine and antagonized by carbachol (Burke *et al.*, 1982; Burke and Fahn, 1988). Other postural asymmetries which have been related to dystonia are produced by manipulation of GABA function in the vestigial nucleus, electrolytic or 6-hydroxydopamine lesions of the nigrostriatal tract and the focal injection of tetanus toxin into basal ganglia (see McGeer and McGeer, 1988). Whether any of these phenomena are truly dystonic in nature remains to be determined.

One further animal model of dystonia is currently causing considerable interest. It stems from the discovery that haloperidol and other neuroleptic drugs which induce acute dystonia in man (see below) act on a population of sigma opiate receptors in brain (Tam and Cook, 1984; Weber *et al.*, 1986). These haloperidol sensitive sigma receptors (σ h) are labelled by ^{3}H-SKF 10,047 or ^{3}H-ditolylguanidine and are distinct in pharmacological characteristics and in distribution within brain from those receptors on which phenyclidine acts. The focal unilateral injection of ditolylguanidine, haloperidol or SKF 10,047, but not phenyclidine, into the σ h-rich red nucleus induces rotation of the head around the sagittal axis so that the eyes ipsilateral to the injection face upwards (Walker *et al.*, 1988). The contralateral paws and limbs may be affected in some cases. The effects are site specific, and do not, for example, occur when ditolylguanidine is injected into substantia nigra. Again there is the problem of whether the behavioural changes are dystonic in nature and the whole question of the relevance of the σ h receptor to dystonia in man. However, of considerable interest is the finding that σ h receptors are increased in number and decreased in affinity in the brain of the dystonic dt rat compared to its normal litter mate (Bowen et al., 1988). Other receptor populations in this mutant strain are normal.

NEUROLEPTIC-INDUCED DYSTONIA

An important clue to the cause of idiopathic dystonia lies in the ability of neuroleptic drugs to induce acute dystonic episodes in man (Marsden et al., 1975; Marsden and Jenner, 1980; Rupniak et al., 1986). Acute dystonia is characterized by muscle spasms, abnormal postures (most affecting the neck and head but also the trunk), retrocollis, forced opening of the jaw, tongue protrusion and oculogyric crises. Acute dystonic reactions affect about 2–5% of patients treated with neuroleptic drugs although the incidence may be higher with more potent neuroleptics and depot preparations. Most acute dystonic reactions occur within a few days of starting therapy, they are rapidly reversible disappearing with the time course of drug clearance, and can be effectively treated by the administration of anticholinergic drugs. The pathophysiology of neuroleptic-induced acute dystonia is uncertain. Previously we have suggested that they occur because of a mismatch between neuroleptic-induced acute dopamine receptor blockade and increased brain dopamine turnover (Marsden and Jenner, 1980; Kolbe et al., 1981). This leads to unoccupied, normal or supersensitive dopamine receptors being activated by excess amounts of the transmitter. However, it is not known why acute dystonic reactions should only occur in a small proportion of patients given neuroleptic treatment or why younger patients appear particularly susceptible.

Acute dystonia can also be produced experimentally in a variety of non-human primate species (see Rupniak et al., 1986). The time of onset, the nature of the movements and their response to drug withdrawal and anticholinergic treatment are strikingly similar to neuroleptic-induced acute dystonia as it occurs in man. Dystonic episodes both in primates and in humans appear to be under central cholinergic control. However, at present it is not known whether cholinergic excess is accompanied by a dopamine deficiency or dopamine excess. All studies agree that an increase in cholinergic activity underlies acute dystonia as is suggested by the following evidence: (1) anticholinergic agents rapidly reverse neuroleptic-induced dystonia (Marsden et al., 1975); (2) cholinergic agonists may induce dystonia in neuroleptic primed monkeys (Meldrum et al., 1977; Casey et al., 1980); (3) focal intrastriatal application of acetylcholine or carbachol in primates induces dystonia (Murphey and Dill, 1972; Cools et al., 1974); (4) acute administration of neuroleptic drugs increases striatal acetylcholine release

(Sethy and van Woert, 1974).

Excessive dopamine release in the aetiology of dystonic movements is indicated by several lines of evidence: (1) administration of the catecholamine precursor L-Dopa to both humans and primates can cause dystonia resembling that induced by neuroleptic drugs (Mones, 1972; Paulson, 1973; Sassin, 1975; Parkes et al., 1976); (2) the intrastriatal application of dopamine in primates induces dystonia (Cools et al., 1974); (3) the pretreatment of baboons with α-methyl-p-tyrosine (AMPT) plus reserpine may prevent neuroleptic-induced acute dystonia (Meldrum et al., 1977).

However, the latter evidence has been challenged. Thus, administration of AMPT to vervet monkeys was found to worsen neuroleptic-induced dystonia and induced acute dystonia in squirrel monkeys (Neale et al., 1984). This suggests dopamine deficiency rather than excess as a cause of acute dystonic episodes. So, at present the concept of excess cholinergic activity in the genesis of acute dystonia is more convincing than the evidence for dopamine involvement.

PURPOSELESS CHEWING MOVEMENTS IN RATS

The neuroleptic treated primate provides an excellent model for studying acute dystonia as it occurs in man. However, it is not always practical to work with monkey species. So, can neuroleptics induce some phenomena in rats which could be utilized to understand the pathophysiology of the disorder and to devise new means of treatment? Acute dystonia has never been reported following the peripheral administration of neuroleptic drugs to rodents. So it may be that rodent species are not able to exhibit the symptoms of acute dystonia seen in humans following neuroleptic treatment. Rats do however exhibit spontaneous purposeless chewing movements that increase in incidence following the administration of neuroleptics (Clow et al., 1979; Iversen et al., 1980; Waddington and Gamble, 1980; Waddington et al., 1983). Such movements are not abnormal but form part of the normal repertoire of behaviour of the animal and neuroleptic treatment serves to increase their intensity.

The increase in purposeless chewing movements was first noted in experiments involving the chronic treatment of rats with neuroleptic drugs for some months. Such treatments with haloperidol, cis-flupenthixol, trifluoperazine or fluphenazine were found to cause an increase in perioral behaviours such as purposeless jaw movements (Clow et al., 1979; Glassman and Glassman, 1980; Gunne et al., 1982; Rupniak et al., 1983, 1985). In our own experiments these take the form of repetitive opening and closing of the jaw which are distinct from yawning and appear as chewing which is not directed towards any specific object. Since these movements were initially observed in long-term neuroleptic studies we initially thought that they might be a manifestation of tardive dyskinesia in rodents. However, the rapid disappearance of the movements on cessation of drug treatment suggested that this was not the case (Clow *et al.*, 1980). Subsequently, we have found that acute administration of neuroleptics provokes an enhanced incidence of purposeless chewing (Rupniak *et al.*, 1986). Such perioral movements may take different forms since others have reported vacuous chewing movements which only appear after chronic neuroleptic treatment and persist for long periods following drug withdrawal (Waddington *et al.*, 1983). It may be that these movements are more akin to tardive dyskinesia than the movements observed in our own studies (Waddington and Molloy, 1987).

Do the movements which we observe provide us with the pharmacological model of acute dystonia? An increased incidence of purposeless chewing movements can be induced by all major classes of neuroleptic drugs (Rupniak *et al.*, 1986; Stewart *et al.*, 1988b). For example, the administration of haloperidol, trifluoperazine and sulpiride to rats for a 4-month period enhanced the incidence of purposeless chewing (Table 29.1; Rupniak *et al.*, 1983). Interestingly, administration of clozapine over the same period did not alter chewing behaviour, a finding which may be related to the high anticholinergic activity of this compound. Purposeless chewing was not induced by the continuous administration of the peripheral dopamine antagonist domperidone, suggesting the movements are central in origin (Rupniak *et al.*, 1985). The nature of the dopamine receptor population involved in the induction of purposeless chewing appears complex. Rosen-

Table 29.1 Spontaneous oral behaviour in rats treated for 4 months with haloperidol, trifluoperazine, sulpiride or clozapine compared to age-matched control animals.

Group	Chewing	Tongue protrusion	Gaping	Grinding teeth chattering
	Mean number of movements/5 min			
Control	18.0 ± 2.7	1.5 ± 0.5	0	0
Haloperidol	$44.7 \pm 5.0^*$	2.4 ± 1.0	0	0
Trifluoperazine	$32.2 \pm 5.0^*$	4.0 ± 1.5	0	0
Sulpiride	$30.6 \pm 2.6^*$	1.7 ± 0.9	0	0
Clozapine	25.6 ± 4.9	3.0 ± 0.8	0	0

Values are mean $\pm$ 1 SEM for observations during a 5-min test period. $n = 8$. Rats received either haloperidol (1.4–$1.6\,mg\,kg^{-1}\,day^{-1}$), trifluoperazine ($4.5$–$5.1\,mg\,kg^{-1}\,day^{-1}$), sulpiride ($102$–$110\,mg\,kg^{-1}\,day^{-1}$) or clozapine ($23$–$26\,mg\,kg^{-1}\,day^{-1}$) for 4 months. Control rats received distilled drinking water alone.

   $^*P < 0.05$ vs control, Mann–Whitney U-test.

   From Rupniak *et al.* (1983).

garten *et al.* (1983) claimed that purposeless chewing is induced by D1 receptors since the D1 agonist SKF 38393 produces purposeless chewing in rats, an effect which was potentiated by the D2 antagonist, sulpiride. However, our own studies suggest that the D1 antagonist SCH 23390 and the D2 agonist quinpirole, also induce purposeless chewing (unpublished data). So it remains unclear precisely how the different dopamine receptor populations are involved in the initiation of purposeless chewing.

The appearance of purposeless chewing is rapid in onset since the administration of an acute bolus of haloperidol to Wistar rats daily for five consecutive days caused an increase in the number of chewing movements on days 3, 4 and 5 of treatment (Fig. 29.1; Rupniak *et al.*, 1986). The increased incidence of chewing was not apparent when the same animals were re-examined 24 h after the last dose. Similarly, like neuroleptic-induced dystonia in primates chewing movements in rats appeared to persist for the duration of neuroleptic treatment, 15 months being the longest period we have examined (Clow *et al.*, 1979; Rupniak *et al.*, 1986).

A critical question in establishing pharmacological relevance of purposeless chewing to acute dystonia is the sensitivity of movements to cholinergic manipulation. In animals treated with haloperidol for 4 months, acute administration of the anticholinergic drugs scopolamine or atropine reduced chewing movements to or below values for control animals (Rupniak *et al.*, 1983, 1985). In contrast, administration of either pilocarpine or physostigmine increased the frequency of purposeless chewing movements. Again, this effect appears to be of central origin since the peripheral anticholinergic agent neostigmine did not enhance haloperidol-induced purposeless chewing and the peripherally acting anticholinergic agent, methylscopolamine, did not reduce chewing. These studies suggest that neuroleptic-induced enhancement of purposeless chewing responds to cholinergic manipulation in a manner identical to that observed for acute dystonic reactions in humans and in other primate species. However, it was clear from these investigations that cholinergic agonist drugs administered alone can themselves enhance perioral chewing in normal rats.

CHOLINERGIC INVOLVEMENT IN PURPOSELESS CHEWING MOVEMENTS

The role cholinergic systems might play in the induction of purposeless chewing warrants further investigation in terms of the ability of anticholinergic drugs to effectively control some symptoms of dystonia.

The administration of a range of cholinergic agonists or cholinesterase inhibitors to rats increases the intensity of purposeless chewing movements

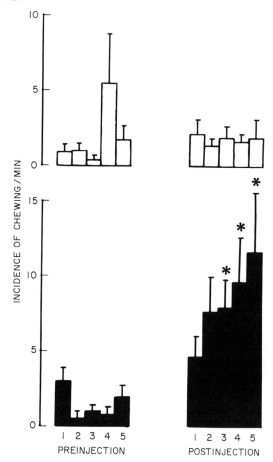

Fig. 29.1   Frequency of purposeless chewing jaw movements in Wistar rats before (left-hand panels) and 3.5 h after (right-hand panels) administration of haloperidol (2 mg kg^{-1} i.p.; solid bars) or saline (open bars) on five consecutive days. *$P < 0.05$ compared to saline-treated control animals. Mann–Whitney U test ($n = 7$–10). From Rupniak et al. (1986).

whether these drugs are administered acutely or over a period of 21 days in animals' drinking water (Fig. 29.2; Rupniak et al., 1985; Salamone et al., 1986; Stewart et al., 1989b). The behaviour is clearly of central origin since the administration of neostigmine, the peripheral cholinergic agonist, for 21 days did not induce purposeless chewing movements. Administration of scopolamine for up to 21 days tended to decrease the number of chewing movements observed in rats. All these results suggest that the administration of cholinomimetic compounds can induce purposeless chewing movements due to an effect on central cholinergic systems.

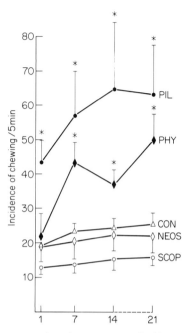

Fig. 29.2   Chewing jaw movements in Wistar rats treated with agents acting on acetylcholine systems for up to 3 weeks compared to control animals. Animals received either physostigmine (PHY; 0.27–0.35 mg kg^{-1}), pilocarpine (PIL; 4.2–5.6 mg kg^{-1}), neostigmine (NEOS; 0.21–0.25 mg kg^{-1}), scopolamine (SCOP; 0.42–0.46 mg kg^{-1}) or no drug (CON) continuously via their daily drinking water. Each point is the mean ± 1 SEM for observations in 6–17 animals. *$P < 0.05$ versus control animals, Mann–Whitney U-test. From Rupniak *et al.* (1985).

The increased incidence of purposeless chewing movements induced by the administration of pilocarpine can be reversed by the administration of scopolamine and other centrally acting anticholinergic drugs but not by the administration of peripheral cholinergic antagonists such as methyl-scopolamine (Rupniak *et al.*, 1985; Salamone *et al.*, 1986; Stewart *et al.*, 1989b). Again this emphasizes the central origin of the increase in perioral movements induced by pilocarpine. Anticholinergic drugs reverse the effects of pilocarpine in a dose-dependent manner but the doses of compounds required are high consistent with the amounts needed to control dystonic symptoms in man (Stewart *et al.*, 1988a).

Interestingly, only neuroleptic drugs with high anticholinergic activity could manipulate pilocarpine induced chewing (Stewart *et al.*, 1988a). Other neuroleptics such as trifluoperazine were without effect. This might suggest that purposeless chewing is a behaviour characterized by cholinergic dominance.

The ability of cholinergic drugs to induce purposeless chewing movements raises interesting questions about the role that acetylcholine systems may play in the treatment and cause of dystonia. However, it is necessary to determine which type of cholinergic receptor may be involved in this phenomenon since cholinergic receptors in brain can be divided into various subclasses, the muscarinic M1 and M2 receptors, and the nicotinic receptor. Nicotinic receptors do not appear involved in the induction of purposeless chewing so we have used a range of selective M1 and M2 muscarinic agonists and antagonists to determine the subtype of receptor responsible for cholinergic involvement in purposeless chewing (Stewart et al., 1989b).

Purposeless chewing was enhanced by the intraperitoneal administration of pilocarpine, RS86, oxotremorine and arecoline. Chewing behaviour was also induced by the i.c.v. injection of carbachol and pilocarpine but not by the putative M1 agonist McN-A-343 or AH 6405. Collectively these data argue in favour of purposeless chewing being an M2-mediated behavioural effect. Chewing behaviour induced by pilocarpine was antagonized in a dose-related manner by the peripheral and i.c.v. administration of a range of muscarinic receptor antagonists. In particular, pilocarpine-induced purposeless chewing was inhibited by the M2 antagonist 4-DAMP and AF-DX 116 but not by the M1 antagonist pirenzepine. So, these studies tentatively suggest that chewing behaviour induced by pilocarpine is not mediated via M1 receptors but rather through central M2 sites.

Where in the brain cholinergic compounds act to induce purposeless chewing is not known. Certainly i.c.v. injection of cholinergic agonists can induce purposeless chewing but it would seem that injection directly into the striatum does not, so sites elsewhere must be examined. The recognition of the type of receptor involved in purposeless chewing and the location of the site involved in the induction of chewing may have important consequences in extending our understanding of the systems responsible for the onset and treatment of neuroleptic-induced dystonia and perhaps even idiopathic dystonia. In addition, the effect of manipulating cholinergic function may have relevance to the treatment of Alzheimer's disease. Cholinergic agonist drugs are currently under development for use in the treatment of this illness. However, there is a danger that such drugs may also induce dystonic reactions. It therefore becomes important to determine the muscarinic subtype producing benefit in Alzheimer's disease compared to that involved in dystonic phenomena.

5HT INVOLVEMENT IN PURPOSELESS CHEWING

Recently another aspect of the purposeless chewing model has been discovered which may lead to the development of new treatments for dystonia and provide a pharmacological test-bed for the study of potential drug treatments in Alzheimer's disease. This arose from a pharmacological assessment of pilocarpine-induced purposeless chewing. The key observation was that the 5HT-2 antagonists ketanserin and spiperone did not alter pilocarpine-induced chewing but that other 5HT antagonists mianserin and methiothepin inhibited the behavioural response (Stewart *et al.*, 1988b). This suggests some form of 5HT involvement in the mediation of pilocarpine-induced chewing. Indeed, pretreatment of rats with reserpine, tetrabenazine or *p*-chlorophenyl-alanine but not AMPT decreased pilocarpine-induced purposeless chewing (Stewart *et al.*, 1987). The only common biochemical change induced by those drugs inhibiting purposeless chewing was a decrease in brain 5HT content. So it appears that 5HT systems may at least modulate cholinergic-induced chewing.

Administration of the 5HT agonists *m*-chlorophenylpiperazine (*m*-CPP), trifluoromethylphenylpiperazine (TFMPP) and quipazine also induced purposeless chewing in rats (Stewart *et al.*, 1989a). However, a range of doses of the 5HT agonists 8-hydroxy-2-(di-n-propylamino)tetralin and 5-methoxy-*N*,*N*-dimethyltryptamine were without effect. The purposeless chewing induced by *m*-CPP was inhibited by the 5HT antagonists methiothepin and mianserin but not by ketaserin, spiperone, ICS 205-930 or (−)-propranolol (Fig. 29.3).

One explanation for the different effects of 5HT agonists and antagonists on purposeless chewing may relate to the existence of different 5HT receptor subtypes (5HT-1A, B, C and D; 5HT-2 and 5HT-3). The lack of effect of the 5HT-2 antagonist ketanserin and the 5HT-2 and 5HT-1A antagonist spiperone and the 5HT-3 antagonist ICS 205-930 suggests 5HT-1A, 5HT-2 and 5HT-3 sites are not involved in *m*-CPP-induced purposeless chewing. Based on the effect of the range of 5HT agonist and antagonist drugs employed, it can be postulated at this stage that 5HT-1B receptor sites appear more likely to be involved than other 5HT-1-like receptors.

Chewing behaviour induced by *m*-CPP was also antagonized by the muscarinic receptor antagonists benzhexol and scopolamine but not methyl-scopolamine (Fig. 29.4). So, both central muscarinic and 5HT receptors appear to be involved in the mediation of purposeless chewing behaviour. This gives support to the previously suggested concept of a central 5HT–acetylcholine interaction.

So, there is a close link between central 5HT and acetylcholine systems in the induction and manipulation of purposeless chewing and this may be a clue to the treatment or even cause of dystonia since Hornykiewicz's post-mortem studies showed changes to occur in 5HT systems. If nothing else, this interaction provides a viable model for 5HT and cholinergic systems which may have implications for the design of treatments for Alzheimer's disease.

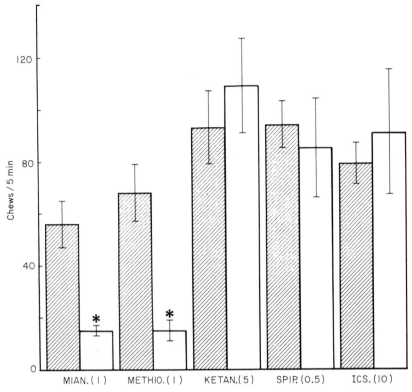

Fig. 29.3 Effect of various 5HT receptor antagonists on *m*-CPP-induced chewing behaviour. Drugs were administered 30 min s.c. prior to *m*-CPP (6 mg kg^{-1} s.c.). Chewing behaviour was assessed for 5 min, 5 min after *m*-CPP administration. Results are expressed as chews/5 min $\pm$ SEM. Hatched columns indicate chewing induced by *m*-CPP in the presence of the various 5HT antagonists. *$P < 0.05$, Mann–Whitney U test. From Stewart *et al.* (1989a).

CONCLUSION

At present dystonia is largely untreatable because its cause remains unknown. Animal models may help in unravelling the cause and in defining treatment. It would seem that the purposeless chewing model has many characteristics akin to those of dystonia as it affects man. This pharmacological model may provide a test-bed in which effective therapies for this crippling disease can be evaluated.

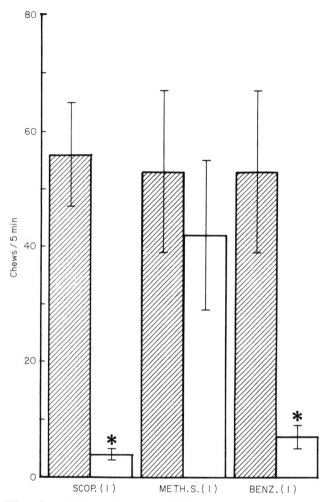

Fig. 29.4 Effect of anticholinergic drugs on *m*-CPP-induced chewing behaviour. Drugs were administered 30 min s.c. prior to *m*-CPP (6 mg kg^{-1} s.c.). Chewing was measured for 5 min, 5 min after *m*-CPP administration. Results are expressed as chews/5 min $\pm$ SEM. Hatched columns indicate chewing induced by *m*-CPP alone. Open columns indicate chewing induced by *m*-CPP in the presence of the various anticholinergic agents. *$P < 0.05$, Mann–Whitney U-test. From Stewart *et al.* (1989a).

ACKNOWLEDGEMENTS

This study was supported by the Dystonia Foundation, the Medical Research Council and the Research Funds of the Bethlem Royal and Maudsley Hospitals and King's College Hospital.

REFERENCES

Abrams, G. M. and Jacquet, Y. F. (1984) *Neurology* **34** (Suppl.), 153.

Adams, L. M. and Foote, S. L. (1988) In *Advances in Neurology, Vol. 50: Dystonia 2* (edited by S. Fahn, C. D. Marsden and D. B. Calne), pp. 313–333. Raven Press, New York.

Bowen, W. D., Walker, J. M., Yashar, A. G., Matsumoto, R. R., Walker, F. O. and Lorden, J. F. (1988) *Eur. J. Pharmacol.* **147**, 153–154.

Burke, R. E. and Fahn, S. (1988) In *Advances in Neurology, Vol. 50: Dystonia 2* (edited by S. Fahn, C. D. Marsden and D. B. Calne), pp. 335–342. Raven Press, New York.

Burke, R. E., Fahn, S., Wagner, H. R. and Smeal, M. (1982) *Brain Res.* **250**, 133–142.

Burke, R. E., Fahn, S. and Marsden, C. D. (1986) *Neurology* **36**, 160–164.

Casey, D. E., Gerlach, J. and Christensson, A. (1980) *Psychopharmacology* **70**, 83–87.

Clow, A., Jenner, P. and Marsden, C. D. (1979) *Eur. J. Pharmacol.* **57**, 365–375.

Clow, A., Theodorou, A., Jenner, P. and Marsden, C. D. (1980) *Eur. J. Pharmacol.* **63**, 145–157.

Cools, A. R., Hendriks, G. and Korten, J. (1974) *J. Neural Transm.* **36**, 91–105.

Fahn, S. (1983) *Neurology* **33**, 1255–1261.

Fahn, S. and Marsden, C. D. (1987) In *Movement Disorders 2* (edited by C. D. Marsden and S. Fahn), pp. 359–382. Butterworths, Kent.

Fahn, S., Marsden, C. D. and Calne, D. B. (1987) In *Movement Disorders 2* (edited by C. D. Marsden and S. Fahn), pp. 332–358. Butterworths, Kent.

Glassman, R. B. and Glassman, H. N. (1980) *Psychopharmacology* **68**, 19–25.

Gunne, L.-M., Growdon, J. and Glaeser, B. (1982) *Psychopharmacology* **77**, 134–139.

Hornykiewicz, O., Kish, S. J., Becker, L. E., Farley, I. and Shannak, K. (1986) *New Engl. J. Med.* **315**, 347–353.

Iversen, S. D., Howells, R. B. and Hughes, R. (1980) In: *Long-term Effects of Neuroleptics. Advances in Biochemical Psychopharmacology, Vol. 24* (edited by F. Catabeni, G. Racagni, P. F. Spano and E. Costa), pp. 305–313. Raven Press, New York.

Jacquet, Y. F. (1984) *Brain Res.* **294**, 144–147.

Jacquet, Y. (1988) In *Advances in Neurology, Vol. 50: Dystonia 2* (edited by S. Fahn, C. D. Marsden and D. B. Calne), pp. 299–311. Raven Press, New York.

Jacquet, Y. F. and Abrams, G. B. (1982) *Science* **218**, 175–177.

Kolbe, H., Clow, A., Jenner, P. and Marsden, C. D. (1981) *Neurology* **31**, 434–439.

Lorden, J. F., McKeon, T. W., Baker, H. J., Cox, N. and Walkley, S. U. (1984) *J. Neurosci.* **4**, 1925–1932.

Lorden, J. F., Oltmans, G. A., Stratton, S. and Mays, L. E. (1988) In *Advances in*

Neurology, Vol. 50: Dystonia 2 (edited by S. Fahn, C. D. Marsden and D. B. Calne), pp. 277–297. Raven Press, New York.

Lutes, J., Lorden, J. F. and Oltmans, G. A. (1985) *Neurosci. Abstr.* **11**, 670.

Marsden, C. D. and Jenner, P. (1980) *Psychol. Med.* **10**, 55–71.

Marsden, C. D., Tarsy, D. and Baldessarini, R. J. (1975) In *Psychiatric Aspects of Neurologic Disease* (edited by D. F. Benson and D. Blumer), pp. 219–266. Grune and Stratton, New York.

Marsden, C. D., Marion, M.-H. and Quinn, N. (1984) *J. Neurol. Neurosurg. Psychiat.* **47**, 1166–1173.

Marsden, C. D., Obeso, J. A., Zarranz, J. J. and Lang, A. E. (1985) *Brain* **108**, 463–483.

McGeer, E. G. and McGeer, P. L. (1988) *Can. J. Neurol. Sci.* **15**, 447–483.

Meldrum, B. S., Anlezark, G. M. and Marsden, C. D. (1977) *Brain* **100**, 313–326.

Mones, R. J. (1972) In *Advances in Neurology, Vol. 1: Huntington's Chorea* (edited by A. Barbeau, T. N. Chase and G. W. Paulson), pp. 665–669. Raven Press, New York.

Murphey, D. L. and Dill, R. E. (1972) *Exp. Neurol.* **34**, 244–254.

Neale, R., Gerhardt, S. and Liebman, J. M. (1984) *Psychopharmacology* **82**, 20–26.

Oltmans, G. A., Beales, M., Lorden, J. F. and Gordon, J. H. (1984a) *Exp. Neurol.* **85**, 216–222.

Oltmans, G. A., Gordon, J. and Beales, M. (1984b) *Neurosci. Abstr.* **9**, 1092.

Parkes, J. D., Bedard, P. and Marsden, C. D. (1976) *Lancet* **i**, 55.

Paulson, G. W. (1973) In *Advances in Neurology, Vol. 1: Huntington's Chorea* (edited by A. Barbeau, T. N. Chase and G. W. Paulson), pp. 647–650. Raven Press, New York.

Rosengarten, H., Schweitzer, J. W. and Friedhoff, A. J. (1983) *Life Sci.* **33**, 2479–2482.

Rotrosen, J., Stanley, M., Kuhn, C., Wazer, D. and Gershon, S. (1980) *Neurology* **30**, 878–881.

Rupniak, N. M. J., Jenner, P. and Marsden, C. D. (1983) *Psychopharmacology* **79**, 226–230.

Rupniak, N. M. J., Jenner, P. and Marsden, C. D. (1985) *Psychopharmacology* **85**, 71–79.

Rupniak, N. M. J., Jenner, P. and Marsden, C. D. (1986) *Psychopharmacology* **88**, 403–419.

Salamone, J. D., Lalies, M. D., Channell, S. L. and Iversen, S. D. (1986) *Psychopharmacology* **88**, 467–471.

Sassin, J. F. (1975) In *Advances in Neurology, Vol. 10: Primate Models of Neurological Disorders* (edited by B. S. Meldrum and C. D. Marsden), pp. 47–54. Raven Press, New York.

Sethy, V. H. and van Woert, M. H. (1974) *Res. Commun. Chem. Pathol. Pharmacol.* **8**, 13–28.

Stewart, B., Rose, S., Jenner, P. and Marsden, C. D. (1987) *Eur. J. Pharmacol.* **142**, 173–176.

Stewart, B. R., Jenner, P. and Marsden, C. D. (1988a) *Psychopharmacology* **96**, 55–62.

Stewart, B. R., Rupniak, N. M. J., Jenner, P. and Marsden, C. D. (1988b) In *Advances in Neurology, Vol. 50: Dystonia 2* (edited by S. Fahn, C. D. Marsden and D. B. Calne), pp. 343–359. Raven Press, New York.

Stewart, B. R., Jenner, P. and Marsden, C. D. (1989a) *Eur. J. Pharmacol.* **162**, 101–107.

Stewart, B. R., Jenner, P. and Marsden, C. D. (1989b) *Psychopharmacology* **97**, 228–234.

Tam, S. W. and Cook, L. (1984) *Proc. Natl. Acad. Sci. USA* **81**, 5618–5622.

Waddington, J. L. and Gamble, S. J. (1980) *Eur. J. Pharmacol.* **68**, 387–388.

Waddington, J. L. and Molloy, A. G. (1987) *Psychopharmacology* **91**, 136–137.

Waddington, J. L., Cross, A. J., Gamble, S. J. and Bourne, R. C. (1983) *Science* **220**, 530–532.

Walker, J. M., Matsumoto, R. R., Bowen, W. D., Gans, D. L., Jones, K. D. and Walker, F. O. (1988) *Neurology* **38**, 961–965.

Weber, E., Sonders, M., Quarum, M., McLean, S., Pou, S. and Keana, J. F. W. (1986) *Proc. Nat. Acad. Sci. USA* **83**, 8784–8788.

PART IV

Other Movement Disorders

30

Degenerative Ataxic Disorders

A. E. Harding

The degenerative ataxias are a complex group of disorders comprising 50 or more distinct syndromes, most of which are genetically determined. Many consider that they defy classification. Until recently attempts at classification have largely been based on pathological grounds; these are not always useful for the clinician and there has been a tendency to ignore clinical and genetic evidence of heterogeneity. A classification based on the clinical and genetic features of over 200 personally studied families with hereditary ataxia is shown in Table 30.1 (Harding, 1983, 1984).

INHERITED ATAXIAS WITH KNOWN METABOLIC DEFECTS

The proportion of patients with hereditary ataxias of adolescent or early adult onset who are recognized to have demonstrable metabolic defects is gradually increasing. A-β-lipoproteinaemia is an important example because the neurological disability associated with this disorder is potentially preventable. It is an autosomal recessive syndrome, possibly caused by failure to synthesize apoprotein B which normally carries lipid from the intestinal cell to the plasma. This results in very low levels of circulating lipids, particularly cholesterol, and absorption of the fat-soluble vitamins A, D, E and K is severely impaired. Symptoms of fat malabsorption are usually mild and may

DISORDERS OF MOVEMENT: CLINICAL, PHARMACOLOGICAL
AND PHYSIOLOGICAL ASPECTS   ISBN 0-12-569685-X

Table 30.1 Classification of degenerative ataxias.

I. Inherited ataxic disorders with known metabolic defects
 1. Progressive ataxias
  (i)   A-β- and hypo-β-lipoproteinaemia
  (ii)   Hexosaminidase deficiency*
  (iii) Cholestanolosis*
  (iv)   Leukodystrophies* (metachromatic, late-onset globoid cell, adrenoleukomyeloneuro-pathy)
  (v)   Mitochondrial encephalomyopathies*
  (vi)   Partial HGPRT deficiency*
  (vii) Wilson's disease*
  (viii) Ceroid lipofuscinosis*
  (ix)   Sialidosis
  (x)   Sphingomyelin storage disorders*

 2. Intermittent ataxias
  (i)   With hyperammonaemia: ornithine transcarbamylase deficiency, argininosuccinate synthetase deficiency (citrullinaemia), argininosuccinase deficiency (argininosuccini-caciduria), arginase deficiency, hyperornithinaemia
  (ii) Aminoacidurias (Hartnup disease, intermittent branched chain ketoaciduria, isovaleric acidaemia)
  (iii) Disorders of pyruvate and lactate metabolism (pyruvate dehydrogenase deficiency, pyruvate carboxylase deficiency, Leigh's syndrome, multiple carboxylase deficiencies)

 3. Disorders associated with defective DNA repair
  (i)   Ataxia telangiectasia
  (ii) Xeroderma pigmentosum*
  (iii) Cockayne syndrome

II. Ataxic disorders of unknown aetiology

 1. Early-onset ataxias (onset usually before 20 years)
  (i)   Friedreich's ataxia
  (ii)   Early-onset cerebellar ataxia with retained tendon reflexes
  (iii) With hypogonadism
  (iv)   With myoclonus (idiopathic Ramsay Hunt syndrome)
  (v)   With pigmentary retinopathy
  (vi)   With optic atrophy $\pm$ mental retardation (including Behr's syndrome)
  (vii) With cataract and mental retardation (Marinesco–Sjögren syndrome)
  (viii) With childhood deafness
  (ix)   With congenital deafness
  (x)   With extrapyramidal features
  (xi)   X-linked recessive spinocerebellar ataxia

 2. Late-onset ataxias (onset usually after 20 years)
  (i)   Autosomal dominant cerebellar ataxia with ophthalmoplegia/optic atrophy/dementia/extrapyramidal features (including Machado–Joseph disease)
  (ii) Autosomal dominant cerebellar ataxia with pigmentary retinopathy $\pm$ ophthalmoplegia/extrapyramidal features
  (iii) "Pure" autosomal dominant cerebellar ataxia of later onset (over 50 years)
  (iv) Autosomal dominant cerebellar ataxia with myoclonus and deafness
  (v)   Autosomal dominant ataxia with essential tremor
  (vi) Periodic autosomal dominant cerebellar ataxia
  (vii) Autosomal recessive late onset ataxia

III. Degenerative late-onset ataxias which may not be inherited

From Harding (1984).
*Ataxia may not be a prominent feature.

be overlooked; patients then present in the second decade of life with a progressive neurological syndrome reminiscent of Friedreich's ataxia comprising ataxia, areflexia and proprioceptive loss. There is now convincing evidence that this syndrome is caused by severe vitamin E deficiency. Treatment with large doses of vitamin E prevents its development and leads to improvement or lack of progression if symptoms are already present (Muller et al., 1983). Patients with a similar neurological syndrome associated with isolated vitamin E deficiency (and otherwise normal serum lipid concentrations), probably due to a specific defect of absorption and inherited as an autosomal recessive trait, have recently been reported (Harding et al., 1985; Harding, 1987a).

A number of enzyme deficiencies which usually give rise to neuro-degenerative disorders in infancy or early childhood have been identified in patients with predominantly ataxic disorders developing in late childhood or early adult life (Adams and Lyon, 1982). These include adrenoleukomyelo-neuropathy, a phenotypic variant of adrenoleukodystrophy (Moser et al., 1987), and galactosylceramide lipidosis (Krabbe's disease). The GM_2 gangliosidoses result from deficiencies of the isoenzymes collectively known as the hexosaminidases; they are genetically heterogeneous, the most common giving rise to the clinical syndrome of Tay–Sachs disease (Johnson, 1981). A progressive neurological disorder with onset in adolescence or early adult life comprising ataxia, tremor, supranuclear ophthalmoplegia, facial grimacing and proximal neurogenic muscle weakness has been described in association with hexosaminidase A deficiency (Harding et al., 1987). Ataxia may be a prominent feature in Niemann Pick disease type C (juvenile dystonic lipidosis), combined with a supranuclear gaze palsy. Sphingomyelin-ase activity is normal but foamy storage cells are found in the bone marrow (Yan-Go et al., 1984).

Cholestanolosis, a rare autosomal recessive disorder caused by defective bile salt metabolism, gives rise to ataxia, dementia, spasticity, peripheral neuropathy, cataract, and tendon xanthomata in the second decade of life. Treatment with chenodeoxycholic acid appears to improve neurological function (Berginer et al., 1984).

Various phenotypes which are classifiable as hereditary ataxias have been described in association with the mitochondrial myopathies. These include late-onset ataxic disorders associated with such features as dementia, deafness and peripheral neuropathy, and the Ramsay Hunt syndrome of ataxia and myoclonus (Petty et al., 1986). The Ramsay Hunt syndrome has a number of causes, including ceroid lipofuscinosis and sialidosis; after intensive investigation a number of cases remain in which no cause can be found, including those with Unverricht–Lundborg disease or so-called Baltic myoclonus (Eldridge et al., 1983).

INHERITED ATAXIAS OF UNKNOWN CAUSE: EARLY ONSET

Degenerative ataxic disorders of unknown cause with onset in childhood or adolescence are nearly always genetically determined; most are autosomal recessive. Friedreich's ataxia is by far the commonest form of hereditary ataxia, comprising over 50% of cases in most large series (Harding, 1983), with a prevalence of 1–2 per 100 000 in Europe. The essential criteria for diagnosis are progressive limb and gait ataxia developing before the age of 25, autosomal recessive inheritance, absent tendon reflexes in the lower limbs, and electrophysiological evidence of an axonal sensory neuropathy. Dysarthria, areflexia, pyramidal weakness of the legs, extensor plantar responses, and distal loss of joint position and vibration sense are not found in every case at the time of presentation but eventually occur in all patients. Less common features include pes cavus, optic atrophy, and distal wasting. Patients lose the ability to walk 15 years after the onset of symptoms on average, and usually die in their forties or fifties. There is an associated heart muscle disease giving rise to electrocardiographic abnormalities in over two-thirds of cases, and diabetes mellitus occurs in 10% (Harding, 1981a, 1984).

Barbeau *et al.* (1984) reviewed the extensive studies attempting to define the basic biochemical defect in Friedreich's ataxia in Quebec. Several unexplained abnormalities were identified in a proportion of patients, such as increased urinary taurine excretion, elevated serum bilirubin concentrations, and abnormal lipid composition; most of these observations have not been confirmed elsewhere. The same applies to reports suggesting that deficiencies of either lipoamide dehydrogenase or the mitochondrial malic enzyme represent the primary defect in Friedreich's ataxia. The Friedreich's ataxia gene has recently been mapped to the centromeric region of chromosome 9 (Chamberlain *et al.*, 1988).

Most of the other early onset ataxias listed in Table 30.1 are rare, with the exception of early onset cerebellar ataxia with retained reflexes, which occurs at a frequency about one-quarter of that of Friedreich's ataxia, and is often confused with it. It is important to distinguish between these two disorders, as the prognosis is much better in the former, and optic atrophy, severe skeletal deformity, and cardiac involvement do not occur. The tendon reflexes are normal or increased, and the gait may have a spastic component (Harding, 1984).

INHERITED ATAXIAS OF UNKNOWN CAUSE: LATE ONSET

The autosomal dominant late-onset cerebellar ataxias (Table 30.1), which are often referred to as the olivopontocerebellar atrophies, are a heterogeneous group of disorders which have provoked most controversy in relation to classification (Harding, 1982). Most patients with dominantly inherited late-onset ataxia do not have a pure cerebellar syndrome; additional features such as supranuclear ophthalmoplegia, optic atrophy, pigmentary retinopathy, dementia, and extrapyramidal dysfunction are common. The clinical syndrome may vary remarkably, even within members of the same family. The pathological findings usually, but not always, include olivopontocerebellar atrophy. Although there is some pathological evidence suggesting that Machado–Joseph disease, described mainly in patients of Azorean descent, is a distinct disorder, it is impossible to distinguish clinically between this syndrome and others included in category II.2.i (Table 30.1).

Patients in this category face the same difficult reproductive decisions as those with Huntington's chorea, because symptoms usually develop between the ages of 25 and 45, i.e. often not until after the reproductive period is over. Genetic linkage studies have shown rather conflicting evidence as to whether the mutant gene causing this disorder is linked to the HLA complex on chromosome 6 (Haines *et al.*, 1984). These results probably reflect further genetic heterogeneity. Further linkage studies, using polymorphic DNA markers on chromosome 6 and elsewhere in the genome, should clarify the issue of classification of the dominant ataxias and possibly provide accurate gene carrier detection and antenatal diagnosis.

DEGENERATIVE LATE-ONSET ATAXIAS WHICH MAY NOT BE INHERITED

In patients who present with slowly progressive cerebellar dysfunction in adult life and do not have similarly affected relatives, it is important to exclude disorders such as hypothyroidism, alcoholism, and malignancy, although paraneoplastic cerebellar degeneration usually gives rise to a subacute ataxic syndrome. About two-thirds of patients with degenerative ataxia developing over the age of 20 years are singleton cases, and they represent a significant clinical problem; it is even difficult to know how to label them. The present author prefers to use the term idiopathic late-onset cerebellar ataxia (Harding, 1981b), rather than the more frequent label of olivopontocerebellar atrophy (OPCA) (Harding, 1987b).

The pathological findings in both inherited and idiopathic late-onset cerebellar degenerations are variable, including cerebello-olivary atrophy and cerebellar cortical atrophy, as well as OPCA (loss of cells in the pontine and olivary nuclei and also of the cerebellar Purkinje cells) which is seen in at least half of sporadic cases. In OPCA there are nearly always accompanying degenerative changes in the basal ganglia, spinal cord, and peripheral nerves, so even as a descriptive pathological term OPCA is rather misleading. There is no evidence that OPCA can be diagnosed with certainty by means of computerized tomography (CT) or magnetic resonance imaging, and it is an inappropriate, albeit comforting, diagnostic label in clinical practice.

The clinical features of idiopathic late-onset cerebellar ataxia are similar to those in the first category of dominant ataxias (Table 30.1), with a wide range of age of onset between 25 and 70 years. Patients with onset after the age of 55 often have a relatively pure midline cerebellar syndrome, with marked gait ataxia but mild cerebellar dysfunction in the limbs. The cause of these disorders is unknown. It is probable that some represent fresh dominant mutations, and others may have an autosomal recessive syndrome, although proven recessive cases of late-onset ataxia are rare.

It has been suggested that some late-onset ataxic disorders, usually developing between the sixth and eighth decades of life, are associated with moderately reduced activity of the enzyme glutamate dehydrogenase (GDH) (Duvoisin et al., 1983). The clinical picture in deficient patients was rather variable, consisting of a combination of atypical parkinsonism with cerebellar ataxia, supranuclear ophthalmoplegia, bulbar palsy, and an axonal peripheral neuropathy in most cases. The CT scan showed striking cerebellar and brain stem atrophy. More recent studies of patients with a variety of neurodegenerative disorders have shown that "low" levels of GDH activity are not disease specific, and therefore probably not of aetiological significance (Aubby et al., 1988)'

DIAGNOSIS

An investigative approach to degenerative ataxias is outlined in Table 30.2. The clinical starting point is termed "probably" degenerative ataxias as it is assumed that other causes of ataxia such as alcoholism, tumour, demyelination, etc. will have been excluded on clinical grounds and imaging studies. Nevertheless, some investigations for acquired causes of cerebellar dysfunction such as malignancy and hypothyroidism are included in this scheme as they present with progressive ataxia and it is important from the point of view of management that they are not overlooked. It should be stressed that taking an adequate family history, and examining relatives if necessary, is essential in the assessment of degenerative ataxic disorders. Nevertheless, neurologists need to bear in mind that most patients with autosomal recessive disorders will not have affected relatives in populations where family size is small, and their absence does not mean that a genetic disorder is unlikely, particularly in early-onset disorders.

In addition to excluding acquired non-genetic cerebellar syndromes, the

Table 30.2 An investigative approach to (probably) degenerative ataxias.

	In all cases if:		
	Early onset (1–20 years)	Late onset (>20 years)	Abnormal in:
Lipids	Yes	If hypo-/areflexic	A-β-lipoproteinaemia
Vitamin E	Yes	If hypo-/areflexic	Vitamin E deficiency
Glucose	Yes	No	Friedreich's ataxia
ECG	Yes	No	Friedreich's ataxia
α-Fetoprotein	Yes	No	Ataxia telangiectasia
Immunoglobulins	Yes	No	Ataxia telangiectasia
VLCFA, ACTH	Males	See below	Adrenol eukodystrophy
Lactate	Yes	Yes	Mitochondrial myopathies
Caeruloplasmin	Yes unless NCS abnormal		Wilson's disease
Thyroxine	Yes	Yes	Hypothyroidism
CXR	Yes	Yes	Parenoplastic syndromes
NCS	Yes	Yes	Many disorders
VEP, SSEP	Yes	Yes	Many disorders
CT/MRI	Yes	Yes	Many disorders

In selected cases with additional clinical features think about:

(1) any age of onset, but usually <50 years: muscle biopsy for mitochondrial myopathy and ceroid lipofuscinosis (fatigable weakness, dementia, stroke-like episodes, myoclonus, retinopathy, short stature, etc.); cholestanol (cataract, tendinous swellings); gonadotrophins (hypogonadism); VLCFA/ACTH (spastic paraparesis, ± axonal neuropathy, radiological evidence of white matter disease, onset earlier in males than females)

(2) particularly (but not exclusively) if early onset (2–20 years): hexosaminidase (gaze palsy, neurogenic weakness, dystonia); aryl sulphatase A, galactocerebrosidase (dementia, psychiatric problems, optic atrophy, demyelinating neuropathy, radiological evidence of white matter disease); bone marrow (gaze palsy, fits, extrapyramidal features, dementia); ammonia/amino acids (early onset, fluctuating course, mental retardation)

(3) in later onset cases (>20 years): anti-Purkinje cell antibodies, pelvic imaging in females (subacute course)

ECG, electrocardiogram; VLCFA, very long chain fatty acids; ACTH, adrenocorticophic hormone; CXR, chest X-ray; NCS, nerve conduction studies; V/SSEPs, visual/somatosensory evoked potentials; CT, computerized tomography; MRI, magnetic resonance imaging.

purposes of the investigations outlined in Table 30.2 are to aid diagnosis of easily recognizable hereditary ataxias (such as nerve conduction studies and electrocardiography in possible cases of Friedreich's ataxia), and to identify biochemically defined degenerative ataxias, some of which are treatable. It is not rational (or economically sound) to investigate all possibilities in every patient, and for this reason the investigative approach outlined here is tailored to the clinical presentation. In practice, only about 5% of "degenerative" ataxias will be found to have a known metabolic basis after intensive investigation. A further 90% can be classified, largely on clinical grounds, into one of the categories in Table 30.1. It is important, as in all degenerative neurological disorders, to make as precise a diagnosis as possible, even if this is merely descriptive, for the purposes of prognosis, genetic counselling, and future investigative studies.

TREATMENT

In view of our poor understanding of the pathophysiology underlying most of the degenerative ataxias, it is not surprising that attempts at treatment have largely been unsuccessful. The drugs which have been used to date are cholinergic agents including physostigmine, lecithin, and choline chloride, serotonergic compounds such as L-5-hydroxytryptophan (L-5HTP) combined with a peripheral decarboxylase inhibitor, the GABAergic drugs baclofen and sodium valproate, and thyrotrophin-releasing hormone (TRH). The rationale for the use of these drugs, and the results of therapy have been reviewed by Plaitakis (1985). Although a small number of cases seem to show objective improvement following treatment, there have been few well-designed trials on homogeneous groups of patients. Double-blind trials have occasionally shown benefit to treated patients, for example following the use of choline chloride and L-5HTP, but this has either not led to functional improvement or has not as yet been achieved in more than one centre. Promising results using TRH in the treatment of ataxia have been reported from Japan (Sobue, 1986), and these observations need to be confirmed.

REFERENCES

Adams, R. D. and Lyon, G. (1982) *Neurology of Hereditary Metabolic Diseases of Children.* McGraw-Hill, New York.

Aubby, D., Saggu, H. K., Jenner, P., Quinn, N. P., Harding, A. E. and Marsden, C. D. (1988) *J. Neurol. Neurosurg. Psychiat.* **51**, 893–902.

Barbeau, A., Melancon, S. B., Lemieux, B. and Roy, M. (1984) *Italian J. Neurol. Sci.* Suppl. 4, 13–26.

Berginer, V. M., Salen, G. and Shefer, S. (1984) *N. Engl. J. Med.* **311**, 1649–1652.

Chamberlain, S., Shaw, J., Rowland, A., Wallis, J., South, S., Nakamura, Y., von Gabain, A., Farrall, M. and Williamson, R. (1988) *Nature* **334**, 248–250.

Duvoisin, R. C., Chokroverty, S., Lepore, F. and Nicklas, W. (1983) *Neurology* **33**, 1322–1326.

Eldridge, R., Iivanainen, M., Stern, R., Koerber, T. and Wilder, B. J. (1983) *Lancet* **ii**, 838–841.

Haines, J. L., Schut, L. J., Weitkamp, L. R., Thayer, M. and Anderson, V. E. (1984) *Neurology* **34**, 1542–1548.

Harding, A. E. (1981a) *Brain* **104**, 589–620.

Harding, A. E. (1981b) *J. Neurol. Sci.* **51**, 259–271.

Harding, A. E. (1982) *Brain* **105**, 1–28.

Harding, A. E. (1983) *Lancet* **i**, 1151–1155.

Harding, A. E. (1984) *The Hereditary Ataxias and Related Disorders*. Churchill Livingstone, Edinburgh.

Harding, A. E. (1987a) *CRC Crit. Rev. Neurobiol.* **3**, 89–103.

Harding, A. E. (1987b) In *Movement Disorders 2* (edited by C. D. Marsden and S. Fahn), pp. 269–271. Butterworths, London.

Harding, A. E., Matthews, S., Jones, S., Ellis, C. J. K., Booth, I. W. and Muller, D. P. R. (1985) *N. Engl. J. Med.* **313**, 32–35.

Harding, A. E., Young, E. P. and Schon, F. (1987) *J. Neurol. Neurosurg. Psychiat.* **47**, 853–856.

Johnson, W. G. (1981) *Neurology* **31**, 1453–1456.

Moser, H. W., Naidu, S., Kumar, A. J. and Rosenbaum, A. E. (1987) *CRC Crit. Rev. Neurobiol.* **3**, 29–88.

Muller, D. P. R., Lloyd, J. K. and Wolff, O. H. (1983) *Lancet* **i**, 225–227.

Petty, R. K. H., Harding, A. E. and Morgan-Hughes, J. A. (1986) *Brain* **109**, 915–938.

Plaitakis, A. (1985) In *Current Therapy in Neurologic Disease* (edited by R. T. Johnson), pp. 254–262. B. C. Decker, Philadelphia.

Sobue, I. (Editor) (1986) *TRH and Spinocerebellar Degeneration*. Elsevier, Amsterdam.

Yan-Go, F. L., Yanagihara, T., Pierre, R. V. and Goldstein, N. P. (1984) *Mayo Clin. Proc.* **59**, 404–410.

31

The Physiology of Myoclonus in Man

J. A. Obeso, J. Artieda and J. M. Martínez-Lage

INTRODUCTION

Myoclonus is defined as sudden, brief, shock-like involuntary movements caused by active muscular contractions (Marsden *et al.*, 1982). Asterixis, which is due to short-lasting pauses of muscular activity causing postural lapses, may be regarded as a form of "negative myoclonus" (Young and Shahani, 1986).

Myoclonus may be a manifestation of a wide variety of pathological conditions affecting the brain, spinal cord or even the peripheral nerves. Myoclonus is a positive sign. The pathological distribution of the lesion(s) causing myoclonus may not coincide with site of the neuronal discharge producing muscle jerking. Detailed clinical and electrophysiological analysis of myoclonus can separate myoclonus from other disorders and distinguish different neuronal discharges by which it is provoked.

Four main pathophysiological categories of myoclonus may be recognized: (1) in *cortical myoclonus* the abnormal activity originates in the sensorimotor cortex and is transmitted down the spinal cord via the pyramidal tract; (2) *subcortical myoclonus* indicates those forms of myoclonus arising in structures located between the cerebral cortex and the spinal cord, for example, reticular reflex myoclonus is a form of subcortical myoclonus

DISORDERS OF MOVEMENT: CLINICAL, PHARMACOLOGICAL
AND PHYSIOLOGICAL ASPECTS   ISBN 0-12-569685-X

originating in the brain stem in which jerks are generalized and produced by external stimuli; (3) *cortical–subcortical myoclonus* is due to an abnormal cortical discharge spreading via cortico-reticulo-spinal pathways to give rise to generalized myoclonus, or when a subcortical focus produces its effect via the motor cortex-pyramidal tract; (4) *spinal myoclonus* is due to abnormal discharges of spinal neurons. Lesions affecting spinal roots, plexi or nerves may produce myoclonus by inducing abnormal firing of spinal or supraspinal neurons.

PHYSIOLOGICAL ASSESSMENT OF MYOCLONUS IN MAN

Pathophysiological classification of myoclonus in man is based on four main sources of information (Marsden *et al.*, 1983): (a) clinical characteristics of the jerks and accompanying neurological signs; (b) EMG recording; (c) EEG–EMG correlation; (d) cortical evoked potentials (Table 31.1).

Table 31.1 Summary of electrophysiological findings and pathophysiological classification of myoclonus in man.

Type	Clinical characteristics	EMG discharge	EEG–EMG correlation	Amplitude of cortical evoked potential
Cortical	Focal and distal Multifocal Action, reflex	10–50 ms Downward activation	Focal EEG wave preceding the jerks by a short latency	Increased ($>12\,\mu$V)
Subcortical	(a) Generalized (b) Spontaneous or reflex and generalized (reticular)	Usually >100 ms Up the brain stem and down the cord 10–50 ms	No EEG correlate on back-averaging	Normal
Cortical–subcortical	Multifocal and/or generalized	10–100 ms Co-contracting	Bilateral brain wave preceding the jerks by >50 ms	Normal or bilaterally enlarged
Spinal	Segmental/focal Spontaneous Rhythmical	>100 ms Synchronous	No correlate	Normal

Clinical data

Focal, stimulus-sensitive jerks, made worse by action, are generally of cortical origin. Multifocal myoclonus may originate from any neuronal pool in the brain, but it is often of the cortical or cortico-subcortical type. Segmental, rhythmical spontaneous myoclonus is suggestive of a spinal cord or, when accompanied by palatal myoclonus, brain stem origin. One must keep in mind however that focal, repetitive myoclonic jerking of one limb may be secondary to lesion of a great variety of structures, including the sensorimotor cortex (epilepsia partialis continua), thalamus, subthalamus, red nucleus, dentorubro-thalamic tract and spinal cord. Subcortical myoclonus usually is generalized and affecting both axial and limb muscles. Generalized stimulus-sensitive myoclonus suggests a brain stem origin ("reticular reflex myoclonus"; Hallett et al., 1977a,b).

Electromyographic activity

Surface EMG recording of antagonist muscles provides useful clues to differentiate myoclonus. In cortical myoclonus (Table 31.1), the EMG bursts are short lasting (< 100 ms), the activation order (when jerks are generalized) follows a rostrocaudal direction, and there is co-contraction of antagonist muscle pairs. Long-lasting EMG discharges (> 100 ms) are seen in conditions such as Creutzfeldt–Jakob disease, torsion dystonia, drug-induced myoclonus, etc., where myoclonus has a subcortical origin (Halliday, 1975; Shibasaki et al., 1981; Obeso et al., 1983a,b). Electromyographic bursts of around 100 ms but showing alternating activity in antagonist muscles have also been described in a family with essential myoclonus (Hallett et al., 1977b) and in two patients with oscillatory myoclonus (Obeso et al., 1983b).

In "reticular reflex myoclonus" the abnormal discharges arise from the pontine reticular formation and propagate up the brain stem and down the cord (Hallett et al., 1977a). Spinal myoclonus usually shows co-contracting, long duration EMG bursts at 2–4 Hz.

EEG–EMG correlation

EEG spikes at the time of jerking can often be observed in patients with myoclonus epilepsy. However, adequate assessment of the relationship between the cortical event and the jerks can only be obtained by applying the technique of back-averaging, which allows one to establish whether or not EEG activity is actually time-locked to the muscle jerks (Marsden et al., 1982).

In cortical myoclonus, a biphasic brain wave precedes the jerks by 20–25 ms for the upper limbs and 30–35 ms for the lower limbs. The cortical origin of this wave is almost always located in the sensorimotor cortex (Shibasaki *et al.*, 1985; Cowan *et al.*, 1986). In rare cases of subcortico-cortical myoclonus, bilateral brain waves may be recorded preceding the jerks by a long latency of 50–200 ms, suggesting a subcortical origin with secondary cortical excitation. Absence of EEG activity preceding the jerks is indicative, but not absolute proof, of a non-cortical origin. Thus, surface EEG recording (even using back-averaging) may fail to detect paroxysmal activity taking place in deep cortical layers (Elger and Speckman, 1980).

A slow, bilateral negativity with fronto-central distribution, preceding the jerks by some 100 ms, has been described in patients with Alzheimer's disease (Wilkins *et al.*, 1985) and Creutzfeldt–Jakob disease (Shibasaki *et al.*, 1981). These two characteristics, the widespread topography and long interval between EEG and EMG discharges, suggest a secondary cortical excitation from subcortical structures.

Cortical evoked potentials

Enhancement of the primary complex of the somatosensory evoked potentials (SEPs) to electric nerve stimulation is a well-documented characteristic of cortical myoclonus. The P_1–N_2 (P_{26}–N_{35}) waves are increased in amplitude ($> 10\,\mu V$), probably reflecting postsynaptic neuronal potentials in the somato-sensory cortex. The large SEP commonly is followed within a short latency by a "reflex" myoclonic discharge in the same limb ("cortical reflex myoclonus"; Hallet *et al.*, 1979). In patients with multifocal or generalized action myoclonus, SEPs often are bilaterally enlarged, suggesting loss of inhibitory influence upon the cortex from other centres (Obeso *et al.*, 1986). Visual evoked potentials (VEPs) to flash or pattern stimulation may also be greatly enhanced and followed by reflex myoclonic discharges in patients with cortical myoclonus. Giant VEPs are particularly frequent in neuronal storage disease and Lafora's disease. Auditory evoked potentials are not usually increased in amplitude, even in patients showing sensitivity to sound (Shibasaki *et al.*, 1988). Cortical evoked potentials are of normal amplitude in subcortical myoclonus.

ANATOMO-PHYSIOLOGICAL BASIS

Myoclonus is a positive sign. One should not expect therefore to find a good correlation between physiological findings and pathological distribution of the lesions. Moreover precise anatomo-physiological correlation is not easy to establish, for myoclonus is rarely due to a single isolated CNS lesion.

(a) *Cortical myoclonus* can be secondary to many different types of focal cortical lesions such as tumours, angiomas, encephalitis or dysplasia (Forster *et al.*, 1949; Kuzniecky *et al.*, 1988; Obeso *et al.*, 1988) affecting the primary sensorimotor cortex. In such cases, myoclonus is usually focal, may be stimulus sensitive, and frequently evolves into repetitive muscle jerking (epilepsia partialis continua), Jacksonian and generalized seizures (Obeso *et al.*, 1985a).

Interestingly, either focal myoclonus or the more common form of multifocal action myoclonus of cortical origin may not be associated with any histological abnormality at the cortical level (Obeso *et al.*, 1985b). In some cases pathology can be found exclusively in subcortical regions as in the case of patients with spino-cerebellar or dentato-rubro-pallido-luysion degenerations (Bird and Shaw, 1978; Bonduelle *et al.*, 1976). Indeed, cortical myoclonus is the rule in patients with the Ramsay Hunt syndrome.

Of subcortical structures, the cerebellum (both dentate and cortex) and thalamus are most often associated with cortical myoclonus, but a single subcortical lesion in patients with multifocal myoclonus of cortical origin is an exceptional finding.

(b) *Subcortical myoclonus* implies that the neuronal discharge producing myoclonus is not originating in the cortex or spinal cord; it should be noted that there is currently no evidence that such discharges actually propagate via subcortical (i.e. reticulo-spinal) pathways totally independently from the cortico-spinal tract. Three categories should be separated for the purpose of this discussion.

(1) Spontaneous, usually synchronous and generalized, jerks may occur in a wide variety of clinical conditions mainly characterized by subcortical pathology, such as Huntington's chorea and Wilson's disease, but are a rare finding in untreated Parkinson's disease. Idiopathic torsion dystonia, supposedly a basal ganglia disorder, is the most common condition causing spontaneous and action myoclonus of subcortical origin.

(2) Reticular reflex myoclonus is characterized by generalized jerks induced by external stimuli or during voluntary movement. Anoxia, herpetic and other acute encephalitides, metabolic encephalopathies (uraemia, liver failure, etc.) are the most frequent causes of reticular reflex myoclonus. The jerks are presumed to originate in the nucleous gigantocellularis of the brain stem reticular formation, but there is as yet no pathological proof. It seems likely that when the brain of a patient with reticular reflex myoclonus comes to post-mortem study, the pathological findings will be rather diffuse.

(3) Palatal myoclonus. Palatal movements, unilaterally or bilaterally, occur at 1.5–3 Hz and are frequently accompanied by synchronous movements of

adjacent muscles such as the external ocular muscles, tongue, larynx, face, neck or diaphragm. Occasionally limb muscles can be involved (Lapresle, 1986). The disorder can be idiopathic, but is also seen after a variety of neurologic lesions including stroke, multiple sclerosis, tumours, trauma and metabolic encephalopathy (Marsden *et al.*, 1982). Common to all the secondary causes appears to be a lesion in the pathway from the dentate nucleus to the inferior olive via the red nucleus (two sides of the "Guillain–Mollaret triangle"). In all cases at post-mortem examination there is an unusual lesion of the inferior olive, described as a hypertrophic degeneration. The recent demonstration (Dubinsky *et al.*, 1987) that the inferior olive shows increased metabolism (measured by PET scan) adds further support to the role of these nuclei in the origin of palatal myoclonus.

(4) *Spinal myoclonus.* Rhythmic myoclonus of spinal origin is much more common than arrhythmic myoclonus. Spinal myoclonus usually occurs spontaneously and may persist during sleep. Involved regions can be one limb, one limb and adjacent trunk, or both legs. Lesions of the spinal cord giving rise to focal movements include infection, degenerative disease, tumour, cervical myelopathy and demyelinating disease, and it may follow spinal anaesthesia or the introduction of contrast media into the CSF (Marsden *et al.*, 1982; Silfverskiold, 1986). Focal and rhythmic myoclonus is not necessarily due to a spinal lesion, but may instead be secondary to a focal lesion of the sensorimotor cortex, subthalamic nucleus, dentate nucleus and superior cerebellar peduncle or associated with peripheral nerve damage. Loss of small neurons in the spinal cord has been described (Davis *et al.*, 1981) in a case of stimulus-sensitive arrhythmic spinal myoclonus apparently due to an ischaemic myelopathy.

There have been a few reports of myoclonus where the lesion has appeared to be in the peripheral nervous system. Cases have been reported with lesions of nerve (Marsden *et al.*, 1984), brachial plexus (Banks *et al.*, 1985) and nerve root (Sotaniemi, 1985). These lesions could cause myoclonus by way of modifying the normal pattern of sensory input to the spinal cord, which in turn could lead to disinhibition of anterior horn motoneurons.

CONCLUSIONS

Myoclonus is a frequent clinical phenomenon with a very diverse pathophysiological basis. Cortical myoclonus is currently considered a fragment of epilepsy, hence due to a paroxysmal depolarization shift in the sensorimotor cortex (Hallett, 1985); whether or not other physiological types of myoclonus have a similar neurophysiological basis is entirely unknown at present. Clinico-physiological classification of myoclonus may serve a practical purpose in the management of patients. For instance, myoclonus of cortical origin responds best to polytherapy with drugs such as clonazepam, sodium valproate and primidone; on the other hand, reticular reflex myoclonus is extremely sensitive to clonazepam or 5-hydroxytryptophan (plus carbidopa), and patients with the combination of myoclonus and dystonia may show an exquisite sensitivity to alcohol. Nevertheless, further dissection of the mechanisms responsible for the different types of myoclonus seen in clinical practice is necessary in order to rationalize their treatment.

REFERENCES

Banks, G., Nielsen, V. K., Short, M. P. and Kowal, C. D. (1985) *J. Neurol. Neurosurg. Psychiat.* **48**, 582–584.

Bird, T. D. and Shaw, C. M. (1978) *J. Neurol. Neurosurg. Psychiat.* **41**, 140–149.

Bonduelle, M., Escourolle, R., Bouygues, P., Lormeau, G. and Gray, F. (1976) *Rev. Neurol.* **132**, 113–124.

Cowan, J. A. A., Rothwell, J. C., Wise, R. J. S. and Marsden, C. D. (1986) *J. Neurol. Neurosurg. Psychiat.* **49**, 796–807.

Davis, S. M., Murray, N. M. F., Diengdoh, J. V., Galea-Debono, A. and Kocen, R. S. (1981) *J. Neurol. Neurosurg. Psychiat.* **44**, 884–888.

Dubinsky, R., Hallett, M. and Schwankhaus, J. (1987) *Neurology* **37** (Suppl. 1) 125.

Elger, C. E. and Speckman, E. J. (1980) *Electroencephalogr. Clin. Neurophysiol.* **48**, 447–460.

Forster, F. M., Penfield, W., Jasper, H. and Madow, L. (1949) *Electroencephalogr. Clin. Neurophysiol.* **I**, 349–356.

Hallett, M. (1985) *Epilepsia* (Suppl. 1) **26**, S67–S77.

Hallett, M., Chadwick, D., Adam, J. and Marsden, C. D. (1977a) *J. Neurol. Neurosurg. Psychiat.* **40**, 253–264.

Hallett, M., Chadwick, D. and Marsden, C. D. (1977b) *Brain* **100**, 299–312.

Hallett, M., Chadwick, D. and Marsden, C. D. (1979) *Neurology* **29**, 1107–1125.

Halliday, A. M. (1975) In *Myoclonic Seizures* (edited by M. H. Charlton), pp 1–29. Excepta Medica, Amsterdam.

Kuzniecky, R., Berkovic, S., Andermann, F., Melanson, D., Olivier, A. and Robitaille, Y. (1988) *Ann. Neurol.* **23**, 317–325.

Lapresle, J. (1986) *Adv. Neurol.* **43**, 265–273.

Marsden, C. D., Hallett, M. and Fahn, S. (1982) In *Movement Disorders.* (edited by C. D. Marsden and S. Fahn) pp. 196–248. Butterworths, London.

Marsden, C. D., Obeso, J. A. and Rothwell, J. C. (1983) In *Motor Control Mechanism in Health and Disease* (edited by J. E. Desmedt), pp. 865–881. Raven Press, New York.

Marsden, C. D., Obeso, J. A., Traub, M. M. *et al.* (1984) *Br. Med. J.* **288**, 173–176.

Obeso, J. A., Rothwell, J. C., Lang, A. E. and Marsden, C. D. (1983a) *Neurology* **33**, 825–830.

Obeso, J. A., Lang, A. E., Rothwell, J. C. and Marsden, C. D. (1983b) *Neurology* **33**, 240–243.

Obeso, J. A., Rothwell, J. C. and Marsden, C. D. (1985a) *Brain* **108**, 193–224.

Obeso, J. A., Artieda, J., Tuñon, T., Luquin, M. R. and Martinez-Lage, J. M. (1985b)

J. Neurol. Neurosurg. Psychiat. **48**, 1277–1283.

Obeso, J. A., Rothwell, J. C. and Marsden, C. D. (1986) *Adv. Neurol.* **43**, 373–384.

Obeso, J. A., Artieda, J. and Marsden, C. D. (1988) In *Parkinson's Disease and Movement Disorders* (edited by J. Jankovic and E. Tolosa). pp. 263–274. Urban and Schwarzenberg, Baltimore.

Shibasaki, H., Motomura, S., Yasmishita, Y. *et al.* (1981) *Ann. Neurol.* **9**, 150–156.

Shibasaki, H., Yamishita, Y., Neshige, R. *et al.* (1985) *Brain* **108**, 225–240.

Shibasaki, H., Kakigi, R., Oda, K. I. and Masukawa, S. I. (1988) *J. Neurol. Neurosurg. Psychiat.* **51**, 572–575.

Silfverskiöld, B. P. (1986) *Adv. Neurol.* **43**, 275–285.

Sotaniemi, K. A. (1985) *J. Neurol. Neurosurg. Psychiat.* **48**, 722–723.

Wilkins, D. E., Hallett, M., Burarcelli, A., Walshe, T. and Alvarez, N. (1984) *Neurology* **34**, 898–903.

Young, R. R. and Shahani, B. T. (1986) *Adv. Neurol.* **43**, 137–156.

32

Animal Models of Myoclonus

G. P. Luscombe

INTRODUCTION

Myoclonus is a descriptive term for rapid, involuntary, single or repetitive contractions of a muscle or a group of muscles. Myoclonus occurs in a wide range of human neurological disorders and its induction has been ascribed to a variety of etiologies, for example, physiological, essential, epileptic and symptomatic (Marsden et al., 1982; Fahn et al., 1986). Myoclonias also vary in their cerebral focus of induction, pathology and electrophysiology (Halliday, 1986; and see Chapter 31 by Obeso et al. in this volume) as well as in their clinical manifestations and treatments (Pranzatelli and Snodgrass, 1985).

Post-anoxic myoclonus is a particular movement disorder resulting from a period of oxygen deprivation to the brain (Lance and Adams, 1963). Muscle jerking may occur either on voluntary movement (action myoclonus) or in response to external stimuli (stimulus-sensitive myoclonus). Post-anoxic myoclonus may be causally related to a deficiency in brain 5-hydroxy-tryptamine (5HT) since the cerebrospinal fluid (CSF) contains reduced concentrations of the 5HT metabolite, 5-hydroxyindoleacetic acid (5HIAA) (Guilleminault et al., 1973; Van Woert and Sethy, 1975; Chadwick et al., 1977). In addition, myoclonus is suppressed by administration of agents

DISORDERS OF MOVEMENT: CLINICAL, PHARMACOLOGICAL AND PHYSIOLOGICAL ASPECTS ISBN 0-12-569685-X

elevating brain 5HT levels such as 5-hydroxytryptophan (5HTP) plus a peripheral decarboxylase inhibitor (e.g. carbidopa), L-tryptophan plus a monoamine oxidase inhibitor and the benzodiazepine, clonazepam (Pranzatelli and Snodgrass, 1985).

Electrophysiological studies of patients with post-anoxic myoclonus have indicated that one of two separate physiological mechanisms is responsible for muscle jerking, namely reticular reflex myoclonus (Hallett *et al.*, 1977) or cortical reflex myoclonus (Hallett *et al.*, 1979).

Reticular reflex myoclonus, the basis for this chapter, typically consists of jerks affecting the whole body with proximal muscles more affected than distal ones, and flexors more active then extensors. Critically, the myoclonic discharge originates in the lower brain stem and then travels both up the brain stem to the cranial nerve muscles and down the spinal cord to limb muscles (Hallett *et al.*, 1977). Cortical EEG spikes are observed during myoclonus but they do not correlate to muscle jerks suggesting that the spike does not originate in the cortex but is a result of a subcortical event and is *not* directly responsible for the myoclonus. The muscle jerk, whether spontaneous or induced by stimuli, is characteristically associated with a brief EMG burst of 10–30 ms (Hallett *et al.*, 1977).

ANIMAL MODELS OF RETICULAR REFLEX MYOCLONUS

The preceding clinical profile of post-anoxic myoclonus provides a blueprint for the development of an appropriate animal model of human reticular reflex myoclonus which must meet the following criteria: (a) myoclonic jerking should be characterized by brief, hypersynchronous muscle action potentials which can be provoked by sensory stimuli; (b) myoclonus should originate in the brain stem, initiate both ascending activation of cranial nerve muscles and descending activation of limb muscles, and exhibit the EEG/EMG characteristics noted above; (c) correlation between myoclonus and changes in cerebral 5HT mechanisms, for example, a decrease in 5HT turnover and antagonism of jerking by 5HT precursors and clonazepam.

The relationships of several potential animal models of myoclonus to the above criteria are summarized in Table 32.1. The most interesting of these models are now reviewed in more detail to evaluate how closely they mimic human reticular reflex myoclonus.

Urea-induced myoclonus in the cat and rat

Patients with renal failure exhibit a variety of abnormal movements including myoclonus (Locke *et al.*, 1961; Tyler, 1968). Chadwick and French (1979)

Table 32.1 Summary of the characteristics of animal models of myoclonus.

Chemical inducing myoclonus	Species	Brief muscle action potential	Brain stem origin of myoclonus	Effect on brain 5HT turnover	Antagonism of myoclonus by 5HT precursors	Antagonism of myoclonus by Clonazepam
Urea	Cat and rat	No	Yes	Unknown	Unknown	Partial
p,p'-DDT	Rat and mouse	No	Yes	Increased	Yes	Equivocal
Chloralose	Cat and guinea-pig	No	Yes	None	No (guinea-pig)	Unknown
Cobalt	Cat (direct application)	Unknown	Yes	Unknown	Unknown	Unknown
Catechol	Rat, mouse and guinea-pig	No	No	None	No (guinea-pig) Inconsistent (mouse)	No
5HTP	Guinea-pig	No	Yes	Increased	Induce myoclonus	No
Muscimol	Mouse	Unknown	No	Unknown	Yes	Yes

References: Urea (page 446); p,p'-DDT (page 449); chloralose (page 450); cobalt (page 451); catechol (Angel and Lemon, 1973, 1975; Angel et al., 1977; Chadwick et al., 1979; Angel, 1986); 5-HTP (Klawans et al., 1973; Weiner et al., 1977; Chadwick et al., 1978; Luscombe et al., 1986); muscimol (Menon and Vivonia, 1981a,b).

reported uraemic-induced myoclonus to be absent at rest but provoked by voluntary movement and a variety of sensory stimuli. Combined with the observation that myoclonus was suppressed by clonazepam, they proposed that uraemic myoclonus closely resembled post-anoxic reticular reflex myoclonus (Chadwick and French, 1979).

Increasing plasma urea concentrations in animals induced a syndrome remarkably similar to that seen in man during clinical uraemia (Zuckermann and Glaser, 1972; Muscatt et al., 1986). Detailed studies in the cat by Zuckermann and Glaser (1972) showed that the spontaneous and stimulus-sensitive myoclonus induced by urea infusion was associated with paroxysmal activity in the brain stem, particularly the reticular formation. These changes were most evident in the bulbar reticular formation, especially at the level of the nucleus reticularis gigantocellularis and nucleus reticularis caudalis. In addition, the response to urea was essentially unchanged following subcollicular brain stem section but markedly decreased by upper cervical cord transection. Since the latter transection depressed but did not completely abolish paroxysmal activity in the reticular formation, such activity appeared to be largely dependent on sensory input and the involvement of a spinoreticular–spinal loop which relays in the nucleus reticularis gigangto-cellularis (Zuckermann and Glaser, 1972).

This detailed approach has recently been extended to the rat to determine if it might provide a more viable model for pharmacological investigations (Muscatt et al., 1986). Electrophysiological studies, including an electrode implanted over the brain stem at the level of the nucleus reticularis gigantocellularis, suggested that urea-induced myoclonus in rats resembled reticular reflex myoclonus in humans. Thus, spontaneous muscle jerks were preceded by a time-locked discharge in the lower brain stem and ipsilateral cerebellum. Stimulus-sensitive muscle jerks, provoked by electrical stimulation of the hindlimb, arose by activation of an excitable focus in the hindbrain which provoked a myoclonic discharge that travelled up the brain stem to the fifth cranial nerve nucleus of the masseter muscle and down the spinal cord to the limb muscles (Muscatt et al., 1986).

The behavioural response of the rat to urea infusion, however, was fairly erratic and it proved impossible to reproducibly evoke a prolonged period of stimulus-sensitive myoclonus (Muscatt et al., 1986). Clonazepam administration at the onset of stimulus-sensitive myoclonus delayed the development of the myoclonic syndrome but it did not prevent myoclonus. The EEG responses of the brain stem and cerebellum, and the EMG response in the left tibilais anterior, to urea infusion were all markedly reduced by clonazepam administration (Muscatt et al., 1986).

In conclusion, urea-induced spontaneous and stimulus-sensitive myoclonus in the cat and rat provides a good electrophysiological model of reticular reflex myoclonus in man. Although the response in the rat is sensitive to clonazepam, the general instability and complexity of the behavioural response to urea infusion suggests that urea-induced myoclonus in the rat is not a simple and reliable screen for detecting drugs that might be beneficial in human myoclonic disorders.

p,p′-DDT-induced myoclonus in rodents

p,p′-DDT (1,1,1-trichloro-2,2-bis-(*p*-chlorophenyl)ethane) induces stimulus-sensitive myoclonus in mice and rats (Chung Hwang and Van Woert, 1978, 1979; Pratt *et al.*, 1985, 1986). The difference in effects of pharmacological manipulations on *p,p′*-DDT-evoked jerking in these two species has led to controversy over the suitability of *p,p′*-DDT-induced myoclonus in rodents as a suitable model of human post-anoxic myoclonus (Pratt *et al.*, 1986).

Electrophysiological studies suggested that *p,p′*-DDT-induced myoclonus in the rat may be similar to reticular reflex myoclonus in humans. The sequence of muscle bursting in spontaneous jerks indicated a discharge origin in the lower brain stem activating first the neck muscles, then subsequently spreading upwards to the cranial nerves and downwards to the spinal cord innervating the limbs (Pratt *et al.*, 1985). Each jerk consisted of an asynchronous burst of EMG activity lasting about 50–100 ms (Pratt *et al.*, 1985), considerably longer than recorded in human reticular reflex myoclonus.

p,p′-DDT-induced myoclonus in rats was abolished by transection of the spinal cord at the lower thoracic level but spared by transection through the anterior midbrain, suggesting initiation of jerking in either the brain stem or cerebellum (Chung Hwang *et al.*, 1980). Direct infusion of *p,p′*-DDT into the medullary reticular formation of the rat induced stimulus-sensitive myoclonus comparable to that evoked by intragastric administration of *p,p′*-DDT (Chung and Van Woert, 1984). Anatomical studies suggested that the region where *p,p′*-DDT infusion produced myoclonus included the nucleus reticularis gigantocellularis (Chung and Van Woert, 1986). This nucleus receives a major input of spinoreticular fibres and is also the origin of the majority of medullary reticulospinal neurones. These fibres appear to be involved in the spinoreticular–spinal reflex, accentuation of which by somatic stimuli may produce myoclonic responses (Halliday, 1975, 1986). *p,p′*-DDT may enhance this spinoreticular reflex by causing repetitive firing of reticulospinal neurons in response to stimuli from spinoreticular neurons, thus producing this stimulus-sensitive type of myoclonus (Chung and Van Woert, 1986).

Lowered concentrations of 5HIAA in the CSF of patients with post-anoxic myoclonus suggests reduced 5HT turnover. Following administration of *p,p′*-

DDT to both rats and mice, however, cerebral 5HIAA levels were elevated indicating increased 5HT turnover (Chung Hwang and Van Woert, 1978; Pratt *et al.*, 1985). The elevation of brain 5HIAA levels did not appear to be associated with the onset of myoclonus, becoming most apparent once *p,p'*-DDT-induced jerking was well established suggesting that changes in 5HT turnover may be a consequence rather than a cause of myoclonus (Pratt *et al.*, 1985).

Further investigation of the role of 5HT in *p,p'*-DDT-induced myoclonus has involved analysis of the effect of a range of 5HT-active drugs on jerking. Different effects of drugs have been observed in the rat and the mouse. Thus, although the 5HT precursor 5HTP reduced the intensity of *p,p'*-DDT-induced myoclonus in both species (Chung Hwang and Van Woert, 1978; Magnussen, 1985; Pratt *et al.*, 1985), the 5HT reuptake blocker Org 6582 only attenuated myoclonus in the mouse (Chung Hwang and Van Woert, 1978) and not in the rat (Pratt *et al.*, 1985). In addition, non-selective 5HT antagonists such as methergoline and methysergide potentiated *p,p'*-DDT-induced myoclonus in mice (Chung Hwang and Van Woert, 1978) but not in rats (Pratt *et al.*, 1985).

Of particular interest was the inhibition by clonazepam of *p,p'*-DDT-induced myoclonus in mice (Chung Hwang and Van Woert, 1979) and rats (Chung Hwang *et al.*, 1981), a further observation that could not be replicated in rats by Pratt and colleagues (Pratt *et al.*, 1985). The discrepancies over the effects of drugs on *p,p'*-DDT-induced myoclonus may be due to the different assessment techniques employed by the two laboratories to measure the severity of myoclonus, namely, activity meters (Chung Hwang and Van Woert, 1979; Chung Hwang *et al.*, 1981) versus observer rating scales (Pratt *et al.*, 1985; see Pratt, *et al.*, 1986).

Thus electrophysiological studies in the rat suggest that *p,p'*-DDT-induced myoclonus may be a suitable model of reticular reflex myoclonus, with a brain stem origin in the region of the nucleus reticularis gigantocellularis. However, biochemical studies in rats and mice show opposite changes in 5HT metabolism to that observed in human myoclonus. In addition, some pharmacological studies suggest that the response in the rat is not a suitable screen for drugs to treat post-anoxic myoclonus. It is uncertain whether the mouse is a more appropriate model as the pharmacological studies have been questioned and the electrophysiology has not been evaluated.

Chloralose-induced myoclonus

Chloralose induces stimulus-sensitive myoclonus (Adrian and Moruzzi, 1939; Alvord and Fuortes, 1954; Alvord and Whitlock, 1954; Ascher *et al.*, 1963).

Adrian and Moruzzi (1939) showed that in cats an abrupt stimulus produced a jerk associated with a motor cortical response and an impulse discharge in the pyramidal tract. Myoclonus was still observed, however, after either decortication (Alvord and Whitlock, 1954) or bilateral section of the medullary pyramids (Ascher *et al.*, 1963). Stimulus-induced myoclonus in chloralose-treated cats was abolished by high spinal cord transections (Alvord and Fuortes, 1954). Chloralose-induced myoclonus thus appears dependent upon brain stem structures, probably the reticular formation, with a role again suggested for the nucleus reticularis gigantocellularis in the spinoreticular–spinal reflex (Angel, 1986).

The EMG correlate of the chloralose-evoked muscle jerk in guinea-pigs, however, consisted of a prolonged polyphasic burst of 20–50 ms (Chadwick *et al.*, 1980) unlike the brief, hypersynchronous muscle action potential in human reticular reflex myoclonus. In addition, chloralose did not appear to alter regional 5HT metabolism in guinea-pig brain, nor was chloralose-induced myoclonus in guinea-pigs modified by 5HT precursors (Chadwick *et al.*, 1980).

In conclusion, chloralose-induced myoclonus is of brain stem origin, possibly associated with the nucleus reticularis gigantocellularis, but the EMG and the role of 5HT in the response differ from that observed in human reticular reflex myoclonus.

Cobalt-induced myoclonus in cats

Metallic cobalt powder applied bilaterally or in the midline of the lower brain stem induced bilateral, generalized, stimulus-sensitive myoclonus in cats which appeared as early as 24 h after cobalt implantation and continued for 4–5 days (Cesa-Bianchi *et al.*, 1967). The cobalt was particularly effective when applied in the medullary reticular formation, an area including the nucleus reticularis gigantocellularis. The resultant marked EEG changes, characterized by hypersynchronous potentials, high voltage spikes, spike and wave complexes, were frequently associated with diffuse muscular jerks of facial, cervical and forelimb flexor muscles (Cesa-Bianchi *et al.*, 1967). Importantly, this generalized myoclonic jerking was in sharp distinction to the localized, contralateral jerking and Jacksonian epilepsy produced by local cobalt implantations into the sensorimotor cortex (Cesa-Bianchi *et al.*, 1967).

Two characteristic features of human reticular reflex myoclonus, namely the order of muscle activation and the length of the EMG, were not stated following cobalt implantation into the lower brain stem. It was suggested, however, that the synchronous appearance of electrical and motor phenomena indicated that the brain stem had ascending and descending actions on the EEG and spinal motoneurons, respectively (Cesa-Bianchi *et al.*, 1967), as observed in human reticular reflex myoclonus (Hallett *et al.*, 1977).

Thus, myoclonus induced in cats by implantation of cobalt into the lower brain stem meets several of the criteria for a model of human reticular reflex myoclonus. However, more accurate studies of the temporal relationship between EEG and EMG events, of the order of muscle activation and of the characteristics of the muscle action potential are all required. If this can be determined in the cat, or more practically in the rat, then this may be a suitable model for studying the biochemical and pharmacological basis of reticular reflex myoclonus.

CONCLUSIONS

An interesting common feature of these animal models of myoclonus is the role of the lower brain stem and in particular the nucleus reticularis gigantocellularis of the medullary reticular formation. This nucleus, or the region surrounding it, has been associated with jerking induced by each of the agents discussed above. The nucleus acts as a relay in the spinoreticular–spinal reflex and it may be that enhancement of this reflex by sensory stimuli, facilitated by chemical action in the nucleus, produces myoclonus in animals. However, the spinoreticular–spinal pathway does not seem to be the prime reflex pathway involved in reticular reflex myoclonus in man since the afferent spinal pathway of the spinorecticular–spinal reflex is rapidly conducting and the efferent pathway slowly conducting; the opposite situation pertains in human reticular reflex myoclonus where afferent conduction time is long and efferent conduction time short (Hallett *et al.*, 1977).

It is clear, therefore, that an animal model of human reticular reflex myoclonus which is both scientifically valid and experimentally practical has not been established. Myoclonus induced by chloralose and by *p,p'*-DDT in rats does not meet all the criteria for a valid animal model, while these criteria have not been fully investigated in the cases of *p,p'*-DDT-induced myoclonus in mice and cobalt-induced myoclonus in cats. Urea-evoked myoclonus in cats and rats provides an interesting model of human reticular reflex myoclonus but it does not allow routine assessment of drug action.

REFERENCES

Adrian, E. D. and Moruzzi, G. (1939) *J. Physiol.* **97**, 153–199.
Alvord, E. C. and Fuortes, M. G. F. (1954) *Am. J. Physiol.* **176**, 253–261.

Alvord, E. C. and Whitlock, D. G. (1954) *Fed. Proc.* **13**, 2.

Angel, A. (1986) *Adv. Neurol.* **43**, 589–609.

Angel, A. and Lemon, R. N. (1973) *Electroenceph. Clin. Neurophysiol.* **35**, 589–601.

Angel, A. and Lemon, R. N. (1975) *J. Physiol.* **248**, 465–488.

Ascher, P., Jassik-Gerschenfeld, D. and Buser, P. (1963) *Electroenceph. Clin. Neurophysiol.* **15**, 246–264.

Angel, A., Clarke, K. A. and Dewhurst, D. G. (1977) *Br. J. Pharmacol.* **61**, 433–439.

Cesa-Bianchi, M. G., Mancia, M. and Mutani, R. (1967) *Electroenceph. Clin. Neurophysiol.* **22**, 525–536.

Chadwick, D. and French, A. T. (1979) *J. Neurol. Neurosurg. Psychiat.* **42**, 52–55.

Chadwick, D., Hallett, M., Harris, R., Jenner, P., Reynolds, E. H. and Marsden, C. D. (1977) *Brain* **100**, 455–487.

Chadwick, D., Hallett, M., Jenner, P. and Marsden, C. D. (1978) *J. Neurol. Sci.* **35**, 157–165.

Chadwick, D., Jenner, P. and Marsden, C. D. (1979) *Br. J. Pharmacol.* **66**, 358–360.

Chadwick, D., Hallett, M., Jenner, P. and Marsden, C. D. (1980) *Br. J. Pharmacol.* **69**, 535–540.

Chung, E. and Van Woert, M. H. (1984) *Exp. Neurol.* **85**, 273–282.

Chung, E. and Van Woert, M. H. (1986) *Adv. Neurol.* **43**, 569–575.

Chung Hwang, E. and Van Woert, M. H. (1978) *Neurology (Minneap.)* **28**, 1020–1025.

Chung Hwang, E. and Van Woert, M. H. (1979) *Eur. J. Pharmacol.* **60**, 31–40.

Chung Hwang, E., Plaitakis, A., Magnussen, I. and Van Woert, M. H. (1980) *Trans. Am. Soc. Neurochem.* **11**, 150.

Chung Hwang, E., Plaitakis, A., Magnussen, I. and Van Woert, M. H. (1981) *Neurosci. Lett.* **24**, 103–108.

Fahn, S., Marsden, C. D. and Van Woert, M. H. (1986) *Adv. Neurol.* **43**, 1–5.

Guilleminault, C., Tharp, B. R. and Cousin, D. (1973) *J. Neurol. Sci.* **18**, 435–441.

Hallett, M., Chadwick, D., Adam, J. and Marsden, C. D. (1977) *J. Neurol. Neurosurg. Psychiat.* **40**, 253–264.

Hallett, M., Chadwick, D. and Marsden, C. D. (1979) *Neurology (Minneap.)* **29**, 1107–1125.

Halliday, A. M. (1975). In *Myoclonic Seizures* (edited by M. H. Charlton), pp. 1–29. Excerpta Medica, Amsterdam.

Halliday, A. M. (1986) *Adv. Neurol.* **43**, 339–355.

Klawans, H. L., Goetz, C. and Weiner, W. J. (1973) *Neurology (Minneap.)* **23**, 1234–1240.

Lance, J. W. and Adams, R. D. (1963) *Brain* **86**, 111–136.

Locke, S., Merrill, J. P. and Tyler, H. R. (1961) *Arch. Int. Med.* **108**, 519–530.

Luscombe, G., Jenner, P. and Marsden, C. D. (1986) *Adv. Neurol.* **43**, 529–543.

Magnussen, I. (1985) *Acta Pharmacol. Toxicol.* **56**, 87–90.

Marsden, C. D., Hallett, M. and Fahn, S. (1982) In *Movement Disorders* (edited by C. D. Marsden and S. Fahn), pp. 196–248. Butterworth Scientific, London.

Menon, M. K. and Vivonia, C. A. (1981a) *Neuropharmacology* **20**, 441–444.

Menon, M. K. and Vivonia, C. A. (1981b) *Eur. J. Pharmacol.* **73**, 155–161.

Muscatt, S., Rothwell, J., Obeso, J., Leigh, N., Jenner, P. and Marsden, C. D. (1986) *Adv. Neurol.* **43**, 553–563.

Pranzatelli, M. R. and Snodgrass, S. R. (1985) *Clin. Neuropharmacol.* **8**, 99–130.

Pratt, J. A., Rothwell, J., Jenner, P. and Marsden, C. D. (1985) *Neuropharmacology* **24**, 361–373.

Pratt, J. A., Rothwell, J., Jenner, P. and Marsden, C. D. (1986) *Adv. Neurol.* **43**, 577–588.
Tyler, H. R. (1968) *Am. J. Med.* **44**, 734–748.
Van Woert, M. H. and Sethy, V. H. (1975) *Neurology (Minneap.)* **25**, 135–140.
Weiner, W. J., Goetz, C., Nausieda, P. A. and Klawans, H. L. (1977) *Eur. J. Pharmacol.* **46**, 21–24.
Zuckermann, E. G. and Glaser, G. H. (1972) *Arch. Neurol.* **27**, 14–28.

33

Chorea

P. D. Thompson

CLINICAL DEFINITION OF CHOREA

Chorea may be defined as the continuous, random flow of muscle activity from one muscle group to another and one part of the body to another. Both proximal and distal limb muscles and cranial and axial muscles may be involved. Flow of movement is the clinical hallmark of chorea and the most important criterion in its recognition. The muscle jerks may become more obvious with concentration, stress or voluntary movement and characteristically interrupt or are even incorporated into voluntary action. Chorea is less pronounced at rest and disappears during sleep. The severity of chorea may vary from an inability to sit still, clumsiness or a tendency to fidget to more vigorous limb or axial movements that interfere with all volitional activity. The gait may be interrupted by lurches, stops and starts producing a "dancing" gait. Bulbar and respiratory muscles may be affected giving rise to dysarthria and respiratory irregularities with grunts and gasps. In severe chorea, involvement of the axial muscles may produce gross arching movements of the back making it impossible for the patient to stand, walk or even sit in a chair.

Ballism (from Greek, "to throw") is considered by many to be a severe form of chorea, with involvement of proximal limb muscles producing wild

DISORDERS OF MOVEMENT: CLINICAL, PHARMACOLOGICAL
AND PHYSIOLOGICAL ASPECTS   ISBN 0-12-569685-X

flinging movements of the limbs. Ballism usually is unilateral (hemiballism) and due to a lesion of the contralateral subthalamic nucleus and/or its connections with the pallidum (Martin, 1927). The terms "hemiballism/hemichorea" often are combined to emphasize that any difference is a matter of degree (Martin and Alcock, 1934). Furthermore, hemichorea often replaces hemiballism during recovery from a vascular lesion of the subthalamic nucleus, while severe unilateral chorea in Sydenham's chorea may be indistinguishable from hemiballism. Additional evidence of the similarities and overlap is derived from studies of experimental dyskinesias which support the notion that both disorders are closely related not only in terms of the lesion site but also the physiological changes within the basal ganglia (Mitchell et al., 1985a,b; Crossman, 1987) (see below).

PATHOPHYSIOLOGICAL STUDIES IN CHOREA

Electromyography of chorea

The recognition of chorea is primarily a clinical task, as might be anticipated from the above clinical definition. There are no specific characteristics that define individual bursts of electromyographic (EMG) activity as chorea, nor is there a single pattern of EMG activity in chorea. The EMG bursts accompanying chorea in Huntington's disease may be of long duration (500 ms to 1 s) as in dystonia or short (less than 50 ms) as in myoclonus (Marsden et al., 1983; Thompson et al., 1988). Thiebaut (1968) examined 22 patients with Sydenham's chorea and found EMG bursts lasting from 20 to 600 ms. Hallett and Kaufman (1981) noted that the EMG bursts accompanying Sydenham's chorea tended to be longer than 100 ms in contrast to the variable duration of the EMG bursts accompanying chorea in Huntington's disease. Short duration EMG bursts were not a feature in their patient with Sydenham's chorea. The wide spectrum of EMG activity and its distribution mirrors the great variability of movement seen in chorea.

The organization of sequences of muscle contraction in chorea also differs from normal voluntary activity. Wilson (1925) first drew attention to the co-contraction of agonist and antagonist muscle groups in a patient with senile chorea, commenting on "... the complete interruption of normal physiological reciprocal innervation during confluent choreic activity". Hoefer and Putman (1940) and Herz (1944) later confirmed Wilson's observations by recording EMG activity in patients with athetosis and Sydenham's chorea. The pattern of activation of agonist and antagonist muscles in chorea is highly variable, particularly in Huntington's disease (Hallett and Alvarez, 1983; Thompson *et al.*, 1988). Chorea in Sydenham's disease, in contrast may exhibit alternating agonist and antagonist muscle contraction (Thiebaut, 1968; Hallett and Kaufman, 1981).

Studies in Huntington's disease

Of the many diseases causing chorea, Huntington's disease is by far the commonest encountered in clinical practice. Accordingly, most studies of chorea in man have been undertaken in Huntington's disease. Many neurophysiological tests have been applied to the study of Huntington's disease, but the relationship of the abnormal findings to the clinical symptoms and signs is not known at the present time.

Voluntary limb movement

It has long been noted that voluntary movements in Huntington's disease are abnormal and often slow to a degree that seemed out of proportion to the chorea (Hamilton, 1908). Even after the abolition of chorea by neuroleptic drugs, there may not necessarily be an improvement in motor function (Shoulson, 1981; Girotti *et al.*, 1984; Quinn and Marsden, 1984).

There may be a delay in the onset of intended voluntary movement, resembling apraxia (Denny-Brown, 1968), while with progression of the disease, hyperkinesia often is replaced by an akinetic rigid syndrome (Hamilton, 1908; Denny-Brown, 1960; Bittenbender and Quadfasel, 1962; Hayden, 1981). In some patients, an akinetic rigid syndrome (without typical chorea) may be the presenting feature (the Westphal variant).

Studies of gait in Huntington's disease have shown similar slowness of movement to that seen in Parkinson's disease (Koller and Trimble, 1985). The walking speed was slower than normal, the stride length was reduced and the patients took fewer steps per minute.

Two recent studies have examined the control of voluntary upper limb movement in patients with Huntington's disease. Hefter *et al.* (1987) found that rapid self-paced extension movements and fast alternating movements of the index finger were slower in the patients than similar movements in age- and sex-matched control subjects. Thompson *et al.* (1988) examined rapid wrist flexion movements through angles of 15° and 60°. There was a greater variability in the end position of each movement and in the patterns of EMG activity of agonist and antagonist muscles, than seen in normal subjects. The angular wrist velocity of such movements was slower than normal. This was most marked in patients who were akinetic and rigid but also was seen in patients with chorea alone. It could not be accounted for by neuroleptic drugs prescribed to decrease chorea, as similar findings were evident in those receiving no medication at all. Complex movements which, unlike simple wrist movements, require the subject to simultaneously or sequentially assemble two motor acts, also were found to be abnormal in this study. Some patients with Huntington's disease were quite unable to combine two movements in a simultaneous or sequential task. Those who could displayed further slowing of the velocity of movement and prolongation of the normal interval between the two movements. Such defects could not be explained by failure of corticomotoneuron conduction, as this has been shown to be normal in patients with Huntington's disease (Thompson *et al.*, 1986). These findings were similar to those seen in Parkinson's disease (Benecke *et al.*, 1986, 1987) and confirmed the clinical impression of underlying bradykinesia in patients with Huntington's disease. They suggest a defect in the planning and programming of complex motor tasks, in addition to the slowed execution of a simple motor act.

As with Parkinson's disease, these motor findings are consistent with the consequences of abnormal basal ganglia function. They are also in keeping with, and provide one explanation for, the observation that pharmacological suppression of chorea in Huntington's disease does not lead to an improvement in motor performance. However, the mechanisms responsible for the defects in voluntary movement in each condition are likely to be different. The EMG patterns accompanying voluntary movement in Huntington's disease are characterized by highly variable patterns of muscle activation and long duration bursts in comparison with the normal sequence of small agonist and antagonist EMG bursts in Parkinson's disease (see Hallett and Khoshbin, 1980).

Ocular movement in Huntington's disease

A supranuclear ocular motor palsy may be a striking feature in some patients with Huntington's disease (Starr, 1967). Several abnormalities of saccade function have been described, including slowing of vertical and horizontal saccades, saccade dysmetria, delayed initiation, or initiation with head thrusts or blinking ("ocular motor apraxia"), difficulty maintaining fixation, excessive distractibility and square wave jerks (Starr, 1967; Leigh *et al.*, 1983; Bollen *et al.*, 1986a; Lasker *et al.*, 1988). The fast phases of rotational and caloric-induced nystagmus also may be slowed while pursuit may be contaminated with saccadic intrusions (Leigh *et al.*, 1983). The pathological basis for these findings is not known. Abnormalities of the brain stem reticular formation are usually held responsible for slowed saccades. However, a recent pathological study of the brain stem from patients with Huntington's disease who slowed vertical saccades, failed to find significant abnormalities in the region of the rostral interstitial nucleus of the medial longitudinal fasciculus, the site where the brain stem generator neurons for vertical saccades are located. This suggests that the defect may lie in the inputs to these areas (Leigh *et al.*, 1985). Disruption of the inputs to the superior colliculus from the frontal eye fields and pars reticulata of the substantia nigra may be responsible for the abnormalities in saccade initiation and excessive distractibility (Leigh *et al.*, 1983; Bollen *et al.*, 1986a; Lasker *et al.*, 1988).

Disordered ocular movements are found in few other causes of chorea and are a valuable clue to the diagnosis of Huntington's disease.

Somatosensory evoked potentials

Loss of the early cortical components of the somatosensory evoked potentials (SEP) are well described although the significance of this is not clear. Abnormalities of sensory perception are not a recognized feature of Huntington's disease. The pathological basis of these SEP changes are poorly understood (Ehle *et al.*, 1984; Noth *et al.*, 1984; Bollen *et al.*, 1985). Both the cortical (Bollen *et al.*, 1985) and thalamic (Ehle *et al.*, 1984) pathologies of Huntington's disease have been implicated. These SEP abnormalities do not appear to correlate with the presence of chorea or the slowing of voluntary movement (Thompson *et al.*, 1988). Furthermore, SEPs are normal in several other conditions causing chorea (see below). This provides further support for the suggestion that changes in SEP components in Huntington's disease are related not so much to the presence of chorea or hyperkinesia (Noth *et al.*, 1984; Thompson *et al.*, 1988) as to the extent of pathological change in Huntington's disease.

Stretch reflexes

Long latency muscle responses to stretch are absent in the intrinsic hand muscles in Huntington's disease (Noth *et al.*, 1985) but may be relatively preserved in the wrist flexor muscles (Thompson *et al.*, 1988). The absence of the long latency muscle stretch reflexes contrasts with their enhancement in Parkinson's disease (Rothwell *et al.*, 1983). No clear correlation between the abnormalities of the long latency stretch reflexes and those of the cortical sensory evoked potentials has been described to date.

Blink reflexes

The latency of the R2 component of the blink reflex may be prolonged in Huntington's disease (Esteban and Gimenez-Roldan, 1975; Bollen *et al.*, 1986b) while the R2 component may show abnormally rapid habituation after paired stimuli (Agostino *et al.*, 1988). Abnormalities of the blink reflex in other basal ganglia diseases have been interpreted as indicating defective control of brain stem interneuron circuits by pallidal efferents (Berardelli *et al.*, 1986).

Positron emission tomography

Studies of cerebral and basal ganglia oxygen and glucose metabolism by positron emission tomography (PET) have shown focal decreases in the striatum and frontal cortex in patients with Huntington's disease (Young *et al.*, 1986). In a detailed single case study by Leenders *et al.* (1986) the capacity of the striatum to store dopamine was normal (assessed by ^{18}F-fluorodopa uptake), but dopamine (D2) receptor binding was reduced consistent with a striatal dopamine preponderance in Huntington's disease. In a study of 15 patients with Huntington's disease Young *et al.* (1986), measured regional basal ganglia and thalamic metabolism by PET scanning with ^{18}F-2-fluoro-2-deoxyglucose. They showed that caudate metabolism correlated with an assessment of the overall functional capacity of the patient and bradykinesia and rigidity, while putamen metabolism correlated with chorea, ocular motor abnormalities and fine manual performances. Thalamic abnormalities correlated with dystonia. The same authors (Young *et al.*, 1987) have also demonstrated normal caudate metabolism in persons at risk for developing Huntington's disease, suggesting that measurement of striatal hypo-metabolism is unlikely to be sensitive enough to detect presymptomatic cases.

Neuropathology and neurochemistry

The pathological changes of Huntington's disease are widely distributed throughout the brain. There may be cortical atrophy affecting the frontal and parietal lobes, but the major abnormalities are found in the striatum with striking atrophy of the caudate nucleus, and to a lesser extent the putamen and external pallidal segment. There is extensive loss of the population of small striatal neurons, particularly the type II spiny projection neurons in the head of the caudate nucleus and the middle and posterior thirds of the putamen (Lange *et al.*, 1976; Roos *et al.*, 1985). Neuronal loss also occurs in the globus pallidus, cerebral cortex, thalamus and subthalamus. Abnormalities in several neurotransmitter systems also are evident in the striatum, corresponding to the neuronal loss. For example, levels of acetylcholine, choline acetyltransferase, γ-aminobutyric acid and glutamic acid decarboxylase are reduced in the striatum while substance P and metenkephalin are reduced in striatopallidal and striatonigral projections (see Marsden, 1982; Martin, 1984; Martin and Gusella, 1986, for reviews).

Studies in other conditions causing chorea

In contrast with Huntington's disease there have been few pathophysiological studies in other causes of chorea.

Voluntary movement

Slowing of voluntary movement has been described in a patient with Sydenham's chorea (Hefter *et al.*, 1987). We have studied the voluntary movements of two patients with vascular chorea (unpublished observations). Both exhibited normal patterns of agonist and antagonist muscle activation during wrist flexion movements. The duration of the EMG bursts was longer than normal in one patient, but the speed of movement was normal. In a single patient with chorea gravidarum (unpublished observations), voluntary movements were slow and the pattern of recruitment of agonist and antagonist muscle activity in simple wrist flexion movements was abnormal. The agonist burst was prolonged and the antagonist was co-contracting or absent.

Somatosensory evoked potentials and long latency stretch reflexes

Normal somatosensory evoked potentials (SEPs) and stretch reflexes have been demonstrated in patients with Sydenham's chorea (Noth *et al.*, 1985; Gledhill and Thompson 1989), benign hereditary chorea (Lange *et al.*, 1987; unpublished observations), neuroacanthocytosis (Kaplan *et al.*, 1986) and chorea gravidarum (unpublished observations). However, since SEPs may be abnormal in Wilson's disease (Chu, 1986) and cerebral vascular disease, both of which may be associated with chorea, such changes are not confined to Huntington's disease.

Positron emission tomography

Studies of choreic syndromes with PET scanning have shown striatal glucose hypometabolism in benign hereditary chorea (Suchowersky *et al.*, 1986), hemichorea (presumed vascular), choreoacanthocytosis, "sporadic progressive chorea and dementia", and the "pseudo-Huntingtonian form of dentato-rubro-pallido-luysian atrophy" (Hosokawa *et al.*, 1987). The reduction of glucose metabolism observed in the striatum of patients with Huntington's disease therefore is not specific. Guttman *et al.* (1987) found normal striatal glucose metabolism in four patients with chorea and systemic lupus erythematosus and suggested the abnormalities present in other choreic syndromes are likely to be the result of striatal disease rather than a correlate of chorea.

PATHOLOGY OF CHOREA

Insight into the pathological anatomy of chorea has long rested on careful clinicopathological studies of focal lesions in man and lesion experiments in non-human primates. The clearest example is the association of lesions of the subthalamic nucleus with hemichorea/hemiballism. Diffuse brain diseases associated with chorea, in contrast, do not permit any such correlation. Either there are many lesions, as in Huntington's disease or lacunar disease of the basal ganglia, or there are no obvious lesions at all, as in Sydenham's chorea or chorea associated with systemic lupus erythematosus.

Chorea in systemic disease

The pathogenesis and the pathological anatomy of chorea in patients with Sydenham's chorea, systemic lupus erythematosus, "lupus-like" disease, and chorea gravidarum, in which no consistent or distinctive pathological change has been found, remains uncertain. The finding in some cases of Sydenham's chorea of tissue-specific antibodies to neurons of the caudate and subthalamic nuclei raises the possibility that this disorder has an immunologic basis. These antibodies were absorbed with the membranes of the group A β-haemolytic streptococcus (Husby et al., 1976), infection with which is a recognized antecedent to Sydenham's chorea. However, analysis of the cerebrospinal fluid is normal in Sydenham's chorea, including tests for oligoclonal immunoglobulin bands (Gledhill and Thompson, 1989).

In a review of clinical and pathological reports of systemic lupus erythematosus and chorea there was no clear association of chorea with vascular lesions involving the basal ganglia (or vice versa) (Bruyn and Padberg, 1984). Immunological factors might therefore be responsible in this situation as well. The possible role of circulating antibodies in patients with systemic lupus erythematosus and lupus-like diseases has recently been discussed (Asherson et al., 1987). These authors comment that even though there is a strong association between thrombosis and antiphospholipid antibodies which might favour small vessel thrombotic occlusion, another possible explanation for central nervous system (CNS) disease is binding of these antibodies to the abundant phospholipid in CNS tissue. The lupus anti-coagulant–antiphospholipid–antibody "lupus-like" syndrome may be a commoner cause of chorea than previously recognized (Asherson et al., 1987).

Both systemic lupus erythematosus and the "lupus-like" syndrome may present as chorea gravidarum (Donaldson and Espiner, 1971; Agrawal and Foa, 1982) or as post-partum chorea (Thomas et al., 1979). However, at least one-third of patients with chorea gravidarum have a past history of

Sydenham's chorea. A similar situation applies to chorea following the use of oral contraceptives. These observations have prompted the suggestion that the hormonal changes induced by pregnancy or oral contraceptives alter dopaminergic sensitivity, particularly in previously damaged basal ganglia (Nausieda et al., 1983).

Other causes of generalized chorea include diffuse cerebrovascular disease, with multiple infarcts in the white matter and basal ganglia (Tabaton et al., 1985; Sethi et al., 1987), and carbon monoxide poisoning (Davous et al., 1986). Isolated cases of thalamic degeneration and chorea (Adams and Malamud, 1971) and chorea (with dystonia) appearing as a paraneoplastic phenomenon (Albin et al., 1988) also have been described. The pathological anatomy of chorea in these situations remains the subject of speculation.

Focal lesions in man producing chorea

Hemiballism and hemichorea are cardinal signs of a structural lesion of the contralateral subthalamic nucleus. Vascular lesions (infarction or haemorrhage) are the most widely recognized (Martin, 1927; Whittier, 1947), but metastatic tumour (Glass et al., 1984), and multiple sclerosis (Riley and Lang, 1988) also have been described.

Lesions outside the subthalamic nucleus also may occasionally result in hemiballism and hemichorea. Such lesions most commonly involve the putamen (Schwarz and Barrows, 1960; Goldblatt et al., 1974; Lang, 1985; Mas et al., 1987), and/or the external segment of the globus pallidus, or the subthalamic efferents to the medial pallidum (Martin, 1957). Abolition of these movements by lesions at other sites has suggested that the involuntary movements are mediated by the pallidal outflow to ventrolateral thalamus, then motor cortex and the corticospinal tract (Martin and McCaul, 1959).

Chorea also has been reported in association with lesions of the caudate nucleus, although many of these cases also appear to have multiple lesions or involvement of the adjacent putamen (Goldblatt et al., 1974; Saris, 1983; Kawamura et al., 1988). This might explain the discrepancy between these observations and the fact that experimental lesions of the caudate nucleus produce behavioural disturbances rather than chorea.

EXPERIMENTAL CHOREA

Destructive lesions

In experimental animals, as in man, lesions in the region of the subthalamic nucleus produce hemiballism or hemichorea (Whittier and Mettler, 1949). In the monkey a destructive lesion involving at least 20% of the volume of the subthalamic nucleus was required before chorea appeared; damage to surrounding structures especially the ansa lenticularis and the internal segment of the globus pallidus abolished the movements. A small lesion involving only part of the pallidal and subthalamic nucleus connections was effective in producing hemiballism. Lesions at other sites also suggested that the hyperkinesia was mediated by similar pathways to those postulated in man (Carpenter et al., 1960).

Pharmacological lesions

A major advance in the study of experimental dyskinesias has been the ability to induce such movements pharmacologically (see Crossman et al., 1984, 1988; Mitchell et al., 1985a,b; Crossman, 1987). These studies have not only confirmed the findings described above but also, when coupled with autoradiographic assessment of cerebral metabolism, have added much new information about basal ganglia circuitry.

Crossman et al. (1984) produced hemiballism in the monkey by injecting bicuculline, a GABA antagonist into the subthalamic nucleus. Of interest was the finding of a decrease in glucose metabolism in the injected subthalamic nucleus, both segments of the globus pallidus and the ventral anterior and ventral lateral nuclei of the thalamus. These changes were consistent with reduced synaptic activity in subthalamopallidal and pallidothalamic projections. Experimental chorea, produced by injections into the lateral segment of the globus pallidus and the adjacent putamen, also resulted in a reduction in metabolic activity of the medial pallidal thalamic projections. Such studies support the clinical impression that chorea and ballism are similar by indicating that they probably share the same neural mechanisms.

SYNTHESIS OF CLINICAL STUDIES IN MAN, EXPERIMENTAL CHOREA AND BASAL GANGLIA FUNCTION

Chorea and ballism principally result from lesions interrupting the striatopallidal–subthalamic–pallidal connections. Striatal lesions also interfere with the performance of simple voluntary movements in man. These observations fit well the current concept of different functional circuits within the basal ganglia (Alexander et al., 1986). One circuit might be primarily involved in the maintenance of normal ongoing motor behaviour and the performance of voluntary motor acts. Another might be primarily engaged in minimizing or suppressing extraneous or unnecessary movement for the task at hand.

Although this model may grossly oversimplify the basal ganglia circuitry (see Kemp and Powell, 1971; Penney and Young, 1983, 1986), it does provide a framework within which the observed defects in voluntary motor control in addition to chorea in certain diseases of the basal ganglia can be explained or at least reconciled. Thus, excitatory corticostriatal projections concerned with the performance of voluntary movement would exert their actions through striato-medial pallidal (inhibitory), pallidothalamic (inhibitory) and thalamocortical (excitatory) connections. The final result would be the facilitation of ongoing motor cortical activity and voluntary motor performance. Concurrently, activity in the striato-lateral pallidal (inhibitory), lateral pallidosubthalamic (inhibitory), subthalamic-medial pallidal (excitatory (Kitai and Kitai, 1987)) and pallidothalamic (inhibitory) pathways would reduce or suppress unwanted motor behaviour (Penney and Young, 1986). These schema also are consistent with the differential effects of dopamine on the striatal output neurons to the different pallidal segments — that is, excitatory to striatal output neurons to the medial pallidum and inhibitory to the lateral pallidum (Penney and Young, 1986). Depriving the striatum of this transmitter could reduce basal ganglia output to motor cortex, providing one mechanism for the bradykinesia seen in Parkinson's disease. Similarly, in Huntington's disease, the loss of striatal output neurons might result in a similar functional effect (i.e. bradykinesia), although the precise mechanism might differ in each condition. The striatal lesion might also deprive the lateral globus pallidus of an inhibitory input leading to excessive inhibition of the subthalamic nucleus, thereby providing an explanation for chorea.

Although an oversimplification, this theory provides one explanation for the combination of bradykinesia and chorea in some patients with Huntington's disease (and also the combination of dyskinesias and bradykinesia in some patients with Parkinson's disease who have been treated with L-dopa).

REFERENCES

Adams, J. E. and Malamud, N. (1971) *Arch. Neurol.* **24**, 101–105.
Agrawal, B. L. and Foa, R. P. (1982) *Arch. Neurol.* **39**, 192–193.
Agostino, R., Berardelli, A., Pauletti, G., Cruccu, G., Stocchi, F. and Manfredi, M. (1988) *Movement Disorders* **3**, 281–289.
Albin, R. L., Bramberg, M. B., Penney, J. B. and Knapp, R. (1988) *Movement Disorders* **3**, 162–169.
Alexander, G. E., Delong, M. R. and Strick, P. L. (1986) *Ann. Rev. Neurosci.* **9**, 357–381.
Asherson, R. A., Derksen, R. H. W. M., Harris, E. N., Bouma, B. N., Gharavi, A. E.,

Kater, L. and Hughes, G. R. V. (1987) *Seminars Arthritis Rheumatism* **16**, 253–259.

Benecke, R., Rothwell, J. C., Dick, J. P. R., Day, B. L. and Marsden, C. D. (1986) *Brain* **109**, 739–758.

Benecke, R., Rothwell, J. C., Dick, J. P. R., Day, B. L. and Marsden, C. D. (1987) *Brain* **110**, 361–379.

Berardelli, A., Rothwell, J. C., Day, B. L. and Marsden, C. D. (1986) *Brain* **108**, 593–608.

Bittenbender, J. B. and Quadfasel, F. A. (1962) *Arch. Neurol.* **7**, 275–288.

Bollen, E., Arts, R. J., Roos, R. A., Van der Velde, E. A. and Buruma, O. J. S. (1985) *Electroencephalogr. Clin. Neurophysiol.* **62**, 235–240.

Bollen, E., Reulen, J. P. H., Den Heyer, J. C., Van der Kamp, W., Roos, R. A. C. and Buruma, O. J. S. (1986a) *J. Neurol. Sci.* **74**, 11–22.

Bollen, E., Arts, R. J. H. M., Roos, R. A. C., Van der Velde, E. A. and Buruma, O. J. S. (1986b) *J. Neurol. Neurosurg. Psychiat.* **49**, 313–315.

Bruyn, G. W. and Padberg, G. (1984) *Eur. Neurol.* **23**, 278–290.

Carpenter, M. B., Correll, J. W. and Hinman, A. (1960) *J. Neurophysiol.* **23**, 288–304.

Chu, N.-S. (1986) *Brain* **109**, 491–507.

Crossman, A. R. (1987) *Neuroscience* **21**, 1–40.

Crossman, A. R., Sambrook, M. A. and Jackson, A. (1984) *Brain* **107**, 579–596.

Crossman, A. R., Mitchell, I. J., Sambrook, M. A. and Jackson, A. (1988) *Brain* **111**, 1211–1234.

Davous, P., Rondot, P., Marion, M. H. and Gueguen, B. (1986) *J. Neurol. Neurosurg. Psychiat.* **49**, 206–208.

Denny-Brown, D. (1960) *Lancet* **ii**, 1099–1105.

Denny-Brown, D. (1968) In *Handbook of Clinical Neurology* (edited by G. Bruyn and P. J. Vinken) Vol. 6, pp. 133–172. North Holland, Amsterdam.

Donaldson, I. M. and Espiner, E. A. (1971) *Arch. Neurol.* **25**, 240–244.

Ehle, A. L., Stewart, R. M., Lellelid, A. and Leventhal, N. A. (1984) *Arch. Neurol.* **41**, 379–382.

Esteban, A. and Gimenez-Roldan, S. (1975) *Acta Neurol. Scand.* **52**, 1145–1157.

Girotti, F., Cavella, F., Scighano, G., Grassi, M. P., Soliveri, P., Giovanninni, P., Paratti, E. and Caraceni, T. (1984) *J. Neurol. Neurosurg. Psychiat.* **47**, 848–852.

Glass, J. P., Jankovic, J. and Borit, B. (1984) *Neurology* **34**, 204–207.

Gledhill, R. F. and Thompson, P. D. (1989) *J. Neurol. Neurosurg. Psychiat.* In press.

Goldblatt, D., Markesbery, W. and Reeves, A. G. (1974) *Arch. Neurol.* **31**, 51–54.

Guttman, M., Lang, A. E., Garnett, E. S., Nahmais, C., Firnau, G., Tyndal, F. Y. and Gordon, A. S. (1987) *Movement Disorders* **2**, 201–210.

Hallett, M. and Alvarez, N. (1983) *J. Neurol. Neurosurg. Psychiat.* **46**, 745–750.

Hallett, M. and Kaufman, C. (1981) *J. Neurol. Neurosurg. Psychiat.* **44**, 329–332.

Hallett, M. and Khoshbin, S. (1980) *Brain* **103**, 301–314.

Hamilton, A. S. (1908) *Am. J. Insanity* **64**, 403–475.

Hayden, M. R. (1981) *Huntington's Chorea.* Springer-Verlag, New York.

Hefter, H., Homberg, V., Lange, H. W. and Freund, H. J. (1987) *Brain* **110**, 585–612.

Herz, E. (1944) *Arch. Neurol. Psychiat.* **51**, 305–318.

Hoefer, P. F. A. and Putnam, T. J. (1940) *Arch. Neurol. Psychiat.* **44**, 517–531.

Hosokawa, S., Ichiya, Y. and Kuwubara, Y. (1987) *J. Neurol. Neurosurg. Psychiat.* **50**, 1284–1287.

Husby, G., Van de Rijn, L., Zabriskie, J. B., Abdin, Z. H. and Williams, R. C. (1976) *J. Exp. Med.* **144**, 1094–1110.

Kaplan, P. W., Erwin, C. W., Bowman, M. H. and Massey, E. W. (1986) *Electroencephalogr. Clin. Neurophysiol.* **63**, 349–352.

Kawamura, M., Takahashi, N. and Hirayama, K. (1988) *J. Neurol. Neurosurg. Psychiat.* **51**, 590–591.

Kemp, J. M. and Powell, T. P. S. (1971) *Phil. Trans. Roy. Soc. Lond. (Biol).* **262**, 411–457.

Kitai, H. and Kitai, S. T. (1987) *J. Comp. Neurol.* **260**, 435–452.

Koller, W. C. and Trimble, J. L. (1985) *Neurology* **35**, 1450–1454.

Lang, A. E. (1985) *Can. J. Neurol. Sci.* **12**, 125–128.

Lange, H., Thorner, G., Hopf, A. and Schroder, K. F. (1976) *J. Neurol. Sci.* **28**, 401–425.

Lange, H. W., Aulich, A., Hefter, H., Homberg, V., Noth, J., Padoll, K. and Strauss, W. (1987) In *Motor Disturbances I* (edited by R. Benecke, B. Conrad and C. D. Marsden), pp. 243–252. Academic Press, London.

Lasker, A. G., Zee, D. S., Hain, T. C., Folstein, S. E. and Singer, H. S. (1988) *Neurology* **38**, 427–431.

Leenders, K. L., Frackowiak, R. S. J., Quinn, N. and Marsden, C. D. (1986) *Movement Disorders* **1**, 69–77.

Leigh, R. J., Newman, S. A., Folstein, S. E., Lasker, A. G. and Jensen, B. A. (1983) *Neurology* **33**, 1268–1275.

Leigh, R. J., Parhad, I. M., Clark, A. W., Buttner-Ennever, J. A. and Folstein, S. E. (1985) *J. Neurol. Sci.* **71**, 247–256.

Marsden, C. D. (1982) *Lancet* ii, 1141–1146.

Marsden, C. D., Obeso, J. A. and Rothwell, J. C. (1983) *Adv. Neurol.* **39**, 865–882.

Martin, J. B. (1984) *Neurology* **34**, 1059–1072.

Martin, J. B. and Gusella, J. F. (1986) *N. Engl. J. Med.* **315**, 1267–1276.

Martin, J. P. (1927) *Brain* **50**, 637–651.

Martin, J. P. and Alcock, N. S. (1934). *Brain* **57**, 504–516.

Martin, J. P. (1957) *Brain* **80**, 1–10.

Martin, J. P. and McCaul, R. (1959) *Brain* **82**, 104–108.

Mas, J. L., Launay, M. and Derousne, C. (1987) *J. Neurol. Neurosurg. Psychiat.* **50**, 104–105.

Mitchell, I. J., Sambrook, M. A. and Crossman, A. R. (1985a) *Brain* **108**, 405–422.

Mitchell, I. J., Jackson, A., Sambrook, M. A. and Crossman, A. R. (1985b) *Brain Res.* **339**, 346–350.

Nausieda, P. A., Bieliauskas, L. A., Bacon, L., Hagerty, M., Voller, W. and Glantz, R. (1983) *Neurology* **33**, 750–754.

Noth, J., Engel, L., Friedmann, H.-H. and Lange, H. W. (1984) *Electroencephalogr. Clin. Neurophysiol.* **59**, 134–141.

Noth, J., Podell, K. and Friedmann, H.-H. (1985) *Brain* **108**, 65–80.

Penney, J. B. and Young, A. B. (1983) *Ann. Rev. Neurosci.* **6**, 73–94.

Penney, J. B. and Young, A. B. (1986) *Movement Disorders* **1**, 3–15.

Quinn, N. and Marsden, C. D. (1984) *J. Neurol. Neurosurg. Psychiat.* **47**, 844–847.

Riley, D. and Lang, A. E. (1988) *Movement Disorders* **3**, 88–94.

Roos, R. A. C., Pruyt, J. F. M., de Vries, J. and Bots, G. Th. A. M. (1985) *J. Neurol. Neurosurg. Psychiat.* **48**, 422–425.

Rothwell, J. C., Obeso, J. A., Traub, M. and Marsden, C. D. (1983) *J. Neurol. Neurosurg. Psychiat.* **6**, 35–44.

Saris, S. (1983) *Arch. Neurol.* **40**, 590–591.

Schwarz, G. A. and Barrows, L. J. (1960) *Arch. Neurol.* **2**, 420–434.

Sethi, K. D., Nichols, F. T. and Yaghami, F. (1987) *Movement Disorders* **2**, 61–66.

Shoulson, I. (1981) *Neurology* **31**, 1333–1335.

Starr, A. (1967) *Brain* **90**, 545–564.

Suchowersky, O., Hayden, M. R., Martin, W. R. M., Stoessl, A. J., Hildebrand, A. M. and Pate, B. D. (1986) *Movement Disorders* **1**, 33–44.

Tabaton, M., Mancardi, G. and Loeb, C. (1985) *Neurology* **35**, 588–589.

Thiebaut, F. (1968) In *Handbook of Clinical Neurology* (edited by P. J. Vinken, and G. W. Bruyn), Vol. 8, pp. 409–436. North Holland, Amsterdam.

Thomas, D., Byrne, P. D. and Travers, R. L. (1979) *Aust. N. Z. J. Med.* **9**, 568–570.

Thompson, P. D., Dick, J. P. R., Day, B. L., Rothwell, J. C., Berardelli, A., Kachi, T. and Marsden, C. D. (1986) *Movement Disorders* **i**, 113–117.

Thompson, P. D., Berardelli, A., Rothwell, J. C., Day, B. L., Dick, J. P. R., Benecke, R. and Marsden, C. D. (1988) *Brain* **iii**, 223–244.

Whittier, J. R. (1947) *Arch. Neurol. Psychiat.* (*Chicago*) **58**, 672–692.

Whittier, J. R. and Mettler, F. A. (1949) *J. Comp. Neurol.* **90**, 319–372.

Wilson, S. A. K. (1925) *Lancet* **ii**, 169–178.

Young, A. B., Penney, J. B., Starosta-Rubenstein, S. Markel, D. S., Berent, S., Giordani, B., Ehrenkavfer, R., Jewett, D. and Hichora, R. (1986) *Ann. Neurol.* **20**, 296–303.

Young, A. B., Penney, J. B., Starosta-Rubenstein, S., Markel, D., Berent, S., Rothley, J., Betley, A. and Hichora, R. (1987) *Arch. Neurol.* **44**, 254–257.

34

The Biochemistry of Huntington's Chorea

C. J. Carter, J. Benavidès and A. Dubois

The mutant gene in Huntington's disease is located on the short arm of chromosome 4 (Gusella *et al.*, 1983, 1985) and its precise location and identity will soon be resolved. There is an enormous data base from post-mortem and *in vivo* neurochemical or biochemical studies in Huntington's disease, and once the primary biochemical deficit is known, this will allow us to trace the development of a neurotoxic process from mutation to expression. This is a unique opportunity for the understanding of neurodegenerative processes in such diseases and the knowledge gained from these studies will be extremely useful in relation to other similar disorders.

Neurochemical studies in Huntington's disease have unfortunately not led to any successful therapeutic rationale, along the lines of L-dopa replacement therapy in Parkinson's disease, but have nevertheless been extremely useful in defining which systems are vulnerable, and as importantly, which systems are not. In addition, certain common features of vulnerable neurons, particularly in relation to their size, position in defined circuits, and dependence upon glucose metabolism, allow the tentative definition of the type of neurotoxic process to expect in Huntington's disease. This in turn may allow the development of rational therapy aimed not at replacing missing circuits, but preventing their destruction.

DISORDERS OF MOVEMENT: CLINICAL, PHARMACOLOGICAL
AND PHYSIOLOGICAL ASPECTS   ISBN 0-12-569685-X

GROSS PATHOLOGY OF HUNTINGTON'S DISEASE

The overall lesion in the terminal stages of Huntington's disease is massive, and not restricted to discrete anatomical areas or a single pathway. At post-mortem, reductions in overall brain weight have been reported to average 200 g (Bruyn *et al.*, 1979), reflecting cortical atrophy (mainly of orbitofrontal and occipital areas), and extensive atrophy of the caudate putamen. The globus pallidus and subthalamic nucleus may be included but cell loss is by no means restricted to these sites. Neuronal loss has also been reported in the thalamus, hypothalamus, substantia nigra and associated midbrain tegmental areas, pontine olivary and dentate nuclei, cerebellar cortex and ventral horn of the spinal cord (Bruyn, 1968; Klintworth, 1973; Lange *et al.*, 1976; Bruyn *et al.*, 1979; Hayden, 1981). In the cerebellum, one study has reported 50% loss of Purkinje cells (Jeste *et al.*, 1984). The vulnerable population in the basal ganglia appear to be characterized by its small to medium size (Lange *et al.*, 1976; Roizin *et al.*, 1976; Bruyn *et al.*, 1979). Although smaller cells are lost, larger neurons may also show abnormal characteristics, such as extensive accumulation of lipofuscin, increased amounts of smooth endoplasmic reticulum, mitochondrial enlargement and disrupted mitochondrial cristae. Mitochondrial numbers may also be increased. Astrocytes show a massive accumulation of lipofuscin, increased deposition of glycogen, proliferation of glial filaments and enlarged mitochondria with disrupted cristae (Tellez Nagel *et al.*, 1974; Roizin *et al.*, 1976).

The disease process thus appears to affect many neurons, regardless of type or position, and as supported by neurochemical data cannot be regarded as a selective neurotoxic process, for example it does not specifically attack a particular type of neuron or any one neurotransmitter system. Despite this apparent generality, it is clear that the lesions are somehow selective in the sense that a characteristic lesion pattern does nevertheless exist in Huntington's disease.

The thesis that we intend to develop in this review is that, if the neurotoxic process is so generalized, the eventual lesion pattern must to a large extent be governed by features of the vulnerable neurons themselves. At first glance, common features of widely differently vulnerable neurons may not seem apparent but they may exist in the form of a common dependence of neurotransmitter synthesis on glucose supply, their position in an extensively used brain circuit, and in many cases their relatively small size. Combinations of these features may perhaps be integrated to provide a "vulnerability index" which eventually defines the anatomy of the lesions.

VULNERABLE NEUROTRANSMITTER SYSTEMS IN HUNTINGTON'S DISEASE

The results of a number of post-mortem studies from various groups who have measured neurotransmitter markers in Huntington's disease brain are summarized in Tables 34.1–34.3.

GABA neurons

It is accepted that there is an extensive loss of GABA neurons in the basal ganglia (Perry et al., 1973a,b; Bird and Iversen, 1974; Iversen et al., 1974; Wu et al., 1979; Spokes, 1980; Butterworth et al., 1983, 1985). These may principally comprise the basal ganglia output systems (striatopallidal, striatonigral) and this loss can be evidenced from large reductions in GABA neuronal markers in these brain regions (see Tables 34.1–34.3). Vulnerability of GABA systems in other areas is less evident, although extrastriatal areas have been less extensively studied. From the data available (see Table 34.1) frontal and motor cortical GABA neurons appear to be spared but there are reports of significantly reduced GABA levels in the temporal (Reynolds and Pearson, 1987) and occipital cortex (Perry et al., 1973b), the cerebellar cortex and the hippocampus and ventrolateral thalamus (Reynolds and Pearson, 1987), although GAD activity in these areas was generally reported as normal. Glutaminase data from cortical or cerebellar regions are equivocal, as this enzyme is extremely sensitive to agonal status, but a clear reduction in glutaminase activity has been suggested in the caudate (Butterworth et al., 1985).

Cholinergic neurons

The cholinergic neuronal marker choline acetyltransferase (CAT) is reduced in the caudate and putamen, but not in the globus pallidus. Cholinergic neuronal loss in the striatum is generally considered as patchy, and may in fact be normal in some Huntington's disease patients. Cortical activity (frontal, temporal, parietal, motor or occipital) appears to be normal (McGeer et al., 1973; Bird and Iversen, 1977; McGeer and McGeer, 1976b; Wong et al., 1981; Sorbi et al., 1983); apart from one report of reduced activity in frontal cortex (Wu et al., 1979). There is evidence for reduced CAT activity in the septum (Spokes, 1980) and hippocampus (Bird and Iversen, 1977) suggesting a lesion of the septo-hippocampal tract, and a small but significant decrease in CAT activity (17%) has been reported in the nucleus accumbens

Table 34.1 The effects of Huntington's disease on various neurotransmitter markers in different brain regions. Data are expressed as a percentage of control values (= 100%). References for each study are superscripted; see footnote to Table 34.3. For each column, values on the left hand side are not significantly different from control values; values on the right hand side are significantly different from control data ($P < 0.05$). FC, frontal cortex; TC, temporal cortex; PC, parietal cortex; OC, occipital cortex; MC, motor cortex; CER, cerebellum; DENT, dentate nucleus; CAU, caudate; PUT, putamen; GP, globus pallidus; ACB, nucleus accumbens; SN, substantia nigra (PC, pars compacta; PR, pars reticulata); HPC, hippocampus; SEP, septum; GABA, γ-aminobutyric acid; GAD, glutamate decarboxylase; GABAT, GABA transaminase; GLNASE, glutaminase; GLUT, glutamate; GLU UP, glutamate uptake; ORNAT, ornithine aminotransferase; CHAT, choline acetyltransferase; GLU BIND, glutamate binding; KAIN, kainate (high- or low-affinity binding sites); TOH, tyrosine hydroxylase; NOR, noradrenaline; 5-HT, 5-hydroxytryptamine; SP, substance P; M-ENK, methionine-enkephalin; CCK, cholecystokinin; ACE, angiotensin converting enzyme; VIP, vasoactive intestinal polypeptide; SOM, somatostatin; TRH, thyrotropin-releasing hormone; NT, neurotensin.

	FC	TC	PC	OC	MC	CER	DENT
GABA	91[1]	63[5]		71[2] 68[1]	97[6]	98[3] 75[1] 85[6] 85[2]	95[3] 108[6]
GAD	85[7] 61[11] 90*[8] 73[12] 90*[9] 111*[9] 118[10]	58[12]	58[12]		105[13]	107*[13]	114*[13]
GLNASE	92*[9] 68[9] 81[11] 142*[9]	58[9] 158*[9]		74[9] 638*[9]	50[9] 159*[9]	61[9] 544*[9]	
GLUT	101[16]	83[5]					

	1	2	3	4	5	6	7	8
GLU UP	98[17]							
ORNAT		66[18]	83[18]	64[18]	105[18]	102[13]	83[13]	95[13]
							84[23]	100[20]
								31[23]
CHAT	83[11]	54[7]	93[18]	128[18]	146[18]			
	97[7]		84[23]	79[23]	62[23]			
	100[18]							
	104[19]							
	116[20]							
	145[23]							
GLU BIND KAIN HIGH	76[22]		93[21]					
			62[22]					
LOW		52[22]	110[22]					
5-HT			131[5]					
SP	70[26]							
CCK	101[28]							
ACE	108[29]			117[29]				
	117[30]							
VIP	87[30]							
SOM	104[32]							
	114[33]							

*In these cases, it was stated that comparative values were controlled for agonal status.

Table 34.2 (*See Table 34.1*)

	CAU	PUT	GP	ACB	SN	HPC	SEP
GABA	60[1] 71[2] 57[4]	38[1] 59[4] 35[5]	32[5] 59[6]	52[6]	20[1] 45[2]	101[6]	71[5]
GAD	27[4] 28*[9] 16[11] 38[12] 50[13]*	35[7] 46*[12] 19[4] 46[13]	19[4] 32[12] 55*[13]		106 PC[13] 26[12] 82* PR[13]	105[8] 81[13]	
GABAT		56[14]					
GLNASE	8[11] 9[9] 7[9] 44*[9]	16[15]			180[9] 1350*[9]		
GLUT	71[16]	68[16] 75[5]	114[5]	86[16]		87[5]	
GLU UP	28[17]	40[17]				68[17]	
CHAT	82[14] 49[11] 46[4] 40[19] 57[13] 38[12] 45[20]	44[4] 50[19] 69[7] 65[13] 41[20] 31[23]	76[9] 167[20] 34[23]	106[12] 82[9]	83[20] 125[23]	97[20] 72[13] 67[20] 76[23]	63[13]
GLU BIND	29[21]	29[21]	33[22]			83[22]	
KAIN HIGH	20[22]	47[22]	550[22]			129[22]	
LOW	71[22]	27[22]					

Table 34.3 (*See Table 34.1*)

	CAU	PUT	GP	ACB	SN	HPC	SEP
DOPAMINE	80(4) 59(24) 132(13)	100(4) 79(24) 169(13)	171(13)	187(13)	105(13) PC 134(13)		
TOH	105(12) 140(25)	79(12) 86(4)	104(12)	118(12)	146(12) 265(25)		
NOR	100(24) 163(13)	50(24)	137(13)	146(13)	140(13)		
5-HT	94(10)	141(24) 197(5)	165(24) 229(5)			81(5)	
SP	95(26) 114(27)	64(26) 68(27)	9(27) 54(26)	68(26)	6(27)		
MENK	134(27)	92(27)	45(27)		40(27) 38(28)		
CCK	90(27)	115(28)	48(28)	70(30)	22(30)		
ACE	40(30)	48(30)	28(30)				
VIP	91(29) 80(30)	108(31)	150(30) 129(31)	147(31) 159(30)			92(30)
SOM	184(33) 468(32)	464(32)	94(30)	290(30) 239(32)	300(32)	118(12) 136(33)	
TRH	233(33) 270(34)	288(34)		78(33)		100(33)	
NT	300(33)			150(33)			

References: (1) Perry *et al.* (1973a); (2) Perry *et al.* (1973b); (3) Kish *et al.* (1983); (4) Bird and Iversen (1974); (5) Reynolds and Pearson (1987); (6) Spokes *et al.* (1980); (7) Wu *et al.* (1979); (8) Bird and Kraus (1981); (9) Butterworth *et al.* (1985); (10) Iversen *et al.* (1974); (11) Butterworth *et al.* (1983); (12) McGeer and McGeer (1976b); (13) Spokes (1980); (14) Carter (1984b); (15) Carter (1985b); (16) Perry (1982); (17) Cross *et al.* (1986); (18) Wong *et al.* (1982); (19) Sorbi *et al.* (1983); (20) Bird and Iversen (1977); (21) Greenmayre *et al.* (1985); (22) London *et al.* (1981); (23) Wastek and Yamamura (1978); (24) Bernheimer and Hornykiewicz (1973); (25) Bird (1980); (26) Kanzawa *et al.* (1979); (27) Emson *et al.* (1980b); (28) Emson *et al.* (1980a); (29) Arregui *et al.* (1979a); (30) Arregui *et al.* (1979b); (31) Emson *et al.* (1979); (32) Aronin *et al.* (1983); (33) Nemeroff *et al.* (1983); (34) Spindel *et al.* (1981).

(Spokes, 1980). There are no other reports of reduced CAT activity outside the basal ganglia.

Glutamatergic neurons

There is recent evidence for damage to corticostriatal glutamate projections and also of afferent glutamatergic inputs to the hippocampus. This was first suggested by Wong and others who measured ornthine aminotransferase activity in Huntington's disease brain and reduced activity of this putative glutamate neuronal marker has been observed in frontal, temporal and parietal (but not occipital) cortex, caudate, putamen and globus pallidus (Wong *et al.*, 1982). Glutamate levels have been reported to be reduced in temporal cortex, hippocampus, caudate and putamen (Perry, 1982; Reynolds and Pearson, 1987) and while there are obvious difficulties in interpreting this because of the myriad metabolic roles of glutamate, the idea of glutamatergic neuronal lesions is supported by reduced glutamate uptake in the caudate, putamen and hippocampus (Cross *et al.*, 1986). Glutamate uptake was normal in frontal cortex and globus pallidus. The concept of destruction of corticofugal glutamate projections is consistent with reports of neuronal damage in largers 3 and 5 of the cortex (Bruyn, 1968; Bruyn *et al.*, 1979), layers which project to subcortical areas (Jones *et al.*, 1977).

Peptidergic neurons

Substance P levels are reduced in the putamen globus pallidus and substantia nigra (Kanazawa *et al.*, 1979; Emson *et al.*, 1980a), while met-enkephalin and cholecystokinin levels are reduced in the globus pallidus and substantia nigra (Emson *et al.*, 1980a,b). Levels of other peptides, for example VIP in the globus pallidus (Emson *et al.*, 1979; Arregui *et al.*, 1979), somatostatin in the caudate putamen, nucleus accumbens and substantia nigra (Nemeroff *et al.*, 1983; Aronin *et al.*, 1983), TRH in the caudate and putamen (Spindel *et al.*, 1981), and neurotensin in the caudate and nucleus accumbens (Nemeroff *et al.*, 1983) are elevated in Huntington's disease. These increases might be explained by tissue shrinkage (for a review, see Martin, 1984). Because of the probability of peptide co-localization with other neurotransmitters, for example GABA in cortical neurons may coexist with CCK, VIP, NPY, substance P or SRIF, and CAT activity with VIP (Jones and Hendry, 1986), the significance of the peptide changes in Huntington's disease is difficult to ascertain. The reduction in SP and met-enkephalin levels in the Huntington's disease basal ganglia might simply reflect their co-localization in vulnerable GABA neurones. This problem of co-localization is an important one as, in the absence of neurons only containing peptides, which seems increasingly likely, the neuronal deficits in Huntington's disease becomes essentially restricted to glutamatergic, GABAergic and cholinergic systems. This then raises the question of whether certain peptides can be used to define subclasses of, for example, GABA neurons and then whether the loss of some peptides but not of others in disease states reflects selective vulnerability of subtypes of GABA neurons defined by the peptide they contain.

NEURONS THAT ARE UNDAMAGED IN HUNTINGTON'S DISEASE

The ascending monoaminergic projections to the basal ganglia in the form of dopaminergic projections from the substantia nigra and serotonergic projections from the raphé nuclei are believed to be intact (McGeer and McGeer, 1976b; Bird and Iversen, 1977; Spokes, 1980). Indeed, 5-HT and 5-HIAA levels are markedly elevated in the putamen and globus pallidus (Reynolds and Pearson, 1987). Noradrenaline levels are not significantly reduced in the caudate putamen, globus pallidus, nucleus accumbens or substantia nigra (Bernheimer and Hornykiewicz, 1973; Spokes, 1980). The lack of damage to these particular catecholamine and indoleamine-containing neurons must be considered an important feature in the pathology of Huntington's disease.

METABOLIC FEATURES OF VULNERABLE NEURONS

If we accept that glutamatergic, GABAergic and cholinergic neurons are vulnerable to the metabolic deficit in Huntington's disease, and that dopaminergic, serotonergic and noradrenergic systems are not, there is a readily apparent difference between the metabolism of the two neurotransmitter groups. Dopamine, 5-HT and noradrenaline are synthesized from blood-borne precursors, phenylalanine, tyrosine and tryptophan, and their metabolites are expelled from the brain as waste products. Oxygen is needed for certain steps of their metabolism, for example by tyrosine or tryptophan hydroxylase and monoamine oxidase (Siesjo, 1978), but there is no dependence upon cerebral glucose metabolism, nor is there any attempt to conserve these neurotransmitter pools by salvaging and reconversion of their metabolites.

The metabolism of glutamate, GABA and acetylcholine is, however, intimately related to glucose and energy metabolism in the brain. Glucose as the sole exogenous respiratory substrate of the brain (Siesjo, 1978) is also supplied in excess, but the margin between supply and oxidative demand is limited by a bottleneck at the entry of pyruvate into the Krebs cycle (pyruvate dehydrogenase). Diversion of glucose-derived products following the entry of pyruvate into this cycle therefore creates no small problem in the brain, and there are extensive metabolic and compartmental modifications to ensure the smooth and energy-efficient coupling between glucose and neurotransmitter metabolism (primarily glutamate, GABA and acetylcholine). The maximal potential activity of pyruvate dehydrogenase may be only 50% above basal metabolic needs (Blass *et al.*, 1980; Sorbi *et al.*, 1983), and any faults in this regulatory apparatus could have severe consequences.

Glutamate and GABA neurons

The metabolic problems that are peculiar to glutamate and GABA neurons and the steps taken to overcome them can be briefly summarized as follows (for reviews, see Balazs *et al.*, 1973; Balazs and Cremer, 1973; Bachelard, 1975; Baxter, 1976; Siesjo, 1978). The synthesis and subsequent release of neurotransmitter from glucose (Ward *et al.*, 1983) or a Krebs cycle intermediate involves the loss of this intermediate from the neuronal compartment. Neurons themselves cannot make up this loss because they lack the ability of *de novo* synthesis of such products, which is a property restricted to glial cells, mainly in the form of CO_2 fixation (pyruvate + CO_2 → oxaloacetate = pyruvate carboxylase) (Yu *et al.*, 1983). Krebs cycle intermediate loss is thus a neuronal problem, while Krebs cycle intermediate synthesis is a glial property. The different Krebs cycles in the glial and neuronal compartments therefore have to be linked in some way to balance the system. This seems to be achieved by the neuronal to glial transfer of glutamate and GABA (Henn and Hamberger, 1971; Schousboe *et al.*, 1977; Schousboe, 1981) which can be fed into and oxidized by the glial Krebs cycle (Yu *et al.*, 1982; Roeder *et al.*, 1984a,b; Tildon and Roeder, 1984). In return, the existence of high affinity uptake site for Krebs cycle intermediates on nerve terminals (for malate and 2-oxoglutarate) (Shank and Campbell, 1981, 1982, 1984; Carter *et al.*, 1986) supports the glial to neuronal transfer of these compounds. Glial cells are themselves able to sustain this loss by CO_2 fixation. The ability of glial cells to oxidize glutamine, the conversion of GABA and glutamate to glutamine in the glial compartment and the high levels of glutamine in brain suggest that glial glutamine may serve as an important amino acid carbon skeleton reservoir or buffer (see also Benjamin and Quastel, 1975; Hertz *et al.*, 1983).

These adaptations allow the metabolism of the amino acid neurotransmitters to be maintained in a closed loop, the purpose of which is to reduce excessive demands an cerebral glucose metabolism. Equally it might be envisaged that malfunctions of this system at any point along this loop, in glia or neurons (release, uptake or enzyme activity), would create severe problems of glucose economy in amino acid neurotransmitter-containing neurons. These adaptive measures are illustrated in Fig. 34.1.

Acetylcholine neurons

Because acetyl-CoA is synthesized from pyruvate (Tucek and Cheng, 1970) by pyruvate dehydrogenase whose activity closely correlates with that of CAT in brain (Reynolds and Blass, 1976; Perry *et al.*, 1980; Sorbi *et al.*, 1983) cholinergic neurons face similar metabolic problems. Acetylcholine, once released, is hydrolysed to choline and acetate, the choline being returned to the nerve terminal (Simon and Kuhar, 1976). It is not yet clear whether there is any linkage between Krebs cycles of the type proposed for amino acid containing neurons, but it is interesting to note that glia are capable of oxidizing acetate (Berl and Frigyesi, 1969; Van den Berg, 1970) and also that they secrete pyruvate in tissue culture (Selak *et al.*, 1985). Thus a similar symbiosis between glia and neurons is theoretically possible.

The important feature of glutamate and GABA and possibly of acetylcholine metabolism, is this type of dovetailing between the glial cell and the neuron. Glial cells are adapted to capture or utilize what is essentially neuronal waste (acetate, GABA, glutamate, CO_2, NH_3), and can store some of this waste in the form of a large buffer pool of glutamine which can serve as a neurotransmitter precursor for glutamate and GABA, can oxidize glutamate, GABA or glutamine, transfer Krebs cycle intermediates to

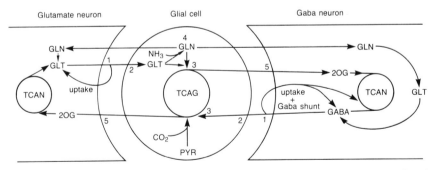

Fig. 34.1 A schematic representation of the metabolic links between glutamate or GABA neurons and glia. (1) Neuronal release of glutamate or GABA; (2) glial uptake of glutamate or GABA; (3) glial oxidation of glutamate, glutamine or GABA; (4) synthesis of glutamine from glutamate and storage in a large pool; (5) glial to neuronal transfer of tricarboxylic acid cycle intermediates. TCAN, neuronal tricarboxylic acid cycle; TCAG, glial tricarboxylic acid cycle; GLT, glutamate; GLN, glutamine; PYR, pyruvate; 2OG, 2-oxoglutarate.

neurons, and sustain this loss by CO_2 fixation. The nerve terminal and glial cell can thus be regarded as a single functional unit dedicated to preserving the economy of cerebral glucose metabolism (see Fig. 34.1). As stated previously this type of metabolic problem is not seen with dopamine, noradrenaline or 5-HT-containing neurons.

GLUTAMATE AND HUNTINGTON'S DISEASE

Although glutamatergic neurons may themselves be damaged in Huntington's disease, a consistent theme running through Huntington's disease research has been that this neurotoxic amino acid, or a related compound, may be responsible for the neurotoxicity seen in Huntington's disease. This stems from the early observation that kainic acid, a rigid analogue of glutamate, produces very similar lesions when injected into the rat striatum (Coyle and Schwartz, 1976; McGeer and McGeer, 1976a; Coyle, 1979; Sanberg and Johnston, 1981). This neurotoxic theory has been extended into many other clinical areas of degenerative diseases, most notably related to brain ischaemia (Meldrum, 1985).

The glutamate theory of Huntington's disease can be expressed in a number of ways. Either the regional distribution of the lesions is influenced by excitatory inputs, which are themselves normal, but whose actions exacerbate the effects of some other deficit, or there is a disturbance in glutamatergic function, leading to overstimulation of glutamate receptors. This second may include (in theory) altered metabolism or uptake, abnormal receptors or channels, or the production of a neurotoxin which acts on these receptors. In relation to this idea, the theory of abnormal production of the NMDA agonist quinolinic acid, a metabolite of tryptophan, has gained some impetus (Schwartz *et al.*, 1983a,b; Beal *et al.*, 1986). Thus far however, there is no direct evidence from studies in Huntington's disease patients to support this idea (Foster *et al.*, 1985; Heyes *et al.*, 1985). In post-mortem brain, glutamate binding is massively reduced in the Huntington's disease caudate and putamen (Greenmayre *et al.*, 1985), as are high- and low-affinity kainate binding. High-affinity kainate binding is also slightly reduced in frontal (-24%) and temporal (-38%) cortex. Although this shows that the missing cells possessed glutamate receptors this does not of course prove that glutamate neurotoxicity is involved in the disease process. However, a contributing role of glutamate, and the very good match of the disease lesion provided by excitotoxic lesions cannot be lightly dismissed. As discussed later there is evidence to suggest that the metabolic lesion in the Huntington's disease brain may be related to disturbances in cerebral energy metabolism. Glutamate receptor-mediated toxicity has been implicated in the pathology of cerebral ischaemia and hypoglycaemia (Simon *et al.*, 1984; Wieloch, 1985) and it is perhaps in this context that a neurotoxic role for glutamate might be considered in Huntington's disease.

FACTORS CONTRIBUTING TO NEURONAL VULNERABILITY

The unifying thread linking vulnerable neurons in Huntington's disease is that they each possess one or more features which would tend to make them metabolically active, and more dependent upon cerebral energy metabolism. This applies to small neurons because of the greater metabolic demands associated with a larger surface area to volume ratio, to neurons supporting long axons for similar reasons (see Kowall *et al.*, 1987), to GABA, acetylcholine and glutamate neurons because of their diversion of glucose into neurotransmitter synthesis, and to neurons with a glutamate input because of the demands associated with excitation.

These factors may each play a contributory role to neuronal vulnerability, and their summation may also create a gradient of vulnerability characterized in the extreme by a small GABAergic neuron supporting a long axon and possessing a strong glutamatergic input (i.e. GABA neurons in the basal ganglia). This concept infers that the responsibility for cell death is shared between the mutant gene and the neurons themselves, which possess features that make them vulnerable to a particular type of metabolic deficit. These factors can be regarded as active (altered metabolism as a result of the mutant gene) or passive in terms of the physical and biochemical characteristics of certain types of neurons. An interplay between these two factors may thus define the eventual anatomical pattern of the lesions.

This approach may have implications for therapy in Huntington's disease, as both active and passive factors might be targeted for drug intervention. The former must await characterization of the mutant protein, but factors influencing the passive features of neuronal vulnerability might be considered.

BIOCHEMICAL AND METABOLIC STUDIES IN HUNTINGTON'S DISEASE

In the previous sections, we have attempted to define certain vulnerability features of the suceptible neurons in the Huntington's disease brain, and suggested that an overdependence upon cerebral glucose or energy metabolism may contribute to such vulnerability. By inference, we are also suggesting that cerebral energy metabolism might be compromised in Huntington's disease, and cells that are more dependent upon this area of metabolism are thus more susceptible to this type of deficit. It is impossible to pinpoint a precise deficit in any particular area of glucose or oxidative metabolism in Huntington's disease patients, or from autopsy experiments, but over the years a number of anomalies have accumulated that suggest that this area of metabolism may not be normal in Huntington's disease. The main conclusions from these studies in post-mortem brain are very briefly summarized below (for a more extensive review, see Carter, 1987).

Glucose metabolism

Hexokinase activity has been reported as being normal in the putamen and fumarase activity as normal in the caudate (Bird *et al.*, 1977; Perry *et al.*, 1980). In the putamen, however, phosphofructokinase (Bird *et al.*, 1977), pyruvate and 2-oxoglutarate dehydrogenase (Carter, 1985a,b) activities are markedly reduced. Succinate dehydrogenase activity is reduced in the caudate (Stahl and Swanson, 1974). The analysis of glycolytic or Krebs cycle enzymes is obviously far from complete but this apparently selective loss of enzymes must be viewed as suspicious. The reductions in pyruvate and 2-oxoglutarate and succinate dehydrogenase activities in the putamen or caudate (66, 69 and 78%) are also much larger than those observed following kainic acid lesions of the rat striatum (25, 25 and 0%) even though these produce similar reductions in GAD activity and greater reductions in CAT activity (Carter, 1987). This comparison may, however, be complicated by the additional loss of glutamatergic input to the Huntington's disease basal ganglia.

These reductions are not global in the Huntington's disease brain, for example, phosphofructokinase and pyruvate dehydrogenase activity are normal in the frontal cortex (Perry *et al.*, 1980; Butterworth *et al.*, 1983). Pyruvate dehydrogenase activity is, however, reduced in the Huntington's disease hippocampus (Sorbi *et al.*, 1983).

It may be pertinent that phosphofructokinase, pyruvate and 2-oxoglutarate dehydrogenase are each key regulatory enzymes whose efficiency is modified according to the needs of the cell by substrates whose levels reflect energy availability (i.e. feedback control via ATP/ADP or NADH/NAD ratios) (Lehninger, 1975). In relation to this, it has been noted that in purified mitochondrial preparations from the Huntington's disease caudate, the oxidation of succinate and cytochrome oxidase activities are markedly decreased. The mitochondrial content of cytochrome aa_3 was reduced by $\geq 50\%$ independently of the degree of tissue atrophy, suggesting abnormal function of the electron transport chain, and inefficient NADH oxidation and ATP generation. This deficit was not, however, observed in frontal cortex (Brennan *et al.*, 1985).

While these data do not permit a definitive conclusion that this area of metabolism is globally compromised in Huntington's disease, they at least demonstrate that enzymes whose localization reflects high energy demand are concentrated in vulnerable striatal neurons.

PET scan studies and *in vivo* detection

These changes can be detected *in vivo* in PET scan studies using ^{18}F-2-deoxyglucose, where large deficits in striatal glucose metabolism have been recorded in early Huntington's disease (Kuhl et al., 1982, 1984; Hayden et al., 1986). For a time it was thought that this technique might be useful for presymptomatic screening, but although striatal hypometabolism may be detected presymptomatically in some individuals, the margin of overlap between control and Huntington's disease patients may not be sufficient to allow precise diagnosis, nor is glucose metabolism in other areas frankly abnormal (Young et al., 1987). Such studies have so far measured glucose metabolism at rest, and perhaps more information could be gained by studying metabolism in functionally activated systems, for example during defined visual or auditory stimulation or the performance of specific cognitive or physical tasks.

A further tool, not yet investigated but potentially useful in PET scan studies, might be the imaging of peripheral benzodiazepine binding (ω_3) sites in Huntington's disease brain. In contrast to the central type receptor found in brain and responsible for the anxiolytic/hypnotic or antiepileptic properties of these drugs the ω_3 binding sites have a predominantly peripheral localization (Anholt et al., 1985). Binding sites in brain exist for this subtype of receptor (on glial cells but not on neurons), but in very low numbers. However, it has been observed that in lesioned brain (excitotoxic or ischaemic lesions), the number of ω_3 sites is dramatically increased within the affected area (Benavidès et al., 1987; Dubois et al., 1988). Either haematogenous cells (e.g. macrophages) bearing this receptor are drafted into the lesioned area, or these receptors are expressed *de novo* in the cells involved in the reactive gliotic events that accompany any type of neuronal injury. The reasons and role for this proliferation of ω_3 sites is unknown, but as is dramatically illustrated in Fig. 34.2, this phenomenon can be used to visualize the distribution of neuronal damage with very good resolution. PET scan ligands for this receptor do exist and have been used to image human heart *in vivo* (Charbonneau et al., 1986). Although no data is yet available from positron emission studies in human brain there is a clear application for such ligands in the study of Huntington's disease and of other human degenerative

A

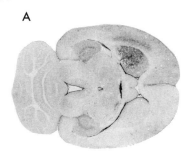

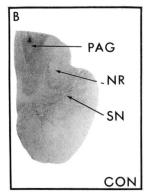

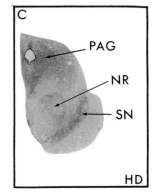

Fig. 34.2   (A) The autoradiographic distribution of ω_3 binding sites (labelled with ^{3}H-PK 11195) in a rat lesioned by intrastriatal injection of kainic acid (left hand side). Note the ipsilateral increase in striatal and deep cortical labelling. There is also a marked bilateral increase in hippocampal labelling possibly reflecting secondary seizure-induced damage (see Benavidès *et al.*, 1987). (B), (C) Autoradiographic images of single control (CON) and Huntington's disease (HD) brain sections at the level of the substantia nigra, illustrating the relative density of ω_3 binding sites labelled with ^{3}H-PK 11195. Note the band of dark labelling in the Huntington's disease substantia nigra. SN, substantia nigra; NR, nucleus ruber; PAG, periaqueductal gray. Quantitative autoradiographic measurements of ^{3}H-PK 11195 binding have so far been made at three different levels in two control and two HD subjects, and are as follows. NR, CON = 668 ± 79; HD = 715 ± 56; PAG, CON = 1137 ± 125; HD = 1607 ± 163 (+41%). SN pars reticulata, CON = 546 ± 69; HD = 871 ± 63 (+60%) (fmol mg protein^{-1}).

disorders. Using such ligands the identification and fine-resolution mapping of brain lesions in patients is now feasible.

Amino acid "Krebs cycle" enzymes

This definition relates to the intimate coupling between glucose and neurotransmitter amino acid metabolism described in previous sections, and includes the enzymes of the GABA shunt (glutamate dehydrogenase, glutamate decarboxylase, GABA transaminase, succinic semialdehyde dehydrogenase) and of glutamate metabolism (glutaminase and glutamine synthetase). Clearly, some of these are reduced because of their selective or concentrated localization in vulnerable neurons (for example, glutamate decarboxylase, GABA transaminase and glutaminase); glutamate dehydrogenase (glutamate - formation) activity is reduced by 32% in Huntington's disease putamen (Carter, 1985b, 1987), while its activity in the direction of 2-oxoglutarate formation has been reported as normal in the caudate and putamen and elevated in the globus pallidus (Bird *et al.*, 1977). This difference may reflect the presence of GDH isoenzymes favouring glutamate synthesis or glutamate oxidation as demonstrated in rat heart (MacDaniel *et al.*, 1984). Logically, the former would be concentrated in neurons and the latter in glia, although this remains to be formally characterized.

Perhaps the most abnormal finding in relation to this area of metabolism is the reduction in glutamine synthetase activity in Huntington's disease frontal and temporal cortex, caudate and putamen and cerebellum (but not in thalamus, olive or hippocampus) (Carter, 1982). This enzyme has a glial localization (Norenberg, 1979) and its activity is conversely increased in kainic acid lesioned rat striata (Nicklas *et al.*, 1979a,b). Apart from its role in linking glutamate and glucose metabolism this enzyme plays the major role in cerebral ammonia assimilation (Cooper *et al.*, 1979). Inhibition of the enzyme has long been known to be neurotoxic (Silver, 1949; Lewey, 1950; Hicks and Coy, 1958), possibly as a result of disrupting cerebral glutamate metabolism, but also because the process of ammonia assimilation is forced upon glutamate dehydrogenase with resultant loss of Krebs cycle intermediates (Cooper *et al.*, 1979; Carter, 1982).

Another clearly abnormal, and perhaps related finding, in Huntington's disease brain is a twofold increase in alanine aminotransferase activity observed in the putamen (Carter, 1984a). Again, this is an effect not observed in the kainic acid lesioned rat striatum. It is known that glutamine synthetase inhibition increases alanine levels in brain (Engelsen and Fonnum, 1985), and there appears to be a correlation between the reduction in glutamine and the increase in alanine levels provoked by such treatment. Alanine aminotransferase catalyses the transamination of glutamate with pyruvate to form 2-oxoglutarate and alanine (or vice versa). It is worth noting that because of the reductions in pyruvate dehydrogenase and glutamine synthetase activity in Huntington's disease (see above), the two substrates for the alanine aminotransferase reaction (glutamate/pyruvate) would be more available. It has also been suggested that alanine aminotransferase is involved in cerebral ammonia assimilation (Fahien *et al.*, 1971; Rucsak *et al.*, 1982), which because of the reduction in glutamine synthetase activity might be compromised in Huntington's disease.

The above changes in enzyme activity in Huntington's disease might be viewed as metabolic changes induced by the primary deficit, although their significance is still far from clear. Certain reductions in activity (for example, phosphofructokinase, pyruvate, 2-oxoglutarate and succinate dehydrogenase) must be regarded as suspiciously large, but this rests but a suspicion while others such on the glutamine synthetase deficit and the increase in alanine aminotransferase seem to reflect a change that is peculiar to Huntington's disease (i.e. such changes are not observed in the kainic acid lesioned striatum and are more than a simple consequence of cell loss).

While it is impossible to define a precise deficit in a cerebral energy metabolism the enzyme changes in Huntington's disease, whether a direct result of cell loss or due to other unknown factors, are clearly localized within an area of interface between carbohydrate and amino acid metabolism as illustrated in Fig. 34.3. This again highlights the dependence of vulnerable neurons upon this area of metabolism.

THERAPEUTIC STRATEGIES IN HUNTINGTON'S DISEASE

The contention of this review is that the damaged neurons in Huntington's disease possess characteristics which tie them closely to cerebral energy metabolism, and furthermore that this must be a feature which dictates their vulnerability. By inference, the primary error leads to deficits which either disrupt the homeostatic measures taken by these neurons to ensure their metabolic economy, or to a deficit to which such active neurons are excessively vulnerable.

In essence this infers that the neurotoxic process must be very closely related to that seen in hypoglycaemia or ischaemia. The effects of such insults have of course been extensively studied in laboratory animals, and while the lesions produced in either situation are not a perfect match for those seen in Huntington's disease there are nevertheless striking similarities. For example, the neurons generally characterized as the most vulnerable to global ischaemia are those in cortical layers III, V and VI (cf. III, V in Huntington's disease), small- to medium-sized striatal neurons (as in Huntington's disease) and cerebellar Purkinje cells (Brierley, 1976). The fate of such neurons in Huntington's disease is contentious but a recent cell count study has revealed severe fall out of these neurons in the Huntington's disease brain (Jeste *et al.*, 1984). Hippocampal pyramidal neurons are affected in ischaemia, and while hippocampal damage is not generally cited as a characteristic feature of Huntington's disease, recent estimates of reduced hippocampal glutamate and GABA levels and of glutamate uptake sites suggest that cell loss may occur in this region (Cross *et al.*, 1986; Reynolds and Pearson, 1987). The

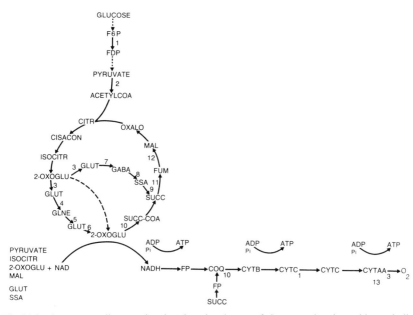

Fig. 34.3 A summary diagram showing the related areas of glucose and amino acid metabolism affected by the Huntington's disease gene. (1) Phosphofructokinase. A rate-limiting glycolytic enzyme whose activity is reduced in the caudate and putamen. (2) Pyruvate dehydrogenase complex. Regulates the flow of pyruvate into the tricarboxyllic acid cycle according to the energy needs of the cell. Its activity is reduced in the caudate, putamen and hippocampus. (3) Glutamate dehydrogenase (reduction amination). Reduced activity in the putamen. (4) Glutamine synthetase. Reduced activity in the frontal and temporal cortex, putamen and cerebellum. (5) Glutaminase. Reduced activity in the caudate and putamen may reflect localization of the enzyme within GABA neurons. (6) Glutamate dehydrogenase (oxidative deamination). Normal activity in the caudate and putamen. Increased activity in the globus pallidus has been interpreted as reflecting the glial localization of the enzyme. Compare with (3). (7) Glutamate decarboxylase. Reduced activity in the basal ganglia is indicative of widespread loss of GABAergic neurons. (8) GABA transaminase. Reduced activity in the putamen probably reflects localization of this enzyme within GABA neurons. (9) Succinic semialdehyde dehydrogenase. Effects of Huntington's disease unknown? (10) 2-Oxoglutarate dehydrogenase complex. Reduced activity in the putamen. (11) Succinate dehydrogenase. Reduced activity in the caudate and putamen. (12) Fumarase. Normal activity in the caudate. (13) Reduced oxidation of succinate, and reduced levels of cytochrome aa_3 in purified mitochondria of the Huntington's disease caudate. F6P, fructose-6-phosphate; FDP, fructose-1,6-diphosphate; CITR, citrate; CIS-ACON, *cis*-aconitate; ISOCITR, isocitrate; 2-OXOGLU, 2-oxoglutarate; GLUT, glutamate; GLNE, glutamine; SUCC-COA, succinyl coenzyme A; SUCC, succinate; FUM, fumarate; MAL, malate; OXALO, oxaloacetate; GABA, γ-aminobutyric acid; SSA, succinic semialdehyde; FP, flavoprotein; COQ_{10}, cytochrome Q_{10} (ubiquinone); CYT, cytochrome.

sensitivity of striatal (but not hippocampal) GABA neurons to global ischaemia has been demonstrated by Francis and Pulsinelli (1982). Striatal cholinergic neurons are more resistant to ischaemia. Following hypoglyca-emic insults, cortical damage is seen more in superficial layers (II > III > IV), in hippocampal pyramidal neurons and in the dentate gyrus. Again in the striatum, small- to medium-sized neurons are more affected (Auer, 1986).

The similarities between the type of damage provoked by these insults and that seen in Huntington's disease is evident. Because of this similarity it would seem logical to discuss the theoretical and practical approaches that have been used in the fields of ischaemia and hypoglycaemia to minimize neuronal damage and to suggest that some of these approaches might be feasible suggestions for the prevention of similar degeneration in Huntington's disease.

The first relates to the general assumption that active neurons will be more vulnerable than quiescent neurons, in which case the idea is that a general dampening of neuronal activity will reduce cell loss. Therapeutically this has involved the use of barbiturates or other inhibitors of neuronal activity such as the benzodiazepines (for reviews, see Hossmann, 1982; MacKenzie *et al.*, 1986). Related to this problem is the idea of reducing neuronal dependence upon cerebral energy metabolism. For glutamate and GABA neurons this entails the provision of a precursor substrate which reduces the necessity to synthesize neurotransmitter from glucose. This approach is at an experimental stage, where it is pertinent to note that glutamine will protect neurons from ischaemic damage *in vitro* (Schurr *et al.*, 1987), and *in vivo* will ameliorate the neurotoxic intrastriatally injected kainic acid upon striatal coinjection (McGeer and McGeer, 1978). An obvious study would be to assess whether systemic treatment with glutamine can provide similar protective effects in ischaemia or hypoglycaemia.

The second relates to the common final routes of neurotoxicity observed following numerous diverse insults, and in particular the cardinal role of calcium entry in cellular toxicity (Schanne *et al.*, 1979; Meldrum *et al.*, 1982). The use of calcium antagonists in Huntington's disease might thus be considered.

An involvement of free radical toxicity in Huntington's disease is suggested by the enormous deposits of lipofuscin in Huntington's disease brain (Siakotos and Munkres, 1982). The free radical scavenger vitamin E (2000 units day^{-1}) has been reported to halt and reverse the disease process in one patient (Hoffer, 1976), but this was not confirmed in a larger study (Caro and Caro, 1978). This is however well worth pursuing.

Finally, we come to the possible use of excitatory amino acid antagonists in neurodegenerative diseases. This was of course the earliest proposal for preventive therapy in Huntington's disease based on the replication of its lesions by the intrastriatal injection of kainic acid. Since then, the idea of excitatory amino acid receptor-mediated toxicity has been linked very closely to the neurotoxic processes involved in both ischaemia and hypoglycaemia (for reviews, see Meldrum, 1985; Wieloch, 1985; Rothman and Olney, 1986). The energy dependence of vulnerable neurons in Huntington's disease, and the parallel with ischaemic and hypoglycaemic brain damage, has been stressed throughout this chapter, and the possible involvement of a glutamatergic neurotoxin in ischaemia and in Huntington's disease reinforces these similarities. It is not certain whether these states are similar because a primary deficit in glutamate metabolism in Huntington's disease mimics the overproduction of excitatory amino acids in ischaemia or hypoglycaemia, or because a fault in the energetic economy of neurons in Huntington's disease produces an intracellular condition akin to hypoglycaemia or ischaemia, resulting in the associated malfunction in excitatory amino acid metabolism. In either case, it is relevant that a biochemical analysis of what actually happens in the Huntington's disease brain independently leads to the conclusion that excitatory amino acids may be involved in the neurotoxic phenomenon.

The search for excitatory amino acid antagonists (mainly for the NMDA receptor) is currently an extremely active field in the pharmaceutical industry. The principal intended goal is to find an agent that will be active in protecting the brain from ischaemic damage following stroke, but it is understood that such drugs might also be useful in other central degenerative disorders where receptor-mediated neurotoxicity is suspected. Many such drugs, whether acting at the NMDA recognition site (CPP) (Murphy et al., 1987) or at the NMDA channel (MK 801) (Wong et al., 1986) are not generally clinically available although they are well advanced in pharmaceutical development. The chronic use of other NMDA channel blockers of the ketamine/phencyclidine family (Kemp et al., 1987), drugs which are already available, is precluded by their hallucinogenic properties. The intense industrial effort now in progress will probably see the development of a number of excitatory amino acid antagonists which hopefully will be useful in degenerative disorders.

At the present time, the most clinically advanced NMDA antagonist is ifenprodil (Vadilex®) which is an anti-ischaemic agent that has recently been demonstrated to have potent NMDA antagonistic potential (Gotti et al., 1988; Carter et al., 1988). The interest in this class of NMDA antagonists is related to their mechanism of action, which although not yet fully understood is clearly distinct from that of either the competitive NMDA receptor antagonists or the psychotomimetic phencyclidine series. Ifenprodil is virtually devoid of behaviourally excitatory effects and will not substitute for phencyclidine in drug discrimination studies (Jackson and Sanger, 1988).

Ifenprodil itself has a limited oral absorption but it is likely that new generation products from this series (e.g. SL 82.0715, Gotti *et al.*, 1988; Carter *et al.*, 1988) will be better adapted. The anti-ischaemic potential of these compounds has been well demonstrated in models of focal ischaemia in rats and cats. Treatment of Huntington's disease with such drugs still remains a future development, but is now a more realistic probability.

As stated in the introduction, the identification of the mutation in Huntington's disease is also feasible in the near future and it will then be possible to define precisely how this provokes neurotoxicity in Huntington's disease, and why the features discussed in this review are manifest. Huntington's disease will then surely serve as a useful base for understanding the phenomenology of neurotoxicity which will have wide applicability in many other degenerative states.

ACKNOWLEDGEMENTS

Some of the work reviewed in this paper was supported by fellowships from the Parkinson's disease Society of Great Britain and the National Huntington's Disease Association (USA), and by research funds from the Association to Combat Huntington's Chorea (to C.J.C.). This work would not have been possible without the help of the MRC Brain Bank at Cambridge, and thanks are especially due to Drs Martin Rossor and Gavin Reynolds for the provision of post-mortem samples.

REFERENCES

Anholt, R. R., De Souza, E. B., Oster-Granite, M. L. and Snyder, S. H. (1985) *J. Pharmacol. Exp. Ther.* **233**, 517–526.

Aronin, N., Cooper, P. E., Lorenz, L. J., Bird, E. D., Sagar, S. M., Leeman, S. E. and Martin, J. B. (1983) *Ann. Neurol.* **13**, 519–526.

Arregui, A., Iversen, L. L., Spokes, F. M. G. and Emson, P. C. (1979a) *Adv. Neurol.* **23**, 517–526.

Arregui, A., Bennett, J. G., Bird, E. D., Tamamura, H. I., Iverson, L. L. and Snyder, S. H. (1979b) *Ann. Neurol.* **2**, 294–298.

Auer, R. N. (1986) *Stroke* **17**, 699–708.

Bachelard, H. S. (1975) In *Brain Work Alfred Benzon Symposium VIII* (edited by D. M. Ingvar and N. A. Lassen), pp. 79–87. Munksgaard, Copenhagen.

Balazs, R. and Cremer, J. E. (1973) In *Metabolic Compartmentation in the Brain.* McMillan Press, London.

Balazs, R., Machiyama, Y. and Patel, A. J. (1973) In *Metabolic Compartmentation in the Brain* (edited by R. Balazs and J. F. Cremer), pp. 57–70. McMillan, London.

Baxter, C. F. (1976) In *GABA in Nervous System Function* (edited by E. Roberts, T. S. Chase and D. B. Tower), pp. 61–87. Excerpta Medica, Amsterdam.

Beal, M. F., Kowall, N. W., Ellison, D. W., Mazurek, M. F., Schwartz, K. J. and Martin, J. B. (1986) *Nature* **321**, 168–171.

Benavidès, J., Fage, D., Carter, C. and Scatton, B. (1987) *Brain Res.* **421**, 167–172.

Benjamin, A. M. and Quastel, J. H. (1975) *J. Neurochem.* **25**, 197–206.

Berl, S. and Frigyesi, T. L. (1969) *Brain Res.* **12**, 444–455.

Bernheimer, H. and Hornykiewicz, O. (1973) *Adv. Neurol.* **1**, 525–531.

Bird, E. D. (1980) *Ann. Rev. Pharmacol. Toxicol.* **20**, 533–551.

Bird, E. D. and Iversen, L. L. (1974) *Brain* **97**, 457–472.

Bird, E. D. and Iversen, L. L. (1977) *Essays Neurochem.* **1**, 177–195.

Bird, E. D. and Kraus, L. J. (1981) In *Transmitter Biochemistry of Human Brain Tissue* (edited by P. Riederer and E. Usdin), pp. 201–220. McMillan, London.

Bird, E. D., Gale, Y. S. and Spokes, E. G. (1977) *J. Neurochem.* **29**, 539–545.

Blass, J. P., Gibson, J. E. and Shimada, M. (1980) In *Biochemistry of Dementia* (edited by P. J. Roberts), pp. 121–134. Wiley, New York.

Brennan, W. A., Bird, E. D. and Aprille, J. R. (1985) *J. Neurochem.* **44**, 1945–1950.

Brierley, J. (1976) In *Greenfields Neuropathology* (edited by W. Blackwood and A. Corsellis). Edward Arnold, London.

Bruyn, E. W. (1968) *Handbook of Clinical Neurology* **6**, 298–348.

Bruyn, E. W., Bots, E. Th. A. M. and Dom, R. (1979) *Adv. Neurol.* **23**, 83–93.

Butterworth, J., Yates, C. M. and Simpson, J. (1983) *J. Neurochem.* **41**, 440–447.

Butterworth, J., Yates, C. M. and Reynolds, G. P. (1985) *J. Neurol. Sci.* **67**, 161–171.

Caro, A. J. and Caro, S. (1978) *Br. Med. J.* **I**, 153–156.

Carter, C. J. (1982) *Life Sci.* **31**, 1151–1159.

Carter, C. J. (1984a) *J. Neurol. Sci.* **66**, 27–32.

Carter, C. J. (1984b) *Neurosci. Lett.* **48**, 339–342.

Carter, C. J. (1985a) *J. Physiol.* **358**, 37P.

Carter, C. J. (1985b) *Biochem. Soc. Trans.* **13**, 958–959.

Carter, C. J. (1987) *Rev. Neurosci.* **1**, 1–34.

Carter, C. J., Savasta, M., Fage, D. and Scatton, B. (1986) *Neurosci. Lett.* **72**, 227–231.

Carter, C. J., Benavides, J., Legendre, P., Vincent, J. P., Noel, F., Thuret, F., Lloyd, K. G., Arbilla, S., Zivkovic, B., MacKenzie, E. T., Scatton, B. and Langer, S. (1988) *J. Pharmacol. Exp. Ther.* **247**, 1222–1232.

Charbonneau, P., Syrota, A., Crouzel, C., Prennat, C. and Crouzel, M. (1986) *Circulation* **73**, 476–483.

Cooper, A. J. L., McDonald, J. M., Gelhard, A. S., Gledhill, R. F. and Duffy, T. E. (1979) *J. Biol. Chem.* **254**, 4982–4992.

Coyle, J. T. (1979) *Biol. Psych.* **14**, 251–276.

Coyle, J. T. and Schwartz, R. (1976) *Nature* **263**, 244–246.

Cross, A. J., Slater, P. and Reynolds, G. P. (1986) *Neurosci. Lett.* **67**, 198–202.

Dubois, A., Benavides, J., Peny, B., Duverger, D., Fage, D., Gotti, B., MacKenzie, E. T. and Scatton, B. (1988) *Brain Res.* **445**, 77–90.

Emson, P. C., Fahrenkrug, J. and Spokes, E. G. S. (1979) *Brain Res.* **173**, 174–178.

Emson, P. C., Arregui, A., Clement-Jones, V., Sandberg, B. E. B. and Rossor, M. (1980a) *Brain Res.* **199**, 147–160.

Emson, P. C., Rehfeld, J. F., Langevin, H. and Rossor, M. (1980b) *Brain Res.* **198**, 497–500.

Engelsen, B. and Fonnum, F. (1985) *Brain Res.* **338**, 165–168.

Fahien, L. A., Lin-Yu, J. H., Smith, S. E. and Happy, J. M. (1971) *J. Biol. Chem.* **246**, 7241–7249.

Foster, A. C., Whetsell, W. O., Bird, E. D. and Schwartz, R. (1985) *Brain Res.* **336**, 207–214.

Francis, A. and Pulsinelli, W. (1982) *Brain Res.* **243**, 271–278.

Gotti, B., Carter, C., Duverger, D., MacKenzie, E. T. and Scatton, B. (1988) *J. Pharmacol. Exp. Ther.* **247**, 1211–1221.

Greenmayre, J. T., Penney, J. B., Young, A. B., D'Amato, C. J., Hicks, S. P. and Shoulson, I. (1985) *Science* **227**, 1496–1499.

Gusella, J. F., Wexler, N. S., Conneally, P. M., Naylor, S. L., Anderson, M. A., Tanzi, R. E., Watkins, P. C., Ottina, K., Wallace, M. R., Sakagnchi, A. Y., Young, A. B., Shenlson, I., Bouilla, E. and Martin, J. B. (1983) *Nature* **306**, 234–238.

Gusella, J. F., Tanzi, R. E., Bader, P. I., Phelan, M. C., Stevenson, R., Hayden, M., Hofman, K. J., Faryniarz, A. G. and Gibbons, K. (1985) *Nature* **318**, 75–78.

Hayden, M. R. (1981) *Huntington's Chorea.* Springer Verlag, Berlin.

Hayden, M. R., Martin, W. R. W., Stoessl, A. J., Clark, C., Hollenberg, S., Adam, M. J., Ammann, W., Harrop, R., Rogers, J., Ruth, T., Sayre, C. and Pate, B. D. (1986) *Neurology* **36**, 888–894.

Henn, F. A. and Hamberger, A. (1971) *Proc. Natl. Acad. Sci.* **68**, 2686–2690.

Hertz, L., Kvamme, E., McGeer, E. G., and Schousboe, A. (1983) In *Glutamine, Glutamate and GABA in the Central Nervous Systems.* Alan R. Liss Inc., New York.

Heyes, M. R., Garnett, E. S. and Brown, R. R. (1985) *Life Sci.* **37**, 1811–1816.

Hicks, S. P. and Coy, M. A. (1958) *Arch. Pathol.* **65**, 378–387.

Hoffer, A. (1976) *J. Orthomolec. Psychiat.* **5**, 169–172.

Hossmann, K. A. (1982) *J. Cereb. Blood Flow Metab.* **2**, 275–297.

Iversen, L. L., Bird, E. D., MacKay, A. V. P. and Rayner, C. N. (1974) *J. Psychiat. Res.* **11**, 255–256.

Jackson, A. and Sanger, D. J. (1988) *Psychopharmacology*, **96**, 87–92.

Jeste, P. V., Barhan, L. and Parisi, J. (1984) *Exp. Neurol.* **85**, 78–96.

Jones, E. G. and Hendry, S. H. C. (1986) *Trends Neurosci.* **8**, 71–76.

Jones, E. G., Coulter, J. D., Burton, H. and Porter, R. (1977) *J. Comp. Neurol.* **173**, 53–80.

Kanazawa, I., Bird, E. D., Gale, J. S., Iversen, L. L., Jessell, T. M. and Muramoto, O. (1979) *Adv. Neurol.* **23**, 495–504.

Kemp, J. A., Foster, A. C. and Wong, E. H. F. (1987) *Trends Neurosci.* **10**, 294–298.

Kish, S. J., Shannak, K. S., Perry, T. L. and Hornykiewicz, O. (1983) *J. Neurochem.* **41**, 1495–1497.

Klintworth, G. K. (1973) *Adv. Neurol.* **1**, 353–368.

Kowall, N. W., Feivante, R. J. and Martin, J. B. (1987) *Trends Neurosci.* **10**, 24–29.

Kuhl, D. E., Phelps, M. E., Markham, C. H., Metter, E. J., Riege, W. H. and Winter, J. (1982) *Ann. Neurol.* **12**, 425–434.

Kuhl, D. E., Metter, E. J., Riege, W. H. and Markham, C. H. (1984) *Ann. Neurol.* **15**, S119–S125.

Lange, H., Thorner, G., Hopf, A. and Schröder, K. F. (1976) *J. Neurol. Sci.* **28**, 401–425.

Lehninger, A. L. (1975) *Biochemistry.* Worth Publishers Inc., New York.

Lewey, F. M. (1950) *J. Neuropath. Exp. Neurol.* **9**, 396–405.

London, E. D., Yamamura, H. I., Bird, E. D. and Coyle, J. T. (1981) *Biol. Psych.* **16**, 155–161.

MacKenzie, E. T., Duverger, D., Gotti, B. and Nowicki, J. P. (1986) In *Pharmacology of Cerebral Ischaemia* (edited by J. Krieglstein), pp. 167–179. Elsevier Science Publishers, Amsterdam.

Martin, J. B. (1984) *Neurology* **34**, 1059–1072.

MacDaniel, H. G., Yeh, M., Jenkins, R., Freeman, B. and Simmons, J. (1984) *Am. J. Physiol.* **246**, 483–490.

McGeer, E. G. and McGeer, P. L. (1976a) *Nature* **263**, 517–519.

McGeer, E. G. and McGeer, P. L. (1978) *Neurochem. Res.* **3**, 501–517.

McGeer, P. L. and McGreer, E. G. (1976b) *J. Neurochem.* **26**, 65–76.

McGeer, P. L., McGeer, E. G. and Fibiger, H. C. (1973) *Neurology* **23**, 912–917.

Meldrum, B. S. (1985) *Clin. Sci.* **68**, 113–122.

Meldrum, B. S., Griffiths, T. and Evans, M. (1982) In *Protection of Tissues Against Hypoxia* (edited by A. Waughler), pp. 275–286. Elsevier Biomedical Press, Amsterdam.

Murphy, D. E., Schneider, J., Boehm, C., Lehmann, J. and Williams, M. (1978) *J. Pharm. Exp. Ther.* **240**, 778–784.

Nemeroff, C. B., Youngblood, W. W., Manberg, P. J., Prange, A. J. and Kizer, J. S. (1983) *Science* **221**, 972–975.

Nicklas, W. J., Duvoisin, R. C. and Berl, S. (1979a) *Brain Res.* **167**, 107–117.

Nicklas, W. R., Nunez, R., Berl, S. and Duvoisin, R. (1979b) *J. Neurochem.* **33**, 839–844.

Norenberg, M. D. (1979) *J. Histochem. Cytochem.* **27**, 756–762.

Perry, E. K., Perry, R. H. and Tomlinson, B. E. (1980) *Neurosci. Lett.* **18**, 105–110.

Perry, T. L. (1982) *Neurosci. Lett.* **28**, 81–85.

Perry, T. L., Hansen, S. and Kloster, M. (1973a) *N. Eng. J. Med.* **288**, 337–342.

Perry, T. L., Hansen, S., Lesk, D. and Kloster, M. (1973b) *Adv. Neurol.* **1**, 609–618.

Reynolds, S. F. and Blass, J. P. (1976) *Neurology* **26**, 625–628.

Reynolds, G. P. and Pearson, S. J. (1987) *Neurosci. Lett.* **78**, 233–238.

Roeder, L. M., Tildon, J. T. and Stevenson, J. H. (1984a) *Biochem. J.* **219**, 125–130.

Roeder, L. M., Tildon, J. T. and Holman, D. C. (1984b) *Biochem. J.* **219**, 131–135.

Roizin, L., Kaufman, M. A., Willson, N., Stellar, S. and Lin, J. C. (1986) *Prog. Neuropathol.* **3**, 447–488.

Rothman, S. M. and Olney, J. W. (1986) *Ann. Neurol.* **19**, 105–111.

Ruscak, M., Orlicky, J., Zubor, V. and Hager, H. (1982) *J. Neurochem.* **39**, 210–216.

Sanberg, P. R. and Johnston, G. A. (1981) *Med. J. Aust.* **2**, 460–465.

Schanne, F. A. X., Kane, A., Young, E. E. and Faber, J. (1979) *Science* **206**, 700–702.

Schousboe, A. (1981) *Int. Rev. Neurobiol.* **22**, 1–45.

Schousboe, A., Svenneby, G. and Hertz, L. (1977) *J. Neurochem.* **29**, 999–1005.

Schuur, A., Changaris, D. G. and Rigor, B. M. (1987) *Brain Res.* **412**, 179–181.

Schwartz, R., Foster, A. C., French, E. D., Whetsell, W. O. and Kohler, C. (1983a) *Life Sci.* **35**, 19–32.

Schwartz, R., Whetsell, O. and Mangane, R. M. (1983b) *Science* **219**, 316–318.

Selak, I., Skaper, S. D. and Varon, S. (1985) *J. Neurosci.* **5**, 23–28.

Shank, R. P. and Campbell, G. LeM. (1981) *Life Sci.* **28**, 843–850.

Shank, R. P. and Campbell, G. LeM. (1982) *Neurochem. Res.* **7**, 601–612.

Shank, R. P. and Campbell, G. LeM. (1984) *J. Neurochem.* **42**, 1153–1161.

Siakotos, A. N. and Munkres, K. D. (1982) In *Ceroid-Lipofluscinosis (Battens Disease)* (edited by D. Armstrong, N. Koppang and J. A. Rider), pp. 167–183. Elsevier Biomedical Press, Amsterdam.

Siesjo, B. K. (1978) *Brain Energy Metabolism.* Wiley.

Silver, M. L. (1949) *J. Neuropath. Exp. Neurol.* **8**, 441–445.

Simon, J. R. and Kuhar, M. J. (1976) *J. Neurochem.* **22,** 93–99.

Simon, R. P., Swan, J. H., Griffiths, T. and Meldrum, B. S. (1984) *Science* **226,** 850–852.

Sorbi, S., Bird, E. D. and Blass, J. P. (1983) *Ann. Neurol.* **13,** 72–78.

Spindel, E. R., Wurtman, R. J. and Bird, E. D. (1981) *N. Engl. J. Med.* **303,** 1235–1236.

Spokes, E. G. S. (1980) *Brain* **103,** 179–210.

Spokes, E. G. S., Garrett, N. J., Rossor, M. and Iversen, L. L. (1980) *J. Neurol. Sci.* **48,** 303–313.

Stahl, W. L. and Swanson, P. D. (1974) *Neurology* **24,** 813–819.

Tellez-Nagel, I., Johnson, A. B. and Terry, R. D. (1974) *J. Neuropath. Exp. Neurol.* **33,** 308–332.

Tildon, J. T. and Roeder, L. M. (1984) *J. Neurochem.* **42,** 1069–1076.

Tucek, S. and Cheng, S. C. (1970) *Biochim. Biophys. Acta* **208,** 538–540.

Van den Berg, C. J. (1970) *J. Neurochem.* **17,** 973–983.

Ward, H. K., Thanki, C. M. and Bradford, H. F. (1983) *J. Neurochem.* **40,** 855–860.

Wastek, G. J. and Yamamura, H. J. (1978) *Mol. Pharmacol.* **14,** 768–780.

Wieloch, T. (1985) *Prog. Brain Res.* **63,** 69–85.

Wong, E. H. F., Kemp, J. A., Priestly, T., Knight, A. R., Woodruff, G. N. and Iversen, L. L. (1986) *Proc. Natl. Acad. Sci.* **83,** 7104–7108.

Wong, P. T. H., McGeer, P. L., Rossor, M. and McGeer, E. G. (1982) *Brain Res.* **231,** 446–471.

Wu, J. Y., Bird, E. D., Chen, M. S. and Huang, W. M. (1979) *Adv. Neurol.* **23,** 527–536.

Young, A. B., Penney, J. B., Starusta-Rubenstein, S., Markel, D., Berent, S., Rothley, J., Bettley, A. and Hichwa, R. (1987) *Arch. Neurol.* **44,** 254–257.

Yu, A. C., Schousboe, A. and Hertz, L. (1982) *J. Neurochem.* **39,** 954–960.

Yu, A. C. H., Drejer, J., Hertz, L. and Schousboe, A. (1983) *J. Neurochem.* **41,** 1484–1487.

35

Tics

A. J. Lees

Tics are abrupt, jerky, purposeless, repetitive movements of discrete muscle groups which last about 100 ms. They are most commonly seen in the head, face and neck. It is possible to suppress them by will power, but this leads to a build-up of internal disquiet. In everyday vernacular a tic refers to a muscular caprice or an eccentric motor habit often associated in the lay mind with a neurotic disposition. In the eighteenth century in France "tique" was used to describe the behavioural vices of domesticated animals and then by the Montpelier physician Boissier de Sauvages who devised a complex classification of spasms and cramps which included 19 distinct varieties of tic. Tics, as we now understand them, were however generally referred to as false chorea or habit chorea or spasm until the publication of Meige and Feindel's influential monograph on tics at the turn of this century. The derivation of the word may be onomatopoeic, creating the feeling of suddenness and abruptness and related to the bookmakers clerk's tick-tack and the tick-tock of a clock; interestingly the word tic is used in almost all European languages and the word "antic" may also have a similar derivation indicating a grotesque gesture or trick.

Tics usually appear for the first time in childhood with a peak mean age of incidence around 7 years. Exceptionally they may occur as early as the first year of life (Burd and Kerbeshian, 1987) or as late as the seventh decade (Klawans and Barr, 1985). They occur three times as commonly in boys as

DISORDERS OF MOVEMENT: CLINICAL, PHARMACOLOGICAL
AND PHYSIOLOGICAL ASPECTS   ISBN 0-12-569685-X

girls and they have been noted in all races and creeds although comparative prevalence figures are not available. They are sudden, stereotyped caricatures of normal voluntary movement which by dint of repetition become less voluntary with time, finally descending to the volitional level of those facial gestures used to signify worry, tiredness or anger. In contrast to these however there is a striking element of compulsive irresistibility to tics which if countered for any period of time inevitably results in a volcanic eruption of inapposite motor energy. Observant ticqueurs say that a vague impulse or feeling precedes the observable movement and it is this which is involuntary, whereas the tic itself remains under a considerable degree of voluntary control.

Tics are characterized by their situation specificity but in general are made worse by boredom, anxiety, excitement or fatigue. They are often at their worst in the home or in the doctors' waiting room whereas in the consulting room itself or at work or school they may be inconspicuous. They may also be triggered by a particular individual towards whom the ticqueur feels strong emotions and there is also a strong element of suggestibility. Engrossing mental or physical activities, alcohol or coitus may temporarily reduce their frequency. Although they generally subside during sleep it has been shown with polysomnographic studies that they may occur in all stages of sleep.

In contrast to abnormal involuntary movements believed to originate in the basal ganglia, such as chorea or dystonia, tics are not invariably exacerbated by voluntary movement of an unaffected body part. Local trauma or irritation such as an episode of conjunctivitis or sinusitis may on occasions be the soil for tic germination. The commonest reported tics are eye winking, head tossing, shoulder shrugging, grimacing and lip and tongue protrusion. Common vocal tics include throat clearing, sniffing, barking and squeaking. Sudden jerks of the arms, trunk and legs closely resembling myoclonus are also quite common and the variety of tics is limited only by the number of voluntary movements the human body can produce. Isolated leg tics are uncommon. An association may exist between tics, frequent blinking and blepharospasm although there is enormous scope for diagnostic confusion in this regard (Elston et al., 1989).

It is estimated in Western Europe and North America that as many as one in ten schoolboys may have a transient tic. The natural history is generally favourable, most ticqueurs remitting spontaneously in adolescence. In one study from Copenhagen (Torup, 1962) half the studied group were free from tics after an average follow-up period of 9 years from diagnosis, a further 46% had improved and only 6% were unchanged. The average age at follow-up was 18 years and the mean age when the tics ceased was 13 years, after a mean duration of 5 years. A second study carried out in the

United Kingdom showed none of 73 children was worse after a mean follow-up period of 5 years, 30 (40%) had completely recovered, 39 (53%) were improved and four (6%) were unchanged, with a mean average age at follow-up of 15 years. In this study a greater proportion of complete remissions occurred in those cases which began between the ages of 6 and 8 years, and vocal tics had a worse prognosis (Corbett *et al.*, 1969).

DIFFERENTIAL DIAGNOSIS

Tics need to be distinguished from a number of other voluntary movements which differ in appearance, but which are frequently seen together with tics and which may stem from similar roots. These can be listed in order of their volitional level in Jacksonian fashion, alternatively on the basis of their complexity or finally with regard to the degree of awareness of an associated non-motor impulse, idea or image.

A *gesture* is a body movement used as a form of silent emotional communication. These range from the voluntary smile or wink through to less voluntary physiognomy.

A *mannerism* is an unusual affected gesture which is more or less peculiar to the individual. Examples would be lifting one leg up or shaking hands with two fingers.

A *compulsion* is a stereotyped motor behaviour preceded or accompanied by an irresistible inner urge which may provoke subjective resistance and acknowledgement of its absurdity. Examples would include touching particular objects, picking up scraps of dust from the floor or constantly sniffing under one's armpits.

A *ritual* is an elaborate, systematized compulsive behaviour. An example of this would be brushing each tooth 33 times, then washing one's mouth out with salt three times before leaving the bathroom.

A *habitual manipulation of the body* is a self-gratifying, compulsive, often socially offensive action which occurs at particular times of anxiety, boredom, tiredness or self-consciousness. Examples would be thumb-sucking, hair twiddling or nose picking.

A *stereotypy* is a purposeless, mechanical movement carried out in a uniformly repetitive way often for very long periods of time and to the exclusion of all other body movements. Examples of this would be body rocking and head banging.

Akathisia is an associated inner tension and mental discomfort with constant, often stereotyped motor restlessness (pacing, marching on the spot, leg crossing). This condition is most commonly seen following sustained neuroleptic treatment or in patients with Parkinson's disease.

In contrast to the abnormal involuntary movements of chorea, tremor, myoclonus and dystonia, mannerisms and compulsions, stereotypies and tics may be regarded as abnormal voluntary movements, and gestures are part and parcel of the normal motor compendium. Irresistibility is a feature of tics, stereotypies, compulsions and rituals and akathisia. Stereotypies in particular seem to be under very little voluntary control. Compulsions and rituals usually occur in direct response to some well-formulated recurring mental intrusion and they have been regarded by behavioural therapists as an avoidance response to an irrational fear. The abnormal feeling is less clearly defined with akathisia and tics and is most commonly described as an inner tension or malaise.

GILLES DE LA TOURETTE SYNDROME

Gilles de la Tourette syndrome or "maladie des tics" is not a separate disease, but simply the most severe end of the spectrum of tics. Its prevalence may be at least as high as 29 per 100000 in childhood (Caine *et al.*, 1988). Diagnosis requires the presence of multiple motor and vocal tics and I have seen cases presenting as late as 35 years. Common associated clinical features include compulsions and rituals, stereotypies, coprolalia, copromimia, echo phenomena, self-injurious behaviour and specific learning disabilities. A few patients exhibit abnormal movements which are clinically indistinguishable from dystonia, chorea or myoclonus and a few others have an exaggerated startle response. There appear to be no major cross-cultural differences in the clinical expression of the condition although the phenomenology of exhibited gestures obviously varies from country to country as does the incidence and nature of copro-phenomena. The Gilles de la Tourette syndrome might be re-defined as a perverse behaviour pattern occurring in response to an inner feeling against the need for self-restraint in situations where a lack of restraint leads to distress or injury, with a spectrum of symptoms ranging from simple tics at one end to complex compulsions at the other. A quote from O, the celebrated patient described in Meige and Feindel's textbook, provides further insight:

> We who tic are consumed with a desire for the forbidden fruit. It is when we are required to keep quiet that we are tempted to restlessness, when silence is compulsory that we feel we must talk... Moreover we abhor a vacuum and fill it as we may. Various are the artifices we might employ such for instance as speaking aloud, but that is much too obvious and does not satisfy. To make a little grunt or sigh, what a comfort in a tic like that.

Individual tics can often be triggered by discussion of the condition. This is one of the reasons why some ticqueurs are reluctant to attend lay meetings. Vocalizations are usually primitive, emotionally charged utterances commonly ejaculated in a loud, staccato, aggressive tone. They have been considered as ontogenic fragments of speech. Some patients have severe vocal tics and very few muscular tics and recurring scatological ideation may occur in isolation or in association with coprolalia.

NEUROPHYSIOLOGICAL ABNORMALITIES

Tics consist of short bursts of muscle activity between 50 and 200 ms, indistinguishable electromyographically from myoclonus and chorea in many instances. Reciprocal activation of antagonist muscles may be present. Marsden and colleagues have reported that simple tics are not preceded by a normal negative premovement EEG potential and suggest that simple tics are physiologically distinct from normal self-paced voluntary movements (Obeso et al., 1981). Shibasaki et al. (1981) also failed to record an early potential shift before EMG onset (Bereitschafts potential), but noted a normal potential shift associated with the final phase of preparation (NS^1) 70–100 ms before EMG onset. Abnormalities have also been reported in the recovery cycle of the blink reflex in some patients similar to those seen in blepharospasm (Tolosa et al., 1986). We have also recently examined the blink reflex in 25 patients with Gilles de la Tourette syndrome using paired supra-orbital shocks with interstimulus intervals of 500 ms and 1 s, eight paired shocks being given per patient and allowing an interval of at least 10 s between each one. The average amplitude of the R2 component of the test stimuli is then expressed as a percentage of the initial shock, giving an R2 per cent for the two interstimulus intervals. In normal subjects when the interval between the initial and test stimulus is 1 s, the amplitude of the test R2 is decreased to about 30% of the initial R2 amplitude and to about 10% when the interval was 500 ms. In blepharospasm the response is unaffected when the interval between test and initial stimulus is 1 s and only moderately decreased to about 70% when the interval is 500 ms (Berardelli et al., 1985). We have found a similar abnormality in nine (25%) of the patients with Gilles de la Tourette syndrome, but with an additional finding that the excitability cycle returns to normal when the patient is asked to voluntarily suppress the eye blink, supporting Tolosa's observations. In addition we have found a further group of seven patients who when compared with age-matched controls appear to have a "super-normal" response with complete suppression of the test R2 at 1 s. Studies are now under way with longer interstimulus intervals between the paired shocks in this group. (Smith and Lees, 1989).

Abnormalities of sleep may occur in patients with Gilles de la Tourette syndrome, including difficulty in getting off to sleep, frequent waking, somnabulism and night terrors. Overnight telemetric sleep studies have shown a significantly increased percentage of stage 3–4 sleep and a reduction in the percentage of REM sleep, and motor tics have been recorded during all stages of sleep, but are less common in stages 3 and 4 (Glaze *et al.*, 1983). At the National Hospital for Nervous Disease 11 patients with Gilles de la Tourette syndrome have so far been studied, three of whom have been shown to have severely disrupted sleep patterns with many arousals and wakenings. It was confirmed that tics do occur during sleep, but are much less frequent than when awake. In contrast to previous claims the occurrence of tics during sleep does not differentiate them from other movement disorders. It is possible that the appearance of tics during sleep may be related to arousal phenomena (Fish and Lees, unpublished observations).

IS GILLES DE LA TOURETTE SYNDROME AN OBSESSIONAL COMPULSIVE DISORDER?

The celebrated ticqueur Samuel Johnson was hidebound by a multitude of obsessions and compulsions and in 1905 Meige wrote:

> The irregularity and insufficiency of cortical control favours the appearance and development of fixed ideas, of impulsions and obsessions. In fact, one finds in ticqueurs all types and severities of these mental disorders. Their co-existence with tics is important to be aware of, as inappropriate movements are very often connected with these mental problems.

Pierre Janet (1903) considered tics to be forced agitations, external manifestations of an underlying phobic anxiety or obsessional disorder. He considered that tics were degraded fragments of a compulsive movement and saw them as systematized movements reproduced regularly, but in an entirely useless, inappropriate fashion in response to some uncontrollable inner obligation. Fifteen of Janet's patients presented with tics only to lose them and experience subsequent depression, obsessional ideas or phobias, although in general he was impressed by the fact that tics and obsessions tended to develop simultaneously. He saw them arising out of a special mental state of incompleteness of psychological insufficiency, often expressing a symbolic meaning such as a need for precision, affection or compensation. In my own experience about 30% of patients with Gilles de la Tourette syndrome have intrusive, senseless repetitive thoughts, images or impulses that are unwanted or unacceptable and give rise to stress and subjective resistance. The content is usually worrying, repugnant, blasphemous, obscene or nonsensical. Hassler also believed that tics were a motor manifestation of an inner obsession and carried out bilateral stereotactic surgery of the intralaminar nuclei of the thalamus and median territory of the thalamus in three patients with Gilles de la Tourette syndrome and severe obsessional compulsive disorder with excellent results (Hassler and Dieckmann, 1970). Fifty-eight per cent of a group of American patients with Gilles de la Tourette syndrome and 45% of a British group had scores on the Leyton obsessional questionnaire above 70, which is the standard cut-off point for diagnosis of obsessional compulsive disorder. This compared with a figure of 82% for primary obsessive compulsive disorder patients and 12% of normal controls (Frankel *et al.*, 1986). In the National Hospital study 33 (35%) of 90 patients with Gilles de la Tourette syndrome admitted to obsessional compulsive behaviour and rituals. Significantly higher scores were obtained on the Crown Crisp experimental index obsessional subtest and the questionnaire of the Leyton obsessional inventory than controls and high scores appeared to be correlated with the presence of coprolalia, compulsive touching and echolalia (Robertson *et al.*, 1988). Common obsessions included fear of causing harm to oneself or others, arithromania, bizarre sexual ideation, matching imagery and doubting. Common compulsions included touching self or objects, checking behaviour and self-injurious rituals. In contrast to primary obsessional compulsive disorder, fear of contamination and cleaning rituals were uncommon which may reflect a sex difference as it is known that these behaviours are particularly common in females (Dowson, 1977). One of my patients had a list of words which produced within him a feeling of fear and which had to be neutralized as soon as they occurred. Examples included "black", "poltergeist", "jack", "stab", "devil" and "six". For example the word "black" could be neutralized by him thinking the word "white". About 10% of patients with primary obsessional compulsive disorder have tics and clumsiness has been reported to occur in both conditions. Furthermore, recent careful analysis of pedigree information from 30 families with Gilles de la Tourette syndrome has suggested that tics may be inherited as an autosomal dominant disorder and that obsessional compulsive disorder may be part of the Gilles de la Tourette spectrum of behaviour (Pauls and Leckman, 1986). It is also recognized that obsessional thoughts and tics may occur together occasionally in the neurobehavioural sequelae of epidemic encephalitis lethargica. I have also seen a patient with Gilles de la Tourette syndrome who has recurrent neuroleptic-induced oculogyric crises associated with obsessions.

    Further recent support for Janet's original notion that obsessional behaviour invades a psychological vacuum comes from a series of case reports in which bradyphrenia has been associated with stereotyped movements in the context of localized structural damage in the basal ganglia. Laplane *et al.*

(1984) reported three patients, two with carbon monoxide poisoning and another with a wasp sting encephalopathy, who had lesions on the CT scan localized to the globus pallidus. These patients exhibited severe psychic akinesia with stereotyped compulsive movements and arithromania. Two similar patients were reported in another study following carbon monoxide poisoning, both of whom had bradyphrenia, one of whom hoarded toilet rolls and the other had coprolalia (Ali-Cherif et al., 1984).

In addition to the basal ganglia, dysfunction of the anterior cingulate cortex may also be involved in tics and obsessions. Bilateral anterior cingulotomies improve refractory obsessional compulsive disorders in 50% of cases and this surgical procedure has recently been carried out in one of my patients with Gilles de la Tourette syndrome and self-injurious compulsions with excellent results. Stimulation of this dopamine-containing area in man has been reported to cause stereotyped movements including licking, rubbing of the fingers, touching of the lips and to and fro movements. In the squirrel monkey electrical stimulation may cause vocalizations. It has also been suggested that some areas of the limbic system may be involved in emotionally charged speech. Extensive connections exist between the limbic system and the striatum and disturbances of these areas may be of particular importance in the mediation of obsessional behaviour and tics.

It can be concluded that tics and obsessions frequently occur together and that often the two phenomena are intimately related although on other occasions there seems to be little connection between them and treatment with neuroleptic drugs may abolish the abnormal movements without improving the underlying intrusive ideas. Tics may be regarded as one external manifestation of an underlying obsessional disorder. Unfortunately neurological understanding of obsessional compulsive behaviour has advanced little since the early formulations of Jackson, Tuke and Janet and the recognization by von Economo in 1929 of the relevance of subcortical areas in these disturbances. The neurological investigation of obsessional behaviour may well provide new insights into the function of the brain and provide greater understanding of phenomena such as perseveration, echolalia and imperative ideation in epilepsy.

REFERENCES

Ali-Cherif, A., Royere, M. L., Gosset, A., Poncet, M., Salamon, G. and Khalil, R. (1984) Rev. Neurol. 140, 401–405.
Berardelli, A., Rothwell, J. C., Day, B. L. and Marsden, C. D. (1985) Brain 108, 593–608.
Burd, L. and Kerbeshian, J. (1987) Am. J. Psychiat. 144, 1066–1067.
Caine, E. D., McBride, M. C., Chiverton, P., Bamford, K. A., Rediess, S. and Shiao, J. (1988) Neurology 38, 472–475.

Corbett, J. A., Matthews, A. N., Connell, P. H. and Shapiro, D. A. (1969) *Br. J. Psychiat.* **115**, 1229–1241.

Dowson, J. H. (1977) *Br. J. Psychiat.* **131**, 75–78.

Elston, J. S., Cas Granje, F. and Lees, A. J. (1989). *J. Neurol. Neurosurg. Psychiat.* **52**, 477–480.

Frankel, M., Cummings, J. L., Robertson, M. M., Trimble, M. J., Hill, M. A. and Benson, D. F. (1986) *Neurology* **36**, 378–382.

Glaze, D. G., Frost, J. D. and Jankovic, J. (1983) *Neurology* **33**, 586–592.

Hassler, R. and Dieckmann, G. (1970) *Rev. Neurol.* **123**, 89–106.

Janet, P. P. (1903) In *Les obsessions et la psychasthénie*. Baillière, Paris.

Klawans, H. L. and Barr, A. (1985) *Arch. Neurol.* **42**, 1079–1080.

Laplane, D., Bavlac, M., Widlocher, D. and Dubois, B. (1984) *J. Neurol. Neurosurg. Psychiat.* **47**, 377–385.

Obeso, J. A., Rothwell, J. C. and Marsden, C. D. (1981) *J. Neurol. Neurosurg. Psychiat.* **44**, 735–738.

Pauls, D. L. and Leckman, J. F. (1986) *New Engl. J. Med.* **315**, 993–997.

Robertson, M. M., Trimble, M. and Lees, A. J. (1988). *Brit J. Psychiat.* **152**, 383–390.

Shibasaki, H., Kuroiwa, Y., Barrett, G., Halliday, E. and Halliday, M. (1981) Abstract 12th World Congress of Neorology, Kyoto, Japan No. 3221.

Smith, S. and Lees, A. J. (1989). *J. Neurol Neurosurg. Psychiat.* In press.

Tolosa, E., Montserrat, L. and Bayes, A. (1986) *Neurology* **36** (suppl. 1), 118–119.

Torup, E. (1962). *Acta Paediatr.* **51**, 261–268.

36

Classification of Tremor

Leslie J. Findley and Lynn Cleeves

DEFINITION OF TREMOR

Tremor is most simply defined as a rhythmical oscillation of a body part. Some authors have included additional terms such as "continuous" and "involuntary" (see review by Brumlik and Yap, 1970) but these do not characterize all types of tremor. For example the classical rest tremor of Parkinson's disease may cease when the patient is completely relaxed and unaroused and tremor may be a "voluntary" phenomenon (see below).

Tremor is a periodic movement about an axis which distinguishes it from other movement disorders such as chorea, myoclonic jerks and tics which may not have a fixed period or which may involve complex movements other than simple oscillations. Certain involuntary movements such as clonus and rhythmical myoclonus may be transient or sustained over long periods of time and may be indistinguishable from tremor. Palatal and ocular myoclonus involve sustained oscillations and should probably be classified as tremor (see Gresty and Findley, 1984).

PROBLEMS IN THE CLASSIFICATION OF TREMOR

Tremor may be a normal phenomenon (physiological tremor) which is not clinically visible and causes no disability except when performing at the

DISORDERS OF MOVEMENT: CLINICAL, PHARMACOLOGICAL
AND PHYSIOLOGICAL ASPECTS   ISBN 0-12-569685-X

limits of manual dexterity (e.g. microsurgery or watch-making). On occasions, for example in moments of intense fear or excitement, physiological tremor may be enhanced to a degree which causes clearly visible shaking. This enhanced physiological tremor may be considered normal insofar as it is transient and not indicative of pathology, or abnormal insofar as it can cause significant disability (e.g. in snooker players or musicians). Tremors indistinguishable from enhanced physiological tremor can be a sign of metabolic disturbance or nervous system pathology. Thus, there is often no clear distinction between normal and pathological tremor.

The classification of tremor is essentially phenomenological and a degree of overlap in the characteristics of tremors encountered in different disorders may give rise to problems in diagnosis. Tremors may be classified on a number of different dimensions, some purely descriptive (e.g. aetiological classification, behavioural types), some having implications for pathophysiology (e.g. frequency, response to drugs). There is, however, no single classification scheme which can be used diagnostically.

PHYSIOLOGICAL TREMOR AND ENHANCED PHYSIOLOGICAL TREMOR

Physiological tremor results from a number of interacting factors including cardioballistic thrust and mechanical resonance, motor neuron firing, synchronization around peripheral reflexes from muscle spindle feedback, supraspinal influences and pharmacological influences (see Marsden, 1984). With the limb completely supported and at rest the major factor in the generation of tremor is the interaction of cardioballistic thrust and resonant properties of the limb. When the limb is in posture the observed tremor is predominantly neurogenic in origin. The initial rate of motor neuron firing (around 8 Hz) is below the tetanic fusion frequency of human muscle (Freund et al., 1975) thus producing an oscillation of the limb (8–12 Hz in the outstretched hands).

Physiological tremor may become enhanced and symptomatic in normal individuals. Enhanced physiological tremor is of greater than normal amplitude but of normal frequency. It is present on maintaining posture and persists during movement. Enhanced physiological tremor commonly results from increased peripheral β-adrenergic activity due to increased levels of circulating catecholamines or alteration in responsiveness of peripheral β-2 receptors (or both). Such receptors are present in extrafusal muscle fibres and muscle spindles. Stimulation of these receptors increases the speed of contraction in human muscle, thus raising the tetanic fusion frequency and resulting in greater oscillation from units firing at their initial rates. The output of muscle spindles is also altered. Both effects increase the burst size

Table 36.1   Common causes of enhanced physiological tremor.

Anxiety States	Methylxanthine administration
Emotional stress	Drug withdrawal states
Thyrotoxicosis	Alcohol
Phaeochromocytoma	Alcohol withdrawal
Catecholamine infusion	

of group 1a spindle impulses that occur during the lengthening phases of the small ripples of contraction of human forearm muscle (Hagbarth and Young, 1979).

Central mechanisms can also influence normal tremor, though these are less clearly understood (Sutton and Sykes, 1967).

Common causes of enhanced physiological tremor are presented in Table 36.1. Factors which exacerbate normal tremor will also exacerbate pathological tremors.

CLASSIFICATION BY AETIOLOGY

The most obvious classification for pathological tremors is that based on underlying aetiology. However, any tremulous disorder may give rise to more than one type of tremor and no single type is pathognomonic of a particular disorder. Thus, the classification of tremor based solely on aetiology may not always be useful. For example, several tremulous phenomena can be identified in Parkinson's disease (Findley et al., 1981) (see below). It is therefore unhelpful, particularly in assessing efficacy of treatment, to refer to "parkinsonian tremor" without more precise specification of the phenomena under consideration.

Common tremors (classified by frequency and behavioural characteristics) encountered in different tremulous disorders are summarized in Table 36.2.

CLASSIFICATION BY BEHAVIOURAL CHARACTERISTICS

Tremors are most commonly classified according to the behavioural circumstances in which they appear. Terminology in the literature has not been consistent. Thus, the following classification has been proposed (see Findley and Capildeo, 1984).

Rest tremor

Tremor occurring when the subject is not voluntarily activating the muscle groups concerned (though involuntary muscle tone is often present).

Table 36.2 Frequency classification of common tremors.

Frequency (Hz)	Disease of process/locus of lesion	Behavioural characteristics
2.5–3.5	Cerebellar/brain stem	Postural, movement
	Multiple sclerosis	Postural, movement
	Alcoholic degeneration	Postural, movement
	Post-traumatic	Postural, movement
4–5	Parkinson's disease	Rest
	Cerebellar disease	Posture, movement
	Rubral	Rest, posture, movement
	Drug-induced (e.g. neuroleptics, ?MPTP)	Rest
5.5–7.5	Essential tremor	Postural, movement
	Clonus	
	Parkinson's disease	Postural, movement
	Drug-induced (e.g. sodium valproate, lithium, tricyclics)	Postural, movement
8–12	Enhanced physiological tremor	Postural, movement
	Drug intoxications	Postural, movement
	Essential tremor	Postural, movement
	? cerebro-cortical	Postural, movement

The term "static tremor" should be avoided as it has been used, by different authors, to refer both to rest tremor and to tremor provoked by maintenance of posture.

Rest tremor is most commonly encountered in Parkinson's disease. It is often complex, involving several muscle groups and producing oscillation at more than one joint (e.g. the classical "pill-rolling" action produced by tremor of the hand, thumb and finger). Typically, the rest tremor disappears with the intention to move and during the movement itself, but may become re-established with the maintenance of posture. When this occurs, spectral analysis shows the tremor in posture to be identical in frequency and waveform characteristics to the tremor at rest and it can be presumed that the two share a common mechanism.

Rubral (or cerebellorubral) lesions may produce a rest tremor indistinguishable from that of Parkinson's disease. However, rubral tremor usually persists, unaltered in frequency and waveform, thoroughout movement and in posture. Amplitude may remain constant or become exacerbated at the termination of goal-directed movement.

Some drugs (e.g. neuroleptics) may produce a rest tremor similar to that seen in Parkinson's disease, though more often an action tremor is produced (Pinder, 1984).

Action tremor

Tremor occurring on voluntary contraction of muscle. This includes *postural tremor* (tremor provoked by the maintenance of posture) and *movement tremor* (tremor provoked by any form of movement). Movement tremor may occur at the initiation of movement (initial tremor), during movement (transitional tremor) or at the termination of movement (terminal tremor). Initial, transition and terminal tremor often occur together and may share a common mechanism. Movement tremor may, however, only become apparent (or exacerbated) during one of these phases. The most common causes of action tremor are enhanced physiological tremor and essential tremor. Essential tremor is predominantly a tremor of posture but it may persist during movement and may become exacerbated at the termination of goal-directed movement. When severe it may persist when the limb is supported and stationary. This results from a degree of residual muscle tonus in the supported limb and can be shown by spectral analysis to be identical in frequency and waveform to the tremor observed in posture.

Almost all drugs acting on the nervous system can produce tremor, usually action tremor, as a side-effect (Table 36.3), but their multiplicity of actions provides few clues as to the pharmacological basis of common tremulous states.

The classical term "intention tremor", though firmly established in the literature, can be misleading since any form of voluntary muscular activity is intentional. The term has typically been applied to the large-amplitude, low-frequency tremor at the termination of goal-directed movement seen in cerebellar disease. The frequency of these tremors is not finely tuned and amplitude varies greatly. In some patients these movements may be more appropriately classified as ataxia or dysmetria (see Gresty and Findley, 1984).

Lesions involving the deep cerebellar nuclei, or their outflow pathways, tend to produce well-organized action tremors with exacerbation at the termination of guided movements. More widespread lesions of the cerebellum and its connections, particularly with involvement of cerebellar cortex,

Table 36.3 Common drugs producing tremor.

Transmitter system affected	Drugs
Central cholinergic	Acetylcholine, muscarinic and nicotinic agonists, anticholinesterases, aminopropanols
Central monoaminergic	Neuroleptics, phenylethylamines, indoles
Peripheral adrenergic	Adrenaline, β-agonists, lithium, caffeine, amphetamine, corticosteroids
Other	Heavy metals, metal chelators, carbon tetrachloride

produce a large dysmetric component (abnormal, corrective movements).

The combination of tremor and dysmetria we would classify as ataxia. This is best seen in patients with multiple sclerosis or in patients with post-traumatic brain stem syndromes in which there are widespread and multiple areas of damage to the cerebellum, brain stem and cerebellum–brain stem connections. In extreme examples, there is total disorganization of voluntary movement designated decomposition of movement by Holmes (1904).

As tremor is defined as a rhythmical phenomenon with a relatively fixed period of oscillation, it is appropriate that certain "passive" physical signs be classified as tremor. Behavioural studies have clearly shown these phenomena to be intimately related to tremorgenic processes (Findley *et al.*, 1981; Salisachs and Findley, 1984).

The *cogwheel phenomenon* refers to a rhythmical, repetitive interruption of passive movement of a limb about a joint. The term "active" cogwheel phenomenon has been used to describe repetitive interruptions to movement carried out by the patient. Some have used the term "cogwheel rigidity", assuming the phenomenon is pathognomonic of parkinsonism. It may, however, occur in patients with a variety of other tremors.

Froments sign is a rhythmical resistance to passive movement of a limb about a joint that can be detected specifically when there is voluntary activity of the contralateral limb. This phenomenon may be seen in tremulous disorders other than Parkinson's disease. For example, in a recent survey of 237 essential tremor patients, Froment's sign was found in 72 (30%) (Cleeves *et al.*, 1988).

CLASSIFICATION BY FREQUENCY

By virtue of its rhythmicity, tremor lends itself well to accurate transduction and quantification. Using a variety of objective techniques (see Brumlik and Yap, 1970; Gresty and Findley, 1984) tremor can be characterized in terms of frequency, amplitude and waveform.

Amplitude determines symptomaticity but is of no value in classification. The amplitude of any tremor can be influenced by a wide range of physiological, psychological and environmental factors (Cleeves *et al.*, 1986). Even when attempts are made to identify and minimize such influences, considerable variability can still be seen in the amplitude of pathological tremors throughout the day and from day to day (Cleeves *et al.*, 1986; Cleeves and Findley, 1987).

The tremor waveform reflects the specific sequence and timing of muscle activity involved in the tremor cycle. A "typical" waveform may be described for tremors of different aetiologies (Fig. 36.1). There is, however, no pattern

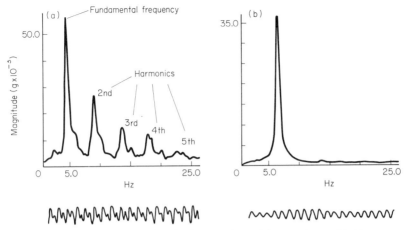

Fig. 36.1 Raw records (lower) and derived spectra (upper) from patients with (a) rest tremor of Parkinson's disease and (b) essential tremor.

which is entirely unique to any one disorder. Bearing this in mind, waveform analysis (see Gresty and Findley, 1984) can nonetheless be a valuable aid in differential diagnosis.

In pathological tremors, frequency is the most stable of the above parameters. Frequency has been presumed to reflect the nature of the "oscillator" and thus has been used as a major criterion of classification.

Holmes (1904) first suggested that the periodicity of tremor may have an important relationship with the nature of the underlying mechanism and be of value in the diagnosis of specific pathophysiology. The frequency of pathological tremors, when recorded over short periods of time under comparable conditions, is relatively constant from recording to recording in individual patients (Findley *et al.*, 1981; Cleeves and Findley, 1987). Considerable changes in frequency have been reported using spectral analysis of EMG activity during long-term monitoring of tremor (Scholz, personal communication). This may, however, reflect artifacts produced by using this method in the freely moving limb.

Frequency bands encountered (in the upper limb) in common tremulous disorders are presented in Table 36.2. These can form the basis of a general guide to diagnosis but it should be borne in mind that the boundaries between bands are somewhat arbitrary and some overlap can occur.

Distinct tremors in different frequency bands can be seen in patients with Parkinson's disease. Common tremor groupings and putative mechanisms are shown in Table 36.4.

Table 36.4 Tremulous phenomena in Parkinson's disease (from Findley and Gresty, 1984).

Classification	Frequency (Hz)	Putative mechanism
Rest tremor	4–5.5	Rhythmical instability in thalamic neurons
Postural tremor, clonus, cogwheel phenomena	ca 6	Spinal mechanisms ? related to clonus
Intermediate-frequency postural tremors	6.5–8.5	? Synchronization from CNS
High-frequency postural tremor, high-frequency cogwheel ("ripple")	9.5–11	Long-term synchronization involving stretch reflexes

"CENTRAL" AND "PERIPHERAL" TREMORS

A gross classification of tremor can be made according to the level, within the nervous system, of the predominant tremorgenic mechanism.

Visible, symptomatic tremor arises from synchronization of discharges of many motor units. Such synchronization results from complex interactions between mechanical properties of the body part, afferent feedback around peripheral reflexes, long-latency reflexes and central "oscillators" (Stein and Lee, 1981). (The term oscillator, as used here, does not presume a single anatomical structure but refers to the sum total of neural influences which achieve expression as tremor through the "final common pathway" of the α-motor neurons.)

The relative influence of these processes may vary but it is possible, in any individual tremor, to assess the contribution of peripheral reflex feedback. If an oscillatory torque, of similar power to that of the tremor, is applied at a freqency slightly above or below the tremor frequency and slowly adjusted to pass through the frequency of the tremor, the tremor may become "entrained" (i.e. the frequency shifts up or down with the frequency of the input but remains regularly sinusoidal). Alternatively the tremor may remain independent of and uninfluenced by the input, resulting in "beating" of the limb as the phase relationship of the input and the ongoing tremor alternately enhances and damps the mechanical response (Fig. 36.2). It can be concluded that, in the former case, the tremor is highly dependent on peripheral drive, whereas in the latter the tremor is generated predominantly by an independent central oscillator (Lee and Stein, 1981).

On the basis of this differential response to peripheral input Marsden has identified subgroups of essential tremor termed type 1 and type 2 (Marsden et al., 1983) (Table 36.5). Type 1 essential tremor is clinically indistinguishable from enhanced physiological tremor (EPT) but occurs in the absence of factors known to produce EPT (Table 36.1). Type 2 is generally of lower

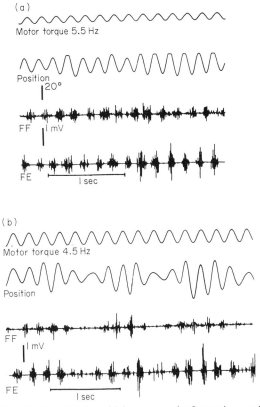

Fig. 36.2 Effect of application of a sinusoidal torque to the fingers in a patient with essential tremor. The traces, from top down, show motor torque, finger position, and EMG recorded from the finger flexors (FF) and finger extensors (FE). In (a) the sinusoidal oscillation was delivered to the finger at 5.5 Hz, the frequency of the patient's spontaneous tremor. The oscillation produced remained more or less sinusoidal as judged from the position trace. In (b) the finger was driven at 4.5 Hz, below its normal tremor frequency. "Beating" can be seen in the position record, suggesting the existence of some central generator. (Reproduced from Marsden, 1984.)

frequency and greater amplitude than type 1. The frequency of type 1 essential tremor can be entrained by the method described above whereas type 2 is resistant to the external input. Longitudinal studies are required to ascertain whether type 1 essential tremor eventually develops into type 2. Elble (1986) has shown that tremor frequencies typical of type 1 and type 2 ET can exist in the same families. Furthermore he has demonstrated, using inertial loading experiments, a frequency-invariant component between 8 and 12 Hz in some normal individuals which is dependent upon synchronous

Table 36.5 Categories of essential tremor (from Marsden *et al.*, 1983).

	Type 1, benign exaggerated physiological tremor	Type 2, benign pathological essential tremor	Type 3, severe pathological essential tremor	Type 4, symptomatic essential tremor
Site	Arms	Arms ± Head ± Legs	Arms Head Legs	Arms
Type	Postural	Postural	Postural	Postural
Frequency (Hz)	8–12	5–7	4–6	5–9
Familial	—	+ + or −	+/−	+/−
Source	Enhanced stretch reflex	? Central oscillator	? Central oscillator	Various
Other causes	Anxiety Alcohol Thyrotoxicosis Drugs			Peripheral neuropathy Torsion dystonia Parkinson's disease

neuronal activity. This component may represent a forme fruste of essential tremor.

Lee and Stein (1981) investigated the relative contributions of peripheral and central factors in essential tremor and parkinsonian rest tremor by evaluating the extent to which the phase of the tremor was reset by mechanical displacement of the wrist joint. The mean resetting index (0 = no resetting; 1 = complete resetting) of essential tremor was 0.64 and for parkinsonian tremor 0.16, indicating a greater central component in the latter.

CLASSIFICATION BY ELECTROMYOGRAPHY (EMG)

Tremor may arise from alternating contractions of antagonist muscle groups, from co-contraction of muscle groups or from rhythmic contractions in a single agonist muscle.

The classical resting tremor of Parkinson's disease invariably arises from alternating contractions of flexor and extensor muscles. All types of EMG activity may be seen in essential tremor and several authors have suggested that this might form the basis of a subgrouping of essential tremor (Sabra and Hallet, 1984; Deuschl *et al.*, 1987a) or for predicting the future development of Parkinson's disease (Shahani and Young, 1978).

Sabra and Hallett (1984) found that essential tremor patients with alternating muscle activity were less responsive to propranolol than those showing co-contraction. Deuschl *et al.* (1987a) described two subgroups of essential tremor patients. Group A was characterized by synchronous tremor bursts in antagonists or in antigravity muscles alone, normal long-latency reflexes and good response to propranolol. Group B was characterized by reciprocal EMG activity in antagonists, abnormal long-latency reflexes and lack of response to propranolol.

In our opinion, however, the pattern of EMG activity is not generally useful for diagnosis, prognosis or predicting response to drugs in essential tremor. In some essential tremor patients monitored over long periods, we have seen the pattern of EMG activity in the same antagonist muscle change from co-contraction to alternating contraction (Findley *et al.*, 1985a).

Shahani and Young (1978) and Young and Shahani (1979) have shown abnormalities in the behaviour of isolated motor units in patients with essential tremor compared to that found in physiological tremor or Parkinson's disease. We ourselves, however, have not found routine concentric needle electrode EMG or surface EMG helpful in the classification or diagnosis of tremulous disorders.

Perhaps the only diagnostic pattern of EMG activity seen in tremulous disorders is that reported in primary orthostatic tremor. Thompson *et al.* (1986) and Deuschl *et al.* (1987b) recorded high frequency (14–18 Hz), bilaterally synchronous EMG bursts which spontaneously halved in frequency during periods of recording. It is not clear whether this disorder should be classified as a distinct neurological entity or a variant of essential tremor.

CLASSIFICATION BY PHARMACOLOGICAL RESPONSE

The response of tremor to drugs is not generally helpful in classification. Most drugs lack absolute specificity, such that any given drug may effect tremors of different aetiologies. Furthermore, not all tremors of similar aetiology will show a response to a particular drug even if this drug is effective in the majority of cases.

All drugs which produce central sedation will attenuate tremor. In general, the degree of improvement is proportional to the degree of central sedation achieved.

The rest tremor of Parkinson's disease exhibits perhaps the most specific pharmacological responsiveness. The majority of patients with early, idiopathic parkinsonian rest tremor show a response, sometimes dramatic, to dopaminergic drugs and also some response to anticholinergics (see Table 36.6). Tremors of other aetiologies, with the exception of some rubral tremors

Table 36.6 Effects of drugs on tremors of different aetiology.

Disease/disorder	Effective drugs
Enhanced physiological tremor	Beta-blockers
	Alcohol
Essential tremor	Beta-blockers
	Primidone
	Phenobarbitone
	Alcohol
"Kinetic-predominant" essential tremor	Clonazepam
Orthostatic tremor	Clonazepam
Parkinson's disease (rest tremor)	L-Dopa
	Dopamine agonists
	Anticholinergics
	Amantadine
Parkinson's disease (postural tremor)	? Beta-blockers
	? Alcohol
Multiple sclerosis	Isoniazid
Wilson's disease	D-Penicillamine

(Findley and Gresty, 1980), are unaffected or worsened by these drugs.

Beta-blockers, in particular propranolol, are effective in the majority of patients with essential tremor although around 20% of patients show no useful response. Propranolol also suppresses enhanced physiological tremor (Shahani and Young, 1976) and has been reported to reduce the amplitude of parkinsonian postural tremor (Schwab and Young, 1971; Rajput *et al.*, 1975; Koller, 1987) and thus cannot be considered specific to essential tremor.

Young and colleagues reported that EPT (Shahani and Young, 1976) but not essential tremor (Young *et al.*, 1975) could be suppressed by intravenous propranolol and suggested that this could be used as a differential diagnostic test. Further studies are required to determine whether there is a differential response to intravenous propranolol between Marsden's type 1 (clinically indistinguishable from EPT) and type 2 essential tremor. Rapid suppression of tremor by intravenous propranolol does not differentiate the many underlying causes of enhanced physiological tremor.

Many patients with essential tremor (40.8% in our series of 130 patients) report some degree of tremorlytic benefit from small quantities of alcohol. This observation has often been mentioned in the literature as diagnositc of essential tremor. However, dramatic response (i.e. complete suppression of tremor by one or two standard units of alcohol) was reported by only six of our patients (4.6%). Other types of tremor (Rajput *et al.*, 1975) and indeed other types of movement disorder such as spasmodic torticollis (Biary and Koller, 1985), essential myoclonus (Korten *et al.*, 1974) and dystonic myoclonus (Quinn and Marsden, 1984), may also show a response to alcohol.

Primidone can be effective, sometimes dramatically, in reducing the amplitude of essential tremor (Findley *et al.*, 1985b; Koller and Royse, 1986). However, not all patients show a response. The effect of primidone on other types of tremor has not been investigated.

The benzodiazepine clonazepam has been reported to be of little value in attenuating classical essential tremor of the hands (Thompson *et al.*, 1984) but to be effective in a "kinetic-predominant" variant of essential tremor (i.e. monosymptomatic tremor, minimal on sustained posture but enhanced during and at the termination of goal-directed movements) (Biary and Koller, 1987). Orthostatic tremor, the high frequency tremor of the trunk and legs which occurs on standing but disappears when walking and sitting (Heilman, 1984), has been reported to respond dramatically to clonazepam but to be unresponsive to propranolol (Heilman, 1984; Kelly and Sharbrough, 1987).

There may, therefore, be a differential response to drugs amongst clinically distinct variants of essential tremor. However, attempts to identify subgroups within the spectrum of "classical" essential tremor on the basis of differential response to drugs have been inconclusive and contradictory. Calzetti *et al.*, (1983) found essential tremor of lower amplitude and higher frequency to be less responsive to propranolol than higher amplitude, lower frequency tremors. In contrast Sabra and Hallett (1984) and Deuschl *et al.* (1987a) found essential tremor generated by alternating antagonist muscle activity (by implication, of higher amplitude and lower frequency) to be *less* responsive to propranolol than (lower amplitude) tremors generated by co-contraction of muscle groups.

HYSTERICAL TREMOR

Tremors may be encountered in the clinic which are difficult to classify on the basis of the above criteria. A number of terms have been used to describe such tremors (e.g. hysterical, functional, psychogenic) but none is entirely satisfactory in describing a wide range of disorders from malingering to classical hysteria. Recently, Marsden (1986) has reviewed the implications of these various terms and suggests the continued use of the term hysterical.

In a retrospective survey of patients attending the National Hospital, Queen Square, in the periods 1951–53–55, 1961–63–65 and 1971–73–75, Trimble (1981) found that tremor represented 1.1%, 3.4% and 4.4%, respectively, of all symptoms documented as hysterical. However, it is not known what percentage of all tremors seen in the clinic are hysterical in origin. In the absence of strict criteria for defining hysterical tremor, misdiagnosis is probably common.

Distinguishing features of hysterical tremors are that they are often not sustained at a fixed frequency, the behavioural characteristics may change from moment to moment, and distraction of the patient's attention may result in abrupt suppression of the tremor. The performance of a manual task with one hand may suppress tremor in the other (or periods of tremor may alternate with non-continuous performance of the task). In our experience, this manoeuvre almost invariably exacerbates parkinsonian rest tremor. Essential tremor may be enhanced, unchanged or slightly diminished but never entirely suppressed.

It should be borne in mind that patients with hysterical tremors may have an underlying tremor of some other aetiology (hysterical "overlay"). Using spectral analysis, we have been able to identify an underlying tremor of essential type in a number of patients diagnosed as having hysterical tremor.

REFERENCES

Biary, N. and Koller, W. C. (1985) *Neurology* **35**, 239–240.
Biary, N. and Koller, W. C. (1987) *Neurology* **37**, 471–474.
Brumlik, J. and Yap, C.-B. (1970) *Normal Tremor: A Comparative Study*. Charles C. Thomas, Springfield, Illinois.
Calzetti, S., Findley, L. J., Perucca, E. and Richens, A. (1983) *J. Neurol. Neurosurg. Psychiat.* **46**, 393–398.
Cleeves, L. and Findley, L. J. (1987) *J. Neurol. Neurosurg. Psychiat.* **50**, 704–708.
Cleeves, L., Findley, L. J. and Gresty, M. A. (1986) *Adv. Neurol.*, **45**, 349–352.
Cleeves, L., Findley, L. J. and Koller, W. C. (1988) *Ann. Neurol.* **24**, 23–26.
Deuschl, G., Lucking, C. H. and Schenk, E. (1987a) *J. Neurol. Neurosurg. Psychiat.* **50**, 1435–1441.
Deuschl, G., Lucking, C. H. and Quintern, J. (1987b) *Z. EEG EMG* **18**, 13–19.
Elble, R. J. (1986) *Neurology* **36**, 225–231.
Findley, L. J. and Capildeo, R. (1984) *Movement Disorders: Tremor*. Macmillan Press, London.
Findley, L. J. and Gresty, M. A. (1980) *Br. Med. J.* **281**, 1043.
Findley, L. J., Gresty, M. A. and Halmagyi, G. M. (1981) *J. Neurol. Neurosurg. Psychiat.* **44**, 534–546.
Findley, L. J., Cleeves, L. and Calzetti, S. (1985a) *Neurology* **34**, 1618–1619.
Findley, L. J., Cleeves, L. and Calzetti, S. (1985b) *J. Neurol. Neurosurg. Psychiat.* **48**, 911–915.
Freund, H. J., Budingen, H. J. and Dietz, V. (1975) *J. Neurophysiol.* **38**, 933–946.
Gresty, M. A. and Findley, L. J. (1984) In *Movement Disorders: Tremor* (edited by L. J. Findley and R. Capildeo), pp. 15–26. Macmillan Press, London.
Hagbarth, K. E. and Young, R. R. (1979) *Brain* **102**, 509–526.
Heilman, K. M. (1984) *Arch. Neurol.* **41**, 880–881.
Holmes, G. (1904) *Brain* **27**, 360–375.
Kelly, J. J. and Sharbrough, F. W. (1987) *Neurology* **37**, 1434.
Koller, W. C. (1987) *Arch. Neurol.* **44**, 921–923.
Koller, W. C. and Royse, V. L. (1986) *Neurology* **36**, 121–124.
Korten, J. J., Nottermans, S. L. H., Frenken, C. W., Garbeels, F. J. and Joosten,

E. M. (1974) *Brain* **97**, 131–138.

Lee, R. G. and Stein, R. B. (1981) *Ann. Neurol.* **10**, 523–531.

Marsden, C. D. (1984) In *Movement Disorders: Tremor* (edited by L. J. Findley and R. Capildeo), pp. 37–84. Macmillan Press, London.

Marsden, C. D. (1986) *Psychol. Med.* **16**, 277–288.

Marsden, C. D., Obeso, J. A. and Rothwell, J. C. (1983) In *Current Concepts of Parkinson's Disease and Related Disorders* (edited by M. D. Yahr), pp. 31–46. Excerpta Medica, Amsterdam.

Pinder, R. (1984) In *Movement Disorders: Tremor* (edited by L. J. Findley and R. Capildeo), pp. 445–461. Macmillan Press, London.

Quinn, N. P. and Marsden, C. D. (1984) *Neurology* **34** (suppl. 1), 236.

Rajput, A. H., Jamieson, H., Hirsh, S. and Quraishi, A. (1975) *Can. J. Neurol. Sci.* **2**, 31–35.

Sabra, A. F. and Hallett, M. (1984) *Neurology* **34**, 151–156.

Salisachs, P. and Findley, L. J. (1984) In *Movement Disorders: Tremor* (edited by L. J. Findley and R. Capildeo), pp. 219–224. Macmillan Press, London.

Schwab, R. S. and Young, R. R. (1971) *Trans. Am. Neurol. Assoc.* **96**, 305–307.

Shahani, B. T. and Young, R. R. (1976) *J. Neurol. Neurosurg. Psychiat.* **39**, 772–783.

Shahani, B. T. and Young, R. R. (1978) In *Progress in Clinical Neurophysiology, Vol. 5* (edited by J. E. Desmedt), pp. 129–137. Karger, Basel.

Stein, R. B. and Lee, R. G. (1981) In *Handbook of Physiology, Section I, Vol. II, Motor Control* (edited by V. B. Brooks), pp. 325–343. American Physiological Society.

Sutton, G. G. and Sykes, K. (1967) *J. Physiol.* **191**, 699–711.

Thompson, C., Lang, A., Parkes, J. D. and Marsden, C. D. (1984) *Clin. Neurolpharmacol.* **7**, 83–88.

Thompson, P. D., Rothwell, J. C., Day, B. L., Berardelli, A., Dick, J. P. R., Kachi, T. and Marsden, C. D. (1986) *Arch. Neurol.* **43**, 584–587.

Trimble, M. R. (1981) *Neuropsychiatry.* Wiley, Chichester.

Young, R. R. and Shahani, B. T. (1979) *Adv. Neurol.* **24**, 175–183.

Young, R. R., Growden, J. H. and Shahani, B. T. (1975) *N. Engl. J. Med.* **293**, 950–953.

37

Orthostatic Tremor

J. C. Rothwell

INTRODUCTION

In 1984, Heilman produced a clinical description of what he believed was a "distinct and treatable neurologic syndrome": orthostatic tremor. The essential features of the syndrome were "rapid, irregular and asynchronous tremors" of the legs and trunk which started when the patient stood and increased progressively to such an extent that the patients were unable to stand still for longer than 1 min. If they tried to do so, they fell. The tremor could be alleviated, or even abolished, if the patients leaned on a support or started to walk. Since the original description, there have been several further reports of patients with similar disorders (Thompson *et al.*, 1986; Wee *et al.*, 1986; Veilleux *et al.*, 1987; Papa and Gershanik, 1988). The question which arises is whether the syndrome of orthostatic tremor is a distinct condition, or whether it is simply a variant of benign essential tremor.

Heilman believed that orthostatic tremor was a separate entity on two grounds. First, only one of his three patients actually had essential tremor. This patient had a familial, postural tremor of the arms. However, the absence of family history and lack of tremor in the arms of the other two patients probably is insufficient evidence to conclude that these patients did not have essential tremor. Second, Heilman found that none of the three patients responded well to treatment with propranolol, although two of them

DISORDERS OF MOVEMENT: CLINICAL, PHARMACOLOGICAL
AND PHYSIOLOGICAL ASPECTS   ISBN 0-12-569685-X

improved dramatically on clonazepam. Although clonazepam has been reported to be ineffective in treating essential tremor of the arms (Thompson *et al.*, 1984), it is not clear whether this is true for all types of essential tremor. For example, there is a report that it may be of some use in low-frequency action tremor (Biary, 1984).

In our own laboratory we have seen three patients with clinical symptoms similar to those described by Heilman (1984). In all three, physiological studies revealed that the tremor was quite different from that usually seen in the arms of patients with benign essential tremor.

PHYSIOLOGICAL STUDIES OF PATIENTS WITH ORTHOSTATIC TREMOR

Clinical details of these three patients can be found in the original papers by Thompson *et al.* (1986) and Van der Kamp *et al.* (1988). Table 37.1 provides a brief summary of relevant details. None of the patients had a family history of tremor, and all, when standing still, developed a violent tremor of the legs and trunk over a period of 30 s or so. This was alleviated, or abolished, by walking or leaning on a support. In two of the three patients, there was no postural tremor of the legs when seated, and only a fine physiological tremor of the outstretched arms. In the other, a rapid, small amplitude postural tremor could be observed in both the arms and legs, None of the three patients had responded well to a variety of medications including propranolol and clonazepam.

The most striking finding in all three patients was the frequency of the leg tremor. EMG recordings showed that there was no tremor in any muscle when the patients were seated and relaxed. However, on standing, or even attempting to stand, a 16 Hz tremor appeared that was alternating between agonist and antagonist muscles. The frequency of the tremor was the same in both legs, and the EMG bursts were synchronous in corresponding muscles (Fig. 37.1). Bursts were prominent in both the quadriceps and hamstring muscles. Yet even though the amplitude of each EMG burst might be large (up to 1 mV), the visible external tremor appeared only as a fine ripple of activity under the skin. Presumably, this was caused by partial fusion of muscle contractions at such a rapid tremor frequency. In two of the patients, the frequency of tremor in quadriceps often slowed abruptly to 8 Hz, while the other muscles continued at 16 Hz. Under these conditions the quadriceps EMG bursts occurred at every other beat in the tremor cycle (Fig. 37.2). At this frequency, the individual bursts of tremor in quadriceps became clearly visible and produced large oscillations of the patella. Both of these patients complained of increased unsteadiness at these times.

There was a remarkably constant relationship between the timing of

Table 37.1 Summary of clinical features, family history, and response to therapy of the patients with orthostatic tremor.

Patient	Age	Sex	Duration (years)	Family history	Posture for leg tremor	Tremor at other sites	Tremor frequency	EMG pattern	Drugs
1	55	M	4	None	Standing	UL physiological	16 Hz	Alternating	None
2	25	F	4	None	Standing	UL physiological	16 Hz	Alternating	None
3	68	M	15	None	Standing postural	UL postural	16 Hz (UL and LL)	Alternating	None

new

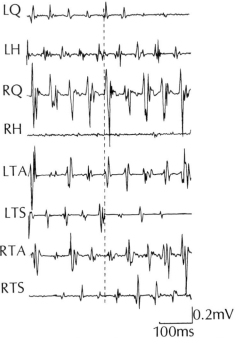

Fig. 37.1 Surface EMG recordings from both legs of patient 1 whilst attempting to stand quietly. Note that activity in corresponding muscles of both legs is synchronous (vertical marker). Agonist and antagonist muscle activity alternates at each joint. Q, quadriceps; H, hamstrings; TA, tibialis anterior; S, soleus. (From Thompson *et al.*, 1986, with permission.)

tremor bursts in different muscles of the leg. Figure 37.3 illustrates the pattern in patient 1. In this figure, 52 periods of tremor have been averaged, aligned to the onset of activity in quadriceps. The bursts in the other muscles are clearly visible: because they were time-locked to the activity in quadriceps, they have not been smoothed out by the averaging process. In this patient, EMG activity in tibialis anterior and extensor hallucis longus preceded that in quadriceps by 18 and 15 ms, respectively. Activity in triceps surae preceded that in hamstrings by 12 ms.

As pointed out above, tremor in the arms was not a prominent feature in any patient. Nevertheless, EMG recordings from the arm and shoulder, revealed that a 16 Hz tremor was present in two of the patients. In one of these (patient 3), the tremor was evident whenever the muscles were active; in the other (patient 1), tremor only appeared during certain tasks. In this patient, arm tremor was prominent when the patient was standing with the arms outstretched in front of, or behind, his body, but there was no tremor when he performed the same tasks when seated, apart from a fine physiological

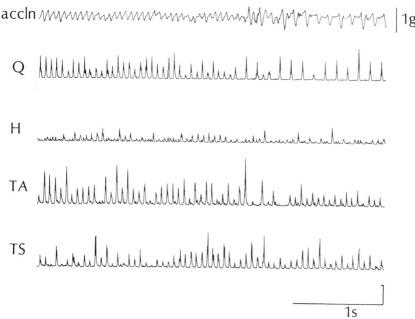

Fig. 37.2 Accelerometer (upper trace) and rectified, smoothed EMG records from the right leg of patient 1 during quiet standing. The accelerometer was attached over the quadriceps muscle. Abbreviations as in Fig. 37.1. Note the change in tremor frequency in the quadriceps muscle from 16 Hz at the start of the record to 8 Hz mid-way through. Tremor frequency remained constant in the other muscles at 16 Hz. Vertical calibration bar is 0.1 mV for Q, H and TA; 0.05 mV for TS (triceps surae). (From Thompson *et al.*, 1986, with permission.)

tremor which did not appear as distinct bursts of activity in the EMG record (c.f. Hagbarth and Young, 1979). When tremor appeared in the arms, it was not time-locked to activity in the leg, nor were the bursts synchronous between corresponding muscles on each side of the body.

It was not possible to influence the frequency or the phase of the tremor in the legs by electrical stimulation of peripheral nerves. Supramaximal stimuli given at random intervals to one tibial nerve while the patients were standing produced a large M-wave, followed by a silent period in the triceps surae. EMG activity resumed as a burst of tremor in phase with the ongoing tremor of the other muscles of the leg.

No patient had any abnormalities of sensory evoked potentials from stimulation of the tibial nerve at the ankle, nor were there any changes in central motor conduction to leg muscles as evaluated using transcranial stimulation of the motor cortex. Stretch reflexes were elicited in the tibialis

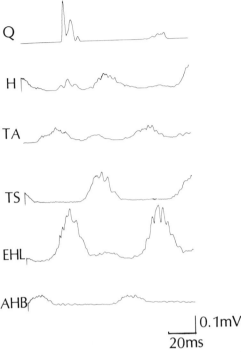

Fig. 37.3 Average of 52 125 ms periods of tremor activity in muscles of the right leg of patient 1. Rectified and smoothed EMG records as in Fig. 37.2. EHL, extensor hallucis longus; AHB, abductor hallucis brevis. The onset of each sweep was triggered by activity in quadriceps, using a 30 ms pretrigger facility. The synchronous nature of the EMG bursts in this average reflects the constant timing of muscle activation in each tremor cycle. Hamstrings trace reveals "pick-up" of EMG activity from the large quadriceps burst preceding each hamstrings burst. (Surface electrodes for H were placed higher up the thigh than those for Q, hence the remote activity appears to begin earlier in H than Q.) Similarly, the TA trace reveals "pick-up" from activity in the triceps surae. (From Thompson *et al.*, 1986, with permission.)

anterior and triceps surae of patient 1 while seated. He had normal latency short and long latency stretch reflexes at about 60 and 110 ms, respectively.

DISCUSSION

The symptoms of these three patients conform to Heilman's original clinical description of orthostatic tremor: i.e. a rapid tremor of the legs and trunk which only appears on standing, and which is alleviated or abolished by walking or leaning against support. However, as pointed out in the introduction, the clinical features alone are inadequate to distinguish whether or not the tremor is a novel entity, or whether it is simply a variant of benign essential tremor. Lack of (i) family history, (ii) response to alcohol or β-blockers and (iii) tremor in other limbs (except in patient 3), are all consistent with, but not conclusive evidence for the suggestion that orthostatic tremor is a separate entity to essential tremor.

However, in addition to satisfying Heilman's clinical criteria, these three patients also shared a unique physiology. Their leg tremor was quite different from that usually described in patients with essential tremor of the arms. The most striking feature was the high frequency. To our knowledge, a 16 Hz tremor has never been described before. It is a far higher frequency than normal 8–10 Hz physiological tremor. One patient thought that it might be similar to a shivering tremor, but previous work (Stuart *et al.*, 1966) has shown that this has a frequency of only 7–8 Hz. Benign essential tremor also has a different frequency (5–8 Hz). In addition, EMG bursts in essential tremor are asynchronous between the limbs, and their phase can be reset by a peripheral nerve stimulus (Bathien *et al.*, 1980). Thus in these three patients, orthostatic tremor had quite a different electrophysiology than essential tremor, at least as normally described in the arms. There are few detailed descriptions of essential tremor in the legs, but our own observations suggest that such tremor is physiologically very similar to that in the arms (Van der Kamp *et al.*, 1988). If so, then we suggest that the tremor of orthostatic tremor is indeed a separate entity to essential tremor.

Table 37.2 summarizes the clinical and electrophysiological features of the present three patients and those originally described by Heilman (1984) and contrasts them with those usually found in patients with benign essential tremor. How do these features of orthostatic tremor compare with those described by other authors? There are three reports available (Wee *et al.*, 1986; Papa and Gershanik, 1987), one of which is only an abstract (Veilleux

Table 37.2   Clinical and physiological differences between orthostatic and essential tremor of the legs.

	Orthostatic tremor	Essential tremor
Clinical	Task-specific	Present in all actions
	Ameliorated by walking	Unaffected by walking
	No tremor elsewhere	Tremor of head or arms
	Family history variable	
Physiological	16 Hz	6–8 Hz
	Synchronous in both legs	Not synchronous
	Phase unaffected by a peripheral nerve shock	Reset by a peripheral nerve shock
Beneficial drugs	Clonazepam and phenobarbitone	Alcohol and β-blockers

Data summarized from present patients and from the article by Heilman (1984).

et al., 1987). In all cases, the predominant complaint of the patients was of leg tremor which was particularly evident when standing, and which was relieved by walking or standing with support. Some patients also had a postural tremor of the head or arms. Clonazepam was an effective treatment in the majority of cases and the family history was variable. Thus the clinical criteria and effective treatment were consistent with the diagnosis of orthostatic tremor. However, several of the patients did not share the physiological features of the three cases documented in this chapter. The patient of Wee *et al.* (1986), and two of the three patients of Papa and Gershanik (1988) had a tremor frequency of only 6–8 Hz. The 18 patients of Veilleux *et al.* (1987) had a mean tremor frequency of 16 Hz, although this varied between patients from 7 to 32 Hz. Resetting the phase of tremor and synchrony of EMG bursts between the legs was not examined in detail in any of the reports.

The conclusion is that several patients who fulfil the clinical criteria for orthostatic tremor have a physiology which is quite different from that seen in the present patients. These individuals lie somewhere between the two extreme categories of Table 37.2. The mechanism of their tremor may be similar to that of essential tremor, yet the clinical presentation is that of orthostatic tremor.

REFERENCES

Bathien, N., Rondot, P. and Toma, S. (1980) *J. Neurol. Neurosurg. Psychiat.* **43**, 713–718.

Biary, N. (1984) *Abstr. Neurol.* **34** (suppl. 1), 128.

Hagbarth, K. E. and Young, R. R. (1979) *Brain* **102**, 509–526.

Heilman, K. H. (1984) *Arch. Neurol.* **41**, 880–881.

Papa, S. M. and Gershanik, O. S. (1988) *Movement Disorders* **3**, 97–108.

Stuart, D. G., Eldred, E. and Wild, W. O. (1966) *J. Appl. Physiol.* **21**, 1918–1924.

Thompson, C., Lang, A., Parkes, J. D. and Marsden, C. D. (1984) *Clin. Neuropharmacol.* **7**, 83–88.

Thompson, P. D., Rothwell, J. C., Day, B. L. *et al.* (1986) *Arch. Neurol.* **43**, 584–587.

Van der Kamp, W., Thompson, P. D., Rothwell, J. C., Day, B. L. and Marsden, C. D. (1989) *Movement Disorders* in press.

Veilleux, M., Sharbrough, F. N., Kelly, J. J., Westmoreland, B. F. and Daube, J. R. (1987) *J. Clin. Neurophysiol.* **4**, 304–305.

Wee, A. S., Subramony, S. H. and Currier, R. D. (1986) *Neurology* **36**, 1241–1245.

38

Long-latency Reflexes

Mark Hallett

In Sherringtonian physiology, the reflex response to stretch is simply the myotatic reflex. The long-latency stretch reflex was discovered by Hammond (1960), but it is clear that its importance and clinical relevance were first recognized by Marsden, Merton and Morton in a search for human evidence of servo control of voluntary movement, a theory postulated by Merton (1953). During the course of a tracking movement made by flexing the top joint of the thumb, the movement was perturbed by stretching the thumb, halting the movement, or unloading (releasing) the thumb and causing it to move faster than intended. EMG activity was monitored in the flexor pollicis longus (FPL) muscle. Consistent with Merton's theory, the response to the perturbation was more rapid than a voluntary reaction time and in a direction that would help restore the thumb to its intended trajectory. Contrary to the theory, however, the latency of the response was not the same as that of the monosynaptic tendon jerk, but was longer. The investigators recognized that this response was indicative of an important physiological process and immediately noted that this new observation might also be helpful in understanding some clinical phenomena in movement disorders.

This work was first presented at a meeting of the Physiological Society in March 1971 (Marsden et al., 1971a). The paper reported the reflex but together with a phenomenon that to this day remains somewhat confusing.

The servo action stretch reflex appeared to depend on peripheral sensation; anaesthesia of the thumb, which did not affect the FPL, reduced the stretch reflex. A second presentation to the Physiological Society in December 1971 (Marsden *et al.*, 1971b), showing that the gain of the stretch reflex increased with force, resulted in the first publication of what the investigators called a "tulip" (Fig. 38.1). Tulips were superimposed records of position or EMG responses coming from the four different experimental conditions: control, stretch, release, and halt.

The first complete report of the subject (Marsden *et al.*, 1972) described the nature of the reflex and pointed out its possible functions, including, for example, that it would be useful "for adjusting to reduced *g* on the moon." Hammond's pioneering observations were noted. Phillips' (1969) suggestion that stretch reflexes could result from the operation of a transcortical pathway (as well as the established spinal cord pathway) was considered as a possible explanation of the long latency of the response. Considering all of the data from the tulips, the halt condition probably gave the strongest support to

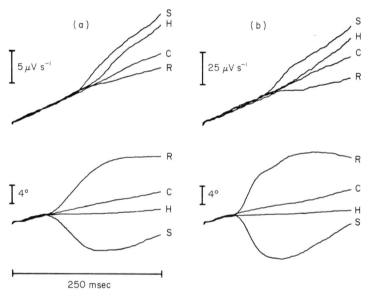

Fig. 38.1 Averaged responses during 20° tracking movements of the top joint of the thumb. The upper records are the integrated EMG responses of flexor pollicis longus; the lower records give the angular position of the terminal phalanx. The average of 16 control records (C) with constant opposition from the motor is compared with the average of 16 trials in which, 50 ms after the start of the sweep, the thumb is either driven back by a sudden increase in the motor current (S), almost brought to a standstill by a simulated spring (H), or allowed to accelerate by a sudden reduction in motor current (R). In (a), the initial resistance to movement was a torque of 150 g cm; in (b), it was 10 times greater. (From Marsden *et al.*, 1971b, with permission.)

the concept that the long-latency reflex was a result of servo action. The halt produced an increase in EMG response at the same latency as the EMG response produced by the stretch, but the muscle was not actually stretched. The EMG response must have been due to a mismatch between the actual and intended position of the thumb. The representation of the intended position of the thumb was in the length of the spindle driven by the fusimotor fibres. With halting, the extrafusal muscle fibres stopped changing length, but the intrafusal fibres continued to shorten. This mismatch increased firing of the Ia afferents, which increased α-motoneuron firing.

Another report (Marsden et al., 1973) stressed the hypothesis about the transcortical loop and further considered the clinical relevance of this work. The first full physiological report was published later (Marsden et al., 1976).

Lee and Tatton (1975) studied responses to sudden limb displacement in monkeys and humans with the goal of developing animal models of human motor system disorders. They planned to perform similar physiological studies in humans and animals and then to apply the findings to animals to learn more about the human disease. Their first interest was Parkinson's disease. In their clinical experiments, they studied stretches of the wrist and recorded EMG from forearm flexor muscles. Like Marsden and colleagues, they found long-latency stretch reflexes before it was possible to respond voluntarily. In the monkeys, two long-latency responses were standard; in humans, there were one or two long-latency reflexes. The short-latency, monosynaptic response was called M1, and the long-latency responses were called M2 and M3. When M2 and M3 could not be differentiated, the response was called the M2–3. This terminology became popular to describe short-latency and long-latency responses. Adam, working in Marsden's laboratory (Marsden et al., 1978a), showed that the long-latency reflex in FPL also often had two components that she called A and B (Fig. 38.2). It seemed that the long-latency reflex was not a unitary phenomenon.

Following these works, long-latency stretch reflexes received increasing attention. In the triceps surae, Melvill Jones and Watt (1971) found a late response they called the functional stretch reflex; Gottlieb and Agarwal (1980) found one they called the postmyotatic response, and Nashner (1976) found one in standing subjects that he also called the functional stretch reflex, but the analogy to Melvill Jones and Watt's (1971) response was not clear. In addition, attention was paid to long-latency responses to electrical stimulation. Upton et al. (1971) demonstrated a long-latency response to mixed nerve stimulation; the short-latency response, believed to be analogous to the H-reflex, was called V1, and the long-latency response was called V2. Caccia et al. (1973) first identified long-latency responses to electrical stimulation of cutaneous nerves.

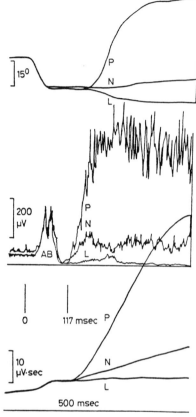

Fig. 38.2 The effect of a normal subject's response to stretch of flexor pollicis longus (FPL) that was held at 10° contracting against a force of 2 N. At 50 ms after the sweep commenced, the thumb was stretched. In separate trials, the subject was instructed either to hold the thumb in a steady position (N), to relax the thumb (L), or to flex the thumb as hard as possible (P) after perceiving the stretch. Each record is the average of 48 trials. The upper record shows the angular position of the thumb. The middle record shows the full-wave rectified EMG responses (see text for explanation of A and B) and the lower record shows the full-wave rectified and integrated EMG responses recorded with surface electrodes from FPL. (From Marsden *et al.*, 1978a, with permission.)

SOME BASIC ISSUES

Terminology in long-latency reflexes is often confusing, and semantics interferes with our understanding of the physiology. Long-latency reflexes have a longer latency than the monosynaptic stretch reflex (the short-latency reflex). Some long-latency reflexes, whose pathways are restricted to the spinal cord, are short-loop reflexes, and some, whose pathways go above the spinal cord, are long-loop reflexes. There are long-latency short-loop reflexes such as the flexor reflex. There may well be long-latency long-loop reflexes, such as some reflexes to stretch, but this is controversial. Referral to a reflex by its latency is unambiguous; referral by its pathway is dangerous until its physiology is established. Confusion arises because of the often unstated equation of long latency and long loop. Names for long-latency reflexes abound, and the designations M1, M2, and M3 are the most popular. Similarities of phenomena found in different situations are not assured. Hence, the second response to one type of stretch at one joint and the second response to another type of stretch at another joint may both be called M2, but they may differ. In general, it is wise to adopt the view that responses found in different experimental conditions are different unless it is demonstrated that they are the same.

Are long-latency reflexes really reflexes? Could they be early voluntary responses? Considerable attention has been paid to this point. They are reflexes in the sense that they depend largely on the mechanical parameters of the stretch and not on the will of the subject. A distinct voluntary response follows the long-latency stretch reflex (Fig. 38.2).

Are long-latency stretch reflexes ever truly long-loop reflexes? Certainly the anatomical pathway exists to permit long-loop reflexes. Primary afferents ascend through the dorsal column, medial lemniscus, and thalamus to sensory cortex. Motor cortex gets input from sensory cortex, and also some direct input from the thalamus which, in turn, has received some of its input from the cerebellum. Motor cortex sends efferents to the α-motoneurons in the spinal cord. The issue seems to be the physiological importance of the connections rather than the connections themselves. The case for a significant long-loop contribution to the long-latency reflex seems strongest for the top joint of the thumb (Marsden et al., 1983). Other mechanisms, including responses to segmented peripheral input resulting from the irregular nature of the stretch (Hagbarth et al., 1981; Berardelli et al., 1982) and responses to group II input (Berardelli et al., 1983b; Mathews, 1984), can contribute to long-latency reflexes.

The physiological role of the long-latency stretch reflex has been debated. The original view, stemming from the motivation of Marsden, Merton and Morton to find human evidence for servo control of voluntary movement, is that the reflex helps to bring a movement back on track after a perturbation. Indeed, it certainly can play that role in many circumstances (Marsden et al., 1983). When the perturbation is "too small" to elicit a long-latency stretch reflex, the muscle properties themselves, perhaps with the short-latency stretch reflex, are sufficient to make corrections. When the perturbation is "too large," the voluntary response that follows the long-latency reflex helps to finish the correction. An alternate hypothesis, proposed by Crago et al.

(1976), stated that the role of the long-latency stretch reflex was to regulate muscle stiffness. Another line of experimentation has produced evidence that long-latency reflexes do indeed correct for perturbations, but not in a simple mechanical way. The reflexes appear able to correct movements in a functional way. The first bit of evidence for functional correction, called the grab reflex, came from an experiment by the Marsden group (Traub et al., 1980). For example, a subject was asked to press the thumb against a lever and the thumb was pulled away from the lever with a tug on the wrist. Instead of an unloading response in FPL, as might be expected in this situation, there was a long-latency increase in the EMG response that had the effect of moving the thumb back to the lever. Subsequent experiments have made the functional aspects of long-latency responses even clearer. For example, during an attempted grasp movement of the thumb and index finger, if the thumb is stretched, the index finger will show the "long-latency stretch reflex", which has the effect of attempting to maintain the functional grasp (Cole and Abbs, 1987).

DERANGEMENTS IN DISEASE

Like virtually all physiological events that are mediated by pathways in the central nervous system, long-latency reflexes can be slowed in multiple sclerosis (Diener et al., 1984). Such investigations, however, have not become as popular as the study of evoked potentials.

Decreases in long-latency stretch reflexes

Part of the initial evidence that long-latency reflexes could be long-loop reflexes was that lesions of the dorsal column (Marsden et al., 1977a), sensorimotor cortex, or corticospinal tract (Marsden et al., 1977b) typically caused decrease or absence of the reflexes. While some authorities argued that this effect was due to a facilitatory influence of the supraspinal pathway and the reflex itself did not use the pathway, the phenomena themselves were clear. In one case, a small lesion of the postcentral gyrus was sufficient to eradicate the long-latency stretch reflex in FPL (Marsden et al., 1977b).

Patients with spasticity typically have lesions somewhere along the long-loop pathway. Additionally, studies of stretch reflexes in spasticity show that the short-latency reflex is enhanced and the magnitude of enhancement is correlated with the clinical impression of increased tone (Berardelli et al., 1983a). There is the expectation that long-latency reflexes should be absent or reduced, and most commonly they are (Berardelli et al., 1983a; Lee et al.,

1983). In some circumstances, however, they may be present. Patients with spasticity can be divided into three groups (Hallett, 1985). The first group has no long-latency reflexes. The second group has apparently normal long-latency reflexes. The third group has enhanced long-latency reflexes characterized by either a single large-amplitude reflex (Fig. 38.3) or a long-duration reflex with abnormal appearance (Fig. 38.4). In some patients, the long-latency reflexes appear to contribute to increased tone. There is no apparent correlation between the patients' clinical appearance and the type of long-latency reflex. Further study is needed in this area.

Noth et al. (1985) found that long-latency reflexes were markedly diminished in patients with Huntington's disease. This was true of long-latency reflexes both to stretch and to electrical stimulation of mixed nerves. The short-latency responses were fully normal. Additionally, the motoneuron pool was responsive to two stretches at a 25 ms interval, suggesting that the loss of the long-latency response was not simply the result of refractoriness of the motoneuron pool. Patients with Huntington's disease also have markedly diminished amplitudes of somatosensory evoked responses. These findings support the idea that long-latency reflexes can be long-loop reflexes.

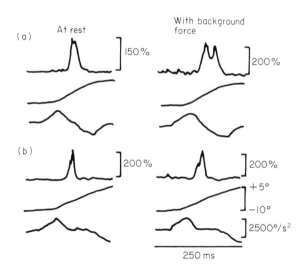

Fig. 38.3 EMG responses of lateral gastrocnemius muscle in two patients, (a) and (b), with multiple sclerosis. Left, responses at rest. Right, responses with background force. In each record, the three traces, from top to bottom, are: rectified EMG, ankle angle, and ankle angular acceleration; and each record is the average of 10 trials. The rectified EMG activity is quantified in relation to EMG of maximal voluntary activity. The record of patient (a) shows an abnormally large long-latency reflex with background force, and the record of patient (b) is abnormal because of the absence of any long-latency reflex. (From Berardelli et al., 1983b, with permission.)

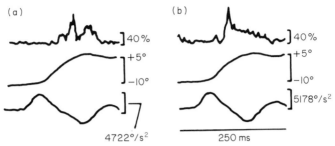

Fig. 38.4 EMG responses of lateral gastrocnemius muscle on the normal side (a) and on the abnormal side (b) of a patient with cervical spondylosis exerting background force. In each record, the three traces, from top to bottom are: rectified EMG, ankle angle, and ankle angular acceleration; and each record is the average of 10 trials. The rectified EMG activity is quantified in relation to EMG of maximal voluntary activity. The normal side (a) shows two distinct long-latency reflexes, and the abnormal side (b) shows a long, continuously decaying response. (From Berardelli *et al.*, 1983b, with permission.)

Increases in long-latency stretch reflexes

Lee and Tatton (1975) first demonstrated that long-latency stretch reflexes were increased in Parkinson's disease (Fig. 38.5), but the evidence was not immediately accepted because of several technical problems. First, long-latency stretch reflexes increase in amplitude with the amount of background contraction. Patients with Parkinson's disease have a difficult time relaxing; hence, their background contraction is higher than that of control subjects at rest, and this feature must be taken into account. Second, the phenomenon turned out not to be present for all muscles. Marsden's group (Rothwell *et al.*, 1983a) studied primarily the FPL, which did not show enhancement, as did most other muscles. From the outset, it was recognized that enhancement of long-latency reflexes might explain rigidity. Long-latency reflexes even appeared when patients were at rest, when normally no long-latency reflexes are seen. There may have been slight enhancement of short-latency reflexes in Parkinson's disease, but this finding was never sufficient to explain increase in tone. Enhanced long-latency reflexes seemed a good candidate. Studies of stretch reflexes in rigidity did show that the magnitude of the enhancement was correlated with the clinical impression of increased tone (Mortimer and Webster, 1978; Berardelli *et al.*, 1983b).

The physiology of enhancement of long-latency reflexes in Parkinson's disease is not clear. One possibility is that the gain of long-loop pathways is increased. Another is that a new long-latency reflex appears in Parkinson's disease that is not seen in normal subjects. Evidence for this second theory was offered by Berardelli *et al.* (1983b). Experiments were done with stretches

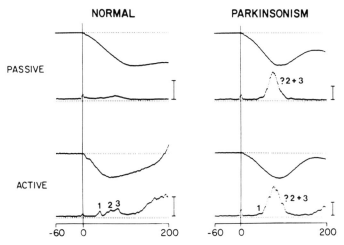

Fig. 38.5 EMG responses from wrist extensors to sudden wrist flexion in a normal subject and in a patient with Parkinson's disease. When the wrist was flexed, the passive task was to do nothing, and the active task was to extend the wrist as quickly as possible. The top trace in each record represents the average handle position, and the bottom trace is the average rectified EMG responses. The normal subject shows an increase in amplitude of the M1, M2 and M3 responses during the active task. The patient shows a markedly increased response without significant modulation between the tasks. Calibration bar = 200 μV. (From Lee and Tatton, 1975, with permission.)

of the triceps surae. In control subjects, the short-latency and long-latency reflexes were abolished by vibration of the muscle. Vibration is known to eradicate the short-latency reflex, and eradication of the long-latency reflex in this circumstance was probably because the long-latency reflex was due to segmented input giving rise to a series of short-latency responses. (Interestingly, vibration does not eradicate normal long-latency responses in all muscles; responses in the wrist flexor muscles are not affected (Hendrie and Lee, 1978).) In the patients with Parkinson's disease, the short-latency response was eradicated, but the long-latency response was maintained. Apparently, in Parkinson's disease there is a long-latency response that is not present in normal subjects and is not vibration sensitive. The authors suggested that this might be due to enhancement of group II mediated reflexes.

The results from spastic and rigid patients taken together make a clinically useful story. Evaluation of stretch reflexes in a patient with increased tone would identify possible increases in short-latency or long-latency mechanisms that would be suggestive of the clinical states of spasticity or rigidity, respectively. Furthermore, the results would be objective quantification of the abnormality.

As with Lee and Tatton's (1975) immediate recognition that long-latency reflexes might increase the understanding of parkinsonian rigidity, Marsden *et al.* (1972, 1973) promptly perceived that long-latency reflexes might be helpful in the understanding of myoclonus. They observed that in some patients with myoclonus a somatosensory stimulus, such as muscle stretch, evoked myoclonic jerking. The latency of the response was long, suggesting that the myoclonic jerk might be an enhanced long-latency reflex. (The enhanced reflexes are more prominent as tonic responses in Parkinson's disease and more prominent as phasic responses in myoclonus.) This observation became an impetus for detailed investigation of patients with myoclonus. The studies showed that there were different types of myoclonus with different types of enhanced long-latency reflexes. The enhanced long-latency reflexes in myoclonus are often called C-reflexes after the pioneering observations of Sutton and Mayer (1974).

The situation that seems best understood, and that supports the idea of long-loop pathways through cerebral cortex, is called cortical reflex myoclonus (Hallett *et al.*, 1979, 1987). Each myoclonic jerk affects only a few adjacent muscles. At about the time of a myoclonic jerk, the EEG shows a focal negative event in sensorimotor cortex somatotopically related to the muscles involved in the jerk (Fig. 38.6). This is demonstrated best by back-averaging the EEG on the occurrence of the myoclonus. The myoclonus that occurs after somatosensory stimulation is also associated with a focal negative event in the EEG that has the same spatial and temporal characteristics as the event preceding the spontaneous jerk. The event is also time-locked to the stimulus and hence is a component of the evoked potential. This component is typically larger than normal and is the so-called giant somatosensory evoked potential. These electrical phenomena are consistent with a hyperactive somatosensory cortex being the source of the myoclonus, and the C-reflex being a hyperactive long-loop reflex.

Rigidity occurs in patients with dystonia, as in patients with Parkinson's disease. Given the findings in Parkinson's disease, patients with dystonia were investigated for abnormalities of long-latency reflexes. Abnormalities in this disorder, however, were not so evident. Tatton *et al.* (1984) suggested that patients with dystonia have long-latency reflexes of abnormally long duration. Rothwell *et al.* (1983b) did not find any abnormality in the amplitude or duration of the stretch reflexes, but noted that the reflexes spread abnormally to other muscles. Since the abnormalities in dystonia were mild, it must be concluded that rigidity in Parkinson's disease and dystonia is not the same and that the constant background muscle activity of dystonia may well be responsible for most of what is clinically appreciated as increased tone.

The issue of long-latency stretch reflexes in patients with cerebellar lesions

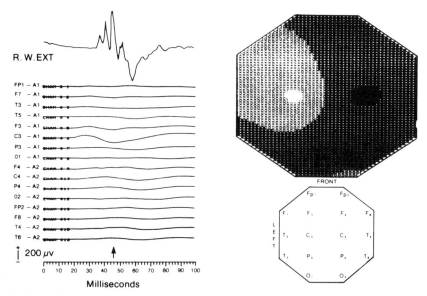

Fig. 38.6   Electrical correlates of myoclonic jerks produced by right median nerve stimulation in a patient with cortical reflex myoclonus. Top left, a single EMG record showing the long-latency reflex, or C-reflex. Bottom left, averaged EEG following the electrical stimulus (the somatosensory evoked potential). Top right, a topographic map of the EEG at the time indicated by the arrow on the bottom left. Light shading is negative, dark shading is positive. R.W.EXT., right wrist extensor. (Modified from Wilkins *et al.*, 1984, with permission.)

is not settled. Marsden *et al.* (1978b) reported results from study of the FPL in a few patients. Long-latency reflexes were absent or delayed, and seemed to underlie the clinical concept of cerebellar hypotonia. The notion of hypotonia itself is a little uncertain, but a reduction in expected response to stretch might well be interpreted clinically as reduced tone. Recent studies in a larger number of patients show increases in long-latency reflexes in the first dorsal interosseous muscle (Friedemann *et al.*, 1987). The clinical correlation of this finding is not immediately apparent.

CONCLUSION

The story about long-latency stretch reflexes is an exciting new chapter in neurophysiology. Almost uniquely, its origin was in human studies and animal studies came afterwards. Investigation of this phenomenon has enlarged our understanding of the normal physiology of the motor system and clarified the pathophysiology of many clinical phenomena in movement disorders. Much has been learned since the first reports from Marsden, Merton and Morton less than 20 years ago, and, of course, much remains to be learned.

ACKNOWLEDGEMENT

I am grateful to B. J. Hessie for skilful editorial assistance.

REFERENCES

Berardelli, A., Hallett, M., Kaufman, C., Fine, E., Berenberg, W. and Simon, S. R. (1982) *J. Neurol. Neurosurg. Psychiat.* **45**, 513–525.

Berardelli, A., Sabra, A. F., Hallett, M., Berenberg, W. and Simon, S. R. (1983a) *J. Neurol. Neurosurg. Psychiat.* **46**, 54–60.

Berardelli, A., Sabra, A. F. and Hallett, M. (1983b) *J. Neurol. Neurosurg. Psychiat.* **46**, 45–53.

Caccia, M. R., McComas, A. J., Upton, A. R. M. and Blogg, T. (1973) *J. Neurol. Neurosurg. Psychiat.* **36**, 960–977.

Cole, K. J. and Abbs, J. H. (1987) *J. Neurophysiol.* **57**, 1498–1510.

Crago, P. E., Houk, J. C. and Hasan, Z. (1976) *J. Neurophysiol.* **39**, 925–935.

Diener, H. C., Dichgans, J., Hulser, P. J., Buettner, U. W., Bacher, M. and Guschlbauer, B. (1984) *Electroencephalogr. Clin. Neurophysiol.* **57**, 336–342.

Friedemann, H. H., Noth, J., Diener, H. C. and Bacher, M. (1987) *J. Neurol. Neurosurg. Psychiat.* **50**, 71–77.

Gottlieb, G. L. and Agarwal, G. C. (1980) *J. Neurophysiol.* **43**, 86–101.

Hagbarth, K. E., Hagglund, J., Wallin, V. and Young, R. R. (1981) *J. Physiol. (Lond.)* **312**, 81–96.

Hallett, M. (1985) In *Clinical Neurophysiology in Spasticity* (edited by P. J. Delwaide and R. R. Young), pp. 83–93. Elsevier, Amsterdam.

Hallett, M., Chadwick, D. and Marsden, C. D. (1979) *Neurology* **29**, 1107–1125.

Hallett, M., Marsden, C. D. and Fahn, S. (1987) In *Handbook of Clinical Neurology, Vol. 5(49): Extrapyramidal Disorders* (edited by P. J. Vinken, G. W. Bruyn and H. L. Klawans), pp. 609–625. Elsevier, Amsterdam.

Hammond, P. H. (1960) In *Proceedings of the III International Conference on Medical Electronics*, pp. 190–199. Institute of Electrical Engineers, London.

Hendrie, A. and Lee, R. G. (1978) *Brain Res.* **157**, 369–375.

Lee, R. G. and Tatton, W. G. (1975) *Can. J. Neurol. Sci.* **2**, 285–293.

Lee, R. G., Murphy, J. T. and Tatton, W. G. (1983) In *Motor Control Mechanisms in Health and Disease* (edited by J. E. Desmedt), pp. 489–508. Raven Press, New York.

Marsden, C. D., Merton, P. A. and Morton, H. B. (1971a) *J. Physiol. (Lond.)* **216**, 21P–22P.

Marsden, C. D., Merton, P. A. and Morton, H. B. (1971b) *J. Physiol. (Lond.)* **222**, 32P–34P.

Marsden, C. D., Merton, P. A. and Morton, H. B. (1972) *Nature* **238**, 140–143.

Marsden, C. D., Merton, P. A. and Morton, H. B. (1973) *Lancet* **i**, 759–761.

Marsden, C. D., Merton, P. A. and Morton, H. B. (1976) *J. Physiol. (Lond.)* **257**, 1–44.

Marsden, C. D., Merton, P. A., Morton, H. B. and Adam, J. (1977a) *Brain* **100**, 185–200.

Marsden, C. D., Merton, P. A., Morton, H. B. and Adam, J. (1977b) *Brain* **100**, 503–526.

Marsden, C. D., Merton, P. A., Morton, H. B., Adam, J. and Hallett, M. (1978a) In *Cerebral Motor Control in Man: Long Loop Mechanisms*, Progress in Clinical Neurophysiology, Vol. 4 (edited by J. E. Desmedt), pp. 167–177. Karger, Basel.

Marsden, C. D., Merton, P. A., Morton, H. B. and Adam, J. (1978b) In *Cerebral Motor Control in Man: Long Loop Mechanisms*, Progress in Clinical Neurophysiology, Vol. 4 (edited by J. E. Desmedt), pp. 334–341. Karger, Basel.

Marsden, C. D., Rothwell, J. C. and Day, B. L. (1983) In *Motor Control Mechanisms in Health and Disease* (edited by J. E. Desmedt), pp. 509–539. Raven Press, New York.

Mathews, P. B. C. (1984) *J. Physiol. (Lond)*. **348**, 383–415.

Melvill Jones, G. and Watt, D. G. D. (1971) *J. Physiol. (Lond.)* **219**, 709–727.

Merton, P. A. (1953) In *The Spinal Cord* (edited by J. L. Malcolm and J. A. B. Gray), pp. 84–91. Churchill Livingstone, London.

Mortimer, J. A. and Webster, D. D. (1978) In *Cerebral Motor Control in Man: Long Loop Mechanisms*, Progress in Clinical Neurophysiology, Vol. 4 (edited by J. E. Desmedt), pp. 342–360. Karger, Basel.

Nashner, L. M. (1976) *Exp. Brain Res.* **26**, 59–72.

Noth, J., Podoll, K. and Friedemann, H.-H. (1985) *Brain* **108**, 65–80.

Phillips, C. G. (1969) *Proc. R. Soc. Lond. (Biol.)* **173**, 141–174.

Rothwell, J. C., Obeso, J. A., Traub, M. M. and Marsden, C. D. (1983a) *J. Neurol. Neurosurg. Psychiat.* **46**, 35–44.

Rothwell, J. C., Obeso, J. A., Day, B. L. and Marsden, C. D. (1983b) *Adv. Neurol.* **39**, 851–863.

Sutton, G. G. and Mayer, R. (1974) *J. Neurol. Neurosurg. Psychiat.* **37**, 207–217.

Tatton, W. G., Bedingham, W., Verrier, M. C. and Blair, R. D. (1984) *Can. J. Neurol. Sci.* **11**, 281–287.

Traub, M. M., Rothwell, J. C. and Marsden, C. D. (1980) *Brain* **103**, 869–884.

Upton, A. R. M., McComas, A. J. and Sica, R. E. P. (1971) *J. Neurol. Neurosurg. Psychiat.* **34**, 699–711.

Wilkins, D. E., Hallett, M., Berardelli, A., Walshe, T. and Alvarez, N. (1984) *Neurology* **34**, 898–903.

39

Cortical Stimulation in Man

B. L. Day

As a neurological tool, cortical stimulation through the intact skull has its most obvious application in the measurement of nerve conduction velocity of central motor fibres. By measuring how long it takes for impulses to travel from the cortex to muscle and subtracting the time taken for conduction in the peripheral nerve, it is possible to estimate the conduction time, and hence velocity, in the central pathway (Marsden et al., 1982). Valuable as this is for detecting central motor lesions in patients (Cowan et al., 1984; Hess et al., 1986; Thompson et al., 1987; Berardelli et al., 1987b; Ingram and Swash, 1987), it would be disappointing if this were the limit of its application. Already a number of other applications have been discovered, for example, as a means of exploring spinal cord and corticospinal circuitry (Rothwell et al., 1984; Cowan et al., 1986; Berardelli et al., 1987a). Hopefully, its greatest potential lies in probing the brain itself. This is the area of research which is only just beginning and which shall be considered here.

BACKGROUND

Bartholow (1874) is credited with being the first to stimulate the human brain directly. He stimulated electrically the brain of a young woman through

DISORDERS OF MOVEMENT: CLINICAL, PHARMACOLOGICAL AND PHYSIOLOGICAL ASPECTS   ISBN 0-12-569685-X

a 2 inch hole in the skull caused by an ulcer and produced muscular contractions in the opposite arm and leg. In the century that followed, stimulation of the human brain remained limited to those relatively rare occasions when the brain was exposed during neurosurgical operations (e.g. Penfield and Jasper, 1954). This situation was altered radically when Merton and Morton (1980) demonstrated a method of stimulating the human cerebral cortex through the scalp and skull. Their method was surprisingly simple and straightforward; a capacitor was discharged through a pair of electrodes held on the head. The brief, but intense, flow of electricity passed mainly along the soft tissue and bone but a small fraction entered the brain. With electrodes placed over the motor strip, twitches in muscles of the opposite side of the body were clearly visible, particularly in the hand and forearm.

Marsden *et al.* (1983) pointed out that responses to a single electrical stimulus could be observed in a wide variety of muscles. In fact, in our laboratory we have yet to find a muscle that cannot be activated in this way. Short-latency responses have been seen even in a vestigial muscle of the ear of a subject who claimed he could not wiggle his ears at will! With bipolar stimulation it is possible to demonstrate a rough somatotopy with activation taking place preferentially under the anode (Rothwell *et al.*, 1987). Thus, with the anode placed at the vertex of the head and the cathode 7 cm in front, muscles in the leg are predominantly activated. With the anode placed 7 cm laterally from the vertex, approximately over the hand area of the motor cortex, then muscles in the arm are activated. The somatotopy suggests that the stimulus activates neurons at some fairly superficial site, probably within the motor cortex. The short latency of the muscle responses (some 20 ms to muscles of the hand) suggests transmission in rapidly conducting descending pathways such as the corticomotoneuronal component of the corticospinal tract (Marsden *et al.*, 1983; Rothwell *et al.*, 1987).

The drawback with the electrical technique is the discomfort of the local sensation caused by the high current density underneath the stimulating electrodes. Although acceptable to motivated researchers experimenting on themselves, this can be a limiting factor when studying some patients. The painless method of magnetic stimulation of the brain, therefore, was a new development of considerable interest to neurologists. Although it has been known for some time that rapidly changing magnetic fields are capable of stimulating nervous tissue (Thompson, 1910), it was not until 1985 that Barker, Jalinous and Freeston introduced the technique of magnetic stimulation of brain. They discharged a capacitor, not directly across the head, but through a flat coil of wire held over the head. The transient flow of electricity produced a rapidly changing magnetic field around the coil which, in turn, induced eddy currents in nearby tissue, including the brain. As with the electrical method, this stimulus caused muscles to contract at short latency.

MULTIPLE SITES OF ACTION

The extent to which cortical stimulation may be used as a tool for unravelling the internal workings of the brain depends primarily upon which neuronal elements are affected by the stimulus. Theoretically at least, the greater the number of different elements the greater the possibilities. As shown below, a single stimulus has the capacity to activate a number of elements depending upon stimulus intensity and mode of stimulation.

An early observation made by Marsden *et al.* (1980) was that the force of a muscle twitch following a cortical shock could match or even exceed that produced by supramaximal stimulation of the peripheral nerve supplying the muscle. To exceed the peripheral twitch force some of the motor units probably fired more than once in response to the cortical shock. Day *et al.* (1987a) produced indirect evidence for such double firing of spinal motorneurons using a collision technique. More recently we have observed directly single motor units firing twice in response to a cortical shock, an example of which is shown in Fig. 39.1 (Day *et al.*, 1989a). The reason for double firing provides an important clue to our understanding of which structures are activated in the brain. Consider the behaviour of a tonically active motor unit from a hand muscle following different intensities of electrical stimulation of the cortex. The behaviour of the unit is best described by a post-stimulus time histogram (PSTH) which plots the probability of the unit firing at different times after the cortical stimulus. Using an intensity of stimulation just above motor threshold, the unit fires with a single period of increased probability of discharge which lasts for around 1.5 ms. Such a discrete peak in the PSTH reflects an excitatory postsynaptic potential (EPSP) in the cell body of the motoneuron, time-locked to the cortical stimulus. The EPSP is produced by the arrival at the motoneuron of a single descending volley in corticospinal fibres. As the stimulus intensity is increased a point is reached when two periods of increased probability of firing are seen; the initial period being accompanied by a second period some 4–5 ms later, reflecting the arrival at the motoneuron of two sequential volleys. A further increase in stimulus intensity produces a third volley at an intermediate latency. Multiple excitations of motoneurons explains why motor units may fire more than once, in quick succession, to a single electrical stimulus to the scalp.

The reason why a single stimulus to the cortex produces more than one excitatory input to spinal motoneurones may be found in the animal literature from the 1950s and 1960s. In the primate, direct recordings from the pyramidal tract following electrical stimulation of the exposed motor cortex revealed multiple descending volleys (Patton and Amassian, 1954; Kernell and Wu, 1967). The earliest descending volley was due to *direct* activation

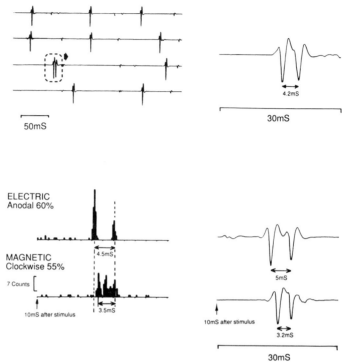

Fig. 39.1 Example of double firing of a single motor unit in FDI muscle. Top: the four rows on the left should be read continuously from left to right and top to bottom. They are part of a continuous record of the unit firing during a weak voluntary contraction. In the third row, there is a spontaneous doublet firing, followed by a long interdischarge interval. The doublet has been expanded in the sweep on the right. Bottom: on the left are two post-stimulus time histograms (PSTH) recorded from this unit following electric anodal stimulation of magnetic clockwise stimulation. Each PSTH starts 10 ms after the stimulus was given. Two peaks are evident in the electric PSTH (P_0 and P_3) and three peaks in the magnetic PSTH (P_1, P_2 and P_3). During construction of these histograms, those trials in which the unit fired twice were excluded. Examples of those occasions on which the unit did fire twice to the two modes of stimulation are shown on the right. Note the correspondence between the intervals separating the peaks in the PSTH and the interval between the double firing. (From Day *et al.*, 1989a)

of the corticospinal neurons (D-wave); the later volleys, which appeared at intervals of about 1.5 ms were thought to be due to *indirect* or synaptic activation of the same descending neurons (I_1, I_2, I_3 etc. waves). The characteristics of these volleys are remarkably similar to those in intact man (as deduced from PSTHs). The intervals between the volleys are the same, as is the recruitment order (Day *et al.*, 1987a, 1989a). It seems that electrical stimulation of the human brain through the scalp is analogous to electrical stimulation of the surface of the exposed primate cortex.

Magnetic stimulation of the cortex appears to act in a similarly complex way but with one crucial difference. Again, using peaks in the PSTH as an indication of the arrival of excitatory volleys at the motoneuron, we find the number of volleys increases with stimulus intensity and that there is some correspondence in time with the volleys produced by electrical stimulation (see Figs 39.2 and 39.3). The main difference between electrical and magnetic stimulation seems to lie in the recruitment order of the individual volleys. The recruitment order of the volleys may be altered further just by reversing the direction of the magnetic field (i.e. by turning the coil over). As a working hypothesis, it has been suggested that the volleys produced by either mode of stimulation are analogous to the descending volleys recorded directly from the baboon's pyramidal tract following electrical stimulation (Day *et al.*, 1987b, 1989a). This hypothesis has received some recent support from the work of Amassian *et al.* (1987) who recorded similar descending volleys in the monkey's pyramidal tract following both electrical and magnetic cortical stimulation.

In summary, it seems that with low to moderate intensities of cortical stimulation, descending volleys arrive at the spinal motoneurons with different latencies depending upon the mode of stimulation. In man the first volley to appear is the D-wave with electrical stimulation and an I-wave with magnetic stimulation. With the magnetic stimulating coil centred on the vertex of the head and with current flowing clockwise in the coil (viewed rom above), the I_1 wave is the first recruited volley to excite motoneurons of right hand muscles. With the current flowing in an anticlockwise direction, an I_3 or an I_2 volley is recruited first. A parsimonious explanation is that each of the descending volleys is produced by primary activation of different neuronal elements within the cortex which act on the spinal apparatus via the fast-conducting corticospinal fibres. Presumably, each element has a different excitability which depends upon the direction of the induced current in the cortex. If this is correct, each of these cortical elements may be isolated physiologically by changing the mode of stimulation, thereby providing an opportunity for studying cortical circuitry. The immediate challenge is to work out the anatomical basis of these cortical elements.

CHANGES IN MOTOR CORTEX EXCITABILITY

Even though we do not know yet precisely which anatomical structures are activated by the cortical stimulus, we can ask the question: during what types of movement are these neurons normally active? One approach is to try to measure the excitability of these neurons during different types of

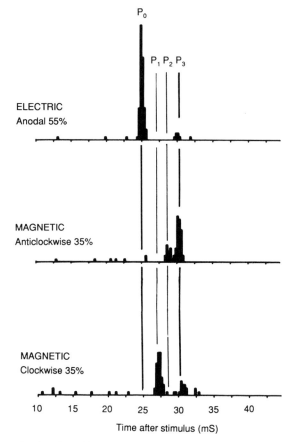

Fig. 39.2 Post-stimulus time histograms (PSTH) from a single motor unit in first dorsal interosseous muscle studied using different modes of cortical stimulation: electric anodal (top), magnetic anticlockwise (middle) and magnetic clockwise (bottom). Each histogram was constructed from the responses to 100 stimuli given 10 ms before the start of the record. Stimulus intensity has been expressed as a percentage of the maximum output of each stimulator and was adjusted to produce two peaks in the PSTH. The histogram peaks have been labelled P_0, P_1, P_2, P_3 (solid vertical lines) according to their latency after the stimulus. These peaks are thought to be caused by the arrival at the motoneurons of D, I_1, I_2 and I_3 waves, respectively (see text). (From Day et al., 1989a.)

muscle activity, the assumption being that their excitability is altered when they are involved in the control of that muscle activity.

A single stimulus delivered to the cortex produces a simple muscle twitch. For a given intensity of stimulation, the size of the muscle response depends upon the excitability of each of the links in the neural chain from cortex to

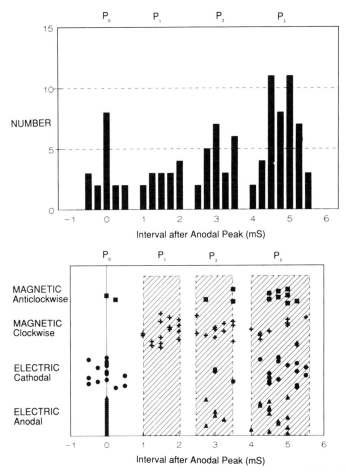

Fig. 39.3 Top: histogram showing relative latencies of PSTH peaks (see Fig. 39.1) in 28 different units from the right FDI muscle, compiled from data using electric anodal, electric cathodal, magnetic clockwise and magnetic anticlockwise stimulation. Peak latencies are expressed relative to the time of the earliest peak ($t = 0$ ms) obtained using electric anodal stimulation in each unit. The ordinate plots the number of units in which a peak occurred at a given interval. Four preferred latencies are visible: at the time of the first anodal peak (-0.5 to 0.5 ms), and at 1–2 ms, 2.5–3.5 ms and 4–4.5 ms later. These are labelled P_0, P_1, P_2 and P_3. These peaks are thought to be caused by the arrival at the motoneurons of D, I_1, I_2 and I_3 waves, respectively (see text). Anodal peaks at P_0 have not been included in the histogram, since they were the standard against which the other peaks were measured. Bottom: the same data displayed according to the mode of stimulation. The abscissa again plots the latency of each peak relative to the first anodal peak (vertical line at $t = 0$ ms). The hatched areas correspond to the time intervals spanned by the peaks in the top picture. Each horizontal row of the graph plots data from the PSTH of a single motor unit grouped according to the method of stimulation. Note that the P_2 and P_3 peaks were produced with all modes of stimulation, whereas only with magnetic clockwise stimulation was the P_1 peak seen and the P_0 peak not seen. (From Day *et al.*, 1989a.)

muscle. For example, changes in spinal cord excitability affect dramatically the size of muscle response (Fig. 39.4). By the same token, changes in excitability of the motor cortex, or other involved sites, would also influence the muscle response. The situation is simplified greatly if the descending volleys from a cortical stimulus activate the spinal motoneurons via a single synapse in the spinal cord, because then only two sites (cortex and spinal cord) need be considered. This is most likely to be true for muscles of the human hand since these muscles are thought to have the densest monosynaptic input from the cortex (the corticomotoneuronal component of the corticospinal tract) (Clough *et al.*, 1968). For this simplest situation, to demonstrate that a change in muscle response is due to a change in cortical excitability, the effects due to spinal cord excitability fluctuations must be ruled out. At present, there is no completely rigorous way of doing this although a few techniques have been tried. Their validity depends ultimately upon the actual pathways activated by the cortical stimulus and the stimulus

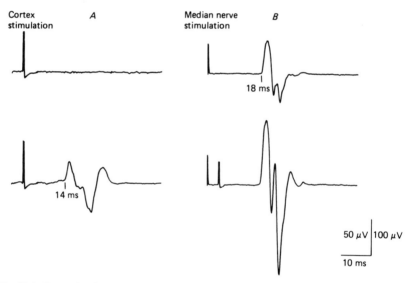

Fig. 39.4   Example of spinal facilitation of the response to an electrical cortical stimulus. In the top left panel a submotor threshold cortical stimulus was given 4 ms after the beginning of the sweep and EMG was recorded from relaxed wrist and finger flexor muscles in the forearm; no response is visible. The bottom left panel shows a synchronous muscle potential to the same stimulus when the subject exerted a small background contraction of the muscles. The top right panel shows an H-reflex in the same subject at rest produced by stimulating the median nerve at the beginning of the sweep. The bottom right panel shows facilitation of the response when both the cortical and median nerve stimuli are given 4 ms apart. All traces are the average of eight trials. Traces in left panels are plotted at twice the gain of those in right panels. (From Cowan *et al.*, 1986.)

which tests the spinal cord. Nevertheless, some interesting results have been obtained which suggest cortical involvement in the genesis of some reflex responses and some movement tasks for muscles controlling the hand.

When a muscle is contracted voluntarily and a stretch stimulus applied, a series of electromyographic responses can be recorded from the muscle. Usually, the reflex responses are separated into short- and long-latency components. It is universally accepted that the short-latency component is produced by the action of the spinal stretch reflex circuit, whereas the mechanism underlying the long-latency component has been a subject of some debate (see Chapter 38 by Hallett in this volume). It has been suggested that the long-latency stretch reflex employs a transcortical mechanism involving muscle spindle afferent signals projecting to and exciting the motor cortex (Marsden *et al.*, 1972). Phillips (1969) had speculated previously that such a circuit may exist and that the efferent limb of this long-loop reflex may use the monosynaptic corticomotoneuronal component of the corticospinal tract; the same output pathway thought to be activated by a cortical stimulus. According to this hypothesis, a stretch stimulus should increase the excitability of motor cortical neurons activated by a cortical stimulus just prior to, and for the duration of, the long-latency stretch reflex. Day *et al.* (1988a) tested this prediction by measuring the size of the muscle response to a magnetic cortical stimulus given at different times after a stretch of a long flexor muscle of the fingers (Fig. 39.5). The muscle response to the cortical stimulus was superimposed on the stretch reflex. The size of the muscle response was larger when it was superimposed on the long-latency component of the stretch reflex compared to when it was superimposed on the spinal short-latency reflex. In addition, when the intensity of cortical stimulation was turned down so that it was just subthreshold and unable to produce a response in the short-latency reflex, a response still appeared when superimposed on the long-latency reflex. Since the spinal motoneurons were excited to a similar degree during both phases of the stretch reflex, it seemed that spinal cord excitability changes could not explain this extra facilitation. Rather, it was more likely that the extra facilitation has occurred at the level of the motor cortex. The timing of this effect was that predicted by the transcortical hypothesis of the long-latency stretch reflex.

Datta *et al.* (1988) used similar reasoning to demonstrate that the stimulated cortical neurons were more active during one of two types of movement. Their experiment focused on the activity of the first dorsal interosseous muscle in the hand. One task was to use this muscle to abduct the index finger on its own; the other task was to perform a power grip with the hand. A magnetic stimulus delivered during the abduction task produced a muscle response, whereas the same stimulus given during the grip task produced no response. They showed electromyographically that the muscle was activated

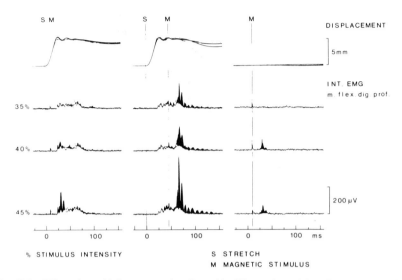

Fig. 39.5 Effect of combining a stretch reflex with different intensities of magnetic cortical stimulation. In the left panel the cortical stimulus was given at a time to produce a response in the middle of the short-latency stretch reflex. In the middle panel the cortical stimulus was timed to produce a response in the middle of the long-latency stretch reflex. The right panel shows the response to a cortical stimulus on its own. Shown are finger displacement traces (top) and rectified EMG traces recorded from wire electrodes inserted into a finger flexor muscle in the forearm. Responses to stretch alone and stretch plus a cortical stimulus have been superimposed. The additional EMG response to the cortical stimulus is shown by the blacked in areas. Note that the EMG response to the cortical stimulus is larger when superimposed on the long-latency component of the stretch reflex, for the three intensities of stimulation. Even when the cortical stimulus was of submotor threshold intensity (35%), a response still appeared when superimposed on the long-latency reflex. (From Day et al., 1988a.)

to a similar degree during two different tasks, suggesting that the difference in response was due to a change in excitability of the stimulated neurons in the cortex. Interestingly, when the experiment was repeated using electrical cortical stimulation, there was little difference in the size of response during the two tasks. Presumably, the change in cortical excitability was "seen" using the magnetic stimulus as it activated the corticospinal neurons transsynaptically, but not with the electrical stimulus as it activated the axons of the corticospinal neurons directly.

STOPPING THE BRAIN

So far, what might be dsecribed as positive effects of cortical stimulation have been considered. It is of interest to consider some negative effects as well. By this I mean the disruptive effects of a cortical stimulus on normal brain activity. It would be no surprise to learn that such a crude stimulus is capable of interrupting normal brain processes, but the question is whether this can teach us anything about those processes. At the present time, two aspects of brain function have been studied in this way: visual perception and movement.

Amassian *et al.* (1988) reported that visual perception can be suppressed by a single stimulus delivered through a magnetic coil placed over the occipital cortex. Their subject looked at a screen upon which was flashed a group of three letters. The magnetic stimulus was given at different times after the appearance of the letters. With an interval of around 100 ms between the stimuli, perception of the letters was impaired. With the coil centred over the midline, the effect usually was to blur the image although on occasions the subject reported seeing nothing. When the interval between stimuli was decreased below 60 ms or increased above 140 ms, there was no effect and the letters were reported correctly. They argued that the site of interference was in the visual cortex because of what happened to the image when the coil was moved. Thus, shifting the coil to the right produced more errors in reporting the letter on the left; with a vertical array of letters, moving the coil rostrally produced interference with the bottom letter. These results suggest that the stimulated neurons in the visual cortex play a role in visual perception during the interval of 60–140 ms after the change in visual scene.

In our laboratory, similar experiments have been undertaken to study the disruptive effect of a cortical stimulus on voluntary movement. It may be expected that a cortical stimulus would suppress movement in the same way as for visual perception, especially as Penfield and Jasper (1954) reported that repetitive 60 Hz stimulation of the exposed human precentral cortex could halt ongoing movement and produce a feeling of paralysis. It turns out that the effect is more interesting. Rather than abolishing movement we find that the movement is executed normally but its initiation may be delayed by up to 150 ms (Day *et al.*, 1989b).

Voluntary movement was studied in the context of a simple reaction task. Subjects were requested to flex their wrist rapidly upon hearing a tone delivered through headphones. Occasionally, in random trials, a cortical stimulus was given at a preset time after the tone. In control trials the subject responded to the tone after a reaction time with the characteristic agonist/antagonist alternating EMG burst pattern associated with rapid limb movements. When a cortical stimulus was interposed between the tone and the *expected* time of movement onset, then the reaction time was prolonged although the movement and its EMG pattern was little affected (Fig. 39.6). This effect was seen with both electrical and magnetic stimulation provided the stimulus was applied to the central areas of cortex. The

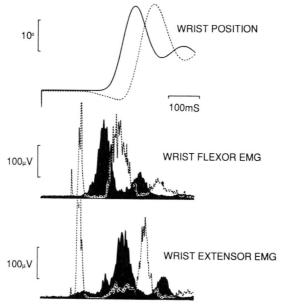

Fig. 39.6 Example of a cortical stimulus delaying the onset of movement during a reaction task. Shown are rapid wrist flexion movements from a normal subject, in response to an audio signal given at the start of the sweep, with (dotted lines) and without (solid traces) magnetic cortical stimulation delivered 100 ms after the start of the sweep. Illustrated are average wrist position (upper traces, flexion upwards), wrist and finger flexor (middle traces) and extensor (lower traces) rectified EMG activity. The control movement (average of 26 trials) is characterized by alternating bursts of activity in the flexor and extensor muscles. In those trials in which a cortical stimulus was given (average of 10 trials) the movement and associated pattern of muscle activities are the same but delayed by some 60 ms. Note the direct muscle responses in the flexor and extensor muscles occurring some 15 ms after the stimulus artefact producing, in this case, a small extension movement seen in the wrist position trace. (From Day *et al.*, 1989b.)

magnitude of the delay was related to the intensity and the timing of the cortical stimulus. The longest delays were obtained with a high-intensity stimulus applied just before the expected onset of movement. The reason for the delay did not seem to be due to refractoriness of the spinal cord since a second cortical stimulus, given just after the first, was capable of producing a large muscle response at a time when the voluntary movement was blocked. Rather, the delaying effect seemed to be due to the cortical stimulus interfering with some brain process. It was possible that the cortical stimulus had interfered with the subject's intention to move, although two lines of evidence suggest this was not the case. First, with long delays the sensation was one of trying to move but being unable to as if transiently paralysed. Second, when the subject was trained to move both limbs at the same time and the stimulus was applied to one hemisphere exclusively, the two limbs were observed to move at different times. Only the limb contralateral to the stimulated hemisphere was delayed to any great extent.

The explanation favoured by Day *et al.* (1989b) was that the brain stimulus caused long-lasting inhibition of strategically placed neurons in the cortex. These cells, which normally would be active at some stage during the preparation and execution of a voluntary movement, would remain unresponsive until they had recovered from the inhibitory process imposed by the shock. Such reasoning explains why a voluntary movement may be *arrested* by a brain stimulus but, on its own, fails to explain why the movement eventually was executed. Presumably, the movement commands upstream from the inhibited cells were retained or stored for the duration of the block; when the cells had recovered from the inhibitory process then the movement commands were allowed to continue their journey. In other words, this experiment provides some evidence that the motor commands, at least for a simple reaction task, may be held in some form of motor memory rather like a buffer store. Also, for the movement to be executed late, not only must the motor commands have been stored but controlling centres in the brain must have had knowledge of the initial failure to move. This knowledge is likely to have arisen from one part of the brain monitoring the status of the brain's motor system, since with small delays (produced by low intensity stimuli) there would have been insufficient time for inspection of signals from peripheral receptors. Thus, during a simple reaction task it seems that motor commands are coded and stored within the brain in a motor memory. Movement occurs upon release of these coded instructions to the motor cortex, under feedback control from an internal monitoring system.

REFERENCES

Amassian, V. E., Quirk, G. J. and Stewart, M. (1987) *J. Physiol.* **394**, 119P.

Amassian, V. E., Cracco, J. B., Eberle, L., Maccabee, P. I. and Rudell, A. (1988) *J. Physiol.* **398**, 40P.

Barker, A. T., Jalinous, R. and Freeston, I. L. (1985) *Lancet* **i**, 1106–1107.

Bartholow, R. (1874). *Am. J. Med. Sci.* **67**, 305–313.

Berardelli, A., Day, B. L., Marsden, C. D. and Rothwell, J. C. (1987a). *J. Physiol.* **391**, 71–83.

Berardelli, A., Inghillieri, M., Formisano, R., Accornero, N. and Manfredi, M. (1987b) *J. Neurol. Neurosurg. Psychiat.* **50**, 732–737.

Clough, J. F. M., Kernell, D. and Phillips, C. G. (1968) *J. Physiol.* **198**, 145–166.

Cowan, J. M. A., Dick, J. P. R., Day, B. L., Rothwell, J. C., Thompson, P. D. and Marsden, C. D. (1984) *Lancet* **ii**, 304–307.

Cowan, J. M. A., Day, B. L., Marsden, C. D. and Rothwell, J. C. (1986) *J. Physiol.* **377**, 333–347.

Datta, A. K., Harrison, L. M. and Stephens, J. A. (1988) *J. Physiol.* **401**, 47P.

Day, B. L., Rothwell, J. C., Thompson, P. D., Dick, J. P. R., Cowan, J. M. A., Berardelli, A. and Marsden, C. D. (1987a) *Brain* **110**, 1191–1209.

Day, B. L., Thompson, P. D., Dick, J. P., Nakashima, K. and Marsden, C. D. (1987b) *Neurosci. Lett.* **75**, 101–106.

Day, B. L., Riescher, H. and Struppler, A. (1988) *Pflugers Arch. Eur. J. Physiol.* Suppl. 1, **411**, R135.

Day, B. L., Dressler, D., Maertens de Noordhout, A., Marsden, C. D., Nakashima, K., Rothwell, J. C. and Thompson, P. D. (1989a) *J. Physiol.* **412**, 449–473.

Day, B. L., Rothwell, J. C., Thompson, P. D., Maertens de Noordhout, A., Nakashima, K., Shannon, K. and Marsden, C. D. (1989b) *Brain.* **112**, 649–663.

Hess, C. W., Mills, K. R. and Murray, N. M. F. (1986) *Lancet* **ii**, 355–358.

Ingram, D. A. and Swash, M. (1987) *J. Neurol. Neurosurg. Psychiat.* **50**, 159–166.

Kernell, D. and Wu, Chien-Ping (1967) *J. Physiol.* **191**, 653–672.

Marsden, C. D., Merton, P. A. and Morton, H. B. (1972) *Nature* **238**, 140–143.

Marsden, C. D., Merton, P. A. and Morton, H. B. (1980) *J. Physiol.* **312**, 5P.

Marsden, C. D., Merton, P. A. and Morton, H. B. (1982) *J. Physiol.* **328**, 6P.

Marsden, C. D., Merton, P. A. and Morton, H. B. (1983) *Adv. Neurol.* **39**, 387–391.

Merton, P. A. and Morton, H. B. (1980) *Nature* **285**, 227.

Patton, H. D. and Amassian, V. E. (1954) *J. Neurophysiol.* **17**, 345–363.

Penfield, W. and Jasper, H. (1954) *Epilepsy and the Functional Anatomy of the Human Brain.* Churchill, London.

Phillips, C. G. (1969) *Proc. Roy. Soc. Lond. (Biol.)* **173**, 141–174.

Rothwell, J. C., Day, B. L., Berardelli, A. and Marsden, C. D. (1984) *Exp. Brain Res.* **54**, 382–384.

Rothwell, J. C., Thompson, P. D., Day, B. L., Kachi, T., Cowan, J. M. A. and Marsden, C. D. (1987) *Brain* **110**, 1173–1190.

Thompson, P. D., Day, B. L., Rothwell, J. C., Dick, J. P. R., Cowan, J. M. A., Asselman, P., Griffin, G. B., Sheehy, M. P. and Marsden, C. D. (1987) *J. Neurol. Sci.* **80**, 91–110.

Thompson, S. P. (1910) *Proc. Roy. Soc. Ser. B* **82**, 396–398.

Index